ARCHIMEDES

ARCHIMEDES

BY

E. J. DIJKSTERHUIS

translated by C. Dikshoorn

With a new bibliographic essay by Wilbur R. Knorr

Summis ingeniis dux et magister fuit

(Heiberg, *Archimedis opera omnia* III,
Prolegomena XCV)

PRINCETON UNIVERSITY PRESS

Princeton, New Jersey

Published by Princeton University Press, 41 William Street,
Princeton, New Jersey 08540
In the United Kingdom: Princeton University Press, Guildford, Surrey

Library of Congress Cataloging in Publication Data will be found
on the last printed page of this book

First printing, 1956 by Ejnar Munksgaard
First Princeton Paperback printing, 1987

LCC 86-43144

ISBN 0-691-08421-1
ISBN 0-691-02400-6 (pbk.)

Reprinted by arrangement with
Uitgeverij Meulenhoff Nederland bv and Ejnar Munksgaard

Chapters I–V of this work were formerly published in Dutch (*Archimedes*, I, P. Noordhoff, Groningen, 1938). The subsequent chapters appeared in the Dutch periodical *Euclides* (XV–XVII, XX; 1938–1944).

Clothbound editions of Princeton University Press books are printed on acid-free paper, and binding materials are chosen for strength and durability. Paperbacks, while satisfactory for personal collections, are not usually suitable for library rebinding.

Printed in the United States of America
by Princeton University Press,
Princeton, New Jersey

CONTENTS

PREFACE

This book is an attempt to bring the work of Archimedes, which is one of the high-water marks of the mathematical culture of Greek Antiquity, nearer to the understanding and the appreciation of the modern reader. Such an attempt has been made twice before, in a way so excellent that I can scarcely hope to equal it: by T. L. Heath in *The Works of Archimedes* and by P. Ver Eecke in *Les Oeuvres Complètes d'Archimède*. My belief that I might be excused for adding to these two excellent editions a new adaptation of the writings of the great Greek mathematician finds its justification in the consideration that the method of treatment here chosen differs fundamentally from the one followed by Heath as well as from that applied by Ver Eecke. As a matter of fact, Heath represents Archimedes' argument in modern notation, Ver Eecke gives a literal translation of his writings. Both methods have their disadvantages: in a representation of Greek proofs in the symbolism of modern algebra it is often precisely the most characteristic qualities of the classical argument which are lost, so that the reader is not sufficiently obliged to enter into the train of thought of the original; the literal translation, on the other hand, which like the Greek text says in words everything that we, spoiled as we have been by the development of mathematical symbolism, can grasp and understand so much more easily in symbols, perhaps helps the present-day reader too little to overcome the peculiar difficulties which are inevitably involved in the reading of the Greek mathematical authors and which certainly are not due exclusively, nay, not even primarily, to the fact that they wrote in Greek.

The method applied in the present book attempts to combine the advantages and avoid the disadvantages of the two methods just outlined. The exposition follows the Greek text closely, but only the propositions are given in a literal translation; after that the proofs are set forth in a symbolical notation specially devised for the purpose, which makes it possible to follow the line of reasoning step by step. This system of notation, which was also used in my work *De Elementen van Euclides* (The Elements of Euclid) (Groningen, 1929, 1931), in long practice has been found a useful aid in the explanation of Greek mathematical arguments.

Apart from the introduction of this aid, I have also tried to meet in another way the difficulties which I know from experience are encountered by present-day mathematicians reading Greek authors. In fact, the Greek mathematicians in their works are wont to give, without a single word of elucidation about the object in view, a dry-as-dust string of propositions and proofs, in which not the slightest distinction is made between lemmas and fundamental theorems, while the general trend of the argument is often very difficult to discover. In order to bring this trend out more clearly I have collected in a separate chapter (Chapter III) all those theorems which in relation to the nucleus of a treatise have the function of elements (*στοιχεῖα*); with each individual work the argument could then be summarized much more briefly, because all the lemmas had already been discussed previously. A decimal classification of Chapter III makes it possible to trace these lemmas quickly, if desired, and to find out how they can be proved. Through this arrangement the additional advantage has been gained that each of Archimedes' treatises can be studied separately.

The publication of this work in English has been rendered possible through the financial aid of the Nederlandse Organisatie voor Zuiver-Wetenschappelijk Onderzoek (The Netherlands Organization for Purely Scientific Research).

The author wishes to express his thanks to Miss C. Dikshoorn, The Hague, for her very careful translation, to Dr Mogens Pihl, Gentofte, for taking the initiative in the publication in Denmark, to the publishing firm, Ejnar Munksgaard, for undertaking the edition, and to the staff of the firm for the care devoted to the work.

CHAPTER I.

THE LIFE OF ARCHIMEDES

1. *The Personality.*

In our scanty knowledge of Greek mathematics the information available about the lives of those who were engaged in its pursuit is one of the weakest points. Reliable reports about the lives and fortunes of hardly any of them are in our possession. The period in which they worked can generally be only roughly approximated; the places where they dwelt are frequently altogether unknown.

To this general rule Archimedes is only an apparent exception. It is true that already in antiquity a great many stories about him were current, which have remained indissolubly bound up with his name down to our own time, but these reports are usually very fanciful and seldom stand the test of historical criticism. It is therefore no more possible to write a coherent and reliable biography[1]) of Archimedes than of any other Greek mathematician. The writer cannot do much beyond arranging the traditional reports and accurately stating the sources from which they originate; only occasionally will it be possible to assess the value of these reports.

The year of his birth already is not established with absolute certainty; it is usually stated to be 287 B.C., because he is reported by the Byzantine polyhistor Tzetzes[2]) to have been seventy-five

[1]) In antiquity there existed a biography of Archimedes written by Heraclides. Eutocius mentions it in his commentary on the *Measurement of a Circle* (*Opera* III, 228). We do not know who this Heraclides was. In the preface to his work *On Spirals* (*Opera* II, 2) Archimedes refers to a man of this name who brought Dositheus a book.

[2]) Joannes Tzetzes lived at Constantinople in the 1st half of the 12th century. He wrote, among other things, a historical work, which is known as the *Chiliades* in view of a posthumous division into books of one thousand lines each. The statement about the age of Archimedes is to be found in *Chil.* II, *Hist.* 35, 105.

when he was killed in 212 B.C. during the Roman conquest of Syracuse.

About his parentage, too, little can be said with certainty. All classical writers agree in calling him a Syracusan by birth, but whereas Cicero[1]) and Silus Italicus[2]) create the impression that he was poor and of humble birth, Plutarch[3]) reports on his family connection and intimate relations with King Hieron II[4]) of Syracuse, which seems to point to quite a different position in the life of the city. Nor need there be any absolute inconsistency between the two possibilities. Indeed, Hieron, who was said to be the illegitimate son of a nobleman by one of his female slaves, apparently owed his brilliant career to his personal merits rather than to hereditary privileges, and his original relations with Archimedes may very well have remained unchanged, the wise mathematician having no desire to share in the power and prosperity of the tyrant.

According to an originally incomprehensible statement by Archimedes himself (a passage in his work *The Sand-reckoner*, which is meaningless in the version of the manuscripts as handed down to us, but has been intelligently emended by F. Blass[5])) he was the son of an astronomer, Phidias, of whose work we know nothing beyond an estimation of the ratio of the diameters of sun and moon, which Archimedes refers to in the passage in question.

Besides these rather vague biographical particulars there are

---

[1]) In his work *Tusculanae Disputationes* (V, 23) Cicero (106–43 B.C.) calls Archimedes *humilem homunculum* (a humble little man), words which may, however, also be meant by way of oratorical contrast with the tyrant Dionysius mentioned just before.

[2]) Silus Italicus, a Roman poet and orator (A.D. 25–100), in his epic *Punica* deals with the second Punic War. He also mentions Archimedes' share in the defence of Syracuse, and calls him *nudus opum* (destitute of means). XIV, 343; ed. L. Bauer (Leipzig 1892). Vol. II, p. 94.

[3]) Plutarch, *Vita Marcelli* XIV, 7 (305) calls him *Ἱέρωνι τῷ βασιλεῖ συγγενὴς καὶ φίλος*.

[4]) Hieron II, illegitimate son of a Syracusan nobleman, became commander-in-chief in 275 B.C. and, after a successful expedition against the Mamertines, in 270 king of Syracuse. After being vanquished in 263 by the Romans, he allied himself with them and continued to support them until his death in 216. Under his peaceful government Syracuse greatly flourished.

[5]) *Opera* II, 220. The manuscripts have: *Φειδία δὲ τοῦ Ἀκούπατρος*, for which F. Blass (*Astr. Nachr.* 104 (1883) No 2488, p. 255) proposed the conjecture: *Φειδία δὲ τοῦ ἁμοῦ πατρὸς* = *our father Phidias.*

some others of a more positive nature, which, however, are not at all worthy of credit[1]); the fact that an Arabian writer calls him a son of Pythagoras is merely mentioned here as an example of the unchecked fancy with which the Arabian mathematicians were wont to write about the lives of their Greek predecessors. And that, as Linceo Mirabello has it, he should have been a pupil of Plato's in his youth at Syracuse must be due to a chronological error, because Plato had been dead for sixty years when Archimedes was born. About his descendants there is a report which is equally strange: Rivault[2]), who in 1615 edited a Latin translation of his collected works, says in his biographical introduction that he has heard from a learned Greek friend that the Sicilian martyr Santa Lucia had been a descendant of the great mathematician.

This much is certain about the life of Archimedes that he spent some time in Egypt on at least one occasion; it is not saying much that Arabian writers mention this, but Diodorus[3]) confirms it in two passages of his *Bibliotheca Historica*[4]), where he refers to the cochlias (a hydraulic machine to which we shall revert in § 4) as an invention of Archimedes, which he was said to have made in Egypt. Moreover it appears from the prefaces to his works that from Syracuse he always maintained very friendly relations with several scholars at Alexandria; it is quite natural to suppose that these ties date from the period when he stayed for his studies at the then centre of Greek science, which must always have exercised a peculiar attraction on mathematicians because at the Museum[5]) the tradition of Euclid lived on. Of his Alexandrian col-

---

[1]) I take these from A. Favaro, *Archimede*. Profili No 21 (Roma 1923), p. 14.

[2]) *Archimedis Opera quae extant novis demonstrationibus commentariisque illustrata* per Davidem Rivaltum a Flurantia. Parisiis. Apud Claudium Morellum. 1615. David Rivault de Flurance (1571–1616) was a mathematician at the court of Louis XIII.

[3]) Diodorus of Agyrium (Sicily) was a historian under the Emperor Augustus. Between 60 and 30 B.C. he wrote his big historical work *Bibliotheca Historica*, the 40 books of which dealt with world history as far as Caesar's Gallic War.

[4]) Diodorus, *Bibliotheca Historica*, rec. F. Vogel (Leipzig 1890), I, 34 (in a description of the Nile delta); V, 37 (in a passage about the draining of silver mines in Spain).

[5]) The Museum was a building at Alexandria where scholars lodged. It was founded in 320 B.C.

leagues he seems to have had the highest regard for the astronomer Conon[1]) of Samos, to whom, until the latter's death, he used to send his mathematical discoveries before their publication, and about whom he always spoke with the greatest admiration[2]). These scholars further included the many-sided Eratosthenes of Cyrene[3]), for whom the *Method* was written, and Conon's pupil, Dositheus of Pelusium[4]), to whom the works *On the Sphere and Cylinder*, *On Conoids and Spheroids*, and *On Spirals* are dedicated.

According to a statement in the biography with which J. Torelli introduces his great Archimedes edition[5]) Archimedes is said after his return from Egypt to have visited other countries as well. There is in particular a story about a voyage to Spain in a note by Leonardo da Vinci[6]), in which the latter mentions that he has read in a history of the Spaniards that the Syracusan Archimedes aided Eclierides, King of the Cilodastri, in a maritime war against the English through the invention of a device for spouting burning pitch on to the ships of his opponents. It is, however, altogether unknown in what work Leonardo can have read this, and authorities on Spanish history ignore both King Eclierides and the people of the Cilodastri. In connection with the possibility of a voyage to Spain, a statement by Diodorus about the use in the Spanish silver mines of the cochlias, which he alleges had been invented by Archimedes[7]), deserves some attention.

---

[1]) Conon of Samos, astronomer and mathematician (3rd century B.C.), wrote a work on astronomy, in which he collected the ancient observations of the Chaldeans; by his geometrical work he laid the foundations for the fourth book of the *Conics* of Apollonius.

[2]) *Opera* I, 4. II, 2; 262.

[3]) Eratosthenes of Cyrene (born about 284 B.C.), when 40 years of age, came to Alexandria at the invitation of Ptolemy III Euergetes, to be tutor to the King's son Philopator; later he became librarian there. He is noted for a measurement of the radius of the earth, for his sieve (*κόσκινον*) for finding prime numbers, and for a solution of the Delian problem.

[4]) About this mathematician no further particulars are known. From the way in which Archimedes starts his correspondence with him after the death of Conon (*Opera* II, 262) one gets the impression that he did not know him personally.

[5]) *Archimedis quae supersunt omnia cum Eutocii Ascalonitae commentariis* ex rec. Josephi Torelli, Veronensis *cum nova versione latina*. Oxonii, 1792, p. XII.

[6]) Quoted in A. Favaro, *Archimede*, p. 19.

[7]) Vide Note 4 to p. 11.

It is possible that Archimedes also returned once more to Egypt, and that on that occasion he constructed the great works concerned with dike and bridge building and regulation of the Nile about which there are some reports in Arabian sources[1]). However, nothing can be said with certainty about this either, and considering the large amount of work of a mathematical and technical kind achieved by him in his native city it is rather more likely that he spent the greater part of his life at Syracuse itself, and that it was from there that his great fame spread in antiquity.

As is understandable from the nature of mathematical work, this fame is not based in the first place on the writings which he sent to his mathematical friends and which have secured him the admiration of the ages, ever since the revival of mathematical science. Even before this, his astronomical work was appreciated in wider circles. He is, however, mainly mentioned in non-mathematical literature on account of the manifestations of his technical ingenuity, and these manifestations appear to have made a great impression in the militaristic Roman Empire chiefly because they made possible the military action—to be discussed more fully in § 5—which he developed during the defence of Syracuse.

If Plutarch is to be believed, he himself seems to have considered the whole of his technical activity as an occupation of a lower order: as it is reported in the *Life of Marcellus*[2]), "he did not deign to leave behind him any written work on such subjects; he regarded as sordid and ignoble the construction of instruments, and in general every art directed to use and profit, and he only strove after those things which, in their beauty and excellence, remain beyond all contact with the common needs of life".

In pure mathematics he found the possibility to satisfy this desire to the full, and if we rely once more on what Plutarch tells us further[3]), there seldom lived anyone who was so much preoccupied with mathematics as he was: "continually bewitched by a Siren who always accompanied him, he forgot to nourish himself and

[1]) A. Favaro, *Archimede* (Roma 1923), p. 21. The report in question occurs, *inter alia,* in the bio-bibliographical work Tar' īkh alhukamā of the Egyptian historian al-Qiftî (1172/73–1248). *Vide* Eilhard Wiedemann, *Beiträge zur Geschichte der Naturwissenschaften* III. Sitzungsber. d. Phys.-med. Soc. in Erlangen **37** (1905), 247–250.

[2]) Plutarch, *Vita Marcelli* XVII, 4 (307).

[3]) Plutarch, *Vita Marcelli* XVII, 6 (307).

omitted to care for his body; and when, as would often happen, he was urged by force to bathe and anoint himself, he would still be drawing geometrical figures in the ashes or with his finger would draw lines on his anointed body, being possessed by a great ecstasy and in truth a thrall to the Muses".

Elsewhere[1]) Plutarch refers once more to the slight importance which Archimedes himself attached to his technical inventions: "most of them were the diversions of a geometry at play which he had practised formerly, when King Hieron had emphatically requested and persuaded him to direct his art a little away from the abstract and towards the concrete, and to reveal his mind to the ordinary man by occupying himself in some tangible manner with the demands of reality".

2. *ΔΟΣ ΜΟΙ ΠΟΥ ΣΤΩ* . . . (*Give me a place to stand on* . . .).

An occasion on which Archimedes was able to comply with Hieron's request seems to have presented itself during the construction of the famous ship Syracusia (later Alexandris), which the king, who was known for his love of display and his propensity to have big projects carried out, had ordered to be built so that after its completion he might present it, loaded with articles of food, to King Ptolemy of Egypt. Detailed reports about the appointments of this famous ship, the size of which is estimated at 4,200 tons and which is said to have been fitted with all conceivable wonders of luxury and technology, are to be found in the *Deipnosophistae* of Athenaeus[2]), who also states that its construction was directed by Archias of Corinth under the superintendence of Archimedes. He further relates that, when no one knew how to launch the ship, Archimedes succeeded in this all alone, with the

[1]) Plutarch, *Vita Marcelli* XIV, 4 (305).

[2]) Athenaeus of Naucratis lived about A.D. 200, first at Alexandria, later in Rome. His dialogue *Δειπνοσοφισταί* contains excerpts from a great many ancient writers on all sorts of subjects. The statements about the ship Syracusia are to be found in *Athenaei Naucratitiae Dipnosophistarum Libri* XV; rec. G. Kaibel (Leipzig 1887). V, 40–44 (206d–209a). Athenaeus seems to have taken his description from an older writer, the physician Moschion (not to be confused with the gynaecologist of the same name from the 5th or 6th century). Other works to be consulted about the ship are A. Holm, *Geschichte Siciliens im Alterthum*, III (Leipzig 1898), pp. 39–41; A. Favaro, *Archimede* (Roma 1923), p. 24.

aid of a few instruments. The same story is told by Proclus[1]), who represents Hieron as operating the device himself and calling out in amazement: "From this day forth Archimedes is to be believed in everything that he may say"[2]).

In Plutarch the description of the launching of a big ship appears in a slightly different form[3]). Archimedes is said to have declared to Hieron that it was possible to move a given weight by a given force. Hieron then invited him to demonstrate this on a ship from the royal fleet, which had been drawn on land with difficulty and had there been loaded with a large crew and the customary freight; upon which Archimedes with quiet movements of his hand put into operation a device which drew the ship to the sea as smoothly as if she were already moving through the water. The story has become famous in particular because of the familiar saying which Archimedes is reported to have uttered on this occasion, viz.: "Give me a place to stand on, and I will move the earth".

In the traditional form in which we have given them here the reports about the setting in motion of a ship are of course purely fantastic; devices capable of achieving what Archimedes must have required of them only exist in the ideal realm of rational mechanics, in which friction and resistance of the air are eliminated. It is, however, hardly to be doubted that the stories about it have a real basis in the invention or demonstration of a device by which force was economized in moving heavy weights. And thus it is after all to the point to speculate with the classical writers on the question what device Archimedes may, theoretically speaking, have used to draw the heavy ship into the sea, and how he proposed to lift the earth off its hinges if only he could have found a place to stand on while doing so.

The writers who deal with this question are anything but unanimous in their answers. Plutarch says[4]) that the device used was a *polyspaston* (πολύσπαστον), i.e. a tackle with a large number of sheaves in each of the two pulley blocks. Tzetzes[5]) speaks in par-

---

1) *Procli Diadochi in primum Euclidis Elementorum librum commentarii*, rec. G. Friedlein (Leipzig 1873), p. 63.

2) *'Από ταύτης τῆς ἡμέρας περὶ παντὸς 'Αρχιμήδει λέγοντι πιστευτέον.* l.c. 63, 24.

3) Plutarch, *Vita Marcelli* XIV, 7 (306).

4) Plutarch, *Vita Marcelli* XIV, 8 (306).

5) Tzetzes, *Chil.* II, *Hist.* 35, 107.

ticular of a *trispaston* (*τρίσπαστον*), i.e. a tackle with two pulley blocks, each of three sheaves. With this apparatus the ideal ratio between effort and load would be 1 : 6, which does not very well agree with his assertion that a weight of 50,000 medimns[1]) of wheat had been moved with it. More casually the *trispaston* is also mentioned by the physician Oribasius[2]); he states that Archimedes used the device to draw ships into the sea, and that the medical profession had begun to apply a small scale model of it for reducing fractures and dislocations.

Heron, on the contrary, for the general solution of the problem To move a given weight by a given force, discusses the so-called *Barulcus*, i.e. an instrument in which a windlass is rotated through a system of toothed wheels, the last of which is driven by an endless screw[3]). Probably Athenaeus is thinking of such a device when in the *Deipnosophistae* he represents Archimedes as using a screw to move the ship Syracusia[4]).

About the moving of the earth we read in Pappus[5]), who mentions the familiar formulation quoted above (*Δός μοι ποῦ στῶ καὶ κινῶ τὴν γῆν*); in Simplicius, in the commentary on the *Physica* of Aristotle[6]), where he says that Archimedes, after having constructed a weighing machine, the so-called *charistion* (*χαριστίων*), is reported to have exclaimed: *πᾶ βῶ καὶ κινῶ τὴν γῆν* (a place to stand on! and I will move the earth); and in two passages in Tzetzes, who on one occassion mentions the *trispaston*[7]), and on the other the *charistion*[8]), and who represents Archimedes as having said in

---

[1]) The medimn is an Attic corn measure of about 54 litres.

[2]) Oribasius was a famous physician in the 4th century, physician in ordinary to the Emperor Julian the Apostate, writer of a big medical encyclopaedia, *Ἰατρικαὶ συναγωγαί*. The reference in question is to be found in *Oribasii Collectionum medicarum reliquiae*, ed. J. J. Raeder IV (Leipzig 1933), 33. The *trispaston* is here dealt with as an invention of Archimedes or Apellis. On p. 12 he speaks of the *polyspaston*.

[3]) *Heronis Opera* II, 1, 256 et seq, and III, 306 et seq.

[4]) Athenaeus, *Deipnosophistae* (vide Note 2 to p. 14) V, 207, a.b.

[5]) Pappus, *Collectio* VIII, 10; 1060.

[6]) Simplicius, *In Aristotelis Physicorum Libros Commentaria*, ed. H. Diels (Berlin 1895), p. 1110.

[7]) Tzetzes, *Chil.* III, *Hist.* 66, 62. The words here are: *ὅπα βῶ καὶ σαλεύσω τὴν χθόνα.*

[8]) Tzetzes, *Chil.* II, *Hist.* 35, 130.

the Dorian dialect of Syracuse: *πᾶ βῶ, καὶ χαριστίωνι τὰν γᾶν κινήσω πᾶσαν* (a place to stand on! and I will move the entire earth by means of a *charistion*).

Now we have in the first place to ask ourselves what is to be understood by the *charistion*, the device which Simplicius calls a weighing machine, which Tzetzes does not define, but which is to be found described in Simon Stevin[1]) as comprising "assen met vijsen" (shafts with screws), so that he seems to imagine something in the style of the *Barulcus*. Now it can hardly be doubted that in this difference of opinion it is Simplicius who is right; indeed, it is fairly natural to suppose with P. Duhem[2]) that the *charistion* is identical with the so-called *charasto* (also called *carasto*, *canisto* or *baracto*), which is discussed in the work *Liber Charastonis*, translated into Latin by Gherard of Cremona from the Arabic version of Thabīt b. Qurra, and which is found to be the balance with unequal arms, which the Romans called *statera*, which is still known in French as *balance romaine*, and in English as *steelyard*. The term *charistion*, about whose origin there has been a good deal of controversy, may originally have been the name of the inventor or of the first describer of the device; this is at least the simplest way to account for the striking fact that a mechanical instrument bears a name derived from a word (*χάρις*) which does not have the slightest connection with technical matters and which on the other hand frequently appears in proper names[3]).

If this interpretation is correct, the hyperbolic words of Archimedes about the moving of the earth must have been inspired by the lever principle manifested in the *charistion*, which consists in the inverse proportion of effort and load to the distances from the fulcrum to their lines of action. And indeed, in view of the fact that in a work by Archimedes dealing with mechanics (*On the Equilibrium*

---

1) Simon Stevin, *De Weeghdaet* (The Practice of Weighing). Voorstel (Prop.) 10. The Principal Works of Simon Stevin. Vol. I. *General Introduction. Mechanics.* Edited by E. J. Dijksterhuis. Amsterdam 1955. p. 354–5.

2) P. Duhem, *Les origines de la Statique* (Paris 1905) I, 79 et seq.

3) For example: *Χαρισθένης, Χαρίστιος, Χαριτώ* et al. The *Liber Charastonis* is now available in an edition of medieval texts on the Science of Weights by E. A. Moody and M. Clagett: *The Medieval Science of Weights* (*Scientia de Ponderibus*). Madison 1952. IV. *Liber Karastonis.* 76–117. The origin of the name is discussed on pp. 79–80.

of Planes) the theory of the lever is discussed, the famous saying was in later days as a rule associated with this theory. Nevertheless it seems doubtful whether this is the right interpretation of the classical tradition; indeed, Plutarch and Heron, who speak about a *polyspaston* or a *trispaston*, and Proclus, who does not mention any particular apparatus at all, were acquainted with the properties of levers just as well as later writers, and there would not have been the slightest reason for their deviating from the tradition, if this had been unambiguous in its reports about the use of a *charistion*.

It would rather seem that the conflicting nature of the various reports corroborates the view that Archimedes did not make his statement about the moving of the earth in connection with any-one particular instrument, but on the basis of a general mechanical insight he had gained by studying several instruments which were already known before his day or had been invented by him (windlass, pulley, lever, toothed wheel, screw, etc.). This insight, which was later to be formulated in the so-called Golden Rule of Mechanics and which was to be fully developed in the principle of virtual displacements, taught the fundamental possibility of reducing at will the effort that has to be applied for moving a given load, provided one takes the trouble to increase proportionally the distance through which it has to be exerted, and thus also indicated in principle the means for moving the earth by a given force from a fixed point outside the earth. It is quite conceivable that Archimedes demonstrated this principle to the Syracusans by moving a large weight by means of tackles, toothed wheels or endless screws, and that in the imagination of later generations this weight grew into the heavily loaded ship which has become an indispensable element in the Archimedes legend.

3. *The Wreath Problem.*

An equally inalienable feature of the legendary figure of Archimedes with the utterance about the moving of the earth is the story of the so-called wreath problem. The story is very familiar; it may, however, be related once more in this context in the form in which the oldest source[1]) mentions it: King Hieron, out of gratitude for the success of one of his enterprises, wants to consecrate a

[1]) Vitruvius, *De Architectura* IX, 3. ed. F. Krohn (Leipzig 1912), p. 198. Vitruvius was a Roman architect under the Emperor Augustus.

gold wreath[1]) to the immortal gods. When it is completed, the wreath is found to have the weight of the gold furnished for it; however, the suspicion arises that a portion of the gold has been replaced by a quantity of silver of the same weight[2]). The king, being unable to force the maker of the wreath to confess, asks Archimedes for a convincing means by which to investigate the charge. One day, when the scholar, still pondering on the problem, steps into a bath, he suddenly becomes aware of the fact that the deeper he descends into the tub, the more water flows over the edge; this suggests to him all at once how he will be able to answer the question, and he is so overjoyed at the discovery that he jumps up and runs home naked, shouting *εὕρηκα, εὕρηκα* (I have found it!).

Thus the story is told by Vitruvius, who then goes on to explain how Archimedes put into practice the newly gained insight. He is reported to have taken a lump of gold and a lump of silver, each having the same weight as the wreath; when immersed successively in a vessel full of water, each of the lumps caused a certain volume of water to overflow, which was measured by ascertaining how much water was required each time to fill the vessel up to the rim again after removal of the lump of metal. When this experiment was repeated with the wreath, the latter proved to cause more water to overflow than the gold and less than the silver. From this, Archimedes could determine the proportion of silver admixed with the gold in the wreath[3]).

Another version of the matter is given in a Latin didactic poem on weights and measures, the *Carmen de Ponderibus*, which used

[1]) It was a wreath (*στέφανος*) and not a crown, as is often mentioned. As Ch. M. van Deventer, *Grepen uit de Historie der Chemie* (Haarlem 1924), p. 114 remarks, this difference is important, because the consecrated wreath, being a sacred object, was not permitted to be subjected to chemical analysis. The whole wreath problem is dealt with in detail by Van Deventer in his Sixth Address (pp. 108–127).

[2]) In the Archimedes translation by Paul Ver Eecke (*Introduction* XLII) it says that the discovery of the fraud took place "*lorsqu'on éprouva l'or par la pierre de touche*". In Vitruvius, however, nothing of the kind is to be found and, as argued by Van Deventer (l.c., p. 115), the dry assay, which is the only one that could have been applied, was unsuitable in this case, both on account of its low degree of accuracy and because technical means already existed at that time for making the surface of an alloy of pure gold.

[3]) If the volumes of the wreath and of equal weights of gold and silver

to be attributed to the grammarian Priscian (5th century)[1]. In this poem, which is of importance for metrology because of the accurate exposition of Roman and Greek metrical systems it contains, the method followed by Archimedes is set forth as follows:

In one of the scales of a balance with equal arms lies a *libra* of pure gold, in the other a *libra* of silver. Both are immersed in water, and the difference in weight is established; suppose this to be three *drachmae*. Now take the wreath to be tested and an equal weight of silver, and establish once more the difference in weight when the same procedure is followed. Let this be 18 *drachmae*; the wreath then contains six *librae* of pure gold. It is easy enough to make sure that the ratio between the difference in weight observed and the difference in weight per unit of weight indicates the number of *librae* of pure gold[2].

The solution of the wreath problem by Archimedes is usually associated with the law of hydrostatics, found by him, concerning the upward thrust experienced by a body immersed in a fluid, his experience in the bath then being looked upon as the origin of the insight formulated in this law. This view appears not to be supported by the report of Vitruvius: the overflowing of the water from the bath does not teach anything about the upward thrust acting on a body immersed in water, and the method for de-

---

are V, V_1, V_2 respectively, and the wreath contains a weight x of gold and a weight y of silver, the following relation is found:

$$\frac{x}{y} = \frac{V_2 - V}{V - V_1}.$$

[1]) This poem was first printed in 1470 in the now very rare work *Volumen de octo partibus orationis*. The lines relating to the wreath problem were later published in the Archimedes edition of Torelli (vide Note 5 to page 12), p. 364. The whole *Carmen de Ponderibus* is to be found in *Metrologicorum Scriptorum reliquiae*, ed. F. Hultsch (Leipzig 1864), II, 88 et seq. The lines referring to Archimedes are lines 124–162. On the reasons for not attributing the poem to Priscian *vide* ibidem II, 24 et seq.

[2]) If the specific gravities of gold and silver are s_1 and s_2 respectively, the difference in weight per unit of weight of the two substances, upon weighing under water, is found to be $\left(\frac{1}{s_2} - \frac{1}{s_1}\right)$. If the wreath weighs G units of weight, of which x is gold, the difference in weight from G units of silver under water becomes $x\left(\frac{1}{s_2} - \frac{1}{s_1}\right)$, from which follows the conclusion mentioned.

termining the specific gravity, which according to Vitruvius was inferred from it, is not therefore based on hydrostatic weighing, like the one in the *Carmen de Ponderibus*, but on the volumetric principle applied in the pyknometer[1]). The familiar anecdote does not therefore conduce to any clarification of our insight into the origin of the famous law of hydrostatics.

4. *Archimedes as a Mechanical Engineer.*

If we may rely on the stories current about it in antiquity, the geometry at play, to which Plutarch referred as the source of so many admired inventions of Archimedes, seems to have occupied his mind to a greater extent than the philosopher of Chaeronea, with his Platonic contempt of technique, is prepared to admit. We will here discuss three mechanical constructions[2]) attributed to him, the *cochlias*, the planetarium, and the hydraulic organ, after which we intend to devote a separate discussion to the engines of war constructed by him.

a) *The Cochlias.*

The *cochlias* or Archimedean screw (also called *limaçon* or hydraulic screw) is a machine for raising water, which according to Diodorus was used in Egypt for the irrigation of fields which were not inundated directly by the water of the Nile[3]), and in Spain for pumping water out of mines[4]). According to Athenaeus[5]), Archimedes himself used it to keep the holds of the ship Syracusia dry. Both writers say that Archimedes was its inventor; the correctness of this assertion is, however, somewhat doubtful. It is not im-

---

[1]) This method of determining the volume is associated with the name of Archimedes by Heron, *Metrica* II, 20 (*Heronis Opera* III, 138); however, he expresses himself very vaguely: "some people say that Archimedes had devised the method".

[2]) For reports about a water-clock constructed by Archimedes, *vide* Chapter II.

[3]) Diodorus, *Bibliotheca Historica* I, 34. Another interpretation is given by A. Favaro, *Archimede* (Rome 1923), p. 23; he holds that the contrivance was used for draining the pools left behind after the water had receded again.

[4]) Diodorus, *Bibliotheca Historica* V, 37. Its use in Spain is confirmed, according to F. M. Feldhaus, *Die Technik der Antike und des Mittelalters* (Potsdam 1931), p. 152, by a statement by Poseidonius.

[5]) Vide Note 2 to page 14.

possible that the machine is of a much older date, and that Archimedes himself became acquainted with it in Egypt[1]). It is also striking that neither Strabo[2]), nor Philo of Byzantium[3]), nor Vitruvius[4]), who all three mention or describe it, associate with it the name of Archimedes.

A detailed description of the *cochlias*, which is not, however, quite clear owing to the absence of any drawing, is found in Vitruvius:

On a wooden cylinder, whose height is as many feet as the diameter finger's breadths, have been drawn eight helices, which start from the vertices of a regular polygon in the upper surface and the pitch of which is equal to the circumference of the basic circle. On these helices have been provided screw threads of flexible branches, which together form the imitation of the snail shell (*κοχλίας*, cochlea) to which the instrument owes its name. The height of these screw threads is such that the diameter of the thickened cylinder is equal to one-eighth of the cylinder axis[5]). The channels thus formed are covered with wood again, so that eight spiral tubes are formed. All the wood has been impregnated with pitch; the whole device is surrounded by iron bands.

The cylinder is now adjusted to be rotatable, so that the axis has an inclination relatively to the horizon which is equal to the smallest angle of a right-angled triangle whose sides are in the ratio 3:4:5. If, at a suitable level of immersion of the base below the surface of the water to be pumped out, it is now rotated in the right direction, i.e. such that the opening of each tube is directed downwards when it enters the liquid, the water is found to rise in the tubes.

It is a phenomenon not without reason termed "not only marvellous, but even miraculous" (*non solo maravigliosa, ma miraco-*

[1]) The Egyptian origin, however, is denied again by F. M. Feldhaus, l.c. (Note 4 to p. 21), p. 137.

[2]) *Strabonis Geographica* XVII, 2, 4; 30. rec. G. Kramer (Berlin 1852), p. 377. Strabo was a Greek geographer (66 B.C.–A.D. 24).

[3]) Philo Judæus: *De confusione linguarum*, cap. 38. In: *Philonis Alexandrini Opera qvae supersunt*. Vol. 2, Ed. Paulus Wendland. (Berlin 1897), p. 236.

[4]) Vitruvius, *De Architectura* X, 6. ed. F. Krohn (Leipzig 1912), p. 237. In other editions: X, 11, e.g. *M. Vitruvii Pollionis de Architectura libri decem* (Amsterdam 1649), p. 217.

[5]) *Uti longitudinis octava pars fiat summa crassitudo*; l.c., p. 238, lines 14–15.

losa) by Galilei[1]; there is indeed something paradoxical in the fact that the water is raised, while under the influence of gravity it continually flows down. We shall not here go into the accurate explanation of this phenomenon, which must be able to account, among other things, for the relation between the magnitudes of the angles of inclination of the cylinder axis relatively to the horizon and of the helices relatively to the base of the cylinder, and also for the level of immersion of the base. In principle it consists in that at the chosen values the first part of each tube inclines backwards at the moment when its mouth leaves the liquid, so that the water it contained as long as the mouth was below the liquid level flows in. The next winding of the tube then lies relatively to this water just as the first lay relatively to the water level, and the water thus reaches higher and higher windings.

Vitruvius' description is in particular obscure where he speaks about the way in which the device is started[2]. It can only be inferred with certainty that this was done by continually treading against spokes which were attached somehow to the cylinder. That this starting method was actually used is confirmed by a discovery made during excavations at Pompeii: a fresco painting there uncovered shows a horizontally placed *cochlias*, which is kept rotating by a regular treadle motion[3] by a slave sitting on a beam over the cylinder.

β) The Planetarium.

In a great many classical writers there are to be found reports about the spheres constructed by Archimedes, in which the motions of the heavenly bodies were imitated by means of a mechanism[4]. According to a report by Cicero[5], Marcellus had kept two such contrivances as sole booty for himself during the sack of Syracuse;

---

[1]) Galilei, *Della Scienza Meccanica e delle utilità che si traggono dagl' Instrumenti di quella*. Bologna 1655, p. 31.

[2]) Vitruvius, *De Architectura* X, 6. ed. F. Krohn (Leipzig 1912), p. 238, 18–22.

[3]) *Notizie degli scavi di antichità, pubblicati della R. Accademia dei Lincei*, 1927. The picture has been reproduced by Gino Loria, *Archimede* (I Curiosi della Natura III, Milan 1928), p. 64.

[4]) Astronomical work of Archimedes is discussed in Chapter XIII, 1.

[5]) Cicero, *De re publica* I, 14. Cf. *Tusculanae Disputationes* I, 25. *De natura deorum* II, 34.

the one, which he had caused to be set up in the temple of Virtus, was apparently a closed star globe, on which the various constellations were shown; the other, however, which later came into the possession of C. Sulpicius Gallus[1]), according to the description must have been a complete, spherical, open planetarium, in which with one revolution the sun, moon, and planets performed the same motions relatively to the sphere of the fixed stars as they do in the sky in one day, and in which one could moreover observe the successive phases and the eclipses of the moon.

Archimedes apparently made a deep impression on the classical world in particular by the construction of the last-mentioned instrument; closed globes which, revolving uniformly, imitated the diurnal motions of the fixed stars had indeed long been known: in the passage cited above, Cicero mentions one of Thales, which is said to have been described later by Eudoxus of Cnidus and had been sung by Aratus in an astronomical poem[2]); by others, Anaximander is held to be the first maker. That Archimedes, however, succeeded in representing the mutually independent and widely different motions of sun, moon, and planets by one mechanism simultaneously with the revolution of the sphere of the stars, seemed to be evidence of a superhuman intelligence. No wonder that literature has preserved many records of this achievement: in a well-known epigram[3]), Claudianus depicts the amazement of Jupiter when he sees the work of his hands imitated in a glass sphere by the art of the old Syracusan scholar; perhaps Ovidius also refers to it when in the *Fasti*[4]), in the description of the temple of Vesta, he speaks of the small reproduction of the immense globe, locked up in metal by Syracusan art; we further find the instrument mentioned, among others, in Martianus Capella[5]) and in Cassiodorus[6]), and in several

---

[1]) C. Sulpicius Gallus, a student of astronomy, was consul in 166 B.C.

[2]) Aratus of Soli (about 275 B.C.) wrote a didactic poem on astronomy, entitled *Phaenomena*. This does not, however, contain any description of the sphere in question. *Arati Phaenomena*, ed. E. Maass (Berlin 1893).

[3]) *Claudii Claudiani Carmina*; rec. J. Koch (Leipzig 1893). *Carmina Minora* LI. Claudianus lived about A.D. 400.

[4]) *Ovidius, Fasti* VI, 277.

[5]) Felix Minaeus Martianus Capella, *De nuptiis Philologiae et Mercurii Libri* VIII, ed. A. Dieck (Leipzig 1925). II, 212. VI, 583. Martianus Capella lived about A.D. 470 at Carthage.

[6]) *Cassiodori Variae*, rec. Th. Mommsen. Monumenta Germ. Hist. Auct.

passages of a philosophical or apologetico-theological nature: Cicero uses it[1]) as an argument against the Epicureans; Sextus Empiricus[2]), in his work against science, was induced by it to argue against the materialists the superiority of the creative intellectual principle to matter; Lactantius[3]) uses it to combat the atheists, in a way which has remained usual up to our days (if man has been able to produce such a thing, could not then God have created the prototype of that which the intelligence of his creature is capable of imitating?).

Archimedes himself seems to have attached greater value to the construction of his planetarium than to any of his other technical achievements. Indeed, for this he apparently made an exception to his custom never to leave any written record of his inventions and mechanical contrivances. At least, among his lost works there is mentioned a book *περὶ σφαιροποιΐας* (*On Sphere-making*)[4]), which can hardly have dealt with anything but the manner in which he constructed contrivances such as his planetaria.

Perhaps this book also furnished information on the much-discussed question how the contrivance described by Cicero was set and kept in motion. The most likely explanation is that this was done by means of one of the hydraulic mechanisms of which Heron describes such a great variety, and the application of which is certain at least for the time of the Emperor Augustus[5]).

γ) *The Hydraulic Organ.*

As already became apparent above, the reports about the technical achievements of Archimedes occur not infrequently in writers where one would least expect them: just as Lactantius wrote about

---

Antiq. XII (Berlin 1895). I, 45; p. 41. Cassiodorus was a high official under the first Ostragothic kings; he lived about 490–580.

[1]) Cicero, *De natura deorum* II, 34.

[2]) Sextus Empiricus, *Adversus Physicos* I, 115. *Opera*; ed. J. A. Fabricius (Leipzig 1718), p. 577. Sextus Empiricus was a sceptic of the 3rd century.

[3]) *L. Caeli Firmiani Lactanti Divinae Institutiones* II, 5, 18. rec. S. Brandt *Opera*, *Pars* I. Corpus Script. Eccl. Lat. XIX (Vienna 1890). Lactantius lived about A.D. 300.

[4]) This work is mentioned, on the authority of Carpus of Antioch, by Pappus, *Collectio* VIII, 3; 1026.

[5]) About this see further A. Favaro, *Archimede* (Rome 1923), pp. 40 et seq.

the planetarium in a theological work for an apologetical purpose, Tertullian[1]) in his book *De Anima* used another wonder of Archimedean technology to explain the essence of the soul. It is the so-called hydraulic organ (in which the air fed to the pipes was compressed above water in an air-chamber), the great variety and complexity of whose different parts, which nevertheless constitute a whole, he praises as a most marvellous work of art, after which he turns it to account to illustrate the unity of the soul in spite of the diversity of its functions.

It may be inferred from this that in antiquity Archimedes was credited with the construction of such an instrument, but not that he was held to be its inventor. This honour is attributed by different writers, specifically Pliny the Elder[2]), Vitruvius[3]), and Athenaeus[4]), to the Alexandrian mechanical engineer Ctesibius[5]).

5. *The Defence of Syracuse.*

The widest fame enjoyed by Archimedes in antiquity, however, is not so much due to the peaceful inventions and contrivances hitherto described as to the active part he took in the defence of Syracuse against the Romans. According to Plutarch[6]) he had already before that time constructed engines of war for offensive and defensive purposes at the request of King Hieron; during the reign of the latter, however, there had never been any occasion to use them. But when in 214 B.C. the Romans laid siege to the town, they were confronted with a formidable and unexpected opponent in the person of the already aged mathematician, who, if the reports are to be trusted, personally superintended the use of his instruments.

Detailed information about this side of Archimedes' work is found

---

[1]) Tertullianus, *De Anima* II, 14; ed. J. H. Waszink (Amsterdam 1933), p. 58.

[2]) *C. Plinii Secundi Naturalis Historiae Libri* XXXVII; ed. C. Mayhoff (Leipzig 1909) VII, 38. Vol. II, 43–44.

[3]) Vitruvius, *De Architectura* IX, 8. ed. F. Krohn (Leipzig 1912), p. 219.

[4]) Athenaeus, *Deipnosophistae* (*vide* Note 2 to p. 14) IV, 174b–d.

[5]) Ctesibius probably lived from about 300 to 230 B.C. The reader may consult on him A. G. Drachmann, *Ktesibios, Philon and Heron. A study in ancient pneumatics.* Copenhagen 1948.

[6]) Plutarch, *Vita Marcelli* XIV, 9 (306).

in three historians, in Polybius[1]), in Livy[2]), and in Plutarch[3]). They tell about powerful ballistic machines, which from the distance discharged heavy blocks of stone on the Roman legions and which, in combination with machines for shorter range, the so-called scorpions, which cast their missiles through holes in the walls, repulsed the enemy on the land side. In detail they further describe the defence on the sea side, where the walls of the suburb Achradina descended sheer into the water, and where the Roman general Marcus Claudius Marcellus personally conducted the assault of the fleet. Here, in addition to the various types of engines of war, cranes were in action which, when turned outwards, dropped large stones or heavy pieces of lead on the approaching Roman ships or which lifted the prows by means of an iron hand, after which they suddenly let the ships fall on the water again. In this way Archimedes even succeeded in baffling the assaults made with the aid of the so-called *sambuca* (*σαμβύκη*[4])), a broad scaling ladder which, rapidly pulled up on two interconnected quinqueremes, was to make possible the scaling of the walls.

On the Roman soldiers all this is said to have made such a deep impression that as soon as they saw a rope or a piece of wood projecting above the walls, they would take to their heels, shouting that Archimedes had invented another engine to destroy them[5]). Marcellus himself, though badly thwarted in his plans, seems to have had the greatest admiration for his mathematical opponent. "Shall we not", Plutarch represents him as saying jestingly to his own engineers[6]), "make an end of fighting against this geometrical

---

[1]) Polybius of Megalopolis, born about 204 B.C., Achaean statesman, writer of the historical work *πραγματική ἱστορία* in 40 books, which have partly been preserved. He was nearest in time to the events at Syracuse. His story about it is to be found in *Polybii Historiae* VIII, 5 et seq.; ed. F. Hultsch (Berlin 1868) I, 623 et seq.

[2]) Titus Livius Patavinus (59 B.C.–A.D. 17) wrote a history of Rome (*Ab urbe condita*) in 142 books, which have partly been preserved. On Archimedes: XXIV, 34.

[3]) Plutarch, *Vita Marcelli* XV–XVII (306–308). There are also reports about the defence of Syracuse by Archimedes in the epic *Punica*, XIV, 292 et seq. by Silus Italicus, ed. L. Bauer (Leipzig 1892) II, 92.

[4]) This engine of war, which owed its name to its resemblance to a musical instrument, is described in detail by Polybius, *Historiae* VIII, 6.

[5]) Plutarch, *Vita Marcelli* XVII, 3 (307).

[6]) Plutarch, *Vita Marcelli* XVII, 1 (307).

Briareus[1]), who uses our ships to ladle water from the sea, who has ignominiously threshed and driven off the *sambuca*, and who by the multitude of missiles that he hurls at us all at once outdoes the hundred-armed giants of mythology?" Actually the Romans did not succeed in conquering the town until after a long siege, about which further details are to be found in Polybius and Livy.

The traditional stories about the defence of Syracuse also refer to the famous burning mirrors, with which Archimedes is said to have focussed the sun's rays so intensely on the ships of the Romans that they caught fire. This story, the technical improbability of which is obvious, is found in none of the three above-mentioned historians, and for this reason already deserves less credit than the reports about the ballistic instruments, which in spite of evident exaggeration may quite well contain a core of truth. The oldest passage speaking about the setting on fire of Roman ships is to be found in Lucian[2]), who does not, however, refer to mirrors and merely states that Archimedes set the enemy's triremes on fire by artificial means; this may also imply that he spouted burning substances on to them in the way he is said to have done this in Spain, according to the report preserved in Leonardo. That he used mirrors is not explicitly stated before the physician Galen[3]) and the Byzantine architect Anthemius of Tralles, who in the 6th century wrote a book on burning mirrors, which is of importance for the history of the focal properties of conic sections. According to him[4]) the contrivance of Archimedes was composed of a large number of small, flat mirrors. In later centuries these mirrors are referred to much more frequently. Eustathius[5]) mentions them in his com-

---

[1]) Briareus or Aegaeon was one of the three *Hecatoncheires*, hundred-armed giants, sons of Uranus and Gaia.

[2]) Lucian of Samosata, rhetorician and philosopher, lived in the 1st century A.D. The passage in question is *Hippias*, cap. 2. ed. N. Nilén (Leipzig 1906), p. 19.

[3]) Galen (A.D. 129–199) was the greatest physician of antiquity after Hippocrates. His statement about Archimedes is to be found in *De Temperamentis* III, 2 *Galeni Opera ex octava Juntarum editione* (Venice 1609), I. 23.

[4]) He was born at Tralles, and died in 534. The statement about Archimedes is to be found in *περὶ παραδόξων μηχανημάτων*, published in *ΠΑΡΑΔΟΞΟΓΡΑΦΟΙ. Scriptores rerum mirabilium graeci*, ed. A. Westermann (Brunsvick 1839), pp. 153, 156.

[5]) *Eustathii Archiepiscopi Thessalonicensis Commentarii ad Homeri Ilia-*

mentary on the Iliad, when he discusses the theory that the inextinguishable blaze shining about the shield and helmet of the hero Diomedes[1]) had been caused by the reflection of sunlight. Naturally Tzetzes does not fail to refer to it[2]), and Zonaras[3]), too, explicitly says on the authority of Dio Cassius[4]) that Archimedes received the sun's rays on a mirror, that he caused the air to ignite through its own density and through the smoothness of the mirror, and that he could thus set fire to any ship lying in the path of this fire. According to him, Proclus used the same weapon at Constantinople in 514, when he helped to defend the city against the fleet of Vitalianus.

In the 17th and 18th centuries the question whether the effect attributed to Archimedes can actually be attained in practice was dealt with by various investigators. We do not intend to go into this, but will only mention the observations of Athanasius Kircher in his *Ars Magna Lucis et Umbrae*[5]), the experimental investigations of Buffon, which are discussed in his *Histoire Naturelle*[6]), and those of Peyrard, which are incorporated in the appendix to his Archimedes translation[7]). It is noteworthy that all three, like Anthemius, try to attain the object by a combination of a large number of small, flat mirrors, which are independently rotatable in every direction[8]).

dem II, 3 (Leipzig 1828). Eustathius was archbishop of Thessalonica in the 12th century.

[1]) *Iliad* V, 4.

[2]) Tzetzes, *Chil.* I, *Hist.* 35, 151. IV, *Epistola,* lines 506–507.

[3]) Joannes Zonaras was a Byzantine monk, who in the first half of the 12th century wrote a historical work on the period from the Creation to 1118, entitled *'Επιτομή ἱστοριῶν. Iohannis Zonarae Annales* IX, 4. ed. M. Pinder (Bonn 1844), II, 210.

[4]) Dio Cassius at the end of the 2nd century A.D. wrote a work on Roman history, of which Zonaras made use very frequently.

[5]) *Athanasii Kircheri Fuldensis Buchonis S.J. Ars magna Lucis et Umbrae in decem Libros digesta* (Rome 1646) Liber X; Pars III. Kircher was a German Jesuit, professor at Würzburg, later at the Collegium Romanum in Rome.

[6]) Buffon, *Histoire naturelle générale et particulière . . . Supplément* I (Paris 1774), *Sixième Mémoire. Expériences sur la lumière et sur la chaleur qu'elle peut produire.* pp. 412 et seq.

[7]) *Oeuvres d'Archimède traduites littéralement, avec un Commentaire* par F. Peyrard . . . *suivies d'un mémoire du Traducteur, sur un nouveau Miroir Ardent* (Paris 1807), p. 543.

[8]) For an exhaustive discussion of the whole question of the burning

6. *The Death of Archimedes.*

During the sack of Syracuse, to which Marcellus, allegedly with tears and not without having made numerous restrictions, had been forced to consent[1]), Archimedes also perished, in spite of the explicit order to spare him, and that under circumstances the main features of which have remained indelibly imprinted on the memory of mankind.

The details of the story indeed have become somewhat corrupted in the course of the ages; this induces us to subject the accounts of classical writers about the story once more to a careful examination; typical anecdotes after all are entitled to historical correctness just as much as demonstrable facts.

The oldest reference is to be found in Livy[2]): according to this account, amid the confusion accompanying the capture of the town, Archimedes while intent on some figures he had drawn in the dust[3]) was killed by a soldier who did not know who he was. Plutarch has several versions: according to two of them[4]) the old mathematician was so much intent on a diagram with his eyes and his mind alike that he had not noticed the incursion of the Romans; a soldier then came up to him, ordering him to follow him to Marcellus; Archimedes refused to do so until he had worked out his problem, upon which the soldier became enraged and killed him. Others say that the soldier already threatened his life while coming up to him, and that Archimedes begged him earnestly to delay the execution of his threat for a moment until the problem was solved and the proof completed. According to a third version he was killed, while on his way to Marcellus with some of his mathematical instruments, by some soldiers who were under the impression that he carried gold.

When we read these accounts, we are chiefly struck by two things: firstly, that none of these oldest sources mentions the familiar saying which is alleged to have been uttered by Archimedes to his assailant and which has become universally known in the form *noli turbare circulos meos*; and secondly, that in none of them is

---

mirrors of Archimedes the reader should consult D. Burger, *Heeft Archimedes de brandspiegels uitgevonden?* Faraday **17** (1947/48) 1–5.

[1]) Plutarch, *Vita Marcelli* XIX, 1 (308).

[2]) Livy, *Ab urbe condita* XXV, 31.

[3]) *intentum formis, quas in pulvere descripserat.*

[4]) Plutarch, *Vita Marcelli* XIX, 4, 5 (308 309).

there any question of the drawing of diagrams in the sand on the ground, to which the accepted tradition of later ages refers. The oldest source putting definite words into the mouth of the threatened mathematician is Valerius Maximus[1]), who makes him say, while stretching out his hands in protection over his diagram: *noli obsecro istum disturbare.* In Tzetzes[2]), even more unrestrainedly and naturally, he snarls at his assailant: *'Απόστηθι, ὦ ἄνθρωπε, τοῦ διαγράμματός μου* (Fellow, stand away from my diagram). The view, however, that Archimedes drew his diagrams in the sand on the ground (in itself already improbable in view of the complexity of his proofs) is based on nothing but the report of Valerius Maximus, which speaks, apparently owing to a mere error, of diagrams "*in terra*" instead of "*in pulvere*"[3]).

Indeed, by this *pulvis*, of which not only Livy, but also Cicero[4]) speaks, is undoubtedly meant something quite different from the sand on the ground. In fact, it is known that the Greek mathematicians drew their diagrams in a surface of smoothed sand or glass dust filling a tray which, just like the familiar calculating instrument, was called abacus[5]). The filling material was called *pulvis*, which also accounts, among other things, for the expression *pulvis eruditus* as a metaphor for mathematics[6]) in Cicero, and for his description of Archimedes as *homunculum e pulvere et radio*[7]).

---

[1]) Valerius Maximus between A.D. 26 and 32 wrote a historical work *Factorum et dictorum memorabilium Libri* 9, where in VIII, 7, 7 ed. C. Kempf (Leipzig 1888), p. 390 the story of the death of Archimedes is told.

[2]) Tzetzes, *Chil.* II, *Hist.* 35, 140.

[3]) *l.c.* (*Note 1*): *at is, dum animo et oculis in terra defixis formas describit, militi qui praedandi gratia domum inreperat strictoque super caput gladio quisnam esset interrogabat, propter nimiam cupiditatem investigandi quod requirebat nomen suum indicare non potuit, sed protecto manibus pulvere "noli", inquit, "obsecro istum disturbare" ac perinde quasi neglegens imperii victoris obtruncatus sanguine suo artis suae lineamenta confudit.*

[4]) Cicero, *De finibus bonorum et malorum* V, 19, 50.

[5]) The most striking reference is the statement of St. Jerome in his commentary on Ezekiel, Chapter 4, where there is question of the portraying of the city of Jerusalem on a tile; he there speaks of a *πλινθεῖον quem nos laterculum et abacum appellare possumus; in cuius pulvere solent geometrae grammas i.e. lineas radiosque describere. Operum D. Hieronymi Quintus Tomus Commentarios in prophetas quos maiores vocant, continet.* Basle 1553. p. 387. col. b.

[6]) Cicero, *De natura deorum* II, 18, 48.

[7]) Cicero, *Tusculanae Disputationes* V, 23.

We shall therefore have to imagine Archimedes in the hour of his death, not out of doors, poring over the ground (he could hardly have failed to notice the tumult of war there), but indoors, sitting at a small table, on which was placed the abacus (which is said to have been rather heavy).

This view is supported in the most striking manner by a mosaic found at Herculaneum, which apparently represents the famous scene of Archimedes' death; a reproduction faces the title-page of this book. It originates from the legacy of Jérôme Bonaparte, later came into the possession of Frau E. Schabell at Wiesbaden, and was published in 1924 by F. Winter[1]). The publisher takes it to be a reproduction of a picture from the beginning of the Hellenistic era, which may have been made while the artist was still under the impression of the scholar's fate. The picture shows a man behind a small table on which is placed a rectangular tray, looking up indignantly at a warrior, who stands beside him with his sword drawn and who apparently orders him to come along with him. The threatened scholar stretches out his hands protectively over his work, and it is hardly to be doubted but he here utters the words which enraged the Roman so much and put an end to his precious life.

According to a statement by Plutarch[2]), Archimedes during his lifetime expressed the wish that upon his tomb there should be placed a cylinder circumscribing a sphere within it, together with an inscription giving the ratio between the volumes of these two bodies[3]), which he had discovered. It seems that Marcellus, who is said to have been greatly distressed at the death of his great opponent, and who conferred honours on his dependents, saw to it that this wish was fulfilled. At least, Cicero relates[4]) that when in 75 B.C. he came as quaestor in Sicily, he succeeded in finding at the gates of Achradina a sepulchral column, overgrown with brushwood and thorns, on which the figure of sphere and cylinder was reproduced.

---

[1]) Franz Winter, *Der Tod des Archimedes.* 82 Winkelmannsprogramm der Archäologischen Gesellschaft zu Berlin (Berlin 1924).

[2]) Plutarch, *Vita Marcelli* XVII, 7 (307).

[3]) *On the Sphere and Cylinder* I, 34; Porism.

[4]) Cicero, *Tusculanae Disputationes* V, 23.

CHAPTER II.

THE WORKS OF ARCHIMEDES MANUSCRIPTS AND EDITIONS

As we already saw, Archimedes was accustomed to send his mathematical writings to his colleagues at Alexandria: Conon, Dositheus, Eratosthenes, who apparently saw to their further distribution. The documents were accompanied by introductory letters, which contained a summary of the propositons to be proved and not infrequently referred back to statements made on earlier occasions. In this way we know some particulars about the order in which he made his discoveries and about the way in which he published them.

It appears that he would usually first send the propositions only, with the request to discuss them with other mathematicians and urge them to find the proofs for themselves; the complete treatises would follow afterwards. On the whole these seem not to have been dispatched until after the death of Conon: in fact, the introduction to one of the oldest works, viz. *Quadrature of the Parabola*, begins with words devoted to the memory of his admired friend, and also contains the suggestion that Dositheus should henceforth act as intermediary with the Alexandrian mathematicians[1]). Afterwards Archimedes more than once states with respect to new treatises that Conon, if he were still alive, would undoubtedly have been capable of appreciating the discoveries and supplying the missing proofs[2]). The question whether Dositheus was less successful in the latter or only took an understandable great interest in the methods applied by the discoverer himself need not occupy us here: again and again, however, he urges Archimedes to send the complete proofs[3]). This desire does not always seem to have been fulfilled very soon; from the introduction to the work *On Spirals* it is evident that many years had elapsed between the first publication of the propositions and problems of *On the Sphere and Cylinder* and the communication of the proofs and solutions; this is justified by

[1]) *Opera* II, 262.

[2]) *Opera* I, 4. II, 2.

[3]) *Opera* I, 168. II, 2.

Archimedes with the argument that he preferred to leave it to mathematicians to find out things for themselves[1]). There was also some malignant design in this, for on the same occasion[2]) he reveals that two of the propositions on the sphere formerly enunciated by him are incorrect, and that he had added them in order to entice those who are always saying of everything that they have found it, without ever giving proofs, into saying that they had discovered something impossible[3]).

The works dispatched by Archimedes seem to have been studied at Alexandria, but never to have been combined into one corpus[4]); that is no doubt the reason why they were partly lost. From statements by Heron, Pappus, and Theon of Alexandria it is, however, certain that in the 3rd and 4th centuries more of his works were extant than are now in our possession. Those other works probably perished in 391 in the fire of the Serapeion[5]). Even those of his works which already had a sufficiently wide circulation in antiquity not to be threatened with extinction through such catastrophes, were early exposed to mutilation and neglect. A proof[6]) announced by Archimedes in *On the Sphere and Cylinder* was no longer present in the 2nd century B.C.[7]); and his treatise *Measurement of the Circle*, which is quite evidently a fragment of or an excerpt from a longer work, in the period which the commentator Eutocius was acquainted with apparently never existed in any other form but that in which we exclusively know it[8]). It also seems that people soon

---

1) *Opera* II, 2.

2) *Opera* II, 4.

3) The propositions in question are as follows:

α) If a sphere is divided by a plane into unequal parts, the ratio of the volumes of the segments is the duplicate ratio of that of their surfaces.

β) Of all segments of spheres which have equal surfaces the segment whose height is the fourth part of the diameter has the greatest volume.

4) The subsequent particulars about the history of the Archimedes manuscripts have been taken from the Prolegomena of the second edition of the Greek text by Heiberg. *Opera* III, i–xcviii.

5) Temple of Serapis of Alexandria with famous library.

6) *Opera* I, 192; lines 5–6.

7) *Opera* III, 130; 1.19. Eutocius here reports that the proof in question was not to be traced in any codex, and that Dionysodorus and Diocles (who both lived in the 2nd century A.D.) consequently tried to supplement the missing argument.

8) *Opera* III, 228.

began to study in particular the more elementary works, specifically *On the Sphere and Cylinder*, *Measurement of the Circle*, and *On the Equilibrium of Planes*, while the profounder investigations fell more or less into oblivion; it is striking to note in what vague terms a writer like Heron of Alexandria can speak about the discoveries of Archimedes[1]), and how slight the influence of his hydrostatic discussions remained throughout antiquity[2]); one even gets the impression that the authoritative and accurate Eutocius, who comments in detail on the three works just mentioned, did not know the other now famous books, notably *Quadrature of the Parabola* and *On Spirals*, or only knew them from hearsay[3]).

Meanwhile the composition of the above-mentioned commentaries does testify to a growing desire to become thoroughly acquainted with at least the more elementary parts of the work of the great mathematician. This desire found expression especially in the school of the Byzantine mathematicians Anthemius[4]) and Isidorus[5]), who in 532 were entrusted by the Emperor Justinian with the re-construction of St. Sophia at Constantinople and who, whether they were urged by the practical problems of architecture to make theoretical studies or were inspired by love of pure mathematics, showed great interest in the work of Archimedes. Isidorus apparently made his pupils study in particular the works *On the Sphere and Cylinder* and *Measurement of the Circle* with the relative commentaries of Eutocius[6]). It is probably also due to the use of

---

[1]) *Vide e.g.* the preface to the second book of Heron's *Metrica*. *Heronis Opera* III, 92. This discussion relates to determinations of volume with regard to spheres, cones, and cylinders, and Heron now says that some people trace the methods for this, as being surprising, to Archimedes, but he does not state whether they were right in doing so. Equally vague is the way in which he speaks about the hydrostatical work of Archimedes in the last chapter of the *Metrica* (II, 20, *Heronis Opera* III, 138).

[2]) In the commentators of Aristotle, Alexander Aphrodisiensis, Themistius, and Simplicius there is no trace of their being acquainted with even the elements of the hydrostatics of Archimedes.

[3]) *Opera* III, xcii.

[4]) Vide Note 4 to p. 28.

[5]) Born at Miletus; after 532 he was charged with superintending the building of St. Sophia.

[6]) These commentaries contain in the manuscripts at the end the statement that they have been "revised by our teacher, the Milesian mechanician Isidorus". This used to be taken for a statement by Eutocius, which is,

such works for instructive purposes that these alone were translated into Attic from the Siculo-Dorian dialect used by Archimedes.

The school of Isidorus achieved considerable merit for the development of mathematics because in an era when the sources of mathematical creative power had dried up they at least realized the value of the great works of the past, and were careful to study and preserve them. Continuing this tradition, the scholars of Constantinople also were the first consciously to make it their object to collect the dispersed works of Archimedes, and it is to these efforts that the most important codices, from which western mathematicians were to derive their knowledge of the work of their great master, owe their existence. In fact, in the 9th century the encyclopaedist Leon of Thessalonica, nicknamed the Iatrosophist or the Philosopher[1]), who through his activity put new life into Byzantine science, provided for the compilation of a manuscript (in the notation of Heiberg: Codex A[2])), which became the archetype of all manuscripts from which prior to 1906 the Greek text of Archimedes could be inferred. It is probably also from Constantinople that the collection of works on mechanics and optics (in Heiberg: Codex 𝔅) originates, which was used in the 13th century in the preparation of the first translation of Archimedes into Latin, and the same applies to the famous manuscript (called Codex C in Heiberg), the discovery of which at the end of the 19th century proved to have constituted a contribution to the knowledge of his works which was as important as it was unexpected.

We will now give a brief outline, based on Heiberg's philological investigations, of the fates of these different manuscripts. The Codices A and 𝔅 seem to have made their way to Western Europe in the days when under the rule of the Norsemen and their suc-

---

however, chronologically impossible, because he was a contemporary of Ammonius, the pupil of Proclus and teacher of Simplicius; he must therefore have been born at the end of the 5th century, and was thus certainly not a pupil of Isodorus. The addition was apparently made by a pupil of the Byzantine school.

[1]) This Leon, who is not to be confused with the Emperor of the same name, Leo VI, surnamed the Philosopher, was archbishop of Thessalonica and professor at the Magnaura University at Constantinople. He lived under the Emperors Theophilus (829–842) and Michael III (842–867). He is known especially for his medical encyclopaedia, the *Σύνοψις ἰατρική*.

[2]) *Opera* I, V et seq. and III, *Prolegomena*.

cessors from the house of Hohenstaufen a new centre of culture began to flourish in Sicily, and the first desire awoke to enrich intellectual life with the treasures of Greek science. After the battle of Benevento (1266), in which Manfred, the last of the German rulers of the Sicilian kingdom, lost his crown and his life, the two Archimedes manuscripts, together with the whole library of the vanquished king, passed into the possession of the Pope. At the end of the 14th century Codex A must have got into private hands; it is further established that in 1491 it was the property of the Italian humanist Giorgio Valla. From this time dates a summary of the contents[1]), from which may be seen that it contained the following works: *On the Sphere and Cylinder*, *Measurement of the Circle*, *On Conoids and Spheroids*, *On Spirals*[2]), *On the Equilibrium of Planes*, *The Sand-reckoner*, *Quadrature of the Parabola*, to which were added the commentaries of Eutocius and the work *De Mensuris* of Heron. Valla seems to have meant to prepare a complete Latin edition of these works, but nothing came of it; fragments from Archimedes only are found in the work of his hand *De expetendis et fugiendis rebus*, published after his death in 1500 (Venice 1501).

After Valla's death the manuscript was bought for 800 gold pieces by Alberto Pio, prince of Carpi, from whom his nephew, Cardinal Rodolfo Pio, inherited it in 1550; the latter died in 1564, but the catalogue of his library, which was then compiled, no longer contains any Archimedes manuscript. Codex A must therefore have been lost between 1550 and 1564 or passed into other hands; it has not been possible to trace it.

Naturally this does not mean that the contents of the precious document were also lost; on the contrary, Codex A had a voluminous offspring in the form of copies and translations which were made of it in the course of more than two centuries and which include the most reliable foundations of the modern editions of the Greek text.

A large part of A was translated into Latin in 1269 by the Flemish Dominican Willem van Moerbeke[3]), who was engaged from 1268

---

[1]) *Opera* III, xi.

[2]) It is apparently this work which is referred to as *De revolutionibus ad Dositheum*.

[3]) He was born about 1215 at Moerbeke-lez-Grammont; he entered the

to 1280 at the Papal Court at Viterbo in making several Greek works accessible in a translation to the scholars of Western Europe (among whom knowledge of Greek was then still exceptional), and who thus had an important share in the spreading of Hellenic culture[1]). The original manuscript of this translation (in Heiberg: Codex B) was found again in Rome in 1884[2]); it contains literal translations (so literal indeed that even where the translator did not understand the text correctly, they are equivalent to a Greek version) of the works *On Spirals*, *On the Equilibrium of Planes*, *Quadrature of the Parabola*, *Measurement of the Circle*, *On the Sphere and Cylinder*, *On Conoids and Spheroids*, *On Floating Bodies*; in addition also the commentaries of Eutocius (except that on *Measurement of the Circle*), a work by Alhazen on burning mirrors, a treatise *De Ponderibus*, and two treatises by Ptolemy. The translation was made between February and 10th December 1269.

Moerbeke's translation contains one work of Archimedes not appearing in Codex A, viz. *On Floating Bodies*. This shows that he must also have used another source for his work. It has been found that this must have been the above-mentioned Byzantine collection of writings on mechanics and optics, which included, besides the work *On Floating Bodies*, also Archimedes' *Quadrature of the Parabola* and *On the Equilibrium of Planes*. It found its way to Western Europe in the same manner as Codex A. Its presence in the Papal library is attested for the years 1295 and 1311; in later years no trace of it is to be found any more.

Codex B itself was in the possession of the German priest Andreas Conerus († 1527) in 1508 in Rome; he took great interest in Greek mathematics and made several corrections in the text. Who were the other owners up to the year 1740, when it got into the library

---

order of St. Dominic, and at Cologne heard the lectures of Albertus Magnus. Then he probably spent a long time in the East, where he acquired a profound knowledge of Greek and of oriental languages. From 1268 to 1276 he was chaplain to Pope Clement IV and his successors, and subsequently archbishop of Corinth. He died before or in 1297. *Vide* H. Bosmans S. J. *Guillaume de Moerbeke et le Traité des corps flottants d'Archimède*. Revue des Questions scientifiques, 1922. Separate.

[1]) Through him, Thomas of Aquinas and other scholastic doctors became acquainted with some of the works of Aristotle in the 13th century.

[2]) Codex Ottobonianus Latinus 1850.

of the Vatican, is to be found accurately enumerated in Heiberg[1]).

Of Codex A several more Greek copies were made. Between 1449 and 1468 Cardinal Bessario[2]) had one made (in Heiberg: E = Codex Marcianus 305, Venice). A second copy (in Heiberg: D = Codex Laurentianus 28, 4to, Florence) dates from the time when Valla was in possession of the original. It was made in 1491 by order of the famous Florentine humanist Angelo Poliziano[3]) for the library of the Medici family; this seems to have taken place not without some protests on the part of the owner, who jealously guarded his treasure, and who refused its inspection by anyone else. Two other important copies (G = Codex Parisiensis 2360, and H = Codex Parisiensis 2361) were made in the time when the original was the property of the Pio family. H was written in 1544 by Christoph Auer, by order of bishop Georges d'Armagnac, ambassador of King Francis I in Rome, for the Royal Library at Fontainebleau. We are passing by in silence several other less important copies.

A second important Latin translation of A, however, which was made in 1450 by order of Pope Nicholas V by Jacobus Cremonensis, a priest of San Cassiano, also has to be recalled. A copy of this, corrected by reference to E, was brought to Germany about 1468 by Joannes Regiomontanus[4]) from his first Italian journey; a plan to publish the work was not carried out, but the manuscript, which is preserved at Nürnberg[5]), was later used in the preparation of the *editio princeps*.

The summary of the events of the Greek manuscripts of works of Archimedes and their Latin translations given above is by no means sufficient to judge of the diffusion of some of these works during the Middle Ages. Recently several Latin translations of Arabic versions were made known[6]). Before 1269 there was a translation

---

1) *Opera* III, lxiii.

2) Bessario lived from 1403 to 1472, and was a cardinal from 1439 onwards. He had an important share in the revival of interest in Greek culture.

3) Angelo Poliziano (1454–1494) was a famous humanist at the court of Lorenzo de' Medici.

4) Johann Müller, from a small place near Königsberg (Franken), called on that account Regiomontanus or Joannes de Monte Regio, was the principal German mathematician and astronomer of the 15th century. He lived from 1436 to 1476.

5) Norimbergensis Cent. V, 15, chartaceus manu Regiomontani scriptus.

6) For more detailed information the reader may consult the valuable articles of Marshall Clagett: *Archimedes in the Middle Ages. The De Mensura*

of *Measurement of the Circle* possibly by Plato of Tivoli, and a second one which in all probability may be ascribed to Gerard of Cremona. It appears that this work was rather popular. Also before 1269 one Johannes de Tinemue wrote a commentary on Book II of *On the Sphere and Cylinder* under the title *De curvis superficiebus Archimenidis*.

Thus far we have only spoken of manuscripts. In the 16th century, however, the growing desire to become acquainted with the works of the great Greek mathematicians gave rise to the publication in printed form of works of Archimedes.

The oldest of these editions is to be found in a now very rare booklet by the Neapolitan mathematician Luca Gaurico on the quadrature of the circle, which appeared in 1503 in Venice[1]); this contains the Latin translation of *Measurement of the Circle* and *Quadrature of the Parabola*, taken from Codex B before its correction by Conerus.

A literal copy of this edition, enlarged with the Latin text of *On the Equilibrium of Planes* and of Book I of *On Floating Bodies*, was published[2]) in 1543 by Nicolo Tartaglia[3]). In his preface the editor boasts of the great difficulties which he had to overcome in the deciphering and the translation of ancient and all but illegible Greek manuscripts, and which he only managed to surmount through his "incredible desire" to accomplish the work. According to Heiberg, however, it is a barefaced lie that he used any Greek text at all; he merely copied the edition of Gaurico and a copy of Codex

---

circuli. Osiris X, 587–618, and a review on George Sarton, *Horus. A Guide to the History of Science*. Waltham (Mass) 1952, in Isis 44: 1,2. Nos 135–136; 92–93.

[1]) *Tetragonismus id est circuli quadratura per Campanum, Archimedem Syracusanum atque Boetium mathematicae perspicacissimos adinventa*. Venetiis 1503. Quoted *Opera* III, lxiii.

[2]) *Opera Archimedis Syracusani Philosophi et Mathematici ingeniosissimi per* Nicolaum Tartaleam Brixianum (*Mathematicarum scientiarum cultorem*) *multis erroribus emendata, expurgata, ac in luce posita, multisque necessariis additis, quae plurimis locis intellectu difficillima erant, commentariolis sane luculentis et eruditissimis aperta, explicata atque illustrata existunt. Appositisque manu propria figuris quae graeco exemplari deformatae ac depravatae erant, ad rectissimam Symetriam omnia instaurata, reducta et reformata elucent.* Venetiis 1543.

[3]) Nicolo Tartaglia (Brescia 1506–Venice 1557) was a well-known Italian mathematician.

B, errors and all[1]). From the legacy of Tartaglia, the entire work *On Floating Bodies* was published in 1565 by Curtius Troianus, a publisher at Venice[2]).

Meanwhile in 1544 the *editio princeps* of the works of Archimedes had been published at Basle by Thomas Gechauff Venatorius[3]); this edition contains all the works known at that time, in Greek with a Latin translation, and also the commentaries of Eutocius. The Greek text was taken from a manuscript[4]) which Wilibald Pirckheymer († 1530) had acquired in Rome; it is mainly a copy of A, but the writer also seems to have consulted B for his work. The Latin text is the one of the translation by Jacobus Cremonensis, corrected by Regiomontanus, which has already been mentioned above.

In 1558 there appeared at Venice a Latin translation of a number of Archimedes' works, prepared by the excellent authority on Greek mathematics, Federigo Commandino[5]); it contains *Measurement of the Circle*, *On Spirals*, *Quadrature of the Parabola*, *On Conoids and Spheroids*, and *The Sand-reckoner*. The text was taken from B or a copy of it; the editor did not use any Greek codex, and did not know Tartaglia's edition. In 1565 he completed his edition with a translation of the work *On Floating Bodies*[6]). Towards the end of the 16th century another new translation of all the works into Latin was published: the *Monumenta* of Francesco Maurolico[7]).

---

[1]) *Opera* III, lxiv.

[2]) *Archimedis de insidentibus aquae* . . . (*ex recensione Nicolai Tartaleae*) Venetiis 1565.

[3]) *Archimedis Syracusani Philosophi ac Geometriae Excellentissimi Opera quae quidem extant, omnia, multis iam seculis desiderata, atque a quam paucissimis visa; nuncque primum et Graece et Latine edita. Adiecta quoque sunt Eutocii Ascalonitae in eosdem Archimedis libros Commentaria item Graece et Latine, nunquam antea excusa.* Basileae. Joannes Hervagius excudi fecit. MDXLIIII.

[4]) Codex Norimbergensis cent. V app. 12 chartaceus s. XVI.

[5]) Federigo Commandino (of Urbino, 1509–1575) edited Latin translations of Euclid, Archimedes, Apollonius, Aristarchus, Ptolemy, Heron, and Pappus. *Archimedis Opera non nulla a* Federico Commandino Urbinate *nuper in latinum conversa, et commentariis illustrata.* Venetiis, MDLVIII.

[6]) *Archimedis de iis quae vehuntur in aqua libri duo a* Federigo Commandino Urbinate *in pristinum nitorem restituti, et commentariis illustrati.* Bononiae, MDLXV.

[7]) *Admirandi Archimedis Syracusani Monumenta omnia mathematica quae extant . . . ex traditione doctissimi viri* D. Francisci Maurolici. Panormi,

The printed editions in Greek and Latin which were published in the 16th century were followed by many more in later days. Of these we mention in the first place the edition by David Rivault[1]) (Paris 1615); this gives the propositions in Greek and in Latin, and the proofs, slightly adapted, in Latin. It is on this edition that the oldest translation into a living language was based, *viz.* the one into German by J. C. Sturm[2]) (Nürnberg 1670). In England Isaac Barrow[3]) prepared a new Latin edition (London 1675), while Wallis[4]) edited *The Sand-reckoner* and *Measurement of the Circle* with the commentary of Eutocius on the latter work (Oxford 1676). At the end of the 18th century appeared, likewise at Oxford, the monumental edition of the Greek text with Latin translation by the Italian mathematician Joseph Torelli[5]) (published after his death by Abram Robertson).

This was followed by translations into living languages again: a translation of the works *On the Sphere and Cylinder* and *Measurement of the Circle* into German by K. F. Hauber[6]) (Tübingen 1798), a French translation of all the works with a commentary by F. Peyrard[7]) (Paris 1807), a German translation of *Quadrature of the*

MDCLXXXV. This is a later reprint of the original edition of 1570, which was lost, but for a few copies, in a shipwreck. Francesco Maurolico (Messina, 1494–1575) published editions and commentaries of several Greek mathematicians.

[1]) Vide Note 2 to p. 11.

[2]) Joh. Chr. Sturm, *Des unvergleichlichen Archimedis Kunstbücher, übersetzt und erläutert* (Nürnberg 1670). Three years before, the same writer had also translated *The Sand-reckoner*. Quoted in Heath, *Archimedes. Introduction* xxx.

[3]) Is. Barrow, *Opera Archimedis, Apollonii Pergaei conicorum libri, Theodosii sphaerica methodo novo illustrata et demonstrata.* Londini 1675. Quoted in Heath, *Archimedes. Introduction* xxx.

[4]) *Archimedis Syracusani Arenarius et Dimensio Circuli. Eutocii Ascalonitae in hanc Commentarius. Cum Versione et Notis* Joh. Wallis. Oxonii 1676. Also in Johannis Wallis *Opera Mathematica tribus voluminibus contenta.* III (Oxoniae 1699), 509, 539. John Wallis (1616–1703) was a well-known English mathematician; he was Savilian Professor of Geometry at Oxford.

[5]) Vide Note 5 to p. 12 Torelli (1721–1781) was an Italian philologist; the quoted edition contains a biography.

[6]) *Archimeds zwey Bücher über Kugel und Cylinder. Ebendesselben Kreismessung. Uebersetz, mit Anmerkungen . . . begleitet* von Karl Friderich Hauber, Tübingen 1798.

[7]) Vide Note 7 to p. 29.

Parabola by J. J. I. Hoffmann[1]) (Aschaffenburg 1817), and a complete German one with critical notes by Ernst Nizze[2]) (Stralsund 1824).

Meanwhile the knowledge of the work of Archimedes had been widened by two new discoveries: in the 17th century Foster[3]) in England and Borelli[4]) in Italy had edited shortly after one another Latin translations of a work by the Arabian mathematician Thâbit b. Qurra[5]), in which at all events certain discoveries of Archimedes are discussed[6]); and in 1773 Lessing[7]) had edited an epigram in which the *Cattle-Problem* attributed to Archimedes is formulated.

Nevertheless there still remained several lacunae. In the first place several works were (and still are) unknown which are quoted by ancient writers (about which more will be said presently), but the Greek text of *On Floating Bodies* also was still missing. In 1828 Cardinal Angelus Maii[8]) based on two manuscripts discovered at

---

[1]) Communication of W. Lorey, Frankfurt am Main.

[2]) *Archimedes von Syrakus vorhandene Werke aus dem Griechischen übersetzt und mit erläuternden und kritischen Anmerkungen begleitet von Ernst Nizze*. Stralsund 1824.

[3]) *Miscellanea sive Lucubrationes Mathematicae Samuelis Foster, Olim Londini in Collegio Greshamensi Astronomiae Professoris Publicae* (sic). *Omnia in lucem edita, et pleraque Latine reddita, opera et Studio* Johannis Twysden. Londini MDCLIX. Tractatus XI. *Lemmata Archimedis apud Graecos et Latinos jam pridem desiderata e vetusto codice M.S.Arabico a* Johanne Gravio *traducta et nunc primum cum Arabum Scoliis publicata* . . . Londini MDCLIX. *Lemmata Archimedis, ex traductione Thebit ibn Corae: cum Commentariis Excellentis Viri, Abi Alhonîn Alî filii Almed Alnasawaei.*

[4]) *Archimedis Liber Assumptorum interprete Thebit Ben-Kora exponente Almochtasso. Ex codice Arabico manuscripto Ser. Magni Ducis Etruriae* Abrahamus Ecchellensis *Latine vertit.* Io. Alfonsus Borellus *Notis illustravit.* This work constitutes an appendix to Borelli's work: *Apollonii Pergaei Conicorum Lib.* V. VI. VII.; ed. Io. Alfonsus Borellus. Florentiae MDCLXI.

[5]) Thâbit b. Qurra (Haran, Mesopotamia; born 826–27 or 835–36; † 901) was one of the important translators of Greek and Syrian works into Arabic; his own work and that of his school greatly helped to preserve the works of the Greek mathematicians.

[6]) The work translated by Thâbit cannot be by Archimedes in the form in which we possess it, because he himself is quoted in it several times.

[7]) Gotthold Ephraim Lessing, *Zur Geschichte der Literatur. Aus den Schätzen der herz. Bibliothek zu Wolfenbüttel.* Zweiter Beitrag. Braunschweig 1773. *Vide* also *Sämtliche Schriften*, ed. Lachmann; 3e Ausgabe (F. Muncker). XIII (Leipzig, 1897), 99.

[8]) The following particulars are taken from H. Bosmans, l.c. (Note 3 to p. 37). Separate, pp. 17 et seq.

the Vatican the publication of a number of Greek fragments of Book I of this work, which were long held to be parts of the original text of Archimedes, and which even still appear as such in the first modern edition of the Greek text of his works, which J. L. Heiberg published in 1881[1]). Afterwards it was, however, established that the fragments published by Maii were nothing but attempts at reconstruction of the Greek text by retranslation from the Latin text prepared by an unknown scholar, not before the 16th century. Besides, the deficiency which they had to supply was made up for definitively by the sensational discovery (in 1899) of a new Archimedes manuscript, which was to prove of inestimable value for the knowledge of his works in other respects as well.

This manuscript (Codex C in Heiberg) was discovered[2]) because Heiberg's attention was drawn to a report by Papadopulos Cerameus about a palimpsest with originally mathematical contents in the library of the monastery of the Holy Sepulchre at Jerusalem. He examined the manuscript at Constantinople in the years 1906 and 1908. It proved to contain an Archimedes text of the 10th century written on parchment, which it had been tried to efface in the 12th, 13th or 14th century, in order to write a Euchologium in its stead. Heiberg has succeeded in deciphering the original text for the greater part, and he thus found, besides fragments of *On the Sphere and Cylinder*, *On Spirals*, *Measurement of the Circle*, and *On the Equilibrium of Planes*, in the first place a considerable part of the Greek text of *On Floating Bodies* and secondly—which was even more important—an almost complete text of an as yet unknown work of Archimedes, the existence of which was known only from quotations by some ancient writers[3]). It is there referred to as

[1]) *Archimedis Opera Omnia cum commentariis Eutocii*, ed. J. L. Heiberg. Leipzig 1880–1881. 2 vols.

[2]) The following particulars on Codex C are taken from Heiberg's report on its discovery: *Eine neue Archimedes-Handschrift.* Hermes XLII (1907), 235 et seq. In this (243–297) the first publication of the Greek text is also found. A German translation with a commentary by H. G. Zeuthen appeared in Bibliotheca Mathematica (3) VII (1906–1907), an English one by T. L. Heath in *The Method of Archimedes, recently discovered by J. L. Heiberg. A Supplement to The Works of Archimedes,* 1897. Cambridge, 1912. A detailed study is given by Enrico Rufini, *Il "Metodo" di Archimede e le origini dell' analisi infinitesimale nell' Antichità* (Per la Storia e la Filosofia delle Matematiche, No 4), Roma 1926.

[3]) Suidas mentions it, adding that Theodosius had written a commentary

'Εφόδιον or *'Εφοδικόν*; Archimedes' own title is *"Εφοδος*, which we may translate by *Method*. As will become evident in the discussion of its contents, this work has furnished us with a truly new insight into the way of thinking of Archimedes. The C manuscript was further found to contain some fragments of the work *Στομαχίον*, in which a kind of puzzle, also known as *loculus Archimedius*, is dealt with, and another fragment of which has been preserved in Arabic[1]). In this way it has also been established that the *Stomachion* is indeed a work by Archimedes (a fact formerly doubted by Heiberg).

The texts newly found in C naturally do not yet appear in the English edition of the works of Archimedes in modern notation published by T. L. Heath in 1897[2]); by the publication of a supplement[3]) this lacuna was filled later, as far as the *Method* is concerned. The first edition containing all the works now known is the second edition of the Greek text by J. L. Heiberg[4]), on which are based all translations subsequently published.

Of these more recent translations we mention in the first place the very reliable, absolutely literal French translation, with detailed comments, by the Belgian engineer Paul Ver Eecke[5]); besides this, there are German translations in the series *Ostwald's Klassiker der exakten Wissenschaften*, by A. Czwalina[6]).

We are concluding this chapter with a summary of the works of Archimedes in the order in which they appear in the Heiberg edition and in which they will also largely be discussed in the present book. As far as the first eight works are concerned, this is the traditional order of the manuscripts derived from A. This is not, however, the same order as that in which the works were written or published. The original order can only be given with any cer-

---

on it (ed. Bekker, Berlin 1854; s.v. Theodosius, 495, col. 1). Heron quotes it in the *Metrica* (*Heronis Opera* III, 80, 84, 130).

[1]) This fragment was published by H. Suter: *Der loculus Archimedius oder Das Syntemachion des Archimedes* ... Abh. z. Gesch. d. Math. 9 (1899), 491–500. It is found in German translation in *Opera* II, 420.

[2]) *Vide* the list of works quoted, p. 417.

[3]) *Vide* Note 2 to p. 44.

[4]) *Vide* the bibliography, p. 417.

[5]) *Vide* the bibliography, p. 417.

[6]) The following translations have appeared: *Ueber Spiralen* (No 201; 1922). *Kugel und Zylinder* (No 202; 1922). *Die Quadratur der Parabel und Ueber das Gleichgewicht ebener Flächen* (No 203; 1923). *Ueber Paraboloide, Hyperboloide und Ellipsoide* (No 210; 1923). *Ueber schwimmende Körper und Die Sandzahl* (No 213; 1925).

tainty in some cases; as far as it is known, we shall indicate it (according to Heath[1]) by the bracketed numbers in Indo-Arabic numerical notation. In the following chapters the works will generally be quoted by the abbreviation given behind each of them.

I. (5) ON THE SPHERE AND CYLINDER. Two books.	*περὶ σφαίρασ καὶ κυλίνδρου α'β'.* *De Sphaera et Cylindro.* **S.C.**
II. (9) MEASUREMENT OF THE CIRCLE.	*κύκλου μέτρησις.* *Dimensio Circuli.* **D.C.**
III. (7) ON CONOIDS AND SPHEROIDS.	*περὶ κωνοειδέων καὶ σφαιροειδέων.* *De Conoidibus et Sphaeroidibus.* **C.S.**
IV. (6) ON SPIRALS.	*περὶ ἑλίκων.* *De lineis spiralibus.* **SPIR.**
V. (1) ON THE EQUILIBRIUM OF PLANES OR CENTRES OF GRAVITY OF PLANE FIGURES. Book I.	*'Επιπέδων ἰσορροπιῶν ἢ κέντρα βαρῶν ἐπιπέδων α'.* *De planorum aequilibriis sive de centris gravitatis planorum I.* **PL.AE. I.**
VI. (3) IDEM. Book II.	**PL.AE. II.**
VII. (10) THE SAND-RECKONER.	*Ψαμμίτης.* *Arenarius.* **AREN.**
VIII. (2) QUADRATURE OF THE PARABOLA.	*τετραγωνισμός παραβολῆς*[2]. *Quadratura Parabolae.* **Q.P.**
IX. (8) ON FLOATING BODIES. Two books.	*'Οχουμένων α'β'.* *De corporibus fluitantibus*[3]. **C.F.**
X. STOMACHION.	*Στομαχίον.* *Loculus Archimedius.*

---

1) T. L. Heath, *Greek Mathematics* II, 22.

2) This title is certainly not authentic, because in Archimedes the word *parabola* does not yet appear in the sense of a conic section. The original title must have been: *τετραγωνισμὸς τῆς τοῦ ὀρθογωνίου κώνου τομῆς.*

3) Formerly usually: *De insidentibus aquae* or *De iis quae in humido vehuntur.*

XI. (4) THE METHOD OF MECHANICAL THEOREMS, ADDRESSED TO ERATOSTHENES.	*περὶ τῶν μηχανικῶν θεωρημάτων προς 'Ερατοσθένην ἔφοδος.* *De mechanicis propositionibus ad Eratosthenem methodus.* **METH.**
XII. LEMMATA.	*Liber Assumptorum.*
XIII. THE CATTLE-PROBLEM.	*πρόβλημα βοεικόν.* *Problema Bovinum.*

About the lost works of Archimedes we have the following reports[1]):

1. Pappus[2]) mentions investigations by Archimedes on semi-regular polyhedra. We will revert to the contents of his report in Chapter XV.

2. In the *Sand-reckoner* Archimedes quotes a few times[3]) an older book by his hand on expressing large numbers, which had been sent to Zeuxippus. The title usually given is *'Αρχαί* (*Principles*), but Hultsch[4]) refers to it as *κατονόμαξις τῶν ἀριθμῶν* (*Naming of Numbers*). The contents of this book are incorporated in the *Sand-reckoner*.

3. The two books of *On the Equilibrium of Planes* do not by any means contain all the writings of Archimedes on the subject of Statics. Several other titles are mentioned, but it is not possible to ascertain with any degree of certainty whether different works are always really meant as such. Book I of *On the Equilibrium of Planes* is probably an excerpt from a larger work, *Elements of Mechanics* (*Στοιχεῖα τῶν μηχανικῶν*), which Archimedes himself quotes by that title[5]). Elsewhere he says that something has been proved in the *Equilibria* (*ἐν ταῖς 'Ισορροπίαις*[6]) or *ἐν τοῖς 'Ισορροπικοῖς*[7])), whilst in the first case it is certainly not the work *On the Equilibrium of Planes* which is meant by this, because the quoted proposition does

[1]) Summarized for the greater part in J. L. Heiberg, *Quaestiones Archimedeae*, Copenhagen 1879, pp. 29–30.

[2]) Pappus, *Collectio* V, 19, 350.

[3]) *Opera* II, 216, lines 17–19; 220, line 11; 236, lines 19–20.

[4]) Hultsch in Pauly-Wissowa, *Real-Encyclopädie der classischen Altertumswissenschaft* s.v. Archimedes, col. 511 *a*.

[5]) **C.F.** II, 2. *Opera* II, 350; lines 21–22.

[6]) **C.F.** II, 2. *Opera* II, 350; line 14.

[7]) **METH.** l. *Opera* II, 438.

not appear in it. Further Pappus[1]) refers to a work *περὶ ζυγῶν* (*On Balances*), while in Heron (in a German translation of an Arabic text) there is question of a *Buch der Stützen*[2]). It is improbable that a statement by Simplicius indicates the existence of a work *κεντροβαρικά*[3]).

4. Theon of Alexandria in his commentary on the *Almagest* attributes to Archimedes a work on optics (*περὶ κατοπτρικῶν*[4])).

This statement is confirmed by Apuleius, who enumerates several optical subjects said to have been treated by Archimedes in a *volumen ingens*[5]). Olympiodorus quotes a remark of Archimedes about *refraction*[6]), and the Scholia to the *Catoptrica* of Euclid contain a proof by his hand on the equality of the angles of incidence and reflection[7]).

5. In Chapter I we already discussed the work *περὶ σφαιροποιίας* (*On Sphere-making*), in which Archimedes is said to have dealt with the construction of his planetaria. As to technical inventions, it is mentioned by Arabian writers that he had written a work on water-clocks[8]). An Arabic treatise on this subject under his name has

---

[1]) Pappus, *Collectio* VIII, 11; 1068.

[2]) *Mechanica* 25. *Heronis Opera* II, 1; 70.

[3]) Simplicius in his commentary on *De Caelo* of Aristotle (*Scholia in Aristotelem*; coll. C. A. Brandis, Berlin 1836; 508 a 30) says that Archimedes and many others have written beautiful *κεντροβαρικά*, but this only means that they wrote on barycentric subjects.

[4]) *Claudii Ptolemaei Magnae Constructionis, id est Perfectae coelestium motuum pertractionis Libri XIII. Theonis Alexandrini in eosdem Commentariorum Libri XI.* Basileae MDXXXVIII. Comm. in I, 3; p. 10.

[5]) *Apulei Apologia sive Pro se de magia liber.* Cap. 16. ed. H. E. Butler and A. S. Owen (Oxford 1914).

Among the subjects dealt with he mentions the questions why in plane, convex, and concave mirrors the image is successively as large as, smaller than, and larger than the object; why right and left are interchanged in the image; why the image is sometimes in and sometimes in front of the same mirror; why it is possible to ignite fuel with concave mirrors on which sunlight falls, etc. Apuleius of Madaura was an African writer of the 2nd century. The passage in question is printed in *Opera* II, 550.

[6]) Olympiodorus *in Aristotelis Meteorologica*, printed in *Opera* II, 550. Olympiodorus was a Greek historian and alchemist (fl. about A.D. 400).

[7]) *Euclidis Opera* VII, 348; No 7. The proof is based on the reversibility of the ray of light.

[8]) This is stated by al-Qiftī, as quoted in E. Wiedemann (l.c., Note 1 to p. 13), p. 249.

even been preserved, in which he gives a detailed description of a waterclock[1]).

6. Finally, several planimetrical works are attributed to him by Arabian writers[2]): *On Circles touching one another*; *On Parallel Lines*; *On Triangles*; *On Properties of Right-angled Triangles*; *On the Assumptions for the Elements of Geometry*; *Book of Data* or *Definitions*; *On the Heptagon in the Circle.*

The existence of the last-mentioned work has been established by the discovery of a treatise by the Arabian mathematician Thâbit b. Qurra on this subject, which was published with a German translation in 1927[3]). Further details about this will be given in Chapter XV.

CHAPTER III

THE ELEMENTS OF THE WORK OF ARCHIMEDES

Owing to the completion of the *Elements* of Euclid, the Greek mathematicians had gained possession of a systematically arranged collection of the fundamental mathematical propositions on which they could base their further investigations. They were thus absolved from the obligation to revert in their works to matters of an elementary, *i.e.* fundamental nature: when a proposition was contained in the *Elements*, it was sufficient, in view of the apparently universal diffusion of the work, merely to mention it.

If the reader was to become acquainted with the work of a writer of Archimedes' level, however, an understanding of the *Elements* of Euclid was—and still is—a prerequisite, though by no means a sufficient preparation; in fact, his works are often found to contain references to the *Elements of Conics* (*τὰ κωνικὰ στοιχεῖα*), by which must be meant one of the works on this subject which had been

---

[1]) E. Wiedemann (l. c., Note 1 to p. 13, p. 257). As a peculiarity of the apparatus described by Archimedes it is mentioned that every hour a raven dropped a sphere in a tray, thus causing a musical note.

[2]) Heiberg (l.c., Note 1 to p. 47), 29–30. E. Wiedemann, l. c., p. 248.

[3]) *Die trigonometrischen Lehren des persischen Astronomen Abu'l-Raihân Muh. ibn Ahmad al-Bîrûnî, dargestellt nach Al-Qânûn al-Mas'ûdî* von Carl Schoy. Nach dem Tode des Verfassers herausgegeben von Julius Ruska und Heinrich Wieleitner (Hannover 1927), pp. 74 et seq.

compiled by Aristaeus[1]) and Euclid[2]). Apparently, however, the various collection of Elements circulating in the days of Archimedes did not yet contain all the fundamental properties which he required for the exposition of his particular investigations; he frequently also derives propositions of an elementary nature in his works, or he formulates them, observing that they are easily proved.

The degree of mathematical knowledge required in the modern reader of Archimedes' works not infrequently constitutes a serious barrier to his appreciation of them, because he acquires his knowledge of the elements of his science in so entirely different a manner from Euclid's disciples. Moreover Archimedes' custom to precede the real nucleus of his treatises by numerous lemmas, the purpose of which does not become apparent until much later, when they are applied, does not make it any the easier to remain aware of the thread of his reasoning.

In order to meet all these difficulties, we intend in this chapter to combine as far as possible the elementary propositions not mentioned by Euclid, but quoted, enunciated or proved by Archimedes, into a system of elements the perusal of which will constitute a sufficient preparation for the study of his investigations proper; in doing so, however, we assume the reader to be acquainted with the main features of the *Elements* of Euclid[3]).

The reader whose only concern is to be informed somewhat quickly about the most essential points of the investigations of Archimedes is advised to skip this chapter and consult it, if desired, during the reading of the following chapters, when reference is made to it in the application of a lemma here dealt with; with a view to such references a decimal notation has been adopted[4]).

We shall first give a short summary of the symbols to be used in this book for the reproduction of the Greek mathematical arguments, and of the chief properties which are in themselves already elementary in relation to the system of Elements to be discussed.

---

[1]) Aristaeus (the elder) was an older contemporary of Euclid. According to Pappus (*Collectio* VII, 30; 672) he wrote a work in five books on loci, connected with the theory of conics (*συνεχῆ τοῖς κωνικοῖς*).

[2]) Pappus (*Collectio* VII, 30; 672) says that Apollonius supplemented the four books of *Conica* of Euclid, and (*ibid.* VII, 34; 676) that Euclid continued the work of Aristaeus.

[3]) See the directions as to quotations in the bibliography, p. 418.

[4]) The mode of quotation is: III, followed by the decimal reference.

0.1 *Notations*:

A rectangle with sides a and b	$\mathbf{O}\ (a, b)$	**O** of *'Ορθογώνιον*
A square with side a	$\mathbf{T}\ (a)$	**T** of *Τετράγωνον*
A circle with diameter d	$\mathbf{K}\ (d)$	**K** of *Κύκλος*

The ratio of two homogeneous[1]) magnitudes A and B

$$(A, B)$$

The ratio of two squares with sides a, b, for example, is written:

$$[\mathbf{T}(a), \mathbf{T}(b)]$$

The above symbols are used in particular in the practice of the so-called Application of Areas or Geometrical Algebra, which is the instrument used in Greek analytical geometry, and the elements of which are dealt with in Euclid[2]). If desired, any line of reasoning expressed by these symbols may at once be translated into the non prevalent algebraic notation by the following substitutions:

$$\mathbf{O}(a, b) = ab \qquad \mathbf{T}(a) = a^2 \qquad (A, B) = A:B\,.$$

0.2. *Fundamental concepts of the application of areas*[2]).

0.21. A plane figure $\mathbf{X}$ is said to be applied *parabolically* to a line segment A when a line segment B is constructed such that

$$\mathbf{O}(A, B) = \mathbf{X}\,.$$

0.22. A plane figure $\mathbf{X}$ is said to be applied *elliptically* to a line segment A with defect of the assigned form (Δ, E) when $\mathbf{X}$ is applied parabolically to a line segment $B < A$, so that

$$\mathbf{X} = \mathbf{O}(B, \Gamma) \qquad \text{and also} \qquad (\Gamma, A-B) = (\Delta, E)\,.$$

0.23. A plane figure $\mathbf{X}$ is said to be applied *hyperbolically* to a line segment A with excess of the assigned form (Δ, E) when $\mathbf{X}$ is applied parabolically to a line segment $B > A$, so that

$$\mathbf{X} = \mathbf{O}(B, \Gamma) \qquad \text{and also} \qquad (\Gamma, B-A) = (\Delta, E)\,.$$

0.24. The Greek names for the three above-mentioned operations

---

[1]) Two magnitudes A and B are called homogeneous when they satisfy the axiom of Eudoxus, *i.e.* when there exist natural numbers m and n such that $m.A > B$ and $n.B > A$.

[2]) *Elements of Euclid* II, 12, 103.

are successively: παραβολή (parabola), ἔλλειψις (ellipse), and ὑπερβολή (hyperbola).

0.3. *Fundamental concepts of the theory of proportions*[1]).

0.31. When $(A, B) = (B, C)$

we call (A, C) the *duplicate ratio* (διπλασίων λόγος) of (A, B).

Symbol: $(A, C) = \mathbf{\Delta\Lambda}(A, B)$.

In this case $(A, C) = [\mathbf{T}(A), \mathbf{T}(B)]$.

The algebraic equivalent of a duplicate ratio is the square of a ratio. In fact, from $a:b = b:c$ it follows that

$$a:c = a^2:b^2 .$$

0.32. When $(A, B) = (B, C) = (C, D)$

we call (A, D) the *triplicate ratio* (τριπλασίων λόγος) of (A, B).

Symbol: $(A, D) = \mathbf{T\Lambda}(A, B)$.

The algebraic equivalent of a triplicate ratio is the cube of a ratio. In fact, from $a:b = b:c = c:d$ it follows that

$$a:d = a^3:b^3 .$$

0.33. When $(A, B) = (M, N)$ and $(B, C) = (P, Q)$

we call (A, C) the *compound ratio* (συγκείμενος λόγος) of the ratios (M, N) and (P, Q).

The algebraic equivalent of a compound ratio is the product of two ratios. In fact, from $a:b = m:n$ and $b:c = p:q$ it follows that

$$a:c = mp:nq .$$

0.4. *Main operations of the theory of proportions*[2]).

0.41. From a proportion $(A, B) = (C, D)$ there result, through the operations to be mentioned below, the proportions given behind them:

[1]) *Elements of Euclid* II, 82–83.

[2]) *Elements of Euclid* II, 71–76.

permutando or *alternando* (ἐναλλάξ)	$(A, C) = (B, D)$	
invertendo (ἀνάπαλιν)	$(B, A) = (D, C)$	
componendo (συνθέντι)	$(A+B, B) = (C+D, D)$	
separando (διελόντι)	$(A-B, B) = (C-D, D)$	provided $A > B$ and therefore $C > D$
convertendo (ἀναστρέψαντι)	$(A, A-B) = (C, C-D)$	

and by a combination of the above:

invertendo componendoque	$(A+B, A) = (C+D, C)$
separando invertendoque	$(B, A-B) = (D, C-D)$
convertendo invertendoque	$(A-B, A) = (C-D, C)$

Contrary to the custom of the Greek mathematicians, we shall as a rule perform these operations without mentioning their names.

0.42. The above-mentioned operations are frequently applied to inequalities as well. In this case the possible change of the sign of inequality is to be noted:

From $(A, B) > (C, D)$[1]) it follows that

permutando	$(A, C) > (B, D)$	
invertendo	$(B, A) < (D, C)$	
componendo	$(A+B, B) > (C+D, D)$	
separando	$(A-B, B) > (C-D, D)$	provided $A > B$ and $C > D$
convertendo	$(A, A-B) < (C, C-D)$	

and by a combination of these:

convertendo invertendoque	$(A-B, A) > (C-D, C)$
separando invertendoque	$(B, A-B) < (D, C-D)$
invertendo componendoque	$(A+B, A) < (C+D, C)$

The way in which these conclusions can be proved is sufficiently apparent from the following example:

Supposition: $(A, B) > (C, D)$

What is required to be proved: $(A+B, B) > (C+D, D)$

Proof: Construct a magnitude $E > C$ such that $(A, B) = (E, D)$.

---

[1]) We remind the reader that this means: there is at least one pair of natural numbers m, n such that

$$m.A > n.B, \quad \text{but} \quad m.C \leqq n.D.$$

Componendo $(A+B, B) = (E+D, D) > (C+D, D)$.

0.43. An inequality of ratios remains true upon duplication, *i.e.* from $(A, B) > (C, D)$ it follows that $\mathbf{\Delta\Lambda}(A, B) > \mathbf{\Delta\Lambda}(C, D)$.

0.44. From $(A, B) > (C, D)$ and $C > D$ it follows that $A > B$.
Proof: Construct E such that $(E, B) = (C, D)$, then $E < A$.
From $C > D$ it follows that $E > B$, hence *a fortiori* $A > B$.

0.45. If $A > B$, and C is any magnitude homogeneous with A and B, we have: $(A, B) > (A+C, B+C)$.
Proof: From $A > B$ it follows that $(A, C) > (B, C)$, hence

$$(A+C, A) < (B+C, B),$$

from which *permutando*

$$(A+C, B+C) < (A, B).$$

0.46. We here remind the reader of the conclusion *ex aequali*, in which it is concluded from

$$(A, B) = (D, E)$$

and

$$(B, C) = (E, F)$$

that

$$(A, C) = (D, F).$$

0.5. *Lemma of Euclid.*

For the operations of the so-called indirect method of passing to a limit (III.8) the so-called lemma of Euclid (*Elements* X, 1) is elementary (*i.e.* functions as an element); in this it is enunciated that if from any magnitude there be subtracted more than its half, from the remainder again more than its half, and so on continually, there will at length remain a magnitude less than any assigned magnitude. As is mentioned in a porism, this proposition is also true in the case when from the remaining magnitude there be continually subtracted its half; in this case it is equivalent to the proposition

$$\lim_{n\to\infty} \frac{1}{2^n} = 0$$

The process of continued bisection is denoted by the term "dichotomy".

0.6. *Numerical system.*

For the expression of numbers Archimedes used the alphabetic system, in which all numbers below 1,000 could be represented by additive juxtaposition of letter symbols for the numbers 1, 2...9, 10, 20...90, 100, 200...900. The symbols in question were:

$\bar{\alpha} = 1$	$\bar{\iota} = 10$	$\bar{\varrho} = 100$
$\bar{\beta} = 2$	$\bar{\varkappa} = 20$	$\bar{\sigma} = 200$
$\bar{\gamma} = 3$	$\bar{\lambda} = 30$	$\bar{\tau} = 300$
$\bar{\delta} = 4$	$\bar{\mu} = 40$	$\bar{\upsilon} = 400$
$\bar{\varepsilon} = 5$	$\bar{\nu} = 50$	$\bar{\varphi} = 500$
$\bar{\varsigma} = 6$	$\bar{\xi} = 60$	$\bar{\chi} = 600$
$\bar{\zeta} = 7$	$\bar{o} = 70$	$\bar{\psi} = 700$
$\bar{\eta} = 8$	$\bar{\pi} = 80$	$\bar{\omega} = 800$
$\bar{\vartheta} = 9$	$\bar{ϙ} = 90$	$\bar{ϡ} = 900$

0.61. The numbers 1,000, 2,000...9,000 were expressed by the symbols for 1, 2...9, with a small dash in front and below the line, the number 10,000 by M (of μύριοι), myriads by $\overset{n}{M}$.
Example:

$$\overset{\lambda\beta}{M}\overline{,\varsigma\varphi\xi\vartheta} = 326{,}569\,.$$

0.62. Unit fractions were expressed by the symbol for the denominator, with an accent affixed. An exception was made for the fraction 1/2, for which the symbol ∟′ was used.
Example:

$$\eta' = \tfrac{1}{8}$$

0.63. General fractions were expressed either in words or as sums or multiples of unit fractions.
Examples: (D.C. 3): *δέκα ἑβδομηκοστομόνα* = ten seventy-oneths

$$\bar{\iota}\ o\alpha' = \tfrac{10}{71}$$

$$∟'\delta' = \tfrac{1}{2} + \tfrac{1}{4} = \tfrac{3}{4}\,.$$

1. *Generation and general properties of conics.*

The way in which conics are introduced and dealt with in Greek geometry, in spite of deeper resemblances, is outwardly so different from their generation and study in our own day that it seems desirable to precede the exposition of the elements on which Archimedes built his work by a short sketch of the evolution of the theory of these curves from the time when it was presumably framed for the first time up to its complete elaboration in the *Conics* of Apollonius.

1.0. The discoverer of conics is considered to be the geometer Menaechmus, who is referred to, for example, in the *Catalogus Geometrarum* of Proclus[1]) on account of his share in the development of geometry after Eudoxus, and who must therefore have flourished at about 350 B.C. The ground on which he is credited with so important a discovery is rather vague: in fact, in a treatise preserved by Eutocius[2]), which is in the form of a letter from the geometer Eratosthenes to King Ptolemy of Egypt, sent with a new solution of the problem of the two mean proportionals, it is stated in a survey of older solutions that the cone should not be cut in the triads of Menaechmus[3]). Since further Eutocius ascribes to Menaechmus a construction of the two mean proportionals[4]), in which use is made of a parabola and an orthogonal hyperbola, the passage in the letter of Eratosthenes is usually considered as an indication that Menaechmus was the discoverer of conics. We are naturally left completely in the dark as to the further particulars of this discovery. We only know through Eutocius that Menaechmus solved the problem of the construction of two line segments B, Γ, which in combination with two given line segments A, Δ satisfy the relations

$$(A, B) = (B, \Gamma) = (\Gamma, \Delta)$$

by a geometrical interpretation of the relations

$$\mathbf{T}(B) = \mathbf{O}(A, \Gamma) \qquad \mathbf{O}(B, \Gamma) = \mathbf{O}(A, \Delta) \qquad \mathbf{T}(\Gamma) = \mathbf{O}(B, \Delta)$$

—in which B and Γ are looked upon as variables—respectively as a parabola, an orthogonal hyperbola, and a parabola (to anticipate for the moment the later names), and that he then determined the two unknowns B and Γ as co-ordinates of a point of intersection of two of these curves. It is obvious that this raises a great many questions: in the first place the question how he then realized that

1) *Procli Diadochi in primum Euclidis Elementorum librum commentarii.* ed. G. Friedlein (Leipzig 1873), p. 67.

2) *Opera* III, 96.

3) *Opera* III, 96; line 17. *μηδὲ Μεναιχμείους κωνοτομεῖν τριάδας.*

4) *Opera* III, 78.

the curves thus defined planimetrically may also be plane sections of a cone; secondly, whether he also determined the other conic sections planimetrically before generating them stereometrically; also, whether his discovery perhaps consisted in his finding that the curves already known as sections of a cone could be made subservient to the solution of the problem of the two mean proportionals. None of these questions, however, can be answered at the present stage of our knowledge of Greek mathematics.

1.1. However this may be, from the fact that as early as about 300 B.C. the above-mentioned special treatises about conics by Aristaeus and Euclid could be written it can be inferred that the theory of these curves must have been very intensively studied in the second half of the fourth century B.C. We have some details about the foundations of the new theory, *i.e.* the manner of generating the conics, from mutually concordant statements by Pappus[1]) and Eutocius[2]) (the latter refers to Geminus). Eutocius relates that by cones "the ancients" exclusively understood the solids generated by a right triangle rotating about one of the sides including the right angle (*i.e.* right circular cones), that they classified these according to the type of the vertical angle of the complete meridian section as right-angled, obtuse-angled, and acute-angled cones, and that they used each kind only for the generation of one type of conic. In fact, they cut every cone by a plane at right angles to a generator, and called the curves thus obtained by the names (which were said by Pappus to originate from Aristaeus):

section of the right-angled cone (*ὀρθογωνίου κώνου τομή*, henceforth to be referred to as *orthotome*).
section of the obtuse-angled cone (*ἀμβλυγωνίου κώνου τομή*, henceforth to be referred to as *amblytome*).
section of the acute-angled cone (*ὀξυγωνίου κώνου τομή*, henceforth to be referred to as *oxytome*).

It is obvious that these are the curves which ever since Apollonius have been known as parabola, hyperbola, and ellipse respectively.

No authentic sources are available to show in what way a symptom (*σύμπτωμα*, *i.e.* a characteristic relation between the co-ordinates of any given point of the curve which, translated into algebraic

1) Pappus, *Collectio* VII, 30; 672.

2) Apollonius, *Conica* II, 168.

symbolism, becomes the equation in respect of the co-ordinate system used) is deduced for each of the curves from these definitions. However, considering the symptomata that were chiefly used prior to Apollonius for characterization purposes, it is highly probable that this was done as follows.

In the following figures let the meridian plane through the generator, to which the cutting plane is at right angles, be chosen for the plane of the paper; the cutting plane is then denoted by AB. A plane at right angles to the axis of the cone intersects the plane of the paper in $\Gamma\Delta$, the cutting plane in $K\Theta$, at right angles to the plane of the paper. Now K is a point of the curve.

Orthotome.

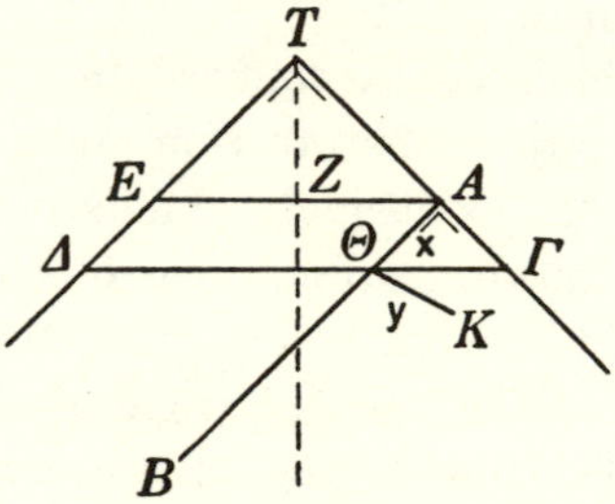

Fig. 1.

Now we have $\quad \mathbf{T}(K\Theta) = \mathbf{O}(\Gamma\Theta, \Delta\Theta)$

$$= \mathbf{O}(\Gamma\Theta, 2AZ)$$

$$\triangle\Gamma\Theta A \sim \triangle TAZ, \text{ thence}$$

$$(\Gamma\Theta, \Theta A) = (TA, AZ), \text{ thence}$$

symptom: $\mathbf{T}(K\Theta) = \mathbf{O}(2\,TA, A\Theta)$.

In algebraic symbolism:

$$(K\Theta = y\,. \qquad A\Theta = x\,. \qquad 2\,TA = p)$$

this amounts to the equation: $y^2 = px$.

Amblytome and Oxytome.

$$\mathbf{T}(K\Theta) = \mathbf{O}(\Gamma\Theta, \Delta\Theta)$$

$$= \mathbf{O}(A\Theta, H\Theta)$$

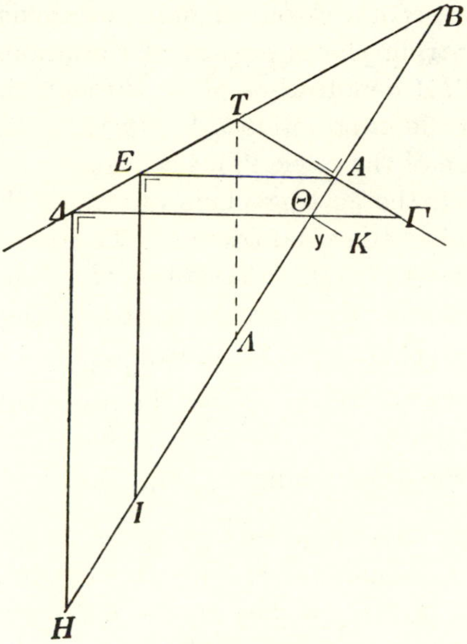

Fig. 2.

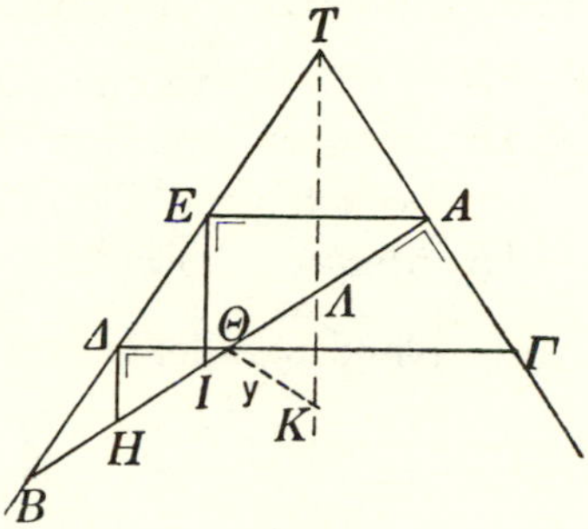

Fig. 3.

$$(H\Theta, IA) = (\Delta\Theta, EA) = (B\Theta, BA)$$

thence $$(H\Theta, B\Theta) = (IA, BA)\,.$$

From this it follows that

$$[\mathbf{O}(A\Theta, H\Theta),\ \mathbf{O}(A\Theta, B\Theta)] = (IA, BA) = (2A\Lambda, BA)$$

or

symptom: $[\mathbf{T}(K\Theta),\ \mathbf{O}(A\Theta, B\Theta)] = (2A\Lambda, AB)$.

In algebraic symbolism:

$$(K\Theta = y,\ A\Theta = x_1,\ B\Theta = x_2,\ 2A\Lambda = p,\ AB = a)$$

the equation:

$$\frac{y^2}{x_1 x_2} = \frac{p}{a}.$$

We shall refer to this form of the equation defining the section as the two-abscissa form.

1.2. For the sake of comparison we shall now describe the way in which, more than a century later, the conics are dealt with by Apollonius; they now appear to be generated on an oblique circular cone, which is intersected by cutting planes of different positions.

In the following figures let $T\Gamma\Delta$ denote that plane through the axis TM[1]) of the cone of which the intersection $\Gamma\Delta$ with the plane of the base of the cone (the plane of the circle M) is at right angles to the straight line EZ, in which the cutting plane intersects the base. Let the triangle $T\Gamma\Delta$ be called the axial triangle. The cutting plane intersects the plane $T\Gamma\Delta$ in AH. AH is thus a line of oblique symmetry for the direction EZ[2]). Now there are three possibilities:

α) AH is parallel to one side of the axial triangle (Fig. 4).

β) AH intersects one side of the axial triangle, and the other side produced beyond T[3]) (Fig. 5).

γ) AB intersects both sides of the axial triangle[4]) (Fig. 6).

Any plane parallel to the base cuts the cone in a circle on $P\Pi$ as diameter, and the cutting plane in $K\Theta$. K lies on the surface of the cone, and is therefore a point of the section. In each of the three cases the following relation applies for the ordinate $K\Theta$[5]):

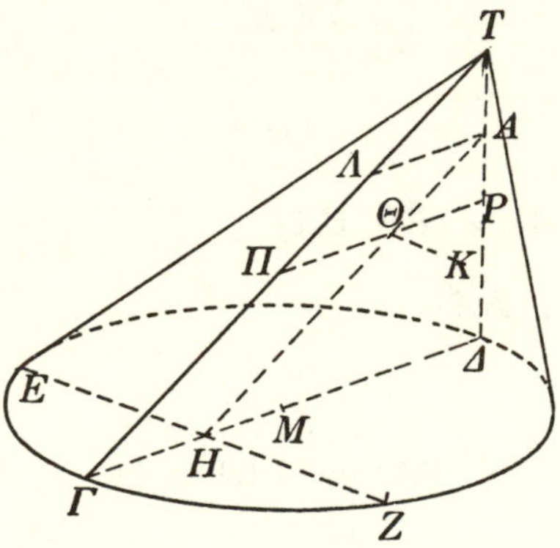

Fig. 4.

$$\mathbf{T}(K\Theta) = \mathbf{O}(\Pi\Theta, P\Theta)\,.$$

Now in Fig. 4 let

$$A\Lambda \parallel \Delta\Gamma, \text{ so that } \Pi\Theta = \Lambda A\,.$$

Now

1) The axis of the cone is the straight line through the vertex and the centre of the base.

2) This symmetry is then, and only then, right symmetry when $EZ \perp AH$, *i.e.* $EZ \perp T\Gamma\Delta$, *i.e.* when $T\Gamma\Delta$ is the plane through the axis at right angles to the base.

3) Apollonius is the first to consider also the part of the curved surface of the cone formed by producing beyond T the half-generators considered by Archimedes. He thus arrives at two branches of the hyperbola.

4) Since the curved surface of the cone is conceived of as extending indefinitely, the base can always be so chosen that this is the case.

5) By the term ordinate, also to be used in discussing Archimedes' works, we translate the standing Apollonian expression τεταγμένως κατηγμένη, *ordinatim ducta*, *i.e.* drawn in the conjugate direction, which Archimedes also uses already on one occasion.

$$(\Lambda A,\, TA) = (\Gamma\Delta,\, T\Delta)$$
$$(P\Theta,\, A\Theta) = (\Delta\Gamma,\, T\Gamma)$$

thence

$$[\mathbf{O}(\Pi\Theta,\, P\Theta),\ \mathbf{O}(TA,\, A\Theta)] = [\mathbf{T}(\Gamma\Delta),\ \mathbf{O}(T\Delta,\, T\Gamma)]\,.$$

Now find a line segment N such that

$$(N,\, TA) = [\mathbf{T}(\Gamma\Delta),\ \mathbf{O}(T\Delta,\, T\Gamma)]$$

then

$$[\mathbf{T}(K\Theta),\ \mathbf{O}(TA,\, A\Theta)] = (N,\, TA) = [\mathbf{O}(N,\, A\Theta),\ \mathbf{O}(TA,\, A\Theta)]$$

therefore symptom: $\mathbf{T}(K\Theta) = \mathbf{O}(N,\, A\Theta)$.

The square on $K\Theta$, applied parabolically (0.21) to N, thus gives a rectangle with side $A\Theta$.

In algebraic symbolism:

$$(K\Theta = y,\ A\Theta = x,\ N = p)$$
$$y^2 = px$$

(equation in respect of the *oblique* system with X axis: AH, origin: A, direction of the ordinates: EZ).

On account of the correlation of the symptoms thus to be derived

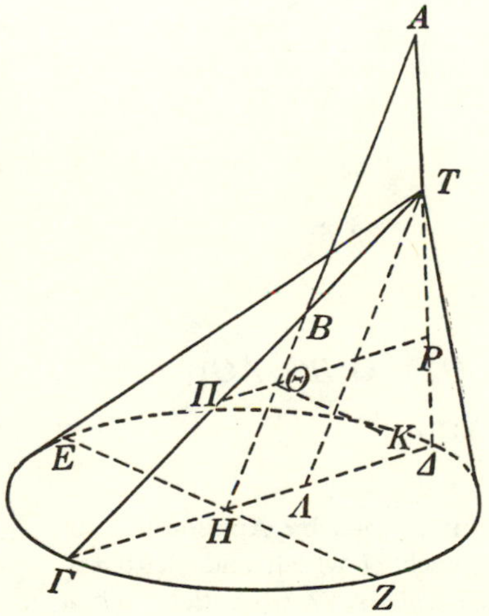

Fig. 5.

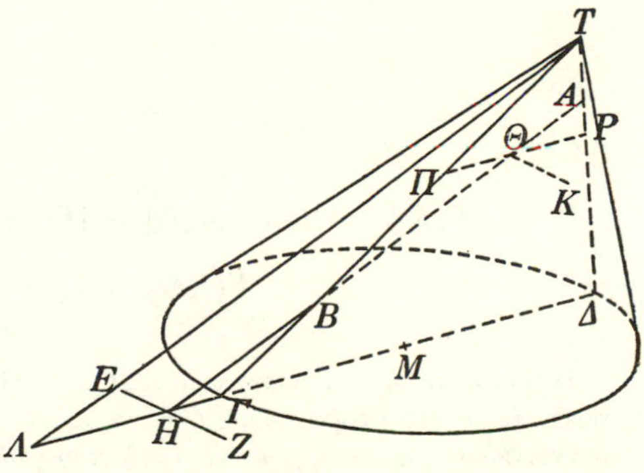

Fig. 6.

with the various forms of application of areas, the curves are called parabola, hyperbola, and ellipse respectively by Apollonius[1]).

In Figs 5 and 6 let further: $T\Lambda \parallel AB$ (Λ on $\Gamma\Delta$).

Now we have

$$(\Pi\Theta, B\Theta) = (\Gamma\Lambda, T\Lambda)$$

$$(P\Theta, A\Theta) = (\Delta\Lambda, T\Lambda)$$

thence

$$[\mathbf{O}(\Pi\Theta, P\Theta),\ \mathbf{O}(B\Theta, A\Theta)] = [\mathbf{O}(\Gamma\Lambda, \Delta\Lambda),\ \mathbf{T}(T\Lambda)]\,.$$

Now find a line segment N such that

$$(N, AB) = [\mathbf{O}(\Gamma\Lambda, \Delta\Lambda),\ \mathbf{T}(T\Lambda)]$$

then we have

$$\text{symptom: } [\mathbf{T}(K\Theta),\ \mathbf{O}(B\Theta, A\Theta)] = (N, AB)\,.$$

If in Figs 7 and 8, $BO = N$, and the straight line AO meets the perpendicular through Θ on AB in Σ, we have:

$$(\Theta\Sigma, A\Theta) = (N, AB)$$

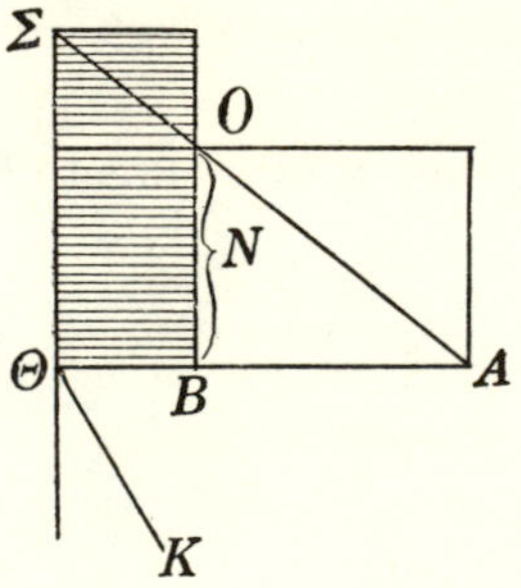

Fig. 7.

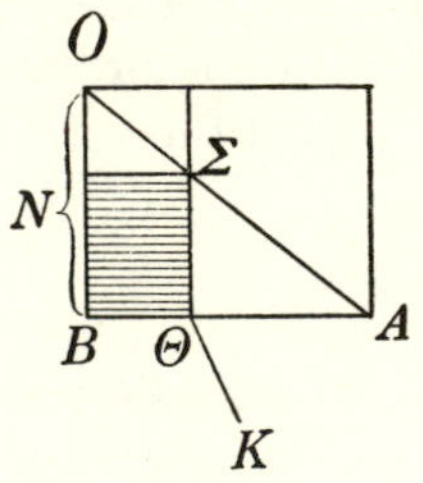

Fig. 8.

whence

$$[\mathbf{T}(K\Theta),\ \mathbf{O}(B\Theta, A\Theta)] = [\mathbf{O}(B\Theta, \Theta\Sigma),\ \mathbf{O}(B\Theta, A\Theta)]$$

or
$$\mathbf{T}(K\Theta) = \mathbf{O}(B\Theta, \Theta\Sigma)\,.$$

---

[1]) That these names were actually newly introduced by Apollonius, appears from the propositions *Conica* I, 11–13, each of which ends with the words καλείσθω δὲ ἡ τοιαύτη τομὴ παραβολή, ὑπερβολή, ἔλλειψις (let such a section be called parabola, hyperbola, ellipse).

The square on $K\Theta$, applied $\left\{\begin{array}{l}\text{hyperbolically}\\ \text{elliptically}\end{array}\right.$ (0.22–0.23)

to N with an $\left\{\begin{array}{l}\text{excess}\\ \text{defect}\end{array}\right.$ the ratio of whose sides is (N, AB), thus yields a rectangle with side $B\Theta$.

In algebraic symbolism:

$$(K\Theta = y,\ B\Theta = x,\ N = p,\ AB = a)$$

$$y^2 = px \pm \frac{p}{a}x^2 \left(\begin{array}{l} + \text{ for amblytome} \\ - \text{ for oxytome}\end{array}\right)$$

(Equation in respect of the *oblique* system with X axis: AB, origin: B, direction of the ordinate: EZ).

1.3. It thus appears that the method of Apollonius differs on two fundamental points from the (presumable) original one of Menaechmus:

a) The symptom is derived in relation to one out of an infinite number of reciprocally equivalent diameters of the section, each of which is a line of oblique symmetry with regard to a conjugate direction; in Menaechmus it referred to a single diameter (later called axis), which was a line of right symmetry. We shall express this henceforth by saying that in Apollonius there is *oblique conjugation* between abscissa and ordinate, in Menaechmus *right conjugation*.

b) The symptom is expressed with the aid of the concepts of the application of areas, as developed in Book VI of Euclid's *Elements*; this does not bring about any alteration, as compared with Menaechmus, in the treatment of the orthotome (*i.e.* no fresh alteration besides that from the main axis to any diameter), but it does with regard to amblytome and oxytome, which only now deserve to be called hyperbola and ellipse. In analytical terms this means that in Apollonius also for amblytome and oxytome the equation is referred to a system of axes consisting of any given diameter and the tangent at one of its extremities.

1.4. The problem that now presents itself when we want to state the elements of the theory of conics in Archimedes consists in that we must try to show what point in the development of this theory, the initial and final stages of which we have now become acquainted with, his method of generating and dealing with conics constitutes.

It has sometimes been considered that this question could at once be answered by pointing out that Archimedes still uses the nomenclature "sections of the right-angled, the obtuse-angled, and the acute-angled cone", which are characteristic of Menaechmus' standpoint, and that he therefore still seems to have generated the conics by cutting right circular cones by planes at right angles to a generator. This argument appears to us to carry little conviction: names are apt to linger on when the meaning of the concepts to which they refer has altered, and we should be equally justified in inferring from the fact that we now still speak of hyperbola and ellipse that we still generate these curves by means of the Euclidean method of application of areas.

There is, however, a more potent—though also terminological—argument which may be advanced in favour of the supposition that the standpoint of Archimedes (at least with regard to the Elements on which he builds) does not yet fundamentally differ from that of Menaechmus. Indeed, when he writes down the symptom of the orthotome for a point with abscissa $A\Theta$ and ordinate $K\Theta$, this is as follows:

$$\mathbf{T}(K\Theta) = \mathbf{O}(N, A\Theta)$$

and he then calls the line segment N "the double of the line as far as the axis" (*ἁ διπλασία τᾶς μέχρι τοῦ ἄξονος*[1])). This is a definition which is perfectly in line with the generating method of Menaechmus, for in Fig. 1 we had

$$\mathbf{T}(K\Theta) = \mathbf{O}(2AT, A\Theta)$$

and thence N was there indeed equal to $2AT$, *i.e.* the double of the part of the generator TA between the curve and the axis of the cone. This definition of the line segment N is not at all in agreement with the generating method of Apollonius, nor is it with other possible methods of generation which may be conceived to have been used between Menaechmus and Apollonius and in which either a right circular cone would be cut by any plane or an oblique one by a plane at right angles to the main section.

It is obvious that this terminological consideration carries much more weight than the preceding one concerning the names of the

---

1) It is to be noted therefore that *ἄξων* refers to the axis of the cone, and not to the straight line which we call the axis of the parabola. In this sense the word is not found in Archimedes.

conics. Indeed, a parabola, a hyperbola, and an ellipse generated according to Apollonius remain nevertheless sections of a right-angled, an obtuse-angled, and an acute-angled cone, and one may therefore continue to call them by these names for the sake of tradition; but the line segment N, to which the squares on the ordinates are applied parabolically (*παρ' ἂν δυνάνται αἱ ἀπὸ τᾶς τομᾶς*, as is the definition in Archimedes as well as in Apollonius), is no longer "the double of the line as far as the axis" as soon as the generating method of Menaechmus is abandoned, and it is difficult to imagine that this expression, which is not a technical abbreviation, but a definition, should yet have continued to be used.

The above formal arguments are supported by others which are even more closely connected with the import of the Archimedean theory of conics. In the first place it is striking that he always speaks of *the* diameter of the orthotome and the amblytome, and of *the* two diameters, the greater and the lesser, of the oxytome[1]); it appears from the context that by the latter he means the principal axes, and he therefore seems to look upon the *symptomata* of the sections (at least in so far as they follow directly from the definition) as relating to right conjugation.

Nor does he ever refer in his discussions on amblytome and oxytome to the magnitude which we denoted by N in our exposition of the method of Apollonius and which is of essential importance for the connection between the theory of conics and the Euclidean application of areas. To put it analytically, Archimedes always uses the equations for ellipse and hyperbola in the form

$$\frac{y^2}{x_1 x_2} = \text{const.},$$

viz. in right conjugation, and never in the Apollonian form

[1]) One may be inclined to observe that it cannot be inferred from the use of the expressions "greater diameter" and "lesser diameter" that only two diameters were being considered, because the axes are also successively the greatest and the least of all diameters. In the Greek text, however, this conclusion is warranted, because it is apparent from the use of the comparative as superlative in the expressions *ἡ μείζων διάμετρος* and *ἡ ἐλάσσων διάμετρος* that only two diameters are involved. As to the amblytome, that here only one diameter is referred to is to be explained by the fact that the amblytome consists of only one branch of a hyperbola.

$$y^2 = px \pm \frac{p}{a} x^2$$

where the conjugation is conceived to be oblique[1]).

All these arguments in combination seem to us to justify the conclusion that the theory of conics of Archimedes does not yet differ on any essential point from that of Aristaeus and Euclid. It will become apparent from the following pages that in some respects he made a step forward, as compared with them, in the direction to be subsequently followed by Apollonius; for the rest, however, a reading of his treatises confirms the impression gained from the *Conica* of Apollonius, *viz.* that the geometer of Perga gave an essentially new foundation to the theory of conics when he based the discussion on the oblique circular cone with any cutting plane and correlated the *symptomata* of all three sections with the concepts of the Euclidean application of areas.

That for the rest the contents of the first four books of the *Conica* were not new in every respect, is known from his own utterances[2]). We shall therefore also, without risking any anachronisms, be able to draw on his work in order to reconstruct the elements of the Archimedean theory of conics, provided we formulate what he finds as oblique conjugation as right conjugation, and provided we do not make use of any propositions which have an essential connection with the concepts of hyperbolic and elliptic application of areas. It may also be said that when speaking about Archimedes, we must not yet look upon the amblytome as a hyperbola and upon the oxytome as an ellipse; an identification of orthotome and parabola is less objectionable, because the symptom of this section is already formulated by Archimedes with the aid of (parabolic) application of areas.

1.5. In Apollonius' *Conica* the definitions of the sections are followed by various propositions relating to the number of points of intersection which they have in common with straight lines of different kinds (*e.g.* parallel or non-parallel to the diameter, drawn in the direction of an ordinate or not), and to the relations "within"

---

[1]) *Vide*, however, III. 5, where it will become apparent that the treatment of ellipse and hyperbola in oblique conjugation after all was already known before Apollonius.

[2]) Apollonius, *Conica* I, 4.

and "without" in respect of the section. Discussions of this nature, which concern the relative position of elements of the figure, are never among the strongest points of Greek mathematicians: the absence of the axiomatic basis on which such discussions would have to be built makes itself felt again and again, and under the guise of an abstract logical argument a tacit appeal to geometrical intuition is yet repeatedly made. It stands to reason that on this point the position of the theory of conics in the days of Archimedes was no better than in the days of Apollonius, and that we are by no means likely to underestimate it by quoting the following propositions from the *Conica* as its probable component elements:

Conica I, 10. *When two points be taken on a conic section, the straight line joining these points will fall within the section and the line produced will fall without it*[1]).

Conica I, 19. *In any conic section a straight line drawn from a point of the diameter in the direction of an ordinate will meet the curve.*

1.6. The statements made above with regard to the—to us rather unsatisfactory—treatment of relations of position in Greek mathematics also apply to the manner in which the tangent concept is introduced. For this, with all curves the method familiar to us for the circle from Euclid's *Elements* is applied unchanged: there (III; Def. 2) the tangent to the circle is defined as the straight line which meets the circle and, when produced, does not intersect the circle, *i.e.* does not meet it again. Upon this, a condition is derived which is necessary and sufficient for a straight line to touch a circle, and finally the close proximity of such a straight line to the circle is further illustrated by the proposition that it is not possible to draw through the point of contact any line all points of which lie between the tangent and the circle.

1.61. For other curves the definition is not even explicitly mentioned again; it is, however, the basis of the derivation of the condition which a straight line has to satisfy in order to touch the curve. In a complete discussion, moreover, the property of the close proximity is proved for each curve anew. With the conics this

[1]) The Greeks do not differentiate between the terms "straight line" and "straight line segment". We are following their custom in the translation; the meaning is always to be inferred from the context.

method does not present any disadvantages. Among the other curves which are certain to have been studied in antiquity, *viz.* the quadratrix of Hippias, the spiral of Archimedes, the conchoid of Nicomedes, and the cissoid of Diocles, it is only the quadratrix which admits of the Euclidean conception without any modification. With the other curves the tangent generally meets the curve elsewhere as well, while in the inflexions of the conchoid there even appears the phenomenon—contrary to the nature of tangency from the Greek point of view—that the tangent intersects the curve in the point of contact.

It does not, however, appear that Greek mathematics concerned itself with the tangent problem for the quadratrix, the conchoid, and the cissoid. And as to the spiral of Archimedes, for this the Euclidean definition can be retained, provided we confine ourselves —as Archimedes will continually be found to do—on each occasion to a consideration of the separate turns which are traced successively by the moving point when the radius vector performs a complete revolution from the zero position.

1.62. In Apollonius' *Conica* we meet with the following fundamental propositions, which are true for all conics:

Conica I, 17. *If a straight line be drawn through the vertex*[1]) *of a conic section in the direction of an ordinate, it will fall without the section.*

This follows at once from the symmetry property of the diameter: if the straight line met the section once more, the vertex of the section would have to be at the same time the middle point and the extremity of the chord obtained. The conclusion is that the straight line therefore touches the section at the vertex.

Conica I, 32. *If a straight line be drawn through the vertex of the section in the direction of an ordinate, it will touch the section, and no other straight line will fall within the area between the section and this straight line.*

In the demonstration of this proposition a distinction is made between the cases where the section is a parabola and those where it is an ellipse, a hyperbola or the circumference of a circle. We will revert to this in our discussion of the different conics.

1.7. We shall find Archimedes speaking repeatedly about similari-

[1]) Vertex (κορυφή) of a conic is any extremity of a diameter.

ty in connection with conics. In all probability he understood by this the same thing as Apollonius, whose conception fits in entirely with the Euclidean view of similarity[1]); his definition may be reproduced, in a slightly abbreviated form, as follows:

We call two conics similar when the lines drawn towards the diameter in the direction of an ordinate are proportional to the parts of the diameter which they determine thereon from the vertex[2]).

The meaning of this is as follows:

Imagine on each of the diameters a series of points with distances to the homologous vertices $x_1, x_2 \ldots; \xi_1, \xi_2 \ldots$, respectively, so that

$$(x_1, \xi_1) = (x_2, \xi_2) = (x_3, \xi_3) \text{ etc.}$$

while we then have for the corresponding ordinates $y_1, y_2 \ldots;$ $\eta_1, \eta_2 \ldots$ etc.

$$(y_1, x_1) = (\eta_1, \xi_1); \qquad (y_2, x_2) = (\eta_2, \xi_2) \text{ etc.}$$

From this it then follows that

$$(y_1, \eta_1) = (y_2, \eta_2) = (y_3, \eta_3) \text{ etc.}$$

Expressed more briefly, this is equivalent to saying that a constant λ can be found such that

$$\text{from } x = \lambda\xi \text{ it follows that } y = \lambda\eta \,.$$

2. *The orthotome.*

2.0. The method of generation has already been mentioned (1.1): a plane is drawn at right angles to a generator of a right circular cone the meridian section of which has a right vertical angle. The orthotome is the section which this plane determines on the surface of the cone. The point where the cutting plane meets the generator to which it is perpendicular is called the *vertex* (κορυφή) of the section; the intersection of the cutting plane with the meridian plane at right angles thereto is for the section a line of right symmetry, which is called the *diameter* (ἡ διάμετρος)[3]). We will already designate the distance of a point of the section to the diameter as *ordinate*[4]), the part of the diameter between the vertex and the

---

[1]) *Elements of Euclid* II, 87–88.

[2]) Apollonius, *Conica* VI, Def. 2.

[3]) By diameter is meant the complete intersection, not the half-line lying within the section.

[4]) *Vide* 1.2; Note 5 on p. 60.

foot of the ordinate as *abscissa.* The double of the line segment bounded by the vertex of the section and the vertex of the cone (ἁ διπλασία τᾶς μέχρι τοῦ ἄξονος, *the double of the line as far as the axis*) we will already call by the Apollonian term *orthia* (ἡ ὀρθία πλευρά = *latus rectum*[1])). The symptom of the section is:

The square on the ordinate of a point of the section is equal to the rectangle contained by the abscissa and the orthia.

In applications it is often used in the following form:

The squares on the ordinates of two points of the section are in the ratio of the abscissae of those points.

2.01. (*Conica* I, 22) *A straight line having two points in common with the orthotome meets the diameter without the section.*

The proof follows from a consideration of the trapezium formed by the line segment between the intersections, their ordinates, and the diameter.

2.02. (*Conica* I, 26) *A straight line parallel to the diameter always meets the section, and only in one point.*

Since the ordinate of the intersection, if any, is known, the abscissa can be unambiguously constructed on the basis of the symptom of the section.

2.03. *Of a straight line parallel to the diameter, the half-line lying on the convex side of the section falls without the section, the other half-line within it.*

The proof of this proposition is difficult to reconstruct in view of the absence of any definition of the terms "within", "without", and "convex"; as will be seen later, other propositions (6.32) are, however, based on the above enunciation, so that they must have occurred in the *Elements*.

2.1. As has already been mentioned in a proposition applying to all conics (1.62), the line drawn through the vertex in the direction of an ordinate is a tangent to the section. The proposition—also enunciated already in a general way—that no other straight line can fall between the section and the tangent is proved as follows for the orthotome (*Conica* I, 32):

In Fig. 9 let the tangent at the vertex of an orthotome with

[1]) It is to be remembered that the classical *latus rectum* is the double of what in analytical geometry is called the parameter; in the equation $y^2 = 2px$ the *latus rectum* is $2p$.

diameter AB and vertex A be $A\Gamma$. Let $A\Delta$, drawn through A within the angle $BA\Gamma$, not meet the section again. Let the straight line, drawn through any given point Δ of this straight line in the direction of the ordinate, meet the section in H, the diameter in E. Then

$$[\mathbf{T}(\Delta E), \mathbf{T}(AE)] > [\mathbf{T}(HE), \mathbf{T}(AE)]$$
$$= (N, AE)\,,$$

when N represents the *orthia* (2.0).

Now find a point Θ on AB such that

$$(N, A\Theta) = [\mathbf{T}(\Delta E), \mathbf{T}(AE)]\,,$$

then

$$A\Theta < AE\,.$$

Let a straight line through Θ, parallel to ΔE, meet the section in Λ, the straight line $A\Delta$ in K.

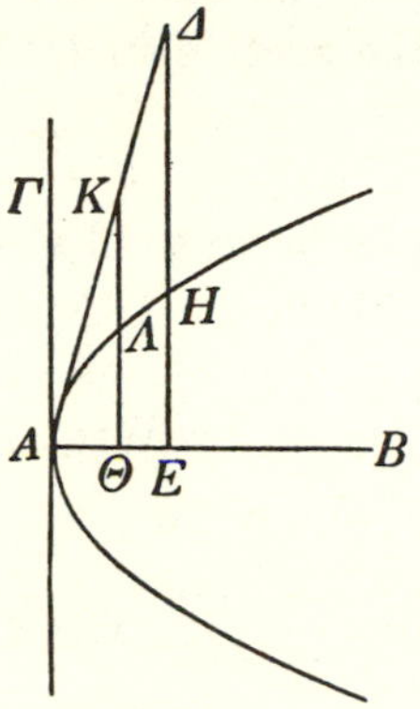

Fig. 9.

Then

$$[\mathbf{T}(K\Theta), \mathbf{T}(A\Theta)] = [\mathbf{T}(\Delta E), \mathbf{T}(AE)] = (N, A\Theta)$$
$$= [\mathbf{O}(N, A\Theta), \mathbf{T}(A\Theta)]\,,$$

whence

$$\mathbf{T}(K\Theta) = \mathbf{O}(N, A\Theta) = \mathbf{T}(\Lambda\Theta)$$

so that K coincides with Λ, and consequently the straight line $A\Delta$ meets the section in K.

2.2. Subsequently a condition may be derived which is sufficient in order that a straight line shall touch the orthotome at a point other than the vertex.

2.20. (*Conica* I, 13) *If* (in Fig. 10) Γ *be a point of the orthotome with abscissa* AB, *and if on the diameter, on the opposite side of the vertex*

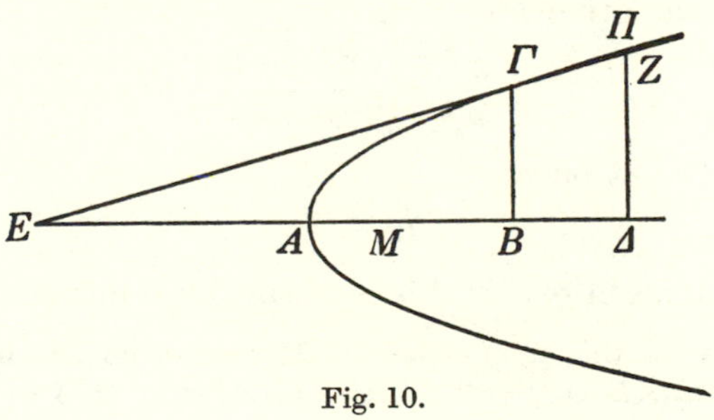

Fig. 10.

A to where the abscissa lies, a line segment AE equal to BA be measured, any point of the straight line $E\Gamma$ other than Γ will fall without the section. $E\Gamma$ will therefore touch the section at Γ.

Proof: Let a point Z of $E\Gamma$ lie within the section. Let the perpendicular through Z to the diameter meet the latter in Δ, the section in Π. Now in view of similarity of triangles

$$[\mathbf{T}(Z\Delta),\ \mathbf{T}(\Gamma B)] = [\mathbf{T}(E\Delta),\ \mathbf{T}(EB)]\,, \qquad (1)$$

and because of the symptom of the section

$$[\mathbf{T}(\Pi\Delta),\ \mathbf{T}(\Gamma B)] = (A\Delta, AB)\,, \qquad (2)$$

thence, because $\Pi\Delta > Z\Delta$,

$$(A\Delta, AB) > [\mathbf{T}(E\Delta),\ \mathbf{T}(EB)]$$

or
$$[\mathbf{O}(4AE, A\Delta),\ \mathbf{T}(EB)] > [\mathbf{T}(E\Delta),\ \mathbf{T}(EB)]$$

or
$$\mathbf{O}(4AE, A\Delta) > \mathbf{T}(E\Delta)\,.$$

This, however, is impossible in view of Euclid II, 5[1]). In fact, A is not the middle point of $E\Delta$, therefore

$$\mathbf{O}(AE, A\Delta) < \tfrac{1}{4}\mathbf{T}(E\Delta)\,.$$

2.21. In order to formulate the given proof algebraically, we write the equation of the section on the system of axes diameter-tangent at the vertex

$$y^2 = 2px\,.$$

If the co-ordinates of Γ are then (x_1, y_1), of $Z(x_1+t, y_2)$, of $\Pi(x_1+t, y_3)$, the two relations given above are:

$$\frac{y_2^2}{y_1^2} = \frac{(2x_1+t)^2}{4x_1^2}\ (1) \text{ and } \frac{y_3^2}{y_1^2} = \frac{x_1+t}{x_1}\ (2),$$

from which we derive

$$\frac{y_3^2}{y_2^2} = \frac{4x_1(x_1+t)}{(2x_1+t)^2},$$

so that we read at once:

$$y_3 < y_2\,.$$

2.22. For a modern reader this is naturally much simpler than the

[1]) According to this proposition, if M represents the middle point of ΔE, we have: $\mathbf{O}(AE, A\Delta) = \mathbf{T}(M\Delta) - \mathbf{T}(MA) < \mathbf{T}(M\Delta) = \tfrac{1}{4}\mathbf{T}(E\Delta)$.

Greek wording. This is partly due to the indirect formulation of the proof of Apollonius, which here, as in many other places, is superfluous. The Greeks have a marked preference for this mode of demonstration: rather than prove that $y_3 < y_2$, they will prove the impossibility of $y_3 > y_2$ (the possibility of $y_3 = y_2$ then being automatically ruled out, because in that case, in view of *Conica* I, 10, other points of the straight line again would fall within the section).

If we now suppose with Apollonius that $y_3 > y_2$, we shall find, as algebraic translation of the Greek proof, from (1) and (2):

$$\frac{x_1 + t}{x_1} > \frac{(2x_1 + t)^2}{(2x_1)^2}$$

or

$$\frac{4x_1(x_1 + t)}{(2x_1)^2} > \frac{(2x_1 + t)^2}{(2x_1)^2},$$

whence

$$4x_1(x_1 + t) > (2x_1 + t)^2 \tag{3}$$

On the ground of our knowledge of familiar simple algebraic identities we now see at once the absurdity of this, and at first view consider it far-fetched when Apollonius, in order to reach the same conclusion, makes use of an identity (Euclid II, 5) which, when formulated algebraically, would read

$$x_1(x_1 + t) = \left(\frac{2x_1 + t}{2}\right)^2 - \left(\frac{t}{2}\right)^2.$$

This example, however, merely shows that by such an algebraical translation we do not realize the essence of the Greek mode of thinking. To a Greek mathematician the identity of Euclid II, 5 is just as fundamental as is one of the familiar algebraic identities to us, and that is why to him in view of the identity

$$\mathbf{O}(AE, A\Delta) = \mathbf{T}(M\Delta) - \mathbf{T}(MA),$$

in which M is the middle point of ΔE, the impossibility of

$$\mathbf{O}(4AE, A\Delta) > \mathbf{T}(E\Delta)$$

is immediately obvious.

2.23. It has now become apparent that the orthotome has a tangent at any point. It still has to be proved that at any point there is no more than one tangent, *i.e.* that the condition that the subtangent shall be double of the abscissa is not only sufficient, but also necessary. The proof of this proposition in Apollonius is as follows (*Conica* I, 35):

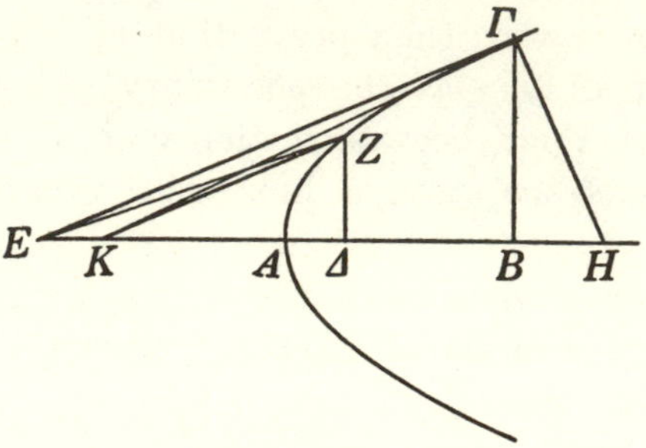

Fig. 11.

In Fig. 11 let a tangent $E\Gamma$ touch the section at Γ, the abscissa being AB. If EA is not equal to AB, then let $A\Delta = EA$. If $A\Delta$ is abscissa of a point Z lying on the same side of the diameter as Γ, then (2.2) EZ will touch the section at Z. EZ will thus have two points in common with $E\Gamma$, and will therefore coincide with $E\Gamma$. Z therefore lies in Γ, Δ in B, and therefore $EA = AB$. The conclusion that, in addition to E, EZ must have another point in common with $E\Gamma$ is based on inspection of the figure, showing that EZ, which must not meet the section again, cannot proceed from the area bounded by $E\Gamma$, the section, and the diameter in any other way but by meeting $E\Gamma$. In the case where Δ lies on the opposite side of B the proof undergoes an obvious modification.

It is further proved that through Γ no other straight line can be drawn between ΓE and the section which meets the section in Γ only (*i.e.* that on the side of ΓE where the section lies no half-line passes through Γ, all other points of which are without the section). Indeed, if ΓK were such a straight line, and if Δ were determined by $A\Delta = AK$, then KZ would be tangent at Z, while (as above) it would coincide with $K\Gamma$, *i.e.* would meet the section again in Γ.

2.24. From the proved proposition about the subtangent it follows at once that the subnormal of the orthotome is constant. Indeed, when the normal of Γ in Fig. 11 meets the diameter in H, and N represents the *orthia*, we have

$$\mathbf{O}(BH, EB) = \mathbf{T}(\Gamma B) = \mathbf{O}(N, AB) = \mathbf{O}(\tfrac{1}{2}N, EB)\,,$$

from which it follows that $BH = \frac{1}{2}N$.

2.3. The property of the pre-Apollonian theory of conics which is most important for our purposes is the one that the symptom of the orthotome applies not only to the diameter, but also to any straight line drawn through a point of the section parallel to the diameter, provided the direction of the tangent at that point be used as the direction of the ordinates. This implies that at any rate the orthotome was also treated in oblique conjugation already before Apollonius, the fundamental difference with the standpoint of the *Conica* remaining that the possiblity of oblique conjugation has to be derived planimetrically, whereas, if the section is generated on an oblique circular cone by any cutting plane, it is present from the very beginning.

There are no indications as to the way in which the property in question may have been proved by the older writers. In the following we will discuss a method which is acceptable on account of its close affinity with the Apollonian mode of thinking.

2.31. (cf. *Conica* I, 43) *If through a variable point of the orthotome straight lines be drawn which are successively parallel to the tangent at the vertex and to the tangent at any given fixed point of the section, these two straight lines together with the diameter include a triangle the area of which is equal to that of the rectangle which has the abscissa of the variable point and the ordinate of the fixed point for its sides.*

In Fig. 12 let A be the vertex of the orthotome, with tangent AK, Γ the fixed point with ordinate ΓB and with tangent ΓE (so that $EA = AB$), Δ the variable point, through which there are drawn towards the diameter $\Delta H \parallel \Gamma E$, $\Delta Z \parallel AK$. Then the proposition is:

$$\triangle\, \Delta ZH = \square\, KZ\,.$$

Proof:

$$[(\triangle\, \Gamma BE,\ \triangle\, \Delta ZH)]$$
$$= [\mathbf{T}(\Gamma B), \mathbf{T}(\Delta Z)] = (AB, AZ)$$
$$= (\tfrac{1}{2}EB, AZ) = [(\triangle \Gamma BE, \square\, ZK)]\,.$$

Therefore

$$\triangle\, \Delta ZH = \square\, ZK\,.$$

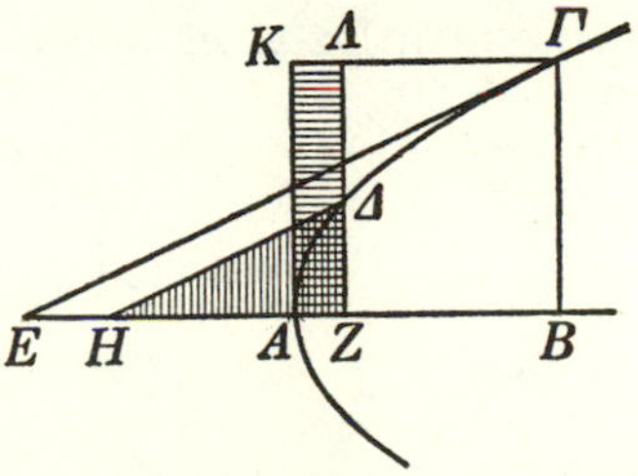

Fig. 12.

2.32. On the ground of this property the symptom of the orthotome may now be derived in oblique conjugation. To this end we prove

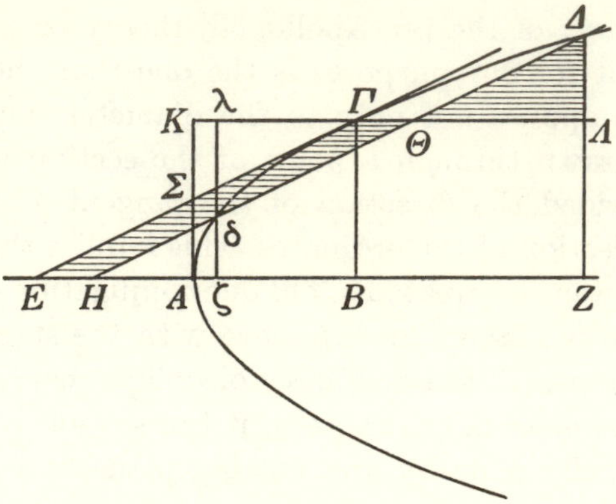

Fig. 13.

(Fig. 13) that between the line segment $\Delta\Theta$, which has been drawn, parallel to the tangent at Γ, towards the straight line through Γ parallel to the diameter, and the line segment $\Gamma\Theta$, which is determined on that straight line by the extremity Θ, there exists the relation

$$\mathbf{T}(\Delta\Theta) = \mathbf{O}(M, \Gamma\Theta)\,,$$

in which M represents a new *orthia*.

Proof: If the tangents at Γ and A meet in Σ, it follows from $AB = AE$ that also $AE = K\Gamma$; therefore

$$\triangle\ A\Sigma E = \triangle\ K\Sigma\Gamma\,.$$

By addition of $A\Sigma\Gamma\Lambda ZA$ to either member it follows from this by 2.31 that

$$E\Sigma\Gamma\Lambda ZE = \square\ AK\Lambda Z = \triangle\ \Delta HZ\,;$$

consequently, after subtraction of $H\Theta\Lambda ZH$ from either member, that

$$\triangle\ \Delta\Lambda\Theta = \square\ \Gamma\Theta HE$$

or $$(\Delta\Theta, \Gamma\Theta) = (2\Gamma E, \Theta\Lambda)\,.$$

Also $$(\Delta\Theta, \Gamma E) = (\Theta\Lambda, 2AB)\,,$$

therefore $$[\mathbf{T}(\Delta\Theta),\ \mathbf{O}(\Gamma\Theta, \Gamma E)] = (\Gamma E, AB)\,.$$

Now take a line segment M such that $(\Gamma E, AB) = (M, \Gamma E)$, then

$$[(\mathbf{T}(\Delta\Theta),\ \mathbf{O}(\Gamma\Theta, \Gamma E)] = (M, \Gamma E) = [\mathbf{O}(M, \Gamma\Theta),\ \mathbf{O}(\Gamma\Theta, \Gamma E)]\,,$$

therefore $$\mathbf{T}(\varDelta\varTheta) = \mathbf{O}(M, \varGamma\varTheta)\,.$$

2.321. In order to correlate the new *orthia* M with the original *orthia* N, which occurred in right conjugation, we write

$$(M, N) = [\mathbf{O}(M, AB),\ \mathbf{O}(N, AB)] = [\mathbf{T}(\varGamma E),\ \mathbf{T}(\varGamma B)] = \\ = [\mathbf{T}(\varDelta\varTheta),\ \mathbf{T}(\varDelta\varLambda)]$$

[in modern notation: $M = \dfrac{N}{\cos^2\varphi}$ if φ represents the angle of the tangents at $\varGamma$ and $\varLambda$].

In this form the property is quoted in the proposition **C.S.** 3 as occurring in the *Elements of Conics*.

2.33. The possibility of treating the orthotome also in oblique conjugation will not be proved completely until it is also established (Fig. 13) that the straight line through $\varGamma$, parallel to the diameter, bisects the chords parallel to the tangent at $\varGamma$. In order to demonstrate this, we consider the chord $\varDelta\delta$ of the direction in question, which meets the straight line through $\varGamma$, parallel to the diameter, in $\varTheta$. Let the ordinate of δ be $\delta\zeta$. Now we have (2.31)

$$\triangle\ \varDelta ZH = \square\ KZ \qquad \text{and} \qquad \triangle\ \delta\zeta H = \square\ K\zeta\,,$$

whence $$\square\ \zeta\delta\varDelta Z = \square\ \lambda Z\,,$$

whence $$\triangle\ \varDelta\varTheta\varLambda = \triangle\ \delta\varTheta\lambda\,,$$

therefore $$\varDelta\varTheta = \delta\varTheta\,.$$

In the case where $\varDelta$ and δ lie on opposite sides of the diameter the proof undergoes an obvious modification.

2.34. Through the proofs given the pre-Apollonian theory of conics has acquired the general concept of the diameters of an orthotome being in affine geometry equivalent to each other, each with the direction of the tangent at the intersection with the section as corresponding direction of the ordinates. We have already seen, however, that even Archimedes does not yet fully draw this conclusion; to him the line of right symmetry remains *the* diameter, and the *orthiae* of the various possible oblique conjugations are not yet directly inter-related, as they are in Apollonius, but they are all correlated with the *orthia* of the diameter through the relation derived in 2.321. It is all the more striking that Archimedes, when discussing segments of the orthotome, *i.e.* figures bounded by the section and a chord, does call the part of the straight line parallel to the diameter, which bisects the chord, diameter of the segment, though he does not look upon the whole of the straight line as the diameter of the section. In practice the general affine standpoint of oblique conjugation has thus also been reached in the terminology used in the treatment of the orthotome. We will, for brevity's sake, call the straight lines parallel to the diameter also diameters, distinguishing the diameter proper therefrom as principal diameter, where necessary.

2.35. Since it is nowhere stated in the formulations and proofs of the propositions mentioned in 2.1; 2.2–2.23 that the direction of the ordinate is at right angles to the direction of the diameter, these propositons are true without any modification for oblique conjugation.

In particular we may now consider as proved the following propositions on the orthotome, which Archimedes quotes as proved in the *Elements of Conics* (Fig. 14; here $AB\Gamma$ is always an orthotome with diameter $B\Delta$).

2.351. **Q.P.** 1. *If* $A\Gamma$, *parallel to the tangent to the section at* B, *meet the straight line* $B\Delta$ *in* Δ, *then* $A\Delta = \Delta\Gamma$. *If inversely* $A\Delta = \Delta\Gamma$, *then* $A\Gamma$ *is parallel to the tangent to the section at* B.

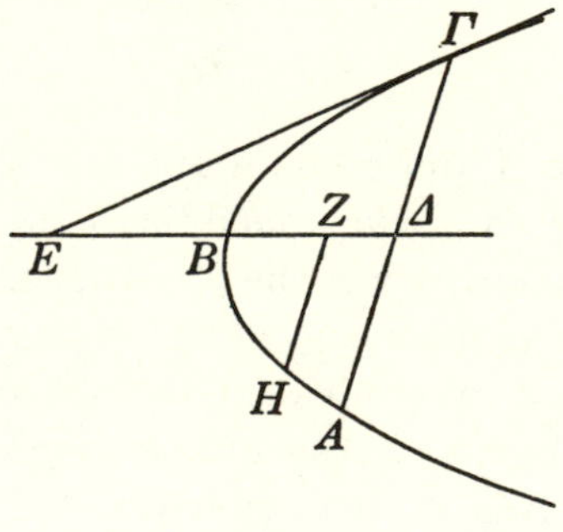

Fig. 14.

2.352. **Q.P.** 2. *If* $A\Gamma$ *be parallel to the tangent at* B, *and the straight line touching the section at* Γ *meet the straight line* $B\Delta$ *in* E, *then* $\Delta B = BE$.

2.353. Finally the property of oblique conjugation can now also be formulated as follows (Fig. 14):

If from two points A and H of the section the straight lines $A\Delta$ and HZ be drawn, parallel to the tangent to the section at B, towards the diameter, then

$$[\mathbf{T}(A\Delta),\ \mathbf{T}(HZ)] = (B\Delta, BZ)\,.$$

Henceforth we shall call $A\Delta$ also ordinate and $B\Delta$ also abscissa.

2.4. Whereas the exposition of the theory of the orthotome so far was somewhat systematic in character, this will no longer be the case in what follows, because we intend to mention only those propositions which are quoted or proved by Archimedes as lemmas.

C.S. 3. *If from the same orthotome two segments be cut off which have equal diameters, the segments are equal to each other, and so are the inscribed triangles.*

By the inscribed triangle of a segment is here understood the triangle which has the boundary chord (base) of the segment for base and the vertex of which lies in the point where the diameter of the segment meets the section (also to be called vertex of the segment).

In Fig. 15 let $B\Gamma\Theta$ be a segment with vertex B, the diameter BH of which falls in the principal diameter of the section, and let ΔEA be a second segment with vertex Δ and diameter ΔZ. Supposition: $BH = \Delta Z$.

Let the *orthia* of the principal diameter be N, then (2.321) the *orthia* M of the diameter ΔZ is determined by the relation

$$(M, N) = [\mathbf{T}(AZ), \mathbf{T}(AK)]\,. \quad (1)$$

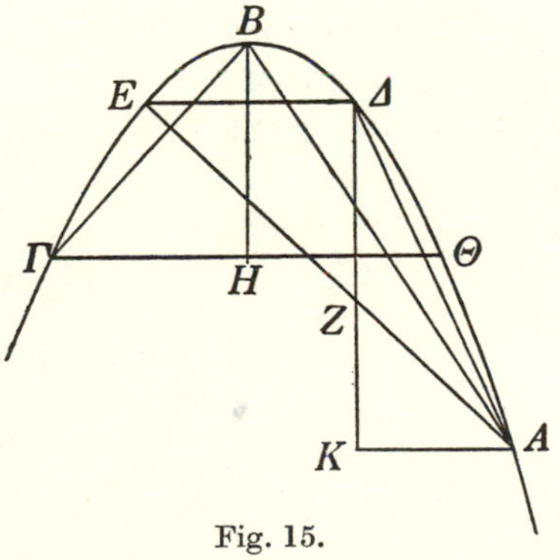

Fig. 15.

In view of the symptom of the section we now have

$$\mathbf{T}(\Theta H) = \mathbf{O}(N, BH)$$
$$\mathbf{T}(AZ) = \mathbf{O}(M, \Delta Z)\,,$$

thence, because $BH = \Delta Z$,

$$[\mathbf{T}(\Theta H, \mathbf{T}(AZ)] = (N, M)\,.$$

In connection with (1) it follows from this that $AK = \Theta H$ whence, because $BH = \Delta Z$, $\triangle\, B\Theta H = \triangle\, \Delta AZ$, and consequently also $\triangle\, B\Theta\Gamma = \triangle\, \Delta AE$.

By the main proposition of the treatise **Q.P.** —to be proved on p. 342, according to which the area of the segment of an orthotome is equal to four-thirds of the area of the inscribed triangle, the areas of the segments $B\Gamma\Theta$ and ΔEA now also are equal.

If the diameter of neither segment falls on the principal diameter of the section, the equality of the areas is proved by comparing both of them with a segment in which this is the case. That Archimedes proceeds in this way, instead of directly comparing any two segments, confirms our earlier impression that right conjugation to him is the primary thing after all, and that he is not acquainted with the general Apollonian proposition (*Conica* I, 49) on the relation between the *orthiae* of different diameters.

2.5. The proposition II, 2 of the treatise **PL.AE.**reads as follows:

If in the segment contained by a straight line and an orthotome a triangle be inscribed which has the same base as the segment and equal height, and triangles be again inscribed in the remaining segments, having the same bases as the segments and equal height, and if in the remaining segments triangles be always inscribed in the same manner, let it be said that the resulting figure is inscribed in the segment "in the recognized manner" (γνωρίμως).

It is now evident that in the figure so inscribed the lines joining the angular points which are nearest to the vertex of the segment and the next in order will be parallel to the base of the segment, and will be bisected by the diameter of the segment, and that they will divide the diameter in the proportions of the successive odd numbers, the number one having reference to the vertex of the segment. These things will be proved in due time.

The proof of this is missing. It may be conceived to have been

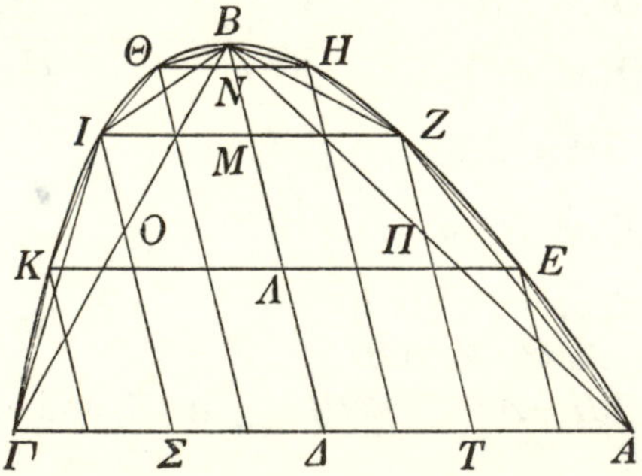

Fig. 16.

furnished as follows. In Fig. 16 let the figure $AEZ \ldots B \ldots \Gamma$ be inscribed in the recognized manner in the segment $AB\Gamma$. The tangent to the section at Z is thus parallel to BA, that at H is parallel to BZ, etc. It is known that the lines drawn through the so defined vertices of the segments, parallel to the diameter, bisect the bases (2.351). If now *e.g.* IO and $Z\Pi$ be produced until they meet $A\Gamma$ successively in Σ and T, then $\Delta\Sigma = \Delta T$. The ordinates of I and Z in respect of $B\Delta$ are therefore equal, and consequently so are the abscissae. The straight lines drawn through I and Z parallel to $A\Gamma$ therefore meet $B\Delta$ in one point. So IZ is parallel to $A\Gamma$. The same proof is true for the pairs of points Θ, H; K, E; etc. At the same time it is plain that $B\Delta$ bisects all these line segments. The ordinates of Θ, I, K, Γ, *i.e.* successively ΘN, IM, $K\Lambda$, $\Gamma\Delta$, are in the proportions of the successive numbers, the abscissae consequently in the proportions of the successive square numbers, and the differences between the abscissae, to which reference is made in the proposition, *viz.* BN, NM, $M\Lambda$, $\Lambda\Delta$, are therefore in the proportions of the successive odd numbers.

2.6. **Q.P.** 4. (Fig. 17a, b) *Let $AB\Gamma$ be a segment which is contained by a straight line and an orthotome, let $B\Delta$ be drawn through the middle*

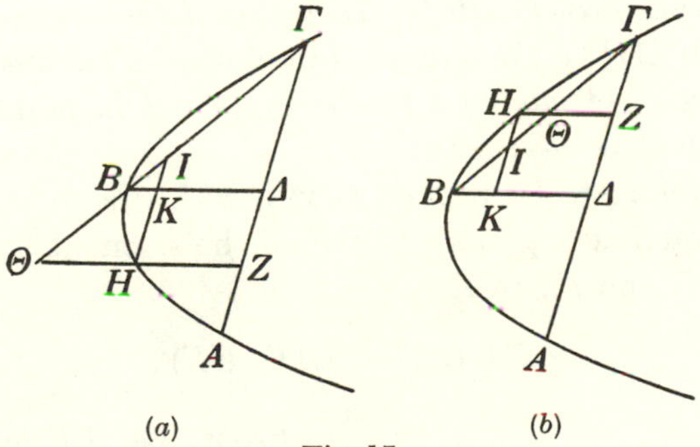

Fig. 17.

point of $A\Gamma$ parallel to the diameter, or be diameter[1]) *itself, and let ΓB be drawn and < if necessary > produced. If now another straight line $Z\Theta$ be drawn parallel to $B\Delta$, which meets the straight line $B\Gamma$, $Z\Theta$ will be to ΘH as ΔA to ΔZ.*

[1]) Here Archimedes again explicitly understands by diameter the principal diameter.

Proof: Let the ordinate of H meet the straight line $B\Gamma$ in I, $B\Delta$ in K. Then

$$[\mathbf{T}(\Gamma\Delta),\ \mathbf{T}(HK)] = (B\Delta, BK)\,,$$

whence $$[\mathbf{T}(\Gamma\Delta),\ \mathbf{T}(Z\Delta)] = (B\Gamma, BI)$$

or $$[\mathbf{T}(B\Gamma),\ \mathbf{T}(B\Theta)] = [\mathbf{T}(B\Gamma),\ \mathbf{O}(BI, B\Gamma)]\,,$$

whence $$\mathbf{T}(B\Theta) = \mathbf{O}(BI, B\Gamma)$$

or $$(BI, B\Theta) = (B\Theta, B\Gamma)\,.$$

From this it follows that

$$(I\Theta, B\Theta) = (\Theta\Gamma, B\Gamma)$$

$$(I\Theta, \Theta\Gamma) = (B\Theta, B\Gamma)$$

or $$(H\Theta, \Theta Z) = (\Delta Z, \Delta\Gamma)\,,$$

therefore $$(\Theta Z, \Theta H) = (A\Delta, Z\Delta)\,.$$

2.7. **Q.P.** 5. (Fig. 18) *Let $AB\Gamma$ be a segment contained by a straight line and an orthotome, and from A let AZ be drawn parallel to the diameter and from Γ the line ΓZ, which touches the section at Γ. If in the triangle $ZA\Gamma$ a straight line be now drawn parallel to AZ, the straight line thus drawn will be divided by the orthotome in the same proportions as $A\Gamma$ by the straight line drawn. The homologous lines will be the part of $A\Gamma$ having A for extremity and the part of the straight line drawn which is adjacent to $A\Gamma$.*

It has therefore to be proved that if $K\Lambda$, drawn parallel to the diameter through any given point K of the segment base $A\Gamma$, meet the curve in Θ and ΓZ in Λ,

$$(K\Theta, \Theta\Lambda) = (AK, K\Gamma)\,.$$

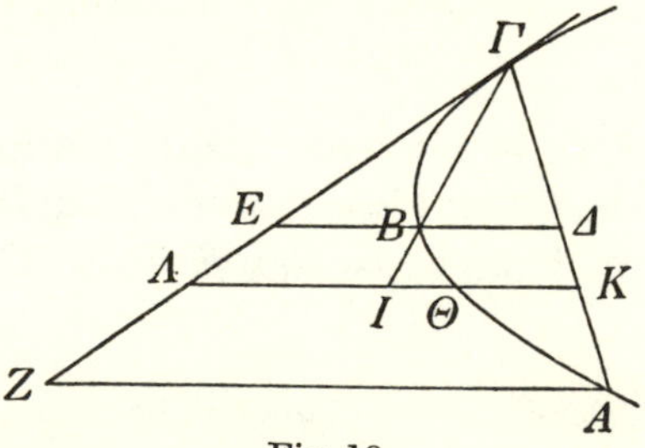

Fig. 18.

Proof: Let ΓB and $K\Lambda$ meet in I. It is known that $EB = B\Delta$, whence $\Lambda I = IK$.

Also (2.6)

$$(KI, \Theta I) = (A\Delta, K\Delta)\,,$$

whence

$$(KI, \Theta K) = (A\Delta, AK)$$

or $$(\Lambda K, \Theta K) = (A\Gamma, AK)$$

or $$(\Lambda\Theta, \Theta K) = (\Gamma K, AK)\,,$$

whence $$(\Theta K, \Theta\Lambda) = (K\Lambda, K\Gamma)\,.$$

2.8. *All orthotomes are similar to one another.*

On the strength of the above definition of similarity of two conics (1.7) this may be proved as follows:

Of two orthotomes let the ordinates and abscissae be successively y and x, η and ξ, the *orthiae* N and M. Then we have

$$\mathbf{T}(y) = \mathbf{O}(N, x)$$
$$\mathbf{T}(\eta) = \mathbf{O}(M, \xi)\,.$$

Now establish between the abscissae of the two curves the relation

$$(x, \xi) = (N, M)\,,$$

then

$$[\mathbf{T}(y), \ \mathbf{T}(\eta)] = [\mathbf{O}(N, x), \ \mathbf{O}(M, \xi)] = [\mathbf{T}(x), \ \mathbf{T}(\xi)]\,,$$

therefore $$(y, \eta) = (x, \xi) \quad \text{or} \quad (y, x) = (\eta, \xi)\,.$$

2.81. *Two segments of orthotomes are called similar when the bases are in the same proportion as the diameters, while in both the angle between diameter and base is the same.*

3. *The oxytome.*

3.0. We already saw (1.1) that the method of generation is the cutting of a right circular cone, the meridian section of which has an acute vertical angle, by a plane at right angles to a generator; the symptom was found to be the relation between the distance of a point of the section to the axis of right symmetry and the two abscissae determined by the foot of the ordinate on the line segment cut off from the axis of symmetry by the sides of the meridian section (Fig. 19):

$$[\mathbf{T}(\Gamma\Delta), \ \mathbf{O}(A\Delta, B\Delta)] = \text{const. (two-abscissa form).}$$

The constant value of this relation is apparently

$$[\mathbf{T}(EK), \ \mathbf{T}(AK)]\,,$$

when EK is the ordinate falling on the straight line bisecting AB

at right angles; from the symptom it also readily follows that this straight line is also an axis of right symmetry. As we already saw, Archimedes calls the line segments determined by the section on the axes of symmetry, *i.e.* AB and EZ, respectively the greater and the lesser diameter (ἡ μείζων διάμετρος and ἡ ἐλάσσων διάμετρος).

By diameter, however, is sometimes also understood the whole of the straight line which is an axis of symmetry. The symptom is also applied in the form

$$[\mathbf{T}(\Gamma\Delta),\ \mathbf{T}(K A)-\mathbf{T}(K\Delta)] = \text{const.} = [\mathbf{T}(EK),\ \mathbf{T}(KA)]\,.$$

3.01. In right conjugation the symptom of the section also holds for the lesser diameter. Indeed, if in Fig. 19 ΓH be at right angles to EZ, then according to Euclid II, 5:

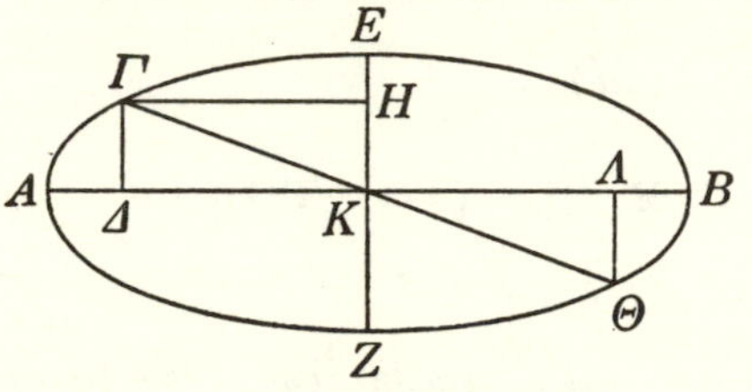

Fig. 19.

$$\mathbf{O}(A\Delta, B\Delta) = \mathbf{T}(AK)-\mathbf{T}(\Delta K)\,,$$

while
$$\mathbf{T}(\Gamma\Delta) = \mathbf{T}(HK)$$
$$= \mathbf{T}(KE)-\mathbf{O}(EH, ZH)\,.$$

The symptom

$$[\mathbf{T}(\Gamma\Delta),\ \mathbf{O}(A\Delta, B\Delta)]$$
$$= [\mathbf{T}(EK),\ \mathbf{T}(KA)]$$

may now be written

$$[\mathbf{T}(KE)-\mathbf{O}(EH, ZH),\ \mathbf{T}(AK)-\mathbf{T}(\Delta K)] = [\mathbf{T}(KE),\ \mathbf{T}(AK)]\,,$$

from which it follows *permutando* and *separando*

$$[\mathbf{O}(EH, ZH),\ \mathbf{T}(\Delta K)] = [\mathbf{T}(KE), \mathbf{T}(AK)]$$

or

$$[\mathbf{T}(\Gamma H),\ \mathbf{O}(EH, ZH)] = [\mathbf{T}(AK),\ \mathbf{T}(EK)]\,.$$

The point of intersection K of the diameters is called *centre* (κέντρον), each of their points of intersection with the section is called *vertex*.

3.02. *When the line joining two points of the oxytome passes through the centre, it is bisected in the centre.*

In Fig. 19 let $\Gamma\Theta$ be the straight line in question, $\Gamma\Delta$ and $\Theta\Lambda$ the ordinates of its extremities.

On account of the symptom of the section we have

$$[\mathbf{T}(\Gamma\Delta),\ \mathbf{O}(A\Delta, B\Delta)] = [\mathbf{T}(\Theta\Lambda),\ \mathbf{O}(A\Lambda, B\Lambda)]$$

or

$$[\mathbf{T}(\Gamma\Delta),\ \mathbf{T}(\Theta\Lambda)] = [\mathbf{T}(KA)-\mathbf{T}(K\Delta),\ \mathbf{T}(KA)-\mathbf{T}(K\Lambda)]\,.$$

From which, in view of similarity of the triangles $K\Delta\Gamma$ and $K\Lambda\Theta$, it follows that

$$[\mathbf{T}(K\Delta),\ \mathbf{T}(K\Lambda)] = [\mathbf{T}(KA)-\mathbf{T}(K\Delta),\ \mathbf{T}(KA)-\mathbf{T}(K\Lambda)]$$

and thence

$$[\mathbf{T}(KA),\ \mathbf{T}(K\Delta)] = [\mathbf{T}(KA),\ \mathbf{T}(K\Lambda)]\,,$$

thence $K\Delta = K\Lambda$. From this it follows again that $K\Gamma = K\Theta$.

3.1. From what has been stated above it readily follows that between the ordinates with a common extremity of an oxytome and of the circle having the greater diameter thereof for its diameter there exists a constant ratio, which is equal to the ratio of the lesser to the greater diameter. This is turned to account in the proposition:

C.S. 4. *Any area contained by an oxytome has to the circle having the greater diameter of the section for its diameter the same ratio as the lesser diameter has to the greater or to the diameter of the circle.*

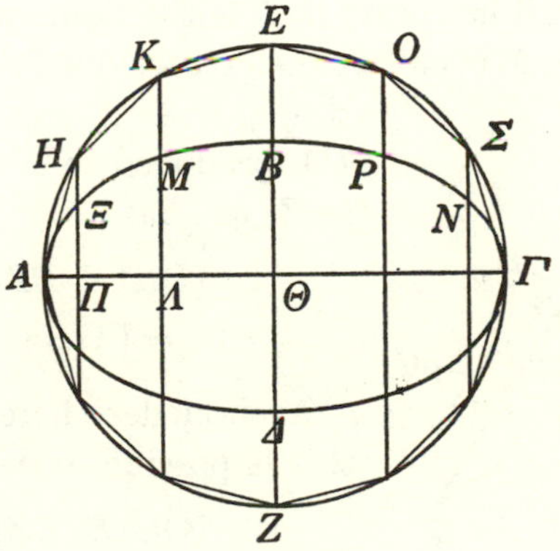

Fig. 20.

The proof of this proposition is furnished by considering in Fig. 20 a regular polygon $AHK\ldots\Gamma\ldots Z\ldots A$ inscribed in the circle and the corresponding polygon $A\Xi\ldots M\ldots\Gamma\ldots\Delta\ldots A$ inscribed in the oxytome, the required passage to the limit being replaced, in the Greek fashion, by an indirect proof. Since the method here to be applied is sufficiently familiar from Euclid XII, we will not here discuss this proof any further.

3.11. From the result obtained it readily follows that the areas of an oxytome and a circle are as the rectangle contained by the diameters of the section to the square on the diameter of the circle (**C.S.** 5), and that the areas of two oxytomes are as the rectangles contained by the diameters in each of the sections (**C.S.** 6).

3.2. What has so far been said about the oxytome fits in completely with the conception we have gained of the original form of the theory of conics. In the treatise **C.S.**, however, Archimedes devotes some elaborate propositions to the possibility of an oxytome lying on an oblique circular cone, apparently a problem newly raised by him, which must have prepared the way for Apollonius' wider conception of the generation of conics.

C.S. 7. *Given an oxytome and a straight line drawn through the centre at right angles to the plane in which the section lies, it is possible to find a cone which has the extremity of the perpendicular for vertex and on whose surface lies the given oxytome.*

In Fig. 21 let the plane of the paper be the plane through the *lesser* diameter AB of the section and the perpendicular $\Delta\Gamma$, erected on its plane through its centre Δ. Γ is the point which is to become vertex of the cone, N is half of the greater diameter. Now construct through A a straight line meeting $\Gamma\Delta$ produced in E and ΓB produced in Z, so that

$$[\mathbf{O}(AE, EZ),\ \mathbf{T}(\Gamma E)] = [\mathbf{T}(N),\ \mathbf{T}(\Gamma\Delta)]\,. \qquad (1)$$

Archimedes here comments that this is possible because

$$[\mathbf{O}(AE, EZ),\ \mathbf{T}(\Gamma E)] > [\mathbf{O}(A\Delta, \Delta B),\ \mathbf{T}(\Gamma\Delta)]\,. \qquad (2)$$

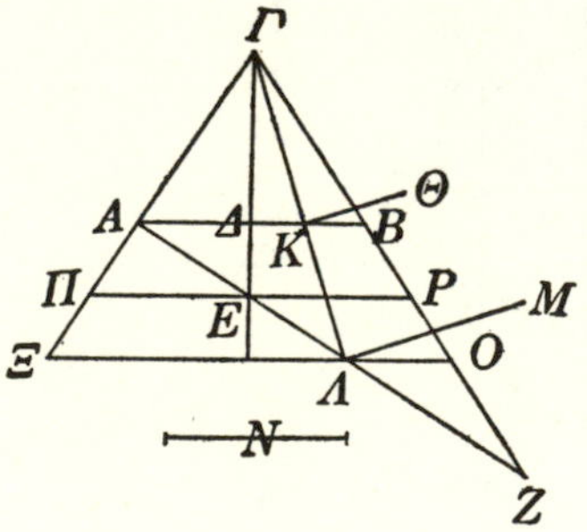

Fig. 21.

We will revert to the significance of this comment in 3.22.

Now draw through AZ a plane at right angles to the plane of the paper, and in this describe the circle on AZ as diameter; this circle is now the curve forming the base of the required cone with vertex Γ.

Proof: Let Θ be a point of the oxytome with ordinate ΘK in respect of AB, let ΓK produced meet AZ in Λ, and let the perpendicular erected through Λ on the plane of the paper (towards the same side as that where Θ lies) meet the circle on AZ as diameter in M. Through Λ let there also be drawn ΞO, through E ΠP, both parallel to AB. It shall now be proved that Θ lies on ΓM.

In fact, we have

$$(AE, A\Lambda) = (\Pi E, \Xi\Lambda)$$

$$(ZE, Z\Lambda) = (PE, O\Lambda),$$

from which it follows that

$$[\mathbf{O}(AE, ZE),\ \mathbf{O}(\Pi E, PE)] = [\mathbf{O}(A\Lambda, Z\Lambda),\ \mathbf{O}(\Xi\Lambda, O\Lambda)].$$

The ratio in the first member is compounded of

$$[\mathbf{O}(AE, ZE),\ \mathbf{T}(\Gamma E)] \quad \text{and} \quad [\mathbf{T}(\Gamma E),\ \mathbf{O}(\Pi E, PE)]$$

i.e. of

$$[\mathbf{T}(N),\ \mathbf{T}(\Gamma\Delta)] \quad \text{and} \quad [\mathbf{T}(\Gamma\Delta),\ \mathbf{O}(A\Delta, B\Delta)]$$

and is therefore equal to

$$[\mathbf{T}(N),\ \mathbf{O}(A\Delta, B\Delta)], \quad \text{thence to} \quad [\mathbf{T}(\Theta K), \mathbf{O}(AK, BK)].$$

Therefore

$$[\mathbf{O}(A\Lambda, Z\Lambda),\ \mathbf{O}(\Xi\Lambda,\ O\Lambda)] = [\mathbf{T}(\Theta K),\ \mathbf{O}(AK, BK)].$$

Also

$$[\mathbf{O}(\Xi\Lambda, O\Lambda),\ \mathbf{T}(\Gamma\Lambda)] = [\mathbf{O}(AK, BK),\ \mathbf{T}(\Gamma K)],$$

whence *ex aequali*

$$[\mathbf{O}(A\Lambda, Z\Lambda),\ \mathbf{T}(\Gamma\Lambda)] = [\mathbf{T}(\Theta K),\ \mathbf{T}(\Gamma K)]$$

or

$$[\mathbf{T}(M\Lambda);\ \mathbf{T}(\Gamma\Delta)] = [\mathbf{T}(\Theta K),\ \mathbf{T}(\Gamma K)]$$

or

$$(M\Lambda, \Gamma\Lambda) = (\Theta K, \Gamma K),$$

from which it follows that Γ, Θ, and M are on one straight line.

Archimedes gives the proof superfluously in the form of a *reductio ad absurdum*, it being initially assumed that Θ does not lie on the surface of the cone; since this assumption is nowhere made use of, this wording is a pure formality.

3.21. The possibility of drawing AZ in such a way that

$$[\mathbf{O}(AE, EZ),\ \mathbf{T}(\Gamma E)] = [\mathbf{T}(N),\ \mathbf{T}(\Gamma\Delta)]$$

will have been proved if it can be shown that through Δ a line is possible which (Fig. 22) meets ΓA in Y and ΓB produced in Φ, so that

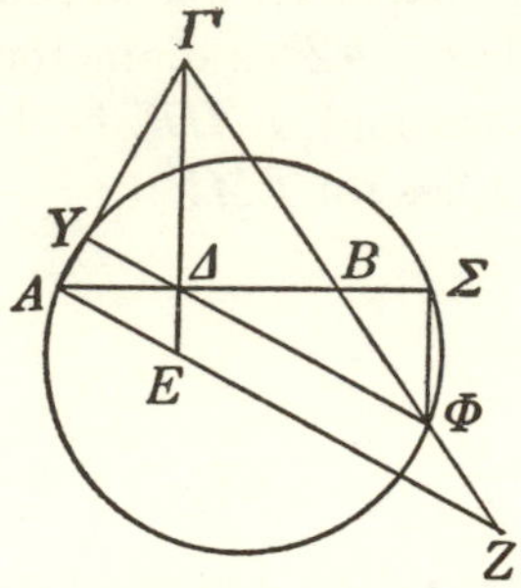

Fig. 22.

$$[\mathbf{O}(Y\Delta, \Delta\Phi), \mathbf{T}(\Gamma\Delta)] = [\mathbf{T}(N), \mathbf{T}(\Gamma\Delta)] .$$

Indeed, the required line through A is then parallel thereto. The line $Y\Phi$ should therefore be so drawn that

$$\mathbf{O}(Y\Delta, \Delta\Phi) = \mathbf{T}(N) . \qquad (3)$$

If now Σ be so taken that

$$\mathbf{T}\ (N) = \mathbf{O}(A\Delta, \Delta\Sigma) ,$$

it is apparently necessary and sufficient for (3) that $AY\Sigma\Phi$ be a cyclic quadrilateral, *i.e.* $\angle\ \Delta\Phi\Sigma = \angle\ YA\Delta$.

Thus Φ is found on ΓB produced as vertex of a triangle with a base $B\Sigma$ and a vertical angle equal to $\angle \Gamma AB$.

For the construction of Φ on ΓB produced to be possible it is therefore necessary and sufficient that Σ shall lie on AB produced, *i.e.* that $\Delta\Sigma > \Delta B$ or $N > A\Delta$. This condition is satisfied because $A\Delta$ is half of the lesser, N half of the greater diameter.

3.22. Since Archimedes explicitly makes use of the inequality (2) of 3.2, it is not likely that the above construction, in which this inequality does not play any part, was applied by him. It is rather to be assumed that, without raising the question as to the construction of AZ, he noted that (Fig. 21) for any straight line AZ passing through a point E on $\Gamma\Delta$ produced the relation (2) has been satisfied, and that he then inferred the possibility of (1) from the inequality

$$[\mathbf{T}(N), \mathbf{T}(\Gamma\Delta)] > [\mathbf{O}(A\Delta, B\Delta), \mathbf{T}(\Gamma\Delta)] ,$$

which amounts to $\mathbf{T}(N) > \mathbf{O}(A\Delta, B\Delta)$ or to $N > A\Delta$.

Formally this inference naturally would not be correct, because from the fact that a variable x (in this case $[\mathbf{O}(AE, EZ), \mathbf{T}(\Gamma E)]$) is greater than a constant a (in this case $[\mathbf{O}(A\Delta, B\Delta), \mathbf{T}(\Gamma\Delta)]$) it cannot be derived that it may become equal to another constant b (in this case $[\mathbf{T}(N), \mathbf{T}(\Gamma\Delta)]$), which is also greater than a. Materially, however, it is correct in the present case when continuity considerations are allowed. In fact, if AZ revolves about A

from the position AB to the position parallel to ΓB, the ratio $[\mathbf{O}(AE, EZ), \mathbf{T}(\Gamma E)]$ increases from the value $[\mathbf{O}(A\Delta, B\Delta), \mathbf{T}(\Gamma\Delta)]$ beyond any pre-assigned limit, and then it must also ultimately attain the value $[\mathbf{T}(N), \mathbf{T}(\Gamma\Delta)]$. It appears probable, in view of the frequent use that Archimedes makes of continuity considerations in his *νεῦσις* constructions (III; 9), that this was the method here applied by him.

3.3 In **C.S.** 8 the same question as in **C.S.** 7 is dealt with, but this time for the case that $\Delta\Gamma$ lies in a plane through one of the diameters of the section at right angles to its plane, without itself being perpendicular to this plane.

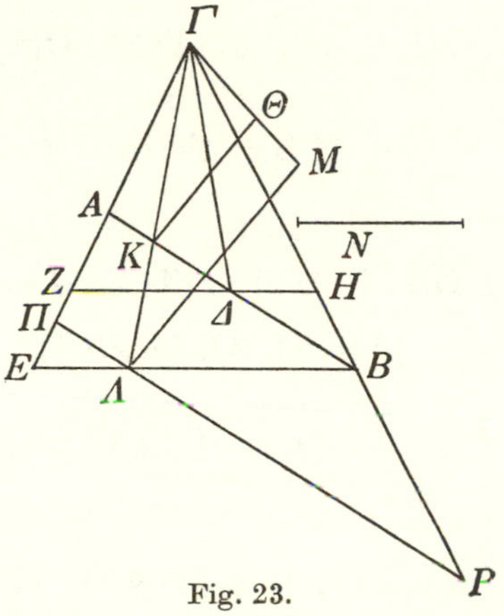

Fig. 23.

In Fig. 23 let AB be the diameter in question, N the half of the other diameter[1]). Measure $\Gamma E = \Gamma B$, and let the straight line drawn through Δ parallel to EB meet ΓA produced in Z, ΓB in H. Now $\mathbf{T}(N)$ is either equal or unequal to $\mathbf{O}(Z\Delta, \Delta H)$.

In the first case describe in a plane through EB at right angles to the plane of the paper a circle on EB as diameter, in the second case an oxytome of which EB is one diameter, while the square of the other diameter, *viz.* Σ, is determined by

$$[\mathbf{T}(N), \mathbf{O}(Z\Delta, \Delta H)] = [\mathbf{T}(\Sigma), \mathbf{T}(EB)]\,. \tag{4}$$

[1]) This other diameter is called *διάμετρος συζυγής*, *conjugate diameter*. This term, however, does not yet have the special connotation which it was to acquire in Apollonius in oblique conjugation and has retained to our own day.

Now the perpendicular, erected through the middle point of EB on the plane of this section, will pass through Γ. According to **C.S.** 7 (3.2) Γ may then be vertex of an oblique circular cone on the surface of which this section lies[1]), while in the case where a circle on EB as diameter has been described, Γ is vertex of a right circular cone which has this circle for its base. It has to be proved that on the surface of this cone lies the given oxytome with diameter AB. Let Θ be a point of the section with ordinate ΘK in respect of AB, M the point of the oxytome with diameter EB, of which the foot Λ of the ordinate is on the same straight line with Γ and K, ΠP being the straight line through Λ parallel to AB.

Now by (4) we have

$$[\mathbf{T}(M\Lambda), \mathbf{O}(E\Lambda, \Lambda B)] = [\mathbf{T}(N), \mathbf{O}(Z\Delta, \Delta H)] .$$

Also

$$[\mathbf{O}(E\Lambda, \Lambda B), \mathbf{O}(\Pi\Lambda, \Lambda P)] = [\mathbf{O}(Z\Delta, \Delta H), \mathbf{O}(A\Delta, \Delta B)] ,$$

therefore *ex aequali*

$$[\mathbf{T}(M\Lambda), \mathbf{O}(\Pi\Lambda, \Lambda P)] = [\mathbf{T}(N), \mathbf{O}(A\Delta, \Delta B)]$$
$$= [\mathbf{T}(\Theta K), \mathbf{O}(AK, BK)]$$

Also

$$[\mathbf{O}(\Pi\Lambda, \Lambda P), \mathbf{T}(\Gamma\Lambda)] = [\mathbf{O}(AK, BK), \mathbf{T}(\Gamma K)] ,$$

therefore *ex aequali*

$$[\mathbf{T}(M\Lambda), \mathbf{T}(\Gamma\Lambda)] = [\mathbf{T}(\Theta K), \mathbf{T}(\Gamma K)]$$

or

$$(M\Lambda, \Gamma\Lambda) = (\Theta K, \Gamma K) ,$$

from which it follows that Γ, Θ, and M are on one straight line, so that Θ lies on the surface of the cone constructed. For the case where a circle has been described on EB as diameter the proof becomes even somewhat simpler.

3.4. In **C.S.** 9 the problem of **C.S.** 8 is discussed once again, but this time for the case where an oblique circular cylinder has to be

[1]) For the application of the construction of **C.S.** 7 (3.2) in the plane of the paper BE would have to be the lesser diameter of the oxytome constructed. This, however, is only the case when

$$\mathbf{T}(N) > \mathbf{O}(Z\Delta, \Delta H) .$$

If, however, EB be the greater diameter, the construction of **C.S.** 7 may be performed in the plane bisecting EB at right angles.

found with axis Γ (Fig. 24). $\Delta\Gamma$ therefore again lies in the plane drawn through one of the diameters of the oxytome (AB) at right angles to the plane of the section, without itself being at right angles to that plane. Now the distance ZH between the two straight lines AZ and BH, drawn parallel to $\Delta\Gamma$, is either equal or unequal to the other diameter Σ of the oxytome.

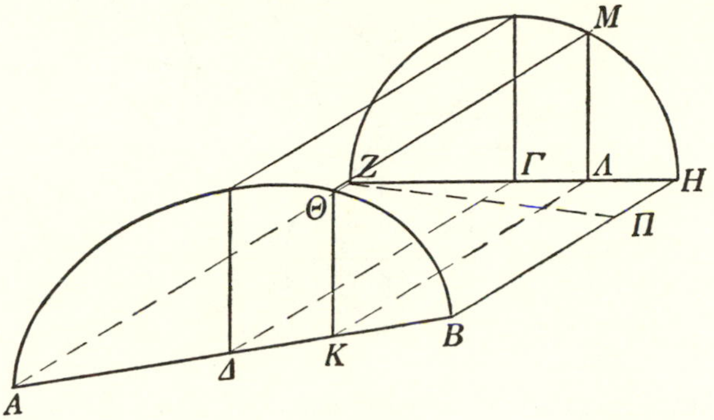

Fig. 24.

α) If $\Sigma = ZH$, describe (Fig. 24) a circle on ZH as diameter in the plane drawn through ZH at right angles to the plane of AZ and BH. The oxytome then lies on the right circular cylinder which has this circle for its base and $\Delta\Gamma$ for its axis. In fact, if Θ be a point of the given oxytome with ordinate ΘK in respect of AB, then

$$[\mathbf{T}(\Theta K),\ \mathbf{O}(AK, BK)] = [\mathbf{T}(\tfrac{1}{2}\Sigma),\ \mathbf{O}(A\Delta, B\Delta)] = [\mathbf{T}(Z\Gamma),\ \mathbf{T}(A\Delta)]\,.$$

If then $K\Lambda \parallel \Delta\Gamma$, we also have

$$[\mathbf{O}(Z\Lambda, \Lambda H),\ \mathbf{O}(AK, KB)] = [\mathbf{T}(Z\Gamma),\ \mathbf{T}(A\Delta)]\,,$$

so that

$$\mathbf{T}(\Theta K) = \mathbf{O}(Z\Lambda, \Lambda H) = \mathbf{T}(M\Lambda)\,.$$

Therefore ΘM is parallel to $K\Lambda$, and consequently to $\Gamma\Delta$, so that Θ lies on the cylinder.

β) If $\Sigma > ZH$, then (Fig. 24) a point Π may be found on BH such that $Z\Pi = \Sigma$. Now describe as above a circle on $Z\Pi$ as diameter, then it will be found in the same way as *sub* α) that the given oxytome lies on the surface of the oblique circular cylinder which has that circle for its base and $\Gamma\Delta$ for axis.

γ) If $\Sigma < ZH$, then find (Fig. 25) on $\Delta\Gamma$ produced a point Ξ such that

$$\mathbf{T}(\Gamma\Xi) = \mathbf{T}(Z\Gamma) - \mathbf{T}(\tfrac{1}{2}\Sigma)$$

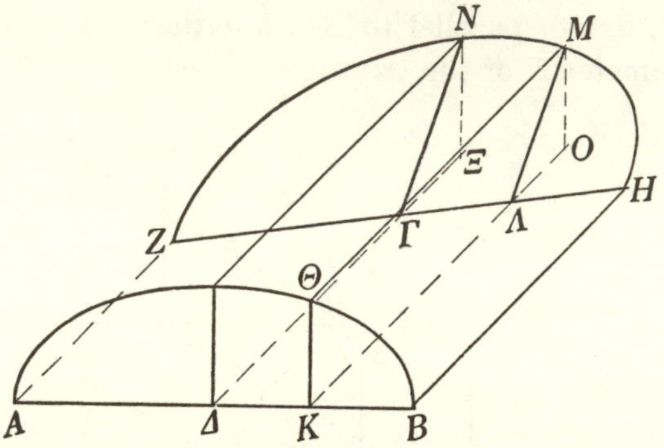

Fig. 25.

and on the perpendicular erected through Ξ on the plane $ABZH$ measure a length $\Xi N = \frac{1}{2}\Sigma$. In the plane ZNH a circle is now described on ZH as diameter; this circle passes through N. Then the given oxytome lies on an oblique circular cylinder which has this circle for its base and $\Delta\Gamma$ for its axis. In fact, if Θ be a point of the oxytome with ordinate ΘK in respect of AB, $K\Lambda$ be parallel to $\Delta\Gamma$, $M\Lambda$ be the ordinate of the circle with foot Λ, and O on $K\Lambda$ produced be the projection of M on the plane $ABZH$, then

$$[\mathbf{O}(Z\Lambda, H\Lambda),\ \mathbf{O}(AK, BK)] = [\mathbf{T}(Z\Gamma),\ \mathbf{T}(A\Delta)]$$

or

$$[\mathbf{T}(M\Lambda,\ \mathbf{O}(AK, BK)] = [\mathbf{T}(\Gamma N),\ \mathbf{T}(A\Delta)]\,,$$

whence, because

$$[\mathbf{T}(M\Lambda),\ \mathbf{T}(N\Gamma)] = [\mathbf{T}(MO),\ \mathbf{T}(N\Xi)]\,,$$

also

$$[\mathbf{T}(MO),\ \mathbf{O}(AK, BK)] = [\mathbf{T}(N\Xi),\ \mathbf{T}(A\Delta)]$$
$$= [\mathbf{T}(\Theta K),\ \mathbf{O}(AK, BK)]\,,$$

therefore

$$MO = \Theta K\,,$$

from which it follows again that Θ lies on the cylinder.

3.5. The propositions proved above are of importance not only for the actual import of the treatise **C.S.**, but also as a means for the evaluation of the Archimedean theory of conics. In fact, when Archimedes proves at such length that any oxytome is to be conceived of in an infinite number of ways as a section generated by cutting an oblique circular cone by a plane at right angles to a principal section, this is apparently a result which did not occur in the *Elements of Conics* and thus constituted a new contribution to the theory. It must naturally have been easy to prove, by an inversion of the argument, that a plane drawn at right angles to the principal section of an oblique circular cone gives, within certain limits as to position, an oxytome as section, which amounts to a step in the direction of the Apollonian generation of conics. Considering the main object of his works (determination of volumes and areas), it is understandable that Archimedes does not go thus far: the theory of conics to him always remains a means to an end, and he therefore does not develop it beyond what is required for his investigations. That he knew it quite well however, appears from a remark in **C.S.** *Definitiones* (*Opera* I, 258): *If a cone be cut by a plane meeting all the sides of the cone, the section will be either a circle or an oxytome.* In the same place (I, 260) it is said that this property is also true of an (oblique) circular cylinder[1]).

3.6. *Two oxytomes are similar, and only then, when the two pairs of diameters form a proportion.*

Let the ordinates in the two sections be called y and η respectively, the greater diameters a and α, the lesser b and β, the abscissae (from one of the vertices of the greater diameter) x and ξ. Establish between the abscissae the relation

$$(x, \xi) = (a, \alpha) . \qquad \text{(cf. 1.7)}$$

It is then derived without difficulty that also

$$[\mathbf{O}(x, a-x),\ \mathbf{O}(\xi, \alpha-\xi)] = [\mathbf{T}(x),\ \mathbf{T}(\xi)] . \qquad (1)$$

Now suppose that

$$(a, \alpha) = (b, \beta) .$$

then

$$[\mathbf{T}(b),\ \mathbf{T}(a)] = [\mathbf{T}(\beta),\ \mathbf{T}(\alpha)]$$

and, in view of the symptom of the section (3.0),

---

1) These two results are already mentioned in Euclid, *Phaenomena* (*Opera* VIII, 6). He probably still refers to right circular cones and cylinders. He says of the oxytome that it resembles a θυρεός, a shield.

$$[\mathbf{T}(y),\ \mathbf{O}(x, a-x)] = [\mathbf{T}(\eta),\ \mathbf{O}(\xi, \alpha-\xi)]\,,$$

from which it follows in connection with (1) that

$$[\mathbf{T}(y),\ \mathbf{T}(\eta)] = [\mathbf{T}(x),\ \mathbf{T}(\xi)]$$

or

$$(y, \eta) = (x, \xi) = (a, \alpha)\,.$$

If inversely it be given that $(y, \eta) = (x, \xi)$, it is found by inversion of the argument that

$$(a, \alpha) = (b, \beta)\,.$$

3.61. With the aid of 3.6 the Corollary of **C.S.** 6 may be derived from the proposition mentioned in 3.11 (**C.S.** 6), according to which the areas of two oxytomes are as the rectangles contained by their diameters, *viz*: *the areas of two similar oxytomes are as the squares on the corresponding diameters.*

4. *The amblytome.*

4.0. We already found (1.1) the method of generation to be the cutting of a right circular cone, the meridian section of which has an obtuse vertical angle, by a plane at right angles to a generator; the symptom was found to be the relation between the distance of a point of the section to the axis of right symmetry and the two abscissae, determined by the foot of the ordinate on the part cut off from the axis of symmetry by the sides of the meridian section (Fig. 26):

$$[\mathbf{T}(\Gamma\Delta),\ \mathbf{O}(A\Delta, B\Delta)] = \text{const. (two-abscissa form).}$$

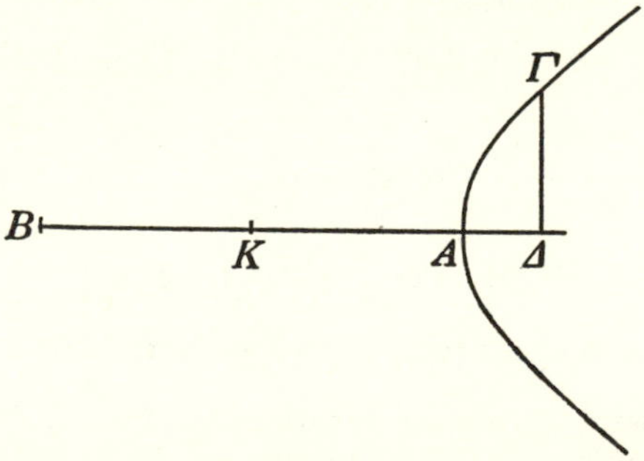

Fig. 26.

Since the discussion of the amblytome before Apollonius remained confined to one of the two branches of which the hyperbola consists, there is only one line of right symmetry; nor can the constant ratio, as with the oxytome, be interpreted as the ratio of the squares of two semi-diameters. The symptom was also frequently applied in the form

$$[\mathbf{T}(\Gamma\Delta),\ \mathbf{T}(K\Delta)-\mathbf{T}(KA)] = \text{const.},$$

in which K is the middle point of the line segment AB. This middle point naturally is not called centre; it is referred to as the *point in which the lines nearest to the section meet* (τὸ σαμεῖον, καθ' ὃ αἱ ἔγγιστα τᾶς τομᾶς συμπίπτοντι).

The existence and the position of these lines may be understood in the following manner, which in principle resembles the method applied by Apollonius at the beginning of the second book of the *Conica*.

4.1. (cf. *Conica* II, 1) *If* (in Fig. 27) *on the line drawn through the vertex A of the section at right angles to the diameter a segment $A\Delta$ be measured such that*

$$[\mathbf{T}(\Delta A),\ \mathbf{T}(KA)]$$

is equal to the constant ratio of the square on an ordinate and the rectangle contained by the two corresponding abscissae, the straight line $K\Delta$ will not meet the section, but any straight line through K lying within the angle ΔKA will do so.

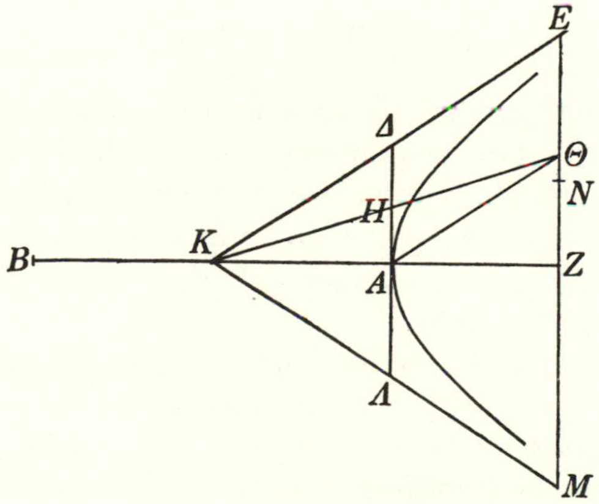

Fig. 27.

α) Suppose that $K\Delta$ met the section in a point E with ordinate EZ, then we should have

$$[\mathbf{T}(EZ),\ \mathbf{O}(AZ, BZ)] = [\mathbf{T}(\Delta A),\ \mathbf{T}(KA)] = [\mathbf{T}(EZ),\ \mathbf{T}(KZ)],$$

whence
$$\mathbf{O}(AZ, BZ) = \mathbf{T}(KZ),$$

which is impossible because, in view of Euclid II, 6,

$$\mathbf{O}(AZ, BZ) = \mathbf{T}(KZ) - \mathbf{T}(KA).$$

β) Let KH be a straight line through K within the angle ΔKA. Draw through A a straight line parallel to $K\Delta$, which shall meet KH in a point Θ. The perpendicular through Θ to the diameter meets $K\Delta$ in E, the diameter in Z, the straight line $K\Lambda$, which lies with $K\Delta$ symmetrically in respect of AB, in M. When KH does not meet the section, Θ lies without the section; therefore a point N of the section must lie between Θ and Z.
Now

$$[\mathbf{T}(NZ),\ \mathbf{O}(AZ, BZ)] = [\mathbf{T}(\Delta A),\ \mathbf{T}(KA)] = [\mathbf{T}(EZ),\ \mathbf{T}(KZ)].$$

From this it follows that

$$[\mathbf{T}(NZ),\ \mathbf{T}(EZ)] = [\mathbf{T}(KZ) - \mathbf{T}(KA),\ \mathbf{T}(KZ)] \text{ and then}$$
$$[\mathbf{T}(EZ) - \mathbf{T}(NZ),\ \mathbf{T}(KA)] = [\mathbf{T}(EZ),\ \mathbf{T}(KZ)],$$

whence
$$[\mathbf{O}(EN, MN),\ \mathbf{T}(KA)] = [\mathbf{T}(\Delta A),\ \mathbf{T}(KA)],$$

therefore
$$\mathbf{O}(EN, MN) = \mathbf{T}(\Delta A).$$

But this is impossible because $EN > E\Theta = \Delta A$ and $MN > \Delta A$.

In view of the proved property the straight lines $K\Delta$ and $K\Lambda$ are called *the straight lines nearest to the section.* We shall further designate them by the Apollonian term *asymptotes*, while we shall also, for brevity's sake, already call their point of intersection, though improperly, *centre*.

4.2. *Two amblytomes are similar, and only then, when the ratio of the square on the ordinate and the rectangle on the two abscissae is the same for both sections.*

The proof can be furnished in the same way as that of 3.6, provided we refrain from interpreting the constant ratio existing in each section between the square on the ordinate and the rectangle on the two abscissae as a ratio of squares of line segments.

4.3. *Mutatis mutandis* the propositions formulated in 2.01–2.03 for the orthotome are also true for the amblytome.

5. *Oxytome and amblytome.*

5.0. In the following paragraphs we will discuss a number of properties which oxytome and amblytome have in common. The proofs for the two sections are entirely analogous; where they differ, the difference generally concerns no more than the sign of an expression. In the following paragraphs by *section* is understood either of the said conic sections, by *diameter* the greater diameter of the oxytome, and the single diameter of the amblytome; the word designates the part of the diameter determined on the axis of symmetry by the sides of the meridian section as well as the whole of the straight line containing this segment. What is called *vertex* in both sections is a point of intersection of a diameter and the section. The straight line determined by the centre of a section and any given point thereof is called *radius vector* of that point; the word denotes this straight line as well as the part of it bounded by the two points.

5.1. We have already seen (1.62) that the straight line drawn through the vertex in the direction of an ordinate touches the section. For the supplementary proposition stating that no other straight line can fall between the section and the tangent we cannot, as for the orthotome (2.1), take the proof from Apollonius' *Conica*, because there use is made of a symptom for oxytome and

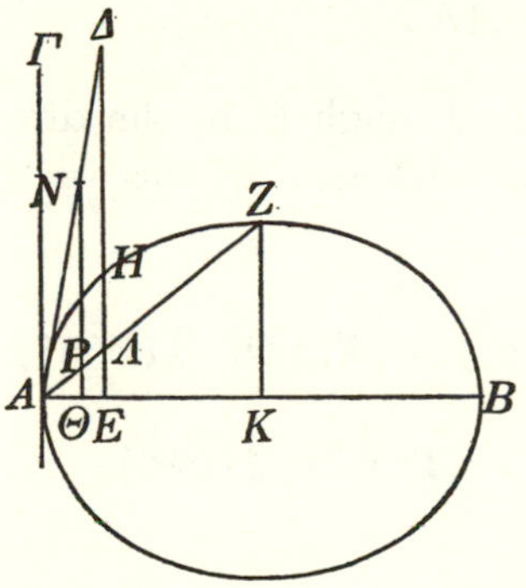

Fig. 28 a.

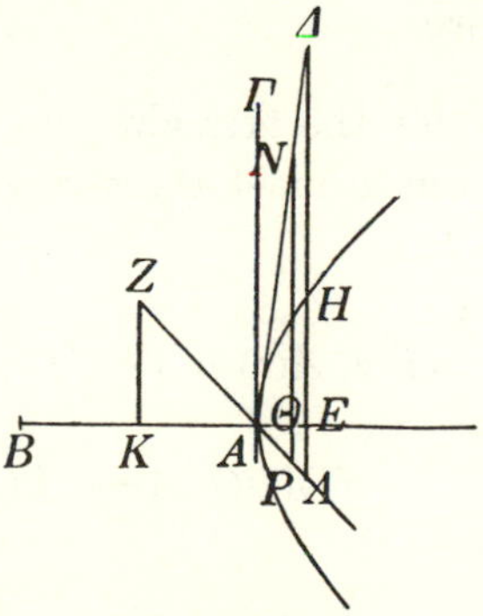

Fig. 28 b.

amblytome not yet occurring in Archimedes. With pre-Apollonian means the proof may be furnished as follows:

In Fig. 28a, b let, for one of the two sections, AB be the diameter, the vertex A, the tangent at the vertex $A\Gamma$. Let $A\Delta$, drawn through A on that sidė of $A\Gamma$ where the section lies, not meet the section. Let the straight line drawn through any point Δ of this straight line in the direction of the ordinate meet the section in H, the diameter in E. Then the following inequality applies:

$$[\mathbf{T}(\Delta E),\ \mathbf{O}(AE, BE)] > [\mathbf{T}(HE),\ \mathbf{O}(AE, BE)] = \text{constant}.$$

For the oxytome the constant is the ratio $[\mathbf{T}(KZ),\ \mathbf{T}(KA)]$, if K be the centre of the section and KZ the half of the other diameter; for the amblytome KZ can be so chosen that the same inequality applies. Now let AZ meet the straight line ΔE in Λ. Then we have

$$[\mathbf{T}(\Delta E),\ \mathbf{O}(AE, BE)] > [\mathbf{T}(\Lambda E),\ \mathbf{T}(AE)]\,,$$

whence

$$[\mathbf{T}(\Delta E),\ \mathbf{T}(\Lambda E)] > (BE, AE)\,.$$

Now find on AB a point Θ such that

$$[\mathbf{T}(\Delta E),\ \mathbf{T}(\Lambda E)] = (B\Theta, A\Theta)\,,$$

then

$$(B\Theta, A\Theta) > (BE, AE)\,,$$

from which it is derived (0.42) that

$$(A\Theta, AB) < (AE, AB)\,,$$

therefore

$$A\Theta < AE\,.$$

Now let the straight line drawn through Θ in the direction of the ordinate meet the straight lines $A\Delta$ and AZ successively in N and in P.

Then

$$(B\Theta, A\Theta) = [\mathbf{T}(\Delta E),\ \mathbf{T}(\Lambda E)] = [\mathbf{T}(N\Theta),\ \mathbf{T}(P\Theta)]\,,$$

whence

$$[\mathbf{O}(B\Theta, A\Theta),\ \mathbf{T}(A\Theta)] = [\mathbf{T}(N\Theta),\ \mathbf{T}(P\Theta)]\,,$$

whence

$$[\mathbf{T}(N\Theta),\ \mathbf{O}(A\Theta, B\Theta)] = [\mathbf{T}(P\Theta),\ \mathbf{T}(A\Theta)] = [\mathbf{T}(KZ),\ \mathbf{T}(KA)]\,,$$

from which it follows that N lies on the section. The straight line $A\Delta$ must therefore meet the section.

5.2. A condition which suffices in order that a straight line shall touch at a point other than the vertex is enunciated in the following proposition (cf. *Conica* I, 34) (Fig. 29a, b):

If Γ be a point of the section with ordinate $\Gamma\Delta$ and abscissae $A\Delta$ and $B\Delta$, and if on the diameter a point E be found such that

$$(A\Delta, B\Delta) = (AE, BE)$$

any point of the straight line $E\Gamma$ other than Γ will fall without the section. The straight line $E\Gamma$ will therefore touch the section at Γ.

Apollonius proves this proposition synthetically and indirectly, which does not make the argument very transparent; for clearness' sake we will formulate it analytically and directly by requiring any point Z of $E\Gamma$ to lie without the section. That is to say: if the perpendicular from Z to AB meet the section in Θ, and the diameter in H, it is required that $ZH > \Theta H$.

Now on account of similarity of triangles we have

$$[\mathbf{T}(ZH),\ \mathbf{T}(\Gamma\Delta)] = [\mathbf{T}(EH),\ \mathbf{T}(E\Delta)]$$

and in view of the symptom of the section

$$[\mathbf{T}(\Theta H),\ \mathbf{T}(\Gamma\Delta)] = [\mathbf{O}(AH, BH),\ \mathbf{O}(A\Delta, B\Delta)\ .$$

Therefore it must be true that

$$[\mathbf{T}(EH),\ \mathbf{T}(E\Delta)] > [\mathbf{O}(AH, BH),\ \mathbf{O}(A\Delta, B\Delta)]$$

or $$[\mathbf{O}(A\Delta, B\Delta),\ \mathbf{T}(E\Delta)] > [\mathbf{O}(AH, BH),\ \mathbf{T}(EH)]\ . \qquad (1)$$

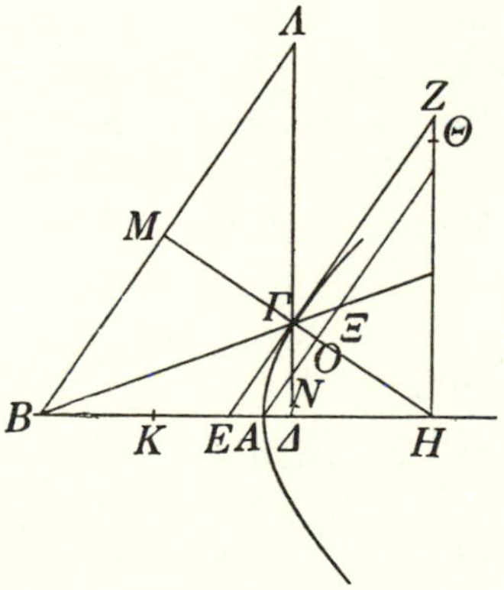

Fig. 29a.

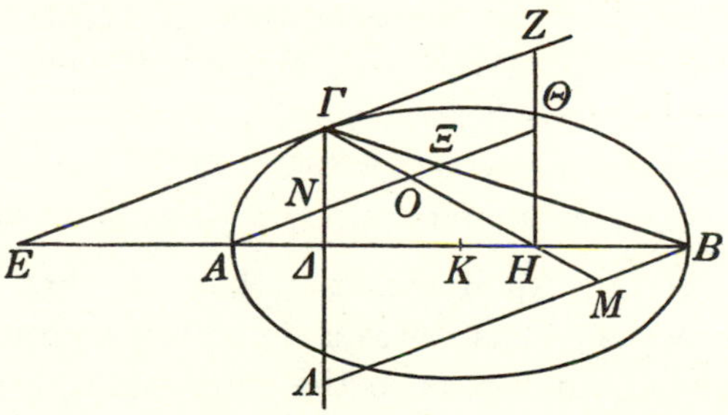

Fig. 29b.

We will now proceed so to transform the ratios of line segments of which the two members are composed that the consequents in the two members are equal. An inequality between ratios of line segments thus results. For this, draw through A a straight line parallel to $E\Gamma$ meeting $\Gamma\Delta$ in N and ΓB in Ξ, through B a straight line parallel to $E\Gamma$ meeting $\Gamma\Delta$ in Λ; further let ΓH meet the straight lines AN and $B\Lambda$ successively in O and M.

Then

$$(A\Delta, E\Delta) = (AN, E\Gamma) \qquad (AH, EH) = (AO, E\Gamma)$$

$$(B\Delta, E\Delta) = (B\Lambda, E\Gamma) \qquad (BH, EH) = (BM, E\Gamma)\,.$$

In view of this the inequality (1) may be written

$$[\mathbf{O}(AN, B\Lambda),\ \mathbf{T}(E\Gamma)] > [\mathbf{O}(AO, BM),\ \mathbf{T}(E\Gamma)]$$

or
$$\mathbf{O}(AN, B\Lambda) > \mathbf{O}(AO, BM)$$

or
$$(AN, AO) > (BM, B\Lambda)$$

or
$$(AN, AO) > (\Xi O, \Xi N)$$

or
$$\mathbf{O}(AN, \Xi N) > \mathbf{O}(AO, \Xi O)\,.$$

When N is a fixed and O a variable point of the line segment $A\Xi$, this will apply then, and only then, for any point O, when N is the middle point of $A\Xi$. This condition is equivalent to

$$(AN, \Lambda B) = (\Xi N, \Lambda B)$$

and consequently to

$$(A\Delta, B\Delta) = (\Gamma\Xi, \Gamma B) = (EA, EB)\,.$$

Since all the stages of the argument can also be reversed, the condition derived is also sufficient. Of course the synthesis may also be given directly, instead of indirectly, as Apollonius does. Starting from the condition $(A\Delta, B\Delta) = (AE, BE)$, he shows the absurdity of the supposition $ZH < \Theta H$; it may just as well be proved that if the condition is satisfied, we always find $ZH > \Theta H$. Here again we meet with one of the numerous examples of the Greek geometers' preference for a *reductio ad absurdum.*

5.21. In Apollonius the theorem that the condition derived is necessary constitutes a separate proposition (I, 36), which is completely analogous to the corresponding proposition for the orthotome (2.23).

5.22. The tangent property derived in 5.2 may also be transformed as follows (cf. *Conica* I, 37):

From
$$(A\Delta, B\Delta) = (AE, BE)$$
it follows

for the oxytome
$$(AB, A\Delta) = (AE+BE, AE),$$

for the amblytome
$$(AB, A\Delta) = (BE-AE, AE),$$

hence, if K be the centre,

$$(2KA, A\Delta) = (2KE, AE)$$

$$(KA, K\Delta) = (KE, KA),$$

therefore
$$\mathbf{T}(KA) = \mathbf{O}(K\Delta, KE).$$

5.23. Another form of the property arises as follows:

$$\mathbf{O}(A\Delta, B\Delta) = \mathbf{T}(KA)-\mathbf{T}(K\Delta) = \mathbf{O}(K\Delta, KE)-\mathbf{T}(K\Delta)$$
$$= \mathbf{O}(K\Delta, \Delta E).$$

5.3. The object of the subsequent propositions consists, just as with the orthotome, in deriving the symptom of the section in oblique conjugation.

5.31. *If tangents to a section be drawn at the vertex and at any given point, these straight lines together with the diameter and the radius vector of the point selected include two triangles of equal area.*

In Fig. 30a, b let AZ be tangent at the vertex, ΓE tangent at the point Γ (E lying on the diameter), K the centre of the section. It has to be proved that

$$\triangle K\Gamma E = \triangle KAZ.$$

Proof:

By 5.22

$$\mathbf{T}(KA) = \mathbf{O}(KE, K\Delta),$$

whence
$$[\mathbf{T}(KA), \mathbf{T}(K\Delta)] = (KE, K\Delta),$$

whence
$$(\triangle KAZ, \triangle K\Delta\Gamma) = (\triangle K\Gamma E, \triangle K\Gamma\Delta),$$

whence
$$\triangle KAZ = \triangle K\Gamma E.$$

5.32. From the proposition proved it follows that

$$\triangle E\Delta\Gamma = \square A\Delta\Gamma Z \quad \text{and} \quad \triangle \Sigma EA = \triangle \Sigma\Gamma Z.$$

5.33. *If through a variable point of a section straight lines be drawn parallel successively to the tangents at the vertex and at any fixed point*

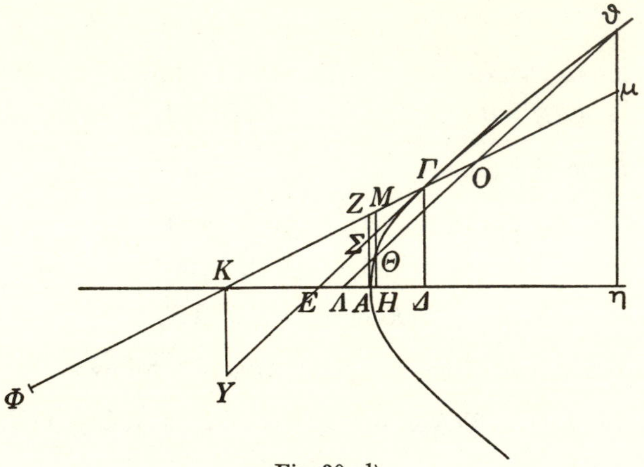

Fig. 30a [1]).

of the section, these two straight lines together with the diameter include a triangle, the area of which is equal to the difference of the areas of the triangles included by the diameter and the radius vector of the fixed point successively with the tangent at the vertex and with the ordinate of the variable point.

In Fig. 30a, b let Θ be the variable point of the section, through which be drawn ΘH parallel to the tangent at the vertex A, $\Theta \Lambda$ parallel to the tangent at Γ. It now has to be proved that

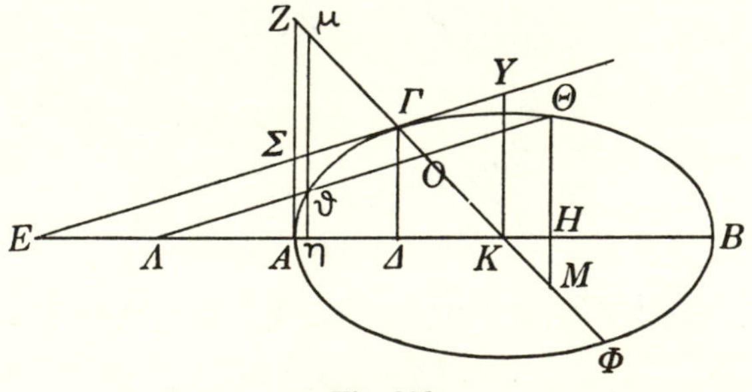

Fig. 30b.

[1]) In Fig. 30a the letter B has been omitted at the point lying on AK produced so that $BK = KA$.

for the oxytome	for the amblytome
$\triangle\, \Theta H\Lambda = \triangle\, KAZ - \triangle\, KHM$	$\triangle\, \Theta H\Lambda = \triangle\, KHM - \triangle\, KAZ$.

Proof:

$$(\triangle\, \Theta H\Lambda,\ \triangle\, \Gamma\Delta E) = [\mathbf{T}(\Theta H),\ \mathbf{T}(\Gamma\Delta)]$$
$$= [\mathbf{O}(AH, BH),\ \mathbf{O}(A\Delta, B\Delta)]$$

$[\mathbf{T}(KA) - \mathbf{T}(KH),\ \mathbf{T}(KA) - \mathbf{T}(K\Delta)] = [\triangle\, KAZ - \triangle\, KHM,\ \triangle\, KAZ - \triangle\, K\Delta\Gamma]$	$[\mathbf{T}(KH) - \mathbf{T}(KA),\ \mathbf{T}(K\Delta) - \mathbf{T}(KA)] = [\triangle\, KHM - \triangle\, KAZ,\ \triangle\, K\Delta\Gamma - \triangle\, KAZ]$.

But we have (5.32)

$\triangle\, \Gamma\Delta E = \triangle\, KAZ - \triangle\, K\Delta\Gamma$,	$\triangle\, \Gamma\Delta E = \triangle\, K\Delta\Gamma - \triangle\, KAZ$,

whence

$\triangle\, \Theta H\Lambda = \triangle\, KAZ - \triangle\, KHM$.	$\triangle\, \Theta H\Lambda = \triangle\, KHM - \triangle\, KAZ$.

5.34. We will now prove the validity of the symptom of the section in oblique conjugation. Since from the proof for the oxytome that for the amblytome can be obtained in the same manner as in 5.33, we will confine ourselves to the oxytome (Fig. 30b).

It is known (5.33) that

$$\triangle\, \Theta H\Lambda = \triangle\, KAZ - \triangle\, KHM$$

or (5.31)
$$\triangle\, \Theta H\Lambda + \triangle\, KHM = \triangle\, K\Gamma E$$

or, after subtraction of $\triangle\, OK\Lambda$ from either member,

$$\triangle\, \Theta OM = \triangle\, K\Gamma E - \triangle\, KO\Lambda\,.$$

Now, if Φ be the second point of intersection of ΓK with the section, we have

$$\mathbf{O}(\Gamma O, \Phi O) = \mathbf{T}(K\Gamma) - \mathbf{T}(KO)\,,$$

while

$$[\mathbf{T}(K\Gamma) - \mathbf{T}(KO),\ \mathbf{T}(K\Gamma)] = [\triangle\, K\Gamma E - \triangle\, KO\Lambda,\ \triangle\, K\Gamma E]\,,$$

whence $[\mathbf{O}(\Gamma O, \Phi O),\ \mathbf{T}(K\Gamma)] = [\triangle\, \Theta OM,\ \triangle K\Gamma E]$ (1)

If the straight line through K parallel to AZ meet the line $E\Gamma$ in Y, we also have

$$[\mathbf{T}(\Gamma\Upsilon),\ \mathbf{T}(\Theta O)] = [\triangle\ \Upsilon K\Gamma,\ \triangle\ \Theta MO]\,, \tag{2}$$

so that the ratio

$$[\triangle\ K\Gamma\Upsilon,\ \triangle\ K\Gamma E],\ \textit{i.e.}\ (\Upsilon\Gamma, E\Gamma)\ \text{or}\ [\mathbf{T}(\Upsilon\Gamma),\ \mathbf{O}(\Upsilon\Gamma, E\Gamma)]\,,$$

is compounded (*vide* (2) and (1)) of

$$[\mathbf{T}(\Upsilon\Gamma),\ \mathbf{T}(\Theta O)] \quad \text{and} \quad [\mathbf{O}(\Gamma O, \Phi O),\ \mathbf{T}(K\Gamma)]\,.$$

Since the ratio $[\mathbf{T}(\Upsilon\Gamma),\ \mathbf{O}(\Upsilon\Gamma, E\Gamma)]$ is also compounded of

$$[\mathbf{T}(\Upsilon\Gamma),\ \mathbf{T}(\Theta O)] \quad \text{and} \quad [\mathbf{T}(\Theta O),\ \mathbf{O}(\Upsilon\Gamma, E\Gamma)]\,,$$

it follows from this that

$$[\mathbf{T}(\Theta O),\ \mathbf{O}(\Gamma\Upsilon, \Gamma E)] = [\mathbf{O}(\Gamma O, \Phi O),\ \mathbf{T}(K\Gamma)]$$

or

$$[\mathbf{T}(\Theta O),\ \mathbf{O}(\Gamma O, \Phi O)] = [\mathbf{O}(\Gamma\Upsilon, \Gamma E),\ \mathbf{T}(K\Gamma)] = \text{const.},$$

by which the symptom has been derived in oblique conjugation.

5.35. The discussion of the sections in oblique conjugation is not complete until it has also been shown that any straight line through the centre is a line of oblique symmetry for a conjugate direction of the ordinate, in this case the direction of the tangent at a point of intersection with the section. This property is enunciated in the following proposition (cf. *Conica* I, 47):

The radius vector of any point of the section bisects the chords parallel to the tangent at that point.

The proof for the amblytome is as follows (Fig. 30a):

Let $\Theta\vartheta$ be a chord parallel to the tangent at Γ, which meets the radius vector $K\Gamma$ in O. Now we have (5.33)

$$\triangle\ \Theta H\Lambda = \triangle\ KHM - \triangle\ KAZ$$

$$\triangle\ \vartheta\eta\Lambda = \triangle\ K\eta\mu - \triangle\ KAZ$$

whence $$\triangle\ KHM - \triangle\ \Theta H\Lambda = \triangle\ K\eta\mu - \triangle\ \vartheta\eta\Lambda$$

or, by subtraction of either member from $\triangle\ \Lambda OK$,

$$\triangle\ \Theta OM = \triangle\ \vartheta O\mu\,,$$

therefore $$O\Theta = O\vartheta\,.$$

5.36. Now that the symptom of the section has been proved in oblique conjugation, all the properties from 5.1 onwards may be considered as proved, when the diameter with perpendicular direction of the ordinate there mentioned is replaced by the radius vector of any given point with oblique conjugate direction of the ordinate. The property 5.33 may therefore be formulated in a general way as follows:

The straight lines, through a variable point of a section parallel to the tangents at two fixed points, together with the radius vector of one of these two points include a triangle the area of which is equal to the difference of the areas of the triangles included by the radius vectors of the two points successively with the tangent at the point selected and with the straight line drawn parallel thereto through the variable point.

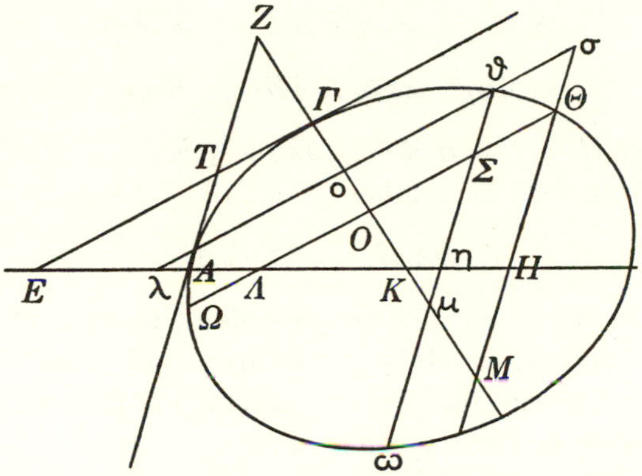

Fig. 31.

Thus, for example, in Fig. 31, where on an oxytome are taken the fixed points A and Γ, and the variable point Θ, we have

$$\triangle\ \Theta\Lambda H = \triangle\ KAZ - \triangle\ KHM$$

$$\triangle\ \Theta OM = \triangle\ K\Gamma E - \triangle\ K\Lambda O$$

5.37. A more symmetrical form of the proposition of 5.36 is the following:

If through any two points of the section there be drawn pairs of straight lines parallel to the tangents at two fixed points, the pair of straight lines parallel to the first tangent, together with one of the straight lines of the second pair and the radius vector of the second point of contact, include a quadrilateral the area of which is equal to that of the quadrilateral included by the pair of straight lines parallel to the second tangent, together with one of the straight lines of the first pair and the radius vector of the first point of contact, provided the two straight lines which in each of the quadrilaterals constitute a side which is not parallel to another side do not meet in any point of the section.

Let A and Γ be the two fixed points, Θ and ϑ the two arbitrarily selected points, then we have

$$1)\quad \square\, \Theta\Sigma\mu M = \square\, \vartheta\Sigma\Lambda\lambda$$

$$2)\quad \square\, \vartheta\sigma M\mu = \square\, \Theta\sigma\lambda\Lambda\,.$$

Indeed, by 5.36 we have

$$\triangle\, \Theta\Lambda H = \triangle\, KAZ - \triangle\, KHM$$

$$\triangle\, \vartheta\lambda\eta = \triangle\, KAZ - \triangle\, K\eta\mu\,,$$

therefore $\triangle\, \Theta\Lambda H + \triangle\, KHM = \triangle\, \vartheta\lambda\eta + \triangle\, K\eta\mu$.

Subtract from either member $\triangle\, \Lambda\Sigma\eta + \triangle\, K\eta\mu$,

then $$\square\, \Theta\Sigma\mu M = \square\, \vartheta\Sigma\Lambda\lambda\,.$$

5.38. At the same time it is now possible to prove a property of all three kinds of conics, which is quoted by Archimedes in **C.S.** 3 and forms the extension to conics of the property of the rectangles under segments of intersecting chords in a circle.

If from a point tangents be drawn to any conic section and inside the conic section straight lines be also drawn, parallel to the tangents and intersecting one another, the rectangles contained by the parts < *into which the point of intersection of the chords divides the latter* > *will be to one another as the squares on the tangents* < *parallel to the chords* >.

We will here state the proof for the oxytome; for the amblytome it proceeds on entirely identical lines, apart from some changes of sign. For the orthotome the principle remains the same; the propositions 2.31 *et seq.* are then used.

In Fig. 31 consider the chords $\Theta\Omega$ and $\vartheta\omega$ (successively parallel to the tangents at Γ and at A), which intersect each other in Σ. Now

$$\mathbf{O}(\Theta\Sigma, \Omega\Sigma) = \mathbf{T}(O\Theta) - \mathbf{T}(O\Sigma)\,.$$

Also

$$[\mathbf{T}(O\Theta) - \mathbf{T}(O\Sigma),\ \mathbf{T}(O\Theta)] = [\triangle\, O\Theta M - \triangle\, O\Sigma\mu,\ \triangle\, O\Theta M]$$

and $$[\mathbf{T}(O\Theta),\ \mathbf{T}(\Gamma T)] = [\triangle\ O\Theta M,\ \triangle\ \Gamma TZ]\,,$$

whence *ex aequali*

$$[\mathbf{T}(O\Theta)-\mathbf{T}(O\Sigma),\ \mathbf{T}(\Gamma T)] = [\triangle\ O\Theta M - \triangle\ O\Sigma\mu,\ \triangle\ \Gamma TZ]$$

or

$$[\mathbf{O}(\Theta\Sigma, \Omega\Sigma),\ \mathbf{T}(\Gamma T)] = (\square\ \Sigma\Theta M\mu,\ \triangle\ \Gamma TZ)\,.$$

Likewise

$$[\mathbf{O}(\vartheta\Sigma, \omega\Sigma),\ \mathbf{T}(TA)] = (\square\ \Sigma\vartheta\lambda\Lambda,\ \triangle\ TAE)\,.$$

By 5.37 and 5.32 the second members are equal, from which follows the equality of the first members, and consequently the correctness of the proposition.

5.4. With the aid of the properties of oblique conjugation, some more properties of oxytome and amblytome used by Archimedes are easily understood.

5.41. *The line joining the points of contact of two parallel tangents to an oxytome passes through the centre.*

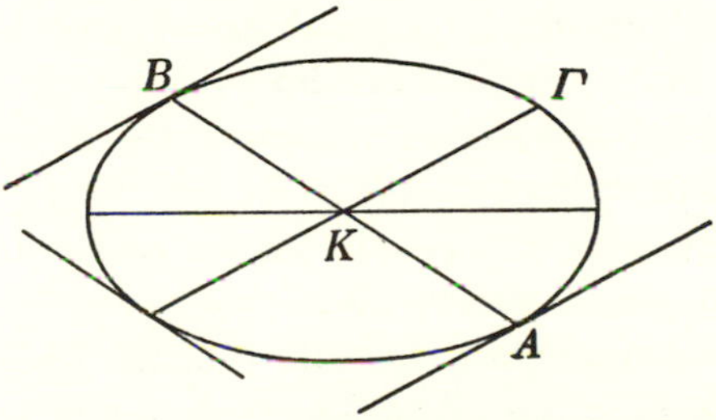

Fig. 32.

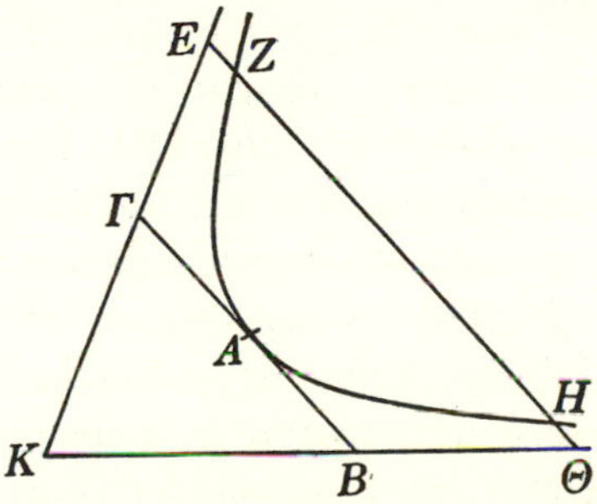

Fig. 33.

If in Fig. 32 A and B be the points of contact of the parallel tangents and K be the centre, the straight line KA as well as the straight line KB bisects all chords parallel to the parallel tangents, so that these straight lines coincide.

5.42. *If through the centre of an oxytome a radius vector be drawn parallel to two parallel tangents, the tangent at the extremity of the radius vector is parallel to the line joining the points of contact of the parallel tangents.*

In Fig. 32 let the straight line through K, parallel to the tangents at A and B, meet the section in Γ. If now the tangent at Γ were not parallel to AB, it would meet AB in a point E, and then we should have (5.2, for oblique conjugation)

$$(AE, BE) = (AK, BK),$$

therefore AE would have to be equal to BE, which is not possible.

5.43. *A tangent to an amblytome intersects the asymptotes in two points which are equidistant from the point of contact.*

The proposition 4.1, when applied for oblique conjugation, shows at once (Fig. 33) that the asymptotes are the lines joining the centre with points B and Γ, which lie on the tangent at A at equal distances from A.

5.431. From 5.43 and 5.35 it also follows that when a straight line meets the section in Z and H, and the asymptotes successively in E and Θ, $ZE = H\Theta$.

6. *Cones, Cylinders, Conoids, and Spheroids.*

6.0. *The Cone and Its Parts.* No definition of a cone is to be found anywhere in Archimedes; this, however, does not imply that he conformed to Euclidean usage, according to which by a cone is understood a figure (XI, Def. 18) generated by the rotation of a right-angled triangle about one of the sides containing the right angle. In the treatise *On the Sphere and Cylinder*, in which only right circular cones occur, he calls—at any rate in the propositions and expositions—the Euclidean cone *isosceles* (*κῶνος ἰσοσκελής*), by which he apparently wishes to denote that the side (*i.e.* the apothem, called *πλευρά* by him) is constant. Another term found is *right cone* (*κῶνος ὀρθός*). In the treatise *On Conoids and Spheroids* the word cone, used without any adjective, denotes a—generally oblique—circular cone. The other terms used are: *vertex* (*κορυφή*); *base* (*βάσις*) for the basic circle and for the plane in which it lies; *axis* (*ἄξων*) for the straight line through the vertex of the cone and the centre of the base, as also for the line segment bounded by these two points. A *segment of a cone* (*ἀπότμαμα κώνου*, **C.S.** *Definitiones*. I, 258) is the portion cut off from a cone on the side of the vertex by a plane which yields an oxytome as section. This plane, just like the section itself, is called *base of the segment*; *axis of the segment* is the line segment bounded by the vertex of the cone and the centre of the section; *height* the distance from the vertex to the base.

It is not explicitly stated that on an oblique circular cone there are also a series of circular sections lying in planes not parallel to the base; the way in which, however, reference is made (**C.S.** *Definitiones*) to the possibility of a plane cutting a cone in a circle (without it being stated that this plane will be parallel to the base) makes it probable that Archimedes was thoroughly aware of this (*vide* 3.5).

6.1. *The Cylinder and Its Parts.* For this, the same things as said with regard to the cone apply *mutatis mutandis*. *Cylinder* (*κύλινδρος*) means an oblique circular cylinder; the right circular cylinder is termed *right* in the propositions and expositions (*κύλινδρος ὀρθός*). A *frustum of a cylinder* (*τόμος κυλίνδρου*) is the portion cut off from a cylinder by two parallel planes which have oxytomes as sections (**C.S.** *Definitiones*). *Axis of the frustum* is the line segment bounded by the centres of the two sections; it lies on the axis of the cylinder. *Height* is the distance between the boundary surfaces.

6.11. **C.S.** 10. On the ground of the same arguments as in Euclid XII it is seen that the ratio of two segments of a cone is compounded of the ratio of their bases and the ratio of their heights, and that any frustum of a cylinder is equal to three times a segment of a cone with the same base and the same height.

6.2. In the letter to Dositheus serving as introduction to the treatise *On Conoids and Spheroids*, the following figures are defined:

6.21. When an orthotome rotates about the diameter, a figure is generated which is called *right-angled conoid* (*ὀρθογώνιον κωνοειδές*; to be translated by *orthoconoid*). When parallel to any tangent plane to this figure a cutting plane is drawn, this plane together with the surface determines a *segment of the conoid* (*τμᾶμα τοῦ κωνοειδέος*), of which the *base* is the section of the cutting plane with the conoid, the *vertex* the point of contact with the tangent plane, the *axis* the part cut off by the two planes from the straight line through the vertex of the segment parallel to the axis of revolution.

The orthoconoid is apparently a paraboloid of revolution.

6.22. Through rotation of an amblytome about the diameter an *obtuse-angled conoid* (*ἀμβλυγώνιον κωνοειδές*; to be translated by *amblyconoid*) is generated. This is therefore one sheet of a two-sheet hyperboloid of revolution. The asymptotes of the section during the rotation generate the *enveloping cone* (κῶνος περιέχων). Of a segment as defined above (6.21) the *axis* is the part cut off by the cutting plane and the tangent plane parallel thereto from the straight line joining the vertex of the enveloping cone with the vertex of the segment. The part of this straight line between these two points themselves is called the *axis produced* (ποτεοῦσα τῷ ἄξονι = *adjacent to the axis*).

6.23. Through rotation of the oxytome about a diameter is generated the *spheroid* (ellipsoid of revolution), which is called *prolate* (παραμᾶκες σφαιροειδές) or *oblate* (ἐπιπλατὺ σφαιροειδές), according as the section has rotated about the greater or about the lesser diameter. *Axis* and *vertex* are defined as above, *centre* is the centre of the rotating section, *diameter* the diameter of the section at right angles to the axis of rotation. A cutting plane together with the spheroid determines two segments, the *vertices* of which are the points where the tangent planes parallel to the cutting plane touch the surface, the parts determined by the cutting plane on the line segment between the points of contact being called the *axes*.

6.3. *The Orthoconoid.*

C.S. 11a. *If an orthoconoid be cut by a plane through, or parallel to, the axis, the section will be the same orthotome as that containing the* < *rotating* > *figure, and its diameter will be the common section of the plane intersecting the figure and the plane erected through the axis at right angles to the cutting plane.*

The proof, which is missing in Archimedes (it is required only for a plane parallel to the axis), may be furnished as follows (Fig. 34).

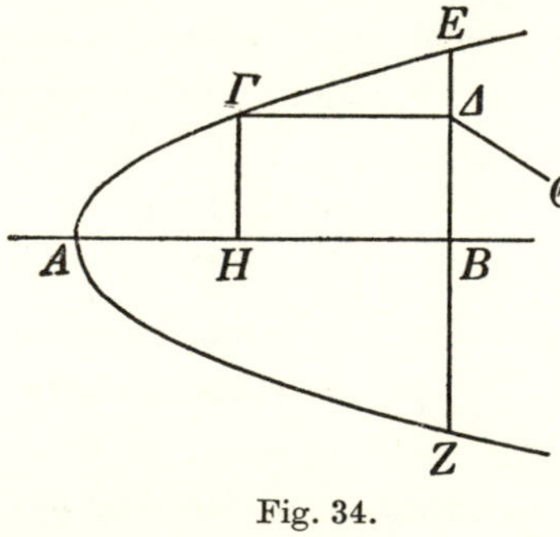

Fig. 34.

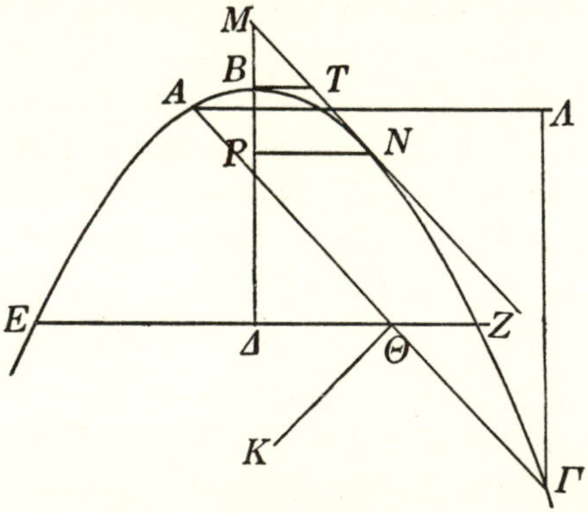

Fig. 35.

Let AB be the axis, $\Gamma\Delta$ the intersection between the plane of the paper and the cutting plane at right angles thereto. Let the straight line drawn through Δ perpendicular to the plane of the paper (*i.e.* in the cutting plane) meet the surface in Θ. Then, if the *orthia* of the meridian section be N, we have

$$\begin{aligned} \mathbf{T}(\Theta\Delta) &= \mathbf{O}(E\Delta, Z\Delta) = \mathbf{T}(BE) - \mathbf{T}(B\Delta) \\ &= \mathbf{O}(N, AB) - \mathbf{T}(\Gamma H) = \mathbf{O}(N, AB) - \mathbf{O}(N, AH) \\ &= \mathbf{O}(N, BH) = \mathbf{O}(N, \Gamma\Delta). \end{aligned}$$

The section therefore is an orthotome with vertex Γ, diameter $\Gamma\Delta$, and the same *orthia* as EAZ, *i.e.* equal and similar thereto.

6.31. **C.S.** 12. *If an orthoconoid be cut by a plane neither passing through the axis, nor parallel to the axis, nor perpendicular to the axis, the section will be an oxytome; the greater diameter of the latter will be the part cut off within the conoid from the intersection between the cutting plane and the plane through the axis at right angles to the cutting plane, the lesser diameter will be equal to the distance between the straight lines drawn from the extremities of the greater diameter, parallel to the axis.*

Proof: In Fig. 35 let $A\Gamma$ be the intersection between the meridian plane, which is the plane of the paper, and the cutting plane at right angles thereto, K a point of the intersection between the cutting plane and the figure, $K\Theta$ the perpendicular from K to $A\Gamma$, EZ the intersection between the plane of the paper and a plane erected through Θ at right angles to the axis; this plane thus intersects the conoid in a circle on EZ as diameter. Now

$$\mathbf{T}(K\Theta) = \mathbf{O}(E\Theta, Z\Theta)\,.$$

On account of the proposition quoted in **C.S.** 3 (*vide* 5.38) we now have

$$[\mathbf{O}(E\Theta, Z\Theta),\ \mathbf{O}(A\Theta, \Gamma\Theta)] = [\mathbf{T}(BT),\ \mathbf{T}(NT)]$$

when the tangent NM parallel to $A\Gamma$ meets the tangent at the vertex (parallel to EZ) in T. Because $NT = MT$ (which follows from $PB = BM$; 2.23), we also have

$$[\mathbf{T}(K\Theta),\ \mathbf{O}(A\Theta, \Gamma\Theta)] = [\mathbf{T}(BT),\ \mathbf{T}(MT)] = [\mathbf{T}(A\Lambda),\ \mathbf{T}(A\Gamma)]\,,$$

therefore $A\Lambda$ and $A\Gamma$ are respectively the lesser and the greater diameter of the oxytome which is the locus of K.

6.32. **C.S.** 15α. *Of the straight lines, drawn from the points of an* *orthoconoid parallel to the axis, the parts which are in the same direction as the convexity* (τὰ κυρτά) *of the surface will fall without the conoid, the parts which are in the other direction within it.*

This is reduced with the aid of **C.S.** 11α (6.3) to the corresponding proposition on the orthotome (2.03).

6.33. **C.S.** *Definitiones. All orthoconoids are similar.* What this means, is not mentioned any further. Probably the reference is simply to similarity of the sections generating the orthoconoids, which are already known to be similar (2.8).

6.34. **C.S.** 15γ. *If a plane meet a conoid without cutting it, it will meet it at one point only, and the plane drawn through the point of contact and the axis will be at right angles to the plane which touches it.*

Suppose the plane to touch the surface at two points A and B. Draw through each of those points a straight line parallel to the axis, and draw a plane through those two straight lines. This plane cuts the conoid in an orthotome on which lie the two points A and B. The points of the line segment AB then fall within the section (1.5), *i.e.* within the conoid.

The second part of the proposition is evident for the tangent plane at the vertex. In fact, the tangents at the vertices to two sections of the conoids in planes through the axis are perpendicular to that axis, and consequently so is the tangent plane (here therefore it is taken into consideration that the tangent plane is determined by two tangents to curves on the surface through the point under review). If the plane touch the conoid at another point, the proposition is understood by noting that the plane contains the tangent to the parallel circle and is consequently at right angles to the meridian section through the point of contact.

6.40. *The Amblyconoid.*

C.S. 11β. *If an amblyconoid be cut by a plane through the axis, or parallel to the axis, or through the vertex of the cone enveloping the conoid, the section will be an amblytome, viz. if the plane pass through the axis, the same as that containing the* < *rotating* > *figure; if it be parallel to the axis, one similar thereto; if it pass through the vertex of the cone enveloping the conoid, one which is not similar thereto; and the diameter of the section will be the common section of the plane cutting the surface and the plane through the axis at right angles to the cutting plane.*

The proof, which is missing in Archimedes, may be furnished as follows:

α) The proposition is evident for a plane through the axis.

β) In Fig. 36 let $\Gamma\Delta$ be the intersection of a plane parallel to the axis BA with a meridian plane chosen as the plane of the paper, at right angles to that plane, Θ any point of the section, ΘE the perpendicular from Θ to $\Gamma\Delta$. Now

$$\mathbf{T}(\Theta E) = \mathbf{O}(ME, NE) = \mathbf{T}(HM) - \mathbf{T}(HE)$$

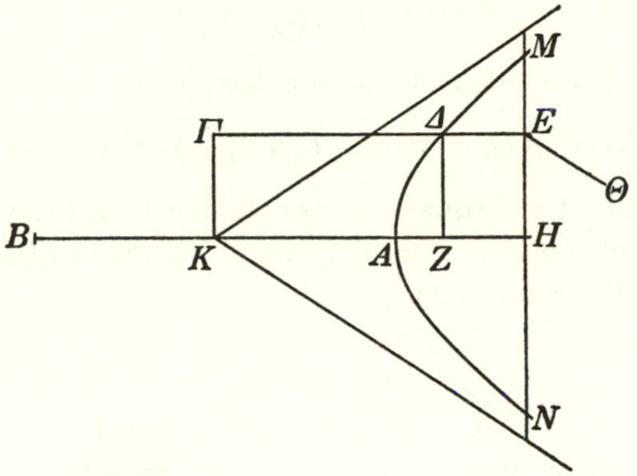

Fig. 36.

On account of the symptom of the section we further have

$$[\mathbf{T}(HM),\ \mathbf{O}(AH, BH)] = [\mathbf{T}(\Delta Z),\ \mathbf{O}(AZ, BZ)], \text{ therefore}$$

$$[\mathbf{T}(HM),\ \mathbf{T}(HE)] = [\mathbf{T}(KH) - \mathbf{T}(KA),\ \mathbf{T}(KZ) - \mathbf{T}(KA)]$$

or (0.41)

$$[\mathbf{T}(\Theta E),\ \mathbf{T}(HM)] = [\mathbf{T}(KH) - \mathbf{T}(KZ),\ \mathbf{T}(KH) - \mathbf{T}(KA)],$$

therefore $[\mathbf{T}(\Theta E),\ \mathbf{T}(\Gamma E) - \mathbf{T}(\Gamma\Delta)] = [\mathbf{T}(HM),\ \mathbf{T}(KH) - \mathbf{T}(KA)]$

The locus of Θ therefore is an amblytome which is similar to the meridian section (4.2).

γ) Now in Fig. 37 let $K\Delta E$ be the intersection with the cutting plane through the vertex of the enveloping cone (centre of the meridian section). We now have

$$\mathbf{T}(\Theta E) = \mathbf{O}(ME, NE) = \mathbf{T}(HM) - \mathbf{T}(HE).$$

Also, if $E\Lambda \parallel \Delta A$,

$$[\mathbf{T}(MH),\ \mathbf{T}(KH)-\mathbf{T}(KA)] = [\mathbf{T}(\varDelta Z),\ \mathbf{T}(KZ)-\mathbf{T}(KA)]$$
$$= [\mathbf{T}(EH),\ \mathbf{T}(KH)-\mathbf{T}(K\varLambda)]\,,$$

therefore $[\mathbf{T}(MH),\ \mathbf{T}(EH)] = [\mathbf{T}(KH)-\mathbf{T}(KA),\ \mathbf{T}(KH)-\mathbf{T}(K\varLambda)]$, from which it follows that

$$[\mathbf{T}(MH)-\mathbf{T}(EH),\ \mathbf{T}(MH)] = [\mathbf{T}(K\varLambda)-\mathbf{T}(KA),\ \mathbf{T}(KH)-\mathbf{T}(KA)]\,,$$

therefore $[\mathbf{T}(\varTheta E),\ \mathbf{T}(K\varLambda)-\mathbf{T}(KA)] = [\mathbf{T}(MH),\ \mathbf{T}(KH)-\mathbf{T}(KA)]$.

The ratio

$$[\mathbf{T}(\varTheta E),\ \mathbf{T}(KE)-\mathbf{T}(K\varDelta)]$$

therefore is compounded of the constant ratios

$$[\mathbf{T}(MH),\ \mathbf{T}(KH)-\mathbf{T}(KA)] \text{ and } [\mathbf{T}(K\varLambda)-\mathbf{T}(KA),\ \mathbf{T}(KE)-\mathbf{T}(K\varDelta)]$$

i.e. of the ratio of the square on the ordinate and the rectangle on the abscissae of the meridian section and the ratio $[\mathbf{T}(KA),\ \mathbf{T}(K\varDelta)]$, which is not the ratio (1,1).

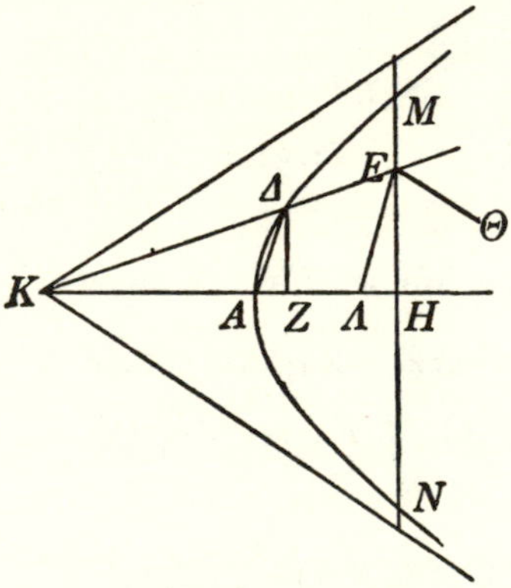

Fig. 37.

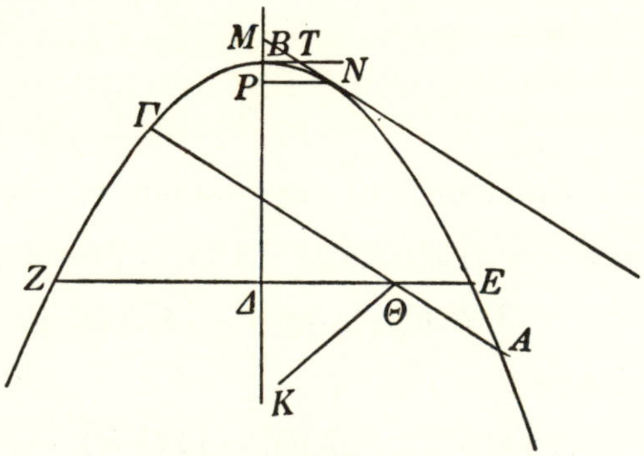

Fig. 38.

The locus of $\varTheta$ therefore is an amblytome, but this is not similar to the meridian section (4.2).

6.41. C.S. 13. *If an amblyconoid be cut by a plane meeting all the generators of the cone enveloping the conoid and not being at right angles to the axis, the section will be an oxytome, and the greater diameter thereof will be the part cut off within the conoid from the intersection of the cutting plane and the plane through the axis at right angles to the cutting plane.*

Just as in 6.31 we find (Fig. 38)

$$[\mathbf{T}(K\Theta),\ \mathbf{O}(A\Theta, \Gamma\Theta)] = [\mathbf{T}(BT),\ \mathbf{T}(NT)]$$

from which it already follows that the locus of K is an oxytome.
Now (5.2) $BP > BM$, whence $TN > TM > BT$, whence $BT < NT$. Therefore $A\Gamma$ is the greater diameter.

6.42. **C.S.** 15β. *Of the straight lines drawn from the points of an amblyconoid parallel to a straight line, drawn to the conoid through the vertex of the cone enveloping the conoid, the parts which are in the same direction as the convexity of the surface will fall without the conoid, the parts which are in the other direction within it.*

6.43. **C.S.** 15γ. For a tangent plane to an amblyconoid, the statement on the orthoconoid in 6.34 is true.

6.44. **C.S.** *Definitiones. Amblyconoids are called similar when the enveloping cones are similar.* This definition is in agreement with the one of 6.33 for orthoconoids. There the criterion of similarity of surfaces of revolution was taken to be similarity of the meridian sections. When, however, two amblytomes are similar (4.2), the asymptotes in both sections will include the same angle with the axis (4.1), and the enveloping cones will therefore be similar; and *vice versa*.

6.5. *The Spheroids.*

C.S. 11γ. *If one of the two spheroids be cut by a plane through the axis, or parallel to the axis, the section will be an oxytome, viz. if the plane pass through the axis, the same as that containing the < rotating > figure; if it be parallel to the axis, one which is similar thereto; and the diameter of the section will be the common section of the plane cutting the surface and the plane drawn through the axis at right angles to the cutting plane.*

The proof, which is missing in Archimedes, can be furnished in the same way as was done for the amblyconoid in 6.40.

6.51. **C.S.** 14α. *If a prolate spheroid be cut by a plane not at right angles to the axis, the section will be an oxytome, and the greater diameter thereof will be the part cut off within the spheroid from the common section of the plane cutting the surface and the plane through the axis at right angles to the cutting plane.*

The proof is entirely identical with that of **C.S.** 13 (6.41), except for the manner in which the conclusion $TB < TN$ is drawn. This follows here (Fig. 39) from the application of the property (5.38) to the pair of chords $O\Pi$, $\Lambda\Sigma$ through the centre. In fact, from this it follows that

$$[\mathbf{T}(TB),\ \mathbf{T}(TN)] = [\mathbf{T}(KO),\ \mathbf{T}(K\Lambda)]$$

and consequently, in view of

$$KO < K\Lambda, \quad \text{also} \quad TB < TN\,.$$

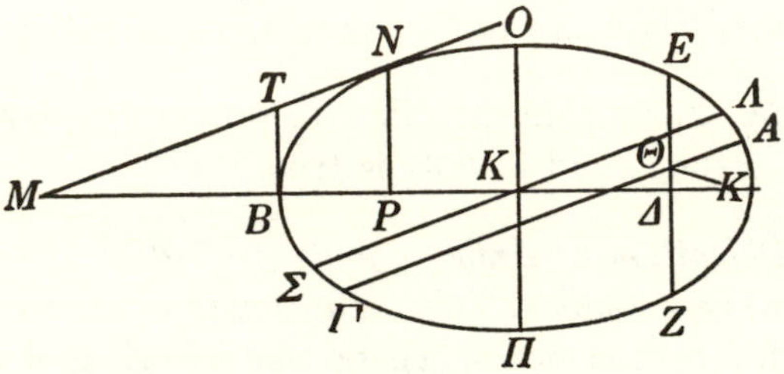

Fig. 39.

6.52. **C.S.** 14β. For the oblate spheroid the same proposition is true, but in this case the part cut off is the lesser diameter.

6.53. **C.S.** 16α. *If a plane meet one of the spheroids without cutting it, it will meet it at one point only, and the plane through the axis and the point of contact will be at right angles to the plane which touches it.*

The proof is entirely identical with that of 6.34.

6.54. **C.S.** 16γ. *If two parallel planes touch one of the spheroids, the straight line joining the points of contact will pass through the centre of the spheroid.*

The proposition is evident for the case that the two planes are at right angles to the axis. If this is not the case, the planes through the axis at right angles to the tangent planes coincide; the axis and the line joining the points of contact therefore lie in a plane cutting the parallel tangent planes in parallel tangents to an oxytome. The proposition then follows from the corresponding one for the oxytome (5.41).

6.55. **C.S.** 17α. *If two parallel planes be drawn which touch a* *spheroid, and through the centre of the spheroid a plane parallel to the tangent planes, the straight lines through the points of the resulting section parallel to the line joining the points of contact will fall without the spheroid.*

In Fig. 40 let the plane of the paper be the plane through the chosen point A of the section in question and the two points of contact B and Δ of the given tangent planes. The section $AB\Gamma\Delta$ is a circle or an oxytome to which have been drawn two parallel tangents ZE and ΘH; from this it follows that if $A\Gamma$ be the diameter, parallel to EZ and ΘH, the lines drawn through A and Γ parallel to $B\Delta$ will touch the section, and any points but the points of contact will fall without the section (5.42).

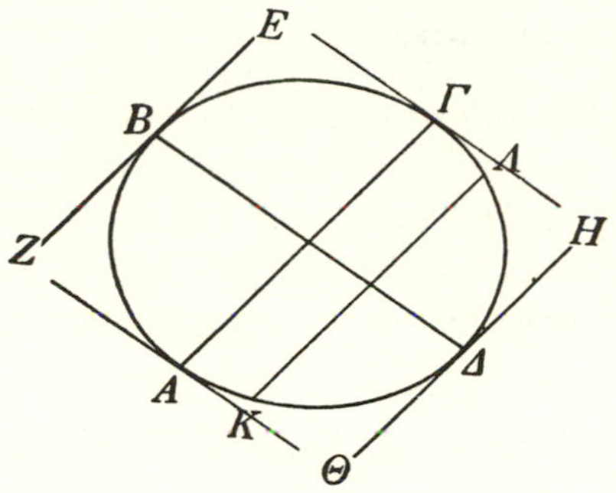

Fig. 40.

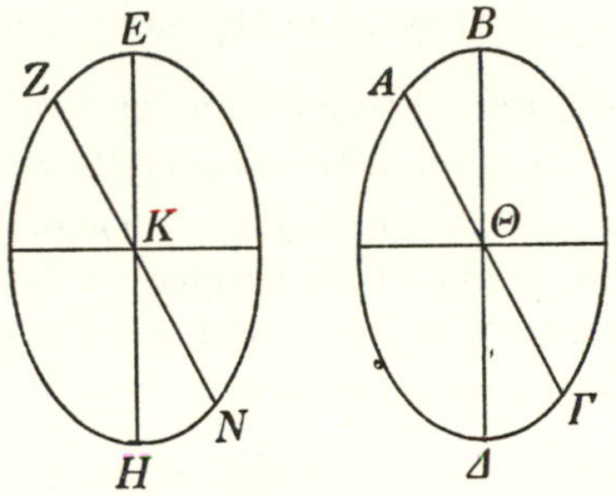

Fig. 41.

6.551. **C.S.** 17β. *If the plane parallel to the tangent planes do not pass through the centre, as $K\Lambda$, of the straight lines, drawn parallel to the line joining the points of contact from the points of the section, the parts which are on the side of the smaller segment will fall without the spheroid; the parts which are on the other side within it.*

6.56. **C.S.** 18. *Any spheroid which is cut by a plane through the centre is divided into two equal parts by this plane, and so is the surface.*

α) If the cutting plane pass through the axis or if it be at right angles to the axis, it is evident that one part of the figure fits on the other, and one part of the surface on the other.

β) If neither the one nor the other be the case, then in Fig. 41, in the plane of the paper through the axis at right angles to the cutting plane, let $B\Delta$ be the axis, Θ the centre, $A\Gamma$ the intersection with the cutting plane. We will now consider a second spheroid with axis EH, equal and similar to the first, which is cut by a plane through the axis in the oxytome $ENHZ$. Through K draw a line ZN, which includes with the axis the same angle as $A\Gamma$ with $B\Delta$, and through ZN draw a plane at right angles to the plane of the paper. The segments ZEN and ZHN of the second spheroid now fit successively on the segments $AB\Gamma$ and $A\Delta\Gamma$ of the first. Now let the second spheroid rotate about the perpendicular in the plane of the paper through K to EH, until H and E have changed places, and subsequently about EH, until N and Z have changed places. Now the segments of the second spheroid can again cover those of the first, *viz.* in such a manner that EZN coincides with $\Delta\Gamma A$, HZN with $B\Gamma A$. Since therefore each of the two segments of the given spheroid may coincide with each of the two segments of the second, they have the same volume and area.

6.60. For all conoids and spheroids the following properties are true:

C.S. 11δ. *When any of the said figures be cut by a plane through the axis, the feet of the perpendiculars dropped on the cutting plane from points of the surface not lying on the section will fall within the section.* The proof of this follows at once from the concept of a surface of revolution.

6.61. **C.S.** 14. *Corollary. Intersections by parallel planes are similar for each of the figures.*

This follows at once from the propositions **C.S.** 12 (6.31), 13 (6.41), 14 (6.51) in connection with the propositions on similarity of conics (2.8; 3.6; 4.2).

6.62. **C.S.** 14β. *A plane drawn through a tangent to an intersection with a plane through the axis, at right angles to that plane, touches the surface in the same point where the tangent touches the conic.*

It cannot touch it anywhere else, for then the foot of the perpendicular from that other point to the plane of the section would fall on a tangent to the section instead of within the section (which is contrary to 6.60).

7. *Lemmas from Arithmetic and Application of Areas.*

7.1. At the end of the introduction to the treatise *On Conoids and Spheroids* Archimedes mentions the following proposition, without proving it.

If any number of magnitudes be given, which exceed one another by an equal amount equal to the least, and also other magnitudes, equal in number to the former, but each equal in quantity to the greatest, all the magnitudes each of which is equal to the greatest will be less than the duplicate of all those exceeding one another by an equal amount and more than the duplicate of these minus the greatest.

The proof will probably have been furnished as follows (Fig. 42):

Conceive the n magnitudes $A, B \ldots H, \Theta$ to form an arithmetical progression with common difference Θ. Produce $B \ldots \Theta$ with $I = \Theta$, $K = H, \ldots O = B$, until they are equal to A. Then one reads at once

$$(n-1)A = 2(B + \ldots + \Theta)$$

thence

$$2(B + \ldots + \Theta) < n.A < 2(A + B + \ldots + \Theta) .$$

In algebraic symbolism this is equivalent to saying that from

$$(n-1)n = 2(1 + 2 + \ldots \overline{n-1})$$

it is derived that

$$2(1 + 2 + \ldots + \overline{n-1}) < n^2 < 2(1 + 2 + \ldots + n) .$$

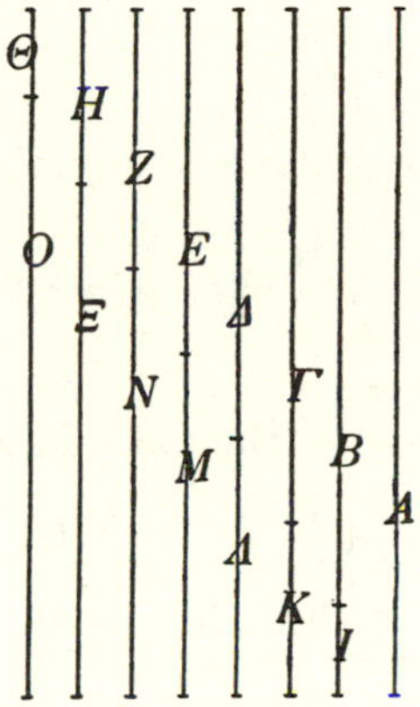

Fig. 42.

7.20 **C.S.** 1. *If of two series of the same number of magnitudes those similarly placed are proportional to one another, and of the magnitudes of the first series all or some bear to other magnitudes any ratios, and the corresponding magnitudes of the second series bear the same ratios to other magnitudes, all the magnitudes of the first series will be to all those of the third as all the magnitudes of the second series to all those of the fourth.*

Archimedes considers two cases:

α) Conceive as given the series

I. $\quad A\ B\ \Gamma\ \Delta\ E\ Z$

II. $\quad H\ \Theta\ I\ K\ \Lambda\ M$,

so that

$$(A, B) = (H, \Theta); \ (B, \Gamma) = (\Theta, I) \text{ etc.} \qquad (1)$$

and the series

III. $N\,\Xi\,O\,\Pi\,P\,\Sigma$

IV. $T\,Y\,\Phi\,X\,\Psi\,\Omega\,,$

so that

$$(A, N) = (H, T);\ (B, \Xi) = (\Theta, Y) \text{ etc.} \tag{2}$$

then it has to be proved that

$$(A + \ldots + Z, N + \ldots + \Sigma) = (H \ldots + M, T + \ldots + \Omega)\,.$$

Proof: From

$$(N, A) = (T, H)$$
$$(A, B) = (H, \Theta)$$
$$(B, \Xi) = (\Theta, Y)$$

it follows *ex aequali* that

$$(N, \Xi) = (T, Y)\,. \tag{3}$$

We likewise find

$$(\Xi, O) = (Y, \Phi) \text{ etc.}$$

We therefore have on the one hand, by (1):

$$(A, H) = (B, \Theta) = (\Gamma, I) \ldots = (A + B + \ldots Z, H + \Theta + \ldots M)$$

on the other hand, by (2) and (3):

$$(A, H) = (N, T) = (\Xi, Y) = (O, \Phi) \ldots = (N + \ldots \Sigma, T + \ldots + \Omega)\,,$$

from which follows that which it was required to prove.

β) Conceive as given the series

I. $A\,B\,\Gamma\,\Delta\,E\,Z$

II. $H\,\Theta\,I\,K\,\Lambda\,M$

with the relations (1)
and the series

III. $N\,\Xi\,O\,\Pi\,P$

IV. $T\,Y\,\Phi\,X\,\Psi$

with, as far as possible, the relations (2).

Following the same line of thought as sub α), we now find:

$$(A + \ldots + Z, N + \ldots + P) = (H + \ldots + M, T + \ldots \Psi)\,.$$

In algebraic symbolism the proposition is evident. Conceive

I.	$A_1, A_2 \ldots A_n$
II.	$\lambda A_1, \lambda A_2 \ldots \lambda A_n$
III.	$\mu_i A_{i_1} \ldots \mu_{i_x} A_{i_x}$
IV.	$\lambda\mu_{i_1} A_{i_1} \ldots \lambda\mu_{i_x} A_{i_x}$

in which $i_1, i_2 \ldots i_x$ are different from one another, ascending in quantity, and all of them not exceeding n.

Now the proposition states that

$$\frac{\sum_1^n i\, A_i}{\sum_1^n i\, \lambda A_i} = \frac{\sum_1^x j\, \mu_{i_j} A_{i_j}}{\sum_1^x j\, \lambda\mu_{i_j} A_{i_j}}.$$

In effect both members are equal to $\frac{1}{\lambda}$.

The cases dealt with by Archimedes refer to $x = n - 1$ and $x = n$.

In the proof it has been assumed that the magnitudes of the four series are all homogeneous; in fact, reference is made to the ratios (A, N), (A, H), and (N, T), and this implies that A is homogeneous with N and with H, N with T, and consequently also A with T.

Of this restriction, however, Archimedes does not take the slightest notice in the applications. We shall find him using the proposition in the case where the magnitudes of the series I and III are volumes, those of the series II and IV lengths, in which case the ratio (A, H) makes no sense. It seems probable that this is a sign of slackening in the strictness of the Euclidean theory of proportions, due to the fact that in applying the propositions of the theory of proportions it was never necessary to take account of the definition of proportion (which explicitly stipulates homogeneity as condition for two magnitudes being in any ratio to each other) and of the way in which these propositions had been derived from the definition. In addition, the custom of representing any magnitudes, of whatever nature, diagrammatically by line segments was bound to conduce to an increasing neglect of the difference in dimension between volumes, areas, and lengths, and to the gradual reaching of a conception which was equivalent to that of positive real numbers.

For the rest it is easy to see that according to the strict conception of the theory of proportions the proposition remains true for the case where the magnitudes of series I are homogeneous only with those of series III, those of series II only with those of series IV.

In fact, from $(A, B) = (H, \Theta)$; $(B, \Gamma) = (\Theta, I)$ etc. it follows, by application of the definition of proportion, that

$$(A + B \ldots + Z, A) = (H + \ldots M, H) .$$

From (3) it follows likewise that

$$(N, N + \Xi + \ldots \Sigma) = (T, T + Y + \ldots \Omega) ,$$

from which, *via*

$$(A, N) = (H, T) ,$$

it follows *ex aequali* that

$$(A + B \ldots + Z, N + \Xi \ldots + \Sigma) = (H + \ldots + M, T + \ldots + \Omega) .$$

7.21. A case of frequent occurrence, in which the proposition 7.20 is applied, consists in that the magnitudes of series III are equal to one another and likewise those of series IV. The proportions (2) then read

$$(A, N) = (H, T); \quad (B, N) = (\Theta, T) \text{ etc.,}$$

while the proportions (1) now result therefrom; in fact, from $(A, N) = (H, T)$ and $(N, B) = (T, \Theta)$ it follows *ex aequali* that $(A, B) = (H, \Theta)$ etc.

The conclusion now is

$$(A + B + \ldots, N + N + \ldots) = (H + \Theta + \ldots, T + T + \ldots)$$

or also

$$(A + B + \ldots, N) = (H + \Theta + \ldots, T) .$$

7.30. **Spir.** 10. *If a series of any number of lines be given, which exceed one another by an equal amount, and the difference be equal to the least, and if other lines be given equal in number to these and in quantity to the greatest, the squares on the lines equal to the greatest, plus the square on the greatest and the rectangle contained by the least and the sum of all those exceeding one another by an equal amount will be the triplicate of all the squares on the lines exceeding one another by an equal amount.*

Proof: In Fig. 42 conceive a number of line segments $A, B, \Gamma, \Delta, E, Z, H, \Theta$ forming an arithmetical progression, the common difference of which is equal to the least, *viz.* Θ.

Produce the segments $\Theta, H, \ldots B$ with the segments $O=B$, $\Xi=\Gamma$, $N=\Delta\ldots$, $I=\Theta$, until they are equal to A.

Then it has to be proved that

$$\mathbf{T}(A)+\mathbf{T}(B+I)\ldots+\mathbf{T}(\Theta+O)+\mathbf{T}(A)+\mathbf{O}(\Theta, A+B+\ldots\Theta)$$
$$= 3[\mathbf{T}(A)+\ldots+\mathbf{T}(\Theta)]\ldots \qquad (1)$$

Proof:

$$\begin{aligned} 2\mathbf{T}(A) &= 2\mathbf{T}(A) \\ \mathbf{T}(B+I) &= \mathbf{T}(B)+\mathbf{T}(I)+2\mathbf{O}(B, I) \\ &\vdots \\ \mathbf{T}(\Theta+O) &= \mathbf{T}(\Theta)+\mathbf{T}(O)+2\mathbf{O}(\Theta, O) \end{aligned} \quad +$$

$$\mathbf{T}(A)+\mathbf{T}(B+I)+\ldots\mathbf{T}(\Theta+O)+\mathbf{T}(A)$$
$$= 2[\mathbf{T}(A)+\mathbf{T}(B)+\ldots\mathbf{T}(\Theta)]+\mathbf{O}(\Theta, 2B+4\Gamma+\ldots 14\Theta)\,.$$

It has therefore still to be proved that:

$$\mathbf{O}(\Theta, A+B+\ldots+\Theta)+\mathbf{O}(\Theta, 2B+4\Gamma+\ldots 14\Theta)$$
$$= \mathbf{T}(A)+\ldots\mathbf{T}(\Theta)$$

or

$$\mathbf{O}(\Theta, A+3B+5\Gamma+\ldots 15\Theta) = \mathbf{T}(A)+\ldots+\mathbf{T}(\Theta)\,.$$

Now

$$\begin{aligned} \mathbf{T}(A) &= \mathbf{O}(8\Theta, A) = [\mathbf{O}\Theta, A+2(B+\Gamma+\ldots\Theta)] \\ \mathbf{T}(B) &= \mathbf{O}(7\Theta, B) = [\mathbf{O}\Theta, B+2(\Gamma+\Delta+\ldots\Theta)] \\ &\vdots \\ \mathbf{T}(H) &= \mathbf{O}(2\Theta, H) = \mathbf{O}[\Theta, H+2\Theta] \\ \mathbf{T}(\Theta) &= \qquad\qquad\quad \mathbf{O}[\Theta, \Theta] \end{aligned} \quad +$$

$$\mathbf{T}(A)+\ldots\mathbf{T}(\Theta) = \mathbf{O}[\Theta, A+3B+\ldots 15\Theta]\,.$$

The numerical example discussed by Archimedes shows the trend of the argument plainly enough; the formulation for a series of n terms is obvious.

7.31. In a Corollary the following inequalities are inferred:

$$3[\mathbf{T}(B)+\ldots\mathbf{T}(\Theta)] < \text{sum of the squares } \mathbf{T}(A)$$
$$< 3[\mathbf{T}(A)\ldots+\mathbf{T}(\Theta)]\,,$$

of which the one on the right follows at once from (1), while the one on the left is derived from (1) by noting that

$$\mathbf{O}(\Theta, A+\ldots+\Theta) < \mathbf{O}[\Theta, A+2(B\ldots\Theta)] = \mathbf{T}(A).$$

7.32. In modern formulation the results obtained are as follows:

Of the arithmetical progression let the common difference v be equal to the first term a, then

$$(n+1)(na)^2+a(a+2a+\ldots+na) = 3[a^2+(2a)^2+\ldots(na)^2],$$

which is equivalent to

$$1^2+2^2+\ldots n^2 = \tfrac{1}{6}n(n+1)(2n+1).$$

Further:

$$3[1^2+2^2+\ldots(n-1)^2] < n^3 < 3[1^2+2^2+\ldots n^2].$$

7.33. In the Corollary mentioned sub 7.31 it is also observed that if on the line segments, referred to in Prop. 7.30, as sides there be described similar figures

$$\mathbf{S}(A),\ \mathbf{S}(B)\ldots\mathbf{S}(\Theta),$$

the inequalities of 7.31 will apply for the areas of these figures:

$$3[\mathbf{S}(B)+\ldots+\mathbf{S}(\Theta)] < \text{Sum of the figures } \mathbf{S}(A) < 3[\mathbf{S}(A)+\ldots+\mathbf{S}(\Theta)].$$

This follows at once from Euclid VI, 20.

7.4. C.S. 2. *If there be lines equal to one another, in any number, and to each of these there be applied an area with a quadratic excess, and the sides of the excesses exceed one another by an equal amount which is equal to the least, and if there be other areas, equal in number to the first-mentioned and in quantity each equal to the greatest, these will be to all the other areas in a less ratio than that of the sum of the side of the greatest excess and one of the equal lines to the sum of one-third of the side of the greatest excess and one-half of one of the equal lines, and to the other areas except the greatest it will be in a greater ratio than this same ratio.*

In Fig. 43 let the line segments A be the equal line segments, to which are hyperbolically applied with quadratic excess the areas $\alpha\vartheta = X_1$, $\beta\iota = X_2$ etc., so that the sides of the excesses $B, \Gamma\ldots H$ form an arithmetical progression with common difference H. It has now to be proved that

$$(nX_1, X_1+X_2\ldots+X_n) < (B+A, \tfrac{1}{3}B+\tfrac{1}{2}A) < (nX_1, X_2+\ldots X_{n-1}).$$

Proof: The rectangles $\Pi_1, \Pi_2 \ldots$ form an arithmetical progression with common difference Π_n; thence by 7.1 we have:

$$2(\Pi_2 + \ldots \Pi_n) < n.\Pi_1 < 2(\Pi_1 + \ldots \Pi_n). \quad (1)$$

For the squares $T_1, T_2 \ldots$ by 7.31 the following applies:

$$3(T_2 + \ldots T_n) < n.T_1 < 3(T_1 + \ldots T_n). \quad (2)$$

Add (1), after division by 2, to (2) after division by 3, then we have:

$$X_2 + \ldots X_n < n(\tfrac{1}{2}\Pi_1 + \tfrac{1}{3}T_1) < X_1 + \ldots X_n \quad (3)$$

or $$X_2 + \ldots X_n < n \cdot \mathbf{O}(B, \tfrac{1}{3}B + \tfrac{1}{2}A) < X_1 + \ldots X_n,$$

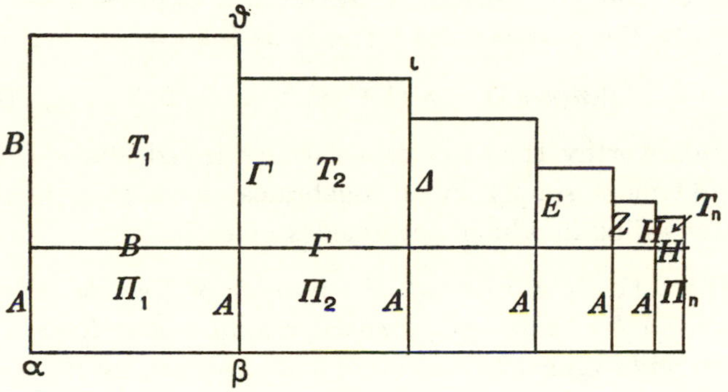

Fig. 43.

thence

$$(nX_1, X_1 + \ldots X_n) < [nX_1, n.\mathbf{O}(B, \tfrac{1}{3}B + \tfrac{1}{2}A)] < (nX_1, X_2 + \ldots X_n),$$

thence

$$(nX_1, X_1 + \ldots X_n) < (B + A, \tfrac{1}{3}B + \tfrac{1}{2}A) < (nX_1, X_2 + \ldots X_n).$$

Algebraically:

Let $A = a$, $H = p$, then $B = np$; $X_1 = np(a + np) \ldots$
$X_n = p(a + p)$,

The proposition is as follows:

$$\frac{n.np(a+np)}{p(a+p) + \ldots np(a+np)} < \frac{np+a}{\tfrac{1}{3}np + \tfrac{1}{2}a}$$
$$< \frac{n.np(a+np)}{p(a+p) + \ldots (n-1)p[a+(n-1)p]}.$$

The proof is furnished by adding together the inequalities

$$ap+2ap+\ldots+nap > \tfrac{1}{2}n.nap > ap+2ap+\ldots+(n-1)ap$$

$$p^2+(2p)^2+\ldots+(np)^2 > \tfrac{1}{3}n(np)^2 > p^2+(2p)^2+\ldots+[(n-1)p]^2$$

from which it follows that

$$p(a+p)+\ldots+np(a+np) > n.np[\tfrac{1}{3}np+\tfrac{1}{2}a)$$
$$> p(a+p)+\ldots+(n-1)p[a+(n-1)p]\ .$$

What was required to be proved follows when $n.np(a+np)$ is divided by the members of this inequality. The wording of this is already entirely adapted to the application in **C.S.** 26. The real contents of the proposition, however, are expressed by the inequality (3). For $p=a=1$ the latter is as follows:

$$1.2+2.3+\ldots+n(n+1) > n^2(\tfrac{1}{2}+\tfrac{1}{3}n) > 1.2+2.3+\ldots+(n-1)n\ .$$

It is noteworthy that the proposition proved above cannot be formulated more simply in an algebraical form than in the geometrical wording in which Archimedes gives it.

7.50. **Spir.** 11. *If a series of any number of lines be given which exceed one another by an equal amount, and also other lines, one fewer in number and in quantity each equal to the greatest, all the squares on the lines equal to the greatest are to the squares on the lines which exceed one another by an equal amount, except the least, in a less ratio than that of the square on the greatest to the sum of the rectangle contained by the greatest and the least and one-third of the square on the difference between the greatest and the least; to the squares, however, on the lines which exceed one another by an equal amount, except the square on the greatest, they will be in a greater ratio than this ratio.*

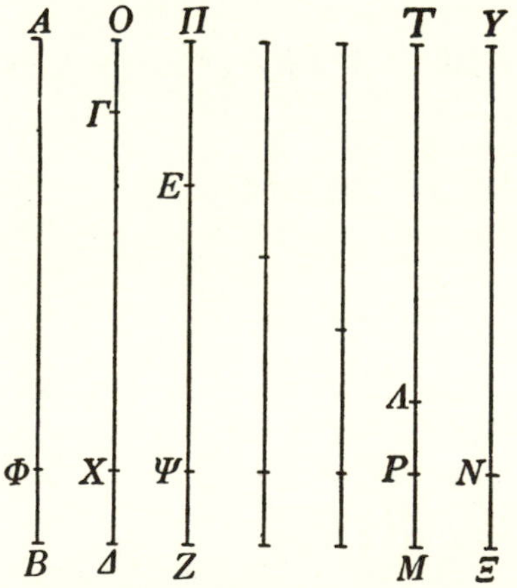

Fig. 44.

In Fig. 44 let there be given the series of line segments AB, $\Gamma\Delta\ldots N\Xi$ forming an arithmetical progression. Archimedes supposes in his proof that the common difference is equal to the least term, $N\Xi$, but he applies the proposition in cases where this condition is not satisfied. We will retain his supposition, but will show subsequently that it is not essential to the validity of the proposition.

Produce all the given line segments by addition of $\Gamma O = N\Xi$, $E\Pi = 2 . N\Xi$ etc., until they are equal to AB. It now has to be proved that

$$[\mathbf{T}(O\varDelta)+\ldots+\mathbf{T}(Y\Xi),\ \mathbf{T}(AB)+\ldots+\mathbf{T}(\varLambda M)] < [\mathbf{T}(AB),\ \mathbf{O}(AB, N\Xi)+\tfrac{1}{3}\mathbf{T}(NY)] < [\mathbf{T}(O\varDelta)+\ldots+\mathbf{T}(Y\Xi),\ \mathbf{T}(\Gamma\varDelta+)\ldots+T(N\Xi)] .$$

If we make $\Phi B = X\varDelta = N\Xi$, then

$$[\mathbf{T}(AB),\ \mathbf{O}(AB, \Phi B)+\tfrac{1}{3}\mathbf{T}(A\Phi)] = [\mathbf{T}(O\varDelta),\ \mathbf{O}(O\varDelta, X\varDelta)+\tfrac{1}{3}\mathbf{T}(OX)]$$

= etc., from which it follows that:

$$[\mathbf{T}(O\varDelta)+\mathbf{T}(\Pi Z)+\ldots+\mathbf{T}(Y\Xi),\ \mathbf{O}(N\Xi, O\varDelta+\Pi Z+\ldots+Y\Xi) + \tfrac{1}{3}\{\mathbf{T}(OX)+\mathbf{T}(\Pi\Psi)+\ldots+\mathbf{T}(YN)\}] = [\mathbf{T}(AB,\ \mathbf{O}(AB, N\Xi)+\tfrac{1}{3}\mathbf{T}(YN)] .$$

If we compare this with what is required to be proved, we find that the proposition will have been proved if we can show that

$$\mathbf{T}(\Gamma\varDelta)+\ldots+\mathbf{T}(N\Xi) < \mathbf{O}(N\Xi, O\varDelta+\ldots+Y\Xi) + \tfrac{1}{3}\{T(OX)+\ldots+\mathbf{T}(YN)\} < \mathbf{T}(AB)+\ldots+\mathbf{T}(\varLambda M) . \qquad (1)$$

Now in the second member of this inequality:

$$\mathbf{O}(N\Xi, O\varDelta\ldots Y\Xi) = \mathbf{O}[N\Xi, (OX+X\varDelta)+\ldots+(YN+N\Xi)] = \mathbf{O}(N\Xi, OX+\ldots+YN)+\underline{\mathbf{T}(X\varDelta)+\ldots+\mathbf{T}(N\Xi)}$$

in the first member:

$$\mathbf{T}(\Gamma\varDelta)+\ldots+\mathbf{T}(N\Xi) = \mathbf{T}(\Gamma X)\ldots+\mathbf{T}(\varLambda P) + 2\mathbf{O}(X\varDelta, \Gamma X+\ldots+P\varLambda)+\underline{\mathbf{T}(X\varDelta)+\ldots+\mathbf{T}(N\Xi)}$$

in the third member:

$$\mathbf{T}(AB)+\ldots+\mathbf{T}(\varLambda M) = \mathbf{T}(A\Phi)+\ldots\mathbf{T}(\varLambda P) + 2\mathbf{O}(\Phi B, A\Phi+\ldots+\varLambda P)+\underline{\mathbf{T}(\Phi B)+\ldots+\mathbf{T}(PM)} .$$

Omitting in all three members the underlined sums, we still have to prove that

$$\mathbf{T}(\Gamma X)+\ldots+\mathbf{T}(\Lambda P)+2\mathbf{O}(X\Delta, \Gamma X+\ldots+\Lambda P) < \tfrac{1}{3}[\mathbf{T}(OX) + \ldots \mathbf{T}(YN)]+\mathbf{O}(N\Xi, OX+\ldots+YN) < \mathbf{T}(A\Phi) + \ldots+\mathbf{T}(\Lambda P)+2\mathbf{O}(\Phi B, A\Phi+\ldots+\Lambda P)\,.$$

If we first compare the sums of squares occurring in the three members, by 7.31 the following applies thereto:

$$\mathbf{T}(\Gamma X)+\ldots+\mathbf{T}(\Lambda P) < \tfrac{1}{3}[\mathbf{T}(A\Phi)+\ldots+\mathbf{T}(TP)] < \mathbf{T}(A\Phi) + \ldots+\mathbf{T}(\Lambda P)\,, \tag{2}$$

in which the second member is equivalent to

$$\tfrac{1}{3}[\mathbf{T}(OX)+\ldots+\mathbf{T}(YN)]\,.$$

A comparison of the rectangles in the three members yields:

$$2\mathbf{O}(X\Delta, \Gamma X+\ldots+\Lambda P) = \mathbf{O}(N\Xi, \Gamma X+\ldots+\Lambda P+T\Lambda\ldots+\Gamma O)=$$

$$\mathbf{O}(N\Xi, OX\ldots+TP) < \mathbf{O}(N\Xi, OX\ldots+YN) < \mathbf{O}(\Phi B, O\Gamma + \Gamma X+\ldots T\Lambda+\Lambda P+2YN) = \mathbf{O}[\Phi B, 2(A\Phi+\ldots\Lambda P)]\,. \tag{3}$$

From addition of the inequalities (2) and (3) follows the validity of (1).

7.51. The following formulation of the proof in algebraic symbols, in which no essential modification has been made in the trend of the argument, may elucidate the demonstration a little better for the present-day reader. At the same time we will abandon the supposition that the common difference of the arithmetical progression is equal to the first term, so that the use that Archimedes will be found to make of the proposition will find its justification.
Suppose

$$N\Xi = t_0 = a;\ \Lambda M = t_1 = a+v;\ \ldots AB = t_n = a+nv\,.$$

We have to prove the inequality (1):

$$\mathbf{t}^2_{n-1}+\ldots+\mathbf{t}_0^2 < \mathbf{a}.\mathbf{n}\mathbf{t}_n+\tfrac{1}{3}\mathbf{n}(\mathbf{n}\mathbf{v})^2 < \mathbf{t}_n{}^2+\ldots+\mathbf{t}_1{}^2\,.$$

In effect:

$$\mathbf{t}_{n-1}{}^2+\ldots\mathbf{t}_0{}^2 = na^2+v^2[1^2+2^2+\ldots+(n-1)^2] + 2av[1+2+\ldots(n-1)] < na^2+\tfrac{1}{3}v^2n^3+n^2.av = na(a+nv)+\tfrac{1}{3}n(nv)^2 = \mathbf{a}.\mathbf{n}\mathbf{t}_n+\tfrac{1}{3}\mathbf{n}(\mathbf{n}\mathbf{v})^2 < na^2 +v^2[1^2+2^2\ldots+n^2]+2av(1+2\ldots+n) = \mathbf{t}_n{}^2+\ldots+\mathbf{t}_1{}^2.$$

7.52. In the same way as in 7.33 it will be seen that the proposition remains true when in the first and in the third member the squares on the given line segments be replaced by other—reciprocally similar—figures, in which these line segments are homologous sides.

7.60 **Q.P.** 23. In *Quadrature of the Parabola* Archimedes proves the following propostion on the sum of a geometrical progression with common ratio $\frac{1}{4}$.

Given a series of magnitudes, each of which is equal to four times the next in order, all the magnitudes and one-third of the least added together will exceed the greatest by one-third.

Let the magnitudes A, B, Γ, Δ, E be given such that

$$B = \tfrac{1}{4}A,\ \Gamma = \tfrac{1}{4}B,\ \text{etc.}$$

It has to be proved that

$$A+B+\ldots+E+\tfrac{1}{3}E = \tfrac{4}{3}A\,.$$

Proof:

Suppose $\quad Z = \frac{1}{3}B,\ H = \frac{1}{3}\Gamma,\ \Theta = \frac{1}{3}\Delta,\ I = \frac{1}{3}E\,,$

then

$$B+Z = \tfrac{1}{3}A;\ \Gamma+H = \tfrac{1}{3}B;\ \Delta+\Theta = \tfrac{1}{3}\Gamma;\ E+I = \tfrac{1}{3}\Delta\,,$$

whence

$$B+\Gamma+\Delta+E+Z+H+\Theta+I = \tfrac{1}{3}(A+B+\Gamma+\Delta)\,.$$

Now

$$Z+H+\Theta = \tfrac{1}{3}(B+\Gamma+\Delta)\,,$$

whence

$$B+\Gamma+\Delta+E+I = \tfrac{1}{3}A$$

or

$$A+B+\Gamma+\Delta+E+\tfrac{1}{3}E = \tfrac{4}{3}A\,.$$

7.61. This result also follows from Euclid XI, 35, for according to this proposition

$$(A-E,\ A+B+\Gamma+\Delta) = (A-B,\ A) = (3,\ 4)\,,$$

whence

$$A+B+\Gamma+\Delta = \tfrac{4}{3}A-\tfrac{4}{3}E$$

or

$$A+B+\Gamma+\Delta+E+\tfrac{1}{3}E = \tfrac{4}{3}A\,.$$

7.62. The general formulation of the method applied by Archimedes is as follows:

Of a geometrical progression $a_1, a_2 \dots a_n$ let the common ratio be $r = \frac{1}{1+b}$ $(b>0)$. Then we have

$$a_i + \frac{1}{b} a_i = a_i \frac{1+b}{b} = \frac{a_{i-1}}{b}$$

whence

$$S_n\left(1+\frac{1}{b}\right) = a_1\left(1+\frac{1}{b}\right) + \frac{a_1}{b} + \dots + \frac{a_{n-1}}{b} = a_1\left(1+\frac{1}{b}\right) + \frac{1}{b}(S_n - a_n)$$

or

$$S_n + \frac{1}{b} a_n = \left(1+\frac{1}{b}\right) a_1 .$$

8. *The Indirect Method for Infinite Processes.*

8. For the discussion of infinite processes, which in present-day mathematics lead to application of the theory of convergent variants, because a magnitude to be calculated is found as the limit of a variant, Archimedes uses a method initiated by Eudoxus, which we will characterize briefly rather than correctly by the expression "indirect passage to the limit". We will avoid the widely used term "exhaustion method"; for a mode of reasoning which has arisen from the conception of the inexhaustibility of the infinite, this is about the worst name that could have been devised.

8.1. The indirect method of Eudoxus, applications of which in the calculation of areas and volumes will be familiar to the reader of Euclid from the twelfth Book of the *Elements*, is encountered in Archimedes in problems of the same kind. It is, however, found in various forms, which are here reduced to two main types.

8.20. I. *The Compression Method.* According to this method the magnitude Σ to be calculated is compressed between a monotonously ascending lower limit I_n and a monotonously descending upper limit C_n; with regard to these limits it is known:

α) that the difference $C_n - I_n$ can, by the choice of n, be made less than any assigned magnitude ε.

or β) that the ratio (C_n, I_n) can, by the choice of n, be made less than the ratio of the greater μ of any two assigned magnitudes to the lesser ν.

The calculation now consists in that a magnitude K is found, which for any value of n lies between I_n and C_n. It is then asserted that

$$\Sigma = K .$$

8.21. The distinction between the cases α) and β) leads to a subdivision of type I into two sub-types:

Iα. *the difference form of the compression method.*

Iβ. *the ratio form of the compression method.*

The proofs of the equality $\Sigma = K$ are somewhat different for these two sub-types.

For both it is given that for any value of n the following inequalities apply:

$$I_n < \Sigma < C_n \quad (1)$$

$$I_n < K < C_n \quad (2)$$

The demonstration now proceeds as follows:

Iα. Difference Form.	Iβ. Ratio Form.

Suppose the equality $\Sigma = K$ not to be true, then either $\Sigma > K$ or $\Sigma < K$.

Case I. Suppose $\Sigma > K$. Now find n so that

$C_n - I_n < \Sigma - K$.	$(C_n, I_n) < (\Sigma, K)$.

Then, because $\Sigma < C_n$ (1), we have *a fortiori*

$\Sigma - I_n < \Sigma - K$,	$(\Sigma, I_n) < (\Sigma, K)$,

therefore $K < I_n$, which is contrary to the supposition (2).

Case II. Suppose $\Sigma < K$. Now find n so that

$C_n - I_n < K - \Sigma$.	$(C_n, I_n) < (K, \Sigma)$.

Then, because $I_n < \Sigma$ (1), we have *a fortiori*

$C_n - \Sigma < K - \Sigma$,	$(C_n, \Sigma) < (K, \Sigma)$,

therefore $C_n < K$, which is contrary to the supposition (2).

Consequently $\Sigma = K$.

8.22. In modern symbolism the conclusion is as follows:

If for any value of n

$$I_n < \Sigma < C_n \qquad \text{and} \qquad I_n < K < C_n$$

with either

$$\lim_{n \to \infty} (C_n - I_n) = 0 \qquad \text{or} \qquad \lim_{n \to \infty} \frac{C_n}{I_n} = 1 ,$$

then $\Sigma = K$.

8.23. The applications of the method naturally differ from one another as regards the findings of K. No general rule can be formulated for this.

However, once the inequalities (1) and (2) have been verified, the demonstration further proceeds automatically; it is, nevertheless, given *in extenso* again in each case.

8.24. Applications of the compression method are to be found: in the difference form:

D.C. 1. **C.S.** 22, 26, 28, 30. **Spir.** 24, 25. **Q.P.** 16. **Meth.** 15.
in the ratio form:

S.C. I, 13, 14, 33, 34, 42, 44.

8.30. II. *The Approximation Method.*

In this method the magnitude Σ to be calculated is approximated from below by the sum S_n of n (positive) terms $a_1, a_2 \ldots a_n$ of a convergent infinite series, where it is known that $\Sigma - S_n$ can, by the choice of n, be made less than any assigned magnitude ε, while the same applies to a_n (*de facto* this follows from the supposition about S_n). The calculation consists in finding a magnitude K, which for any value of n satisfies a relation

$$a_1 + \ldots a_n + R_n = K , \tag{1}$$

in which

$$R_n < a_n$$

It has now to be proved that $\Sigma = K$.

Proof: Suppose this equality not to be true, then either $\Sigma > K$ or $\Sigma < K$.

Case I. Suppose $\Sigma > K$. Now find n so that

$$\Sigma - S_n < \Sigma - K ,$$

then $K < S_n$, which is contrary to (1).

Case II. Suppose $\Sigma < K$. Now find n so that

$$a_n < K - \Sigma .$$

Because of (1) we now have

$$K - S_n < a_n < K - \Sigma ,$$

therefore $\Sigma < S_n$, which is contrary to the supposition with regard to Σ and S_n.

8.31. In the only example of this method occurring in the works of Archimedes (the finding of the area of a parabolic segment, **Q.P.** 18–24) the series $a_1, a_2 \ldots$ is a descending geometrical progression. That $\Sigma - S_n$ can, by the choice of n, be made less than any assigned magnitude, is ensured by Euclid's lemma (X, 1), which states that, if from any magnitude there be subtracted more than its half, from the remainder again more than its half, and so on, there will at length remain a magnitude less than any assigned magnitude. One therefore only has to see to it that

$$\Sigma > a_1 > \tfrac{1}{2}\Sigma$$

$$\Sigma - a_1 > a_2 > \tfrac{1}{2}(\Sigma - a_1)$$

$$\Sigma - (a_1 + a_2) > a_3 > \tfrac{1}{2}[\Sigma - (a_1 + a_2)] , \qquad \text{etc.}$$

which naturally has to be realized geometrically.

That a_n can, by the choice of n, be made less than ε becomes apparent when it is proved that the common ratio of the geometrical progression is less than $\frac{1}{2}$; in this way Euclid's lemma becomes applicable to the successive terms of the geometrical progression. That $R_n < a_n$ becomes apparent when it is proved that $R_n = \lambda a_n$ $(0 < \lambda < 1)$.

8.32. In modern formulation the method amounts to realizing geometrically that

$$\Sigma = \operatorname*{Lim}_{n \to \infty} (a_1 + a_2 + \ldots a_n)$$

and then determining the value K of this limit arithmetically.

9. *Neusis Constructions.*

9. Greek mathematics not infrequently makes use of a construction designated as νεῦσις (neusis) and consisting in that through a given point a straight line has to be drawn from which two given curves

cut off a segment of given length. Although for a literal translation of this term "inclination" would perhaps be the most appropriate (the line segment cut off, when produced, inclines towards, *i.e.* is directed at the given point), the mathematical meaning of the concept is expressed more effectively by the word "insertion": in fact, the construction may be conceived to be performed in such a way that a ruler, on which the given line segment has been measured by means of two marks, is placed through the given point and is then shifted until the two marks lie on the given curves.

We will here mention the lemmas based on a νεῦσις which Archimedes will be found to need in his works. In all of these, let K be the centre of a circle, to which $A\Gamma$ be a tangent or a chord. Through K there now has to be drawn a half-line meeting the circle in B and the straight line $A\Gamma$ in E such that BE in each case shall satisfy an assigned relation. In each case let KP be the diameter parallel to $A\Gamma$, R the radius of the circle.

9.1. **Spir.** 5 (Fig. 45).

$A\Gamma$ touches the circle at Γ. It is required that

$$(BE, R) < (\text{arc}\, B\Gamma, \text{given circumference of a circle}).$$

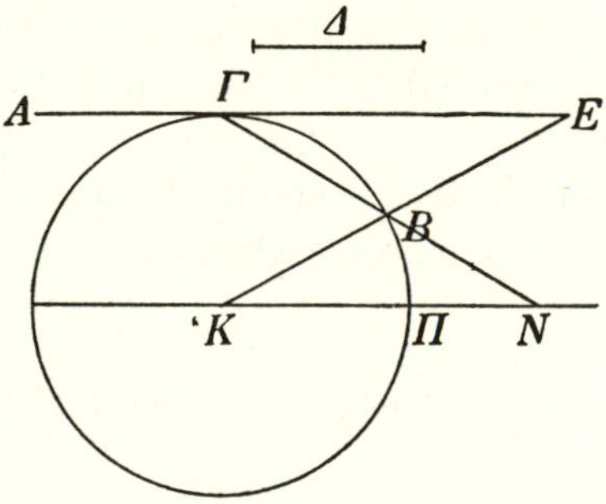

Fig. 45.[1])

If Δ be a line segment greater than the given circumference (the existence of Δ is further argued in **Spir.** 3 by observing that one need merely take a line segment greater than the perimeter of a circumscribed polygon), it is sufficient that

$$(BE, R) < (\text{arc}\, B\Gamma, \Delta)\,.$$

[1]) For Π read P.

Neusis: Insert between the circle and the diameter KP a line segment BN, directed at Γ and equal to Δ, then

$$(BE, R) = (B\Gamma, BN) < (\text{arc}B\Gamma, \Delta) .$$

9.2. **Spir.** 6 (Fig. 46).

$A\Gamma$ is a chord with middle point Θ. E has to lie between A and Γ. Z and H are two line segments satisfying the relation

$$(Z, H) < (\Gamma\Theta, K\Theta) .$$

It is required that

$$(BE, B\Gamma) = (Z, H) .$$

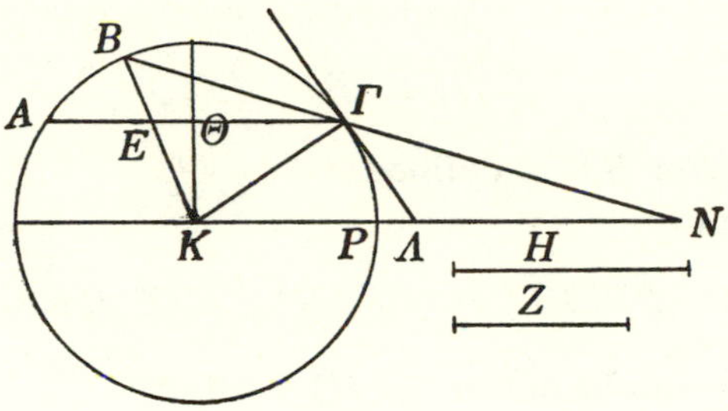

Fig. 46.

Neusis: Insert between the circle and the diameter KP a line segment BN passing through Γ and satisfying the relation

$$(Z, H) = (BK, BN) ,$$

by which relation BN is defined. Apparently we now have

$$(BE, B\Gamma) = (BK, BN) = (Z, H) .$$

If the tangent at Γ meet the straight line KP in Λ, then

$$(Z,H) = (BK, BN) = (K\Gamma, BN) < (K\Gamma, \Gamma\Lambda) = (\Gamma\Theta, K\Theta) . \quad (1)$$

The condition imposed on (Z, H) is therefore necessary. That it is sufficient, is apparently considered by Archimedes to be warranted by the statement that if (1) be satisfied, BN is greater than $\Gamma\Lambda$. This is probably based on a continuity consideration: BN approximates to $\Gamma\Lambda$ when B approximates to Γ along the curve. In fact, it would be conceivable that the difference $BN-\Gamma\Lambda$ would always exceed a given limit, in which case the condition $BN > \Gamma\Lambda$ would not be sufficient to conclude also that $\Gamma N > \Gamma\Lambda$, and consequently that ΓN can be constructed. Such continuity considerations, however, did not appear in the official mathematics of Greek publications; that is why here as well as in some other similar passages there is a gap in the proof.

9.3. Spir. 7 (Fig. 47).

The only difference from 9.2 is that E now has to lie on $A\Gamma$ produced, and that (Z, H) now has to fulfil the condition

$$(Z, H) > (\Gamma\Theta, K\Theta) .$$

Neusis: Insert between the circle and the diameter KP a line segment BN directed at Γ and satisfying the relation

$$(Z, H) = (BK, BN) ,$$

by which relation BN is defined.

Now apparently

$$(BE, B\Gamma) = (BK, BN) = (Z, H) .$$

As necessary condition for (Z, H) we find

$$(Z, H) = (BK, BN) = (K\Gamma, BN) > (K\Gamma, \Gamma\Lambda) = (\Gamma\Theta, K\Theta) .$$

If, inversely, this condition be satisfied, then $BN < \Gamma\Lambda$. That this is sufficient, was probably seen again on the ground of the consideration that BN approximates to 0 if B approximate to P, and to $\Gamma\Lambda$ if B approximate to Γ.

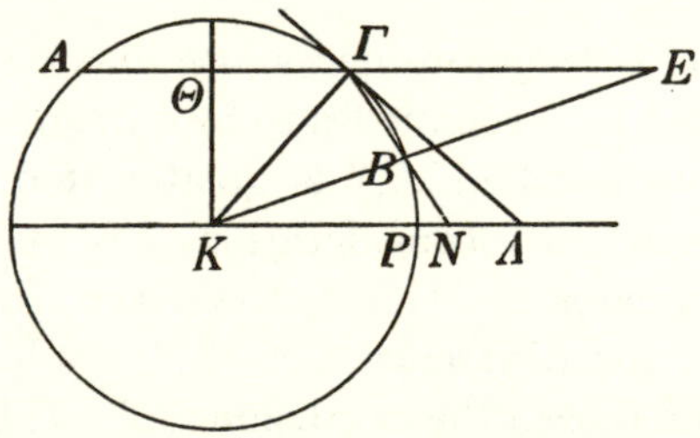

Fig. 47.

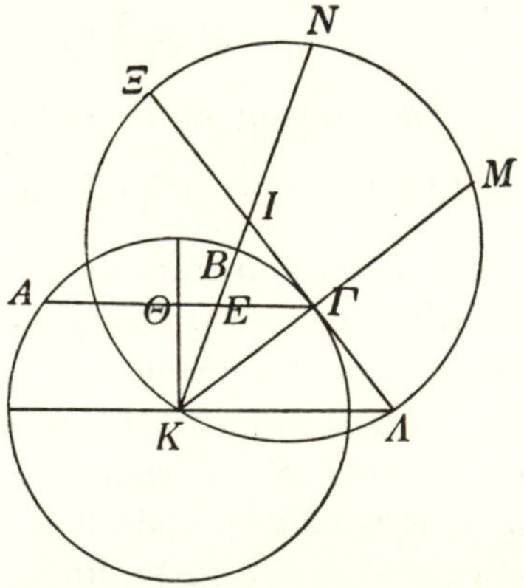

Fig. 48.

9.4. Spir. 8 (Fig. 48).

Now E has to be found on the chord $A\Gamma$ so that when KE meets the tangent to the circle at Γ in I,

$$(BE, I\Gamma) = (Z, H),$$

while

$$(Z, H) < (\Gamma\Theta, \Theta K) = (\Gamma K, \Gamma\Lambda).$$

Now construct a point Ξ such that

$$(Z, H) = (\Gamma K, \Gamma\Xi).$$

From this it follows that

$$\Gamma\Xi > \Gamma\Lambda.$$

Now describe a circle through Λ, K, Ξ. $K\Gamma$ will meet this circle again in M.

Neusis: Insert between $\Lambda\Xi$ and the circle $K\Lambda\Xi$ a line segment IN equal to ΓM and directed at K. Then KI is the required straight line through K.

Proof:

$$(KE, \Gamma\Lambda) = (IK, I\Lambda) = (I\Xi, IN) = (I\Xi, \Gamma M),$$

whence

$$(KE, I\Xi) = (\Gamma\Lambda, \Gamma M) = (\Gamma K, \Gamma\Xi) = (KB, \Gamma\Xi),$$

whence

$$(KE, KB) = (I\Xi, \Gamma\Xi), \text{ therefore } (BE, KB) = (\Gamma I, \Gamma\Xi)$$

or

$$(BE, I\Gamma) = (K\Gamma, \Gamma\Xi) = (Z, H).$$

That the condition imposed on (Z, H) is necessary for the construction to be possible, may be realized as follows: In view of $KI > K\Gamma$

$$\mathbf{O}(KI, IN) > \mathbf{O}(K\Gamma, \Gamma M), \text{ whence}$$

$\mathbf{O}(I\Xi, I\Lambda) > \mathbf{O}(\Gamma\Xi, \Gamma\Lambda)$. If therefore O be the middle point of $\Xi\Lambda$,

then $\mathbf{T}(O\Lambda) - \mathbf{T}(OI) > \mathbf{T}(O\Lambda) - \mathbf{T}(O\Gamma)$, therefore $O\Gamma > OI$.

The middle point O therefore lies on the same side of Γ as I, and since I lies on the opposite side of Γ to Λ,

$\Gamma\Xi > \Gamma\Lambda$, whence $(Z, H) = (\Gamma K, \Gamma\Xi) < (\Gamma K, \Gamma\Lambda) = (\Gamma\Theta, \Theta K)$.

9.5. **Spir.** 9 (Fig. 49).

This proposition is entirely analogous to 8.4. The only difference is that E now has to lie on $A\Gamma$ produced. This is the case when $\Gamma\Xi < \Gamma\Lambda$. In view of this the condition imposed on (Z, H) now becomes

$$(Z, H) > (\Gamma\Theta, K\Theta)\ .$$

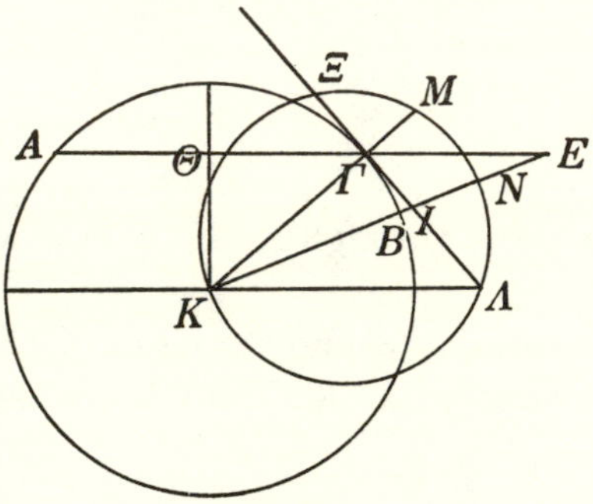

Fig. 49.

9.6. The commentators not infrequently speculate on the question how Archimedes can have performed the neuseis in question. The most obvious answer seems to be: as neuseis. The insertion of line segments of a given length between two given curves does not indeed fit in very well with the traditional prescript that in planimetrical construction only straight lines and circles are to be used, but there is not the slightest evidence that this restriction applied in Greek mathematics to non-elementary problems; there are even many indications to the contrary. Such indications are to be found in the numerous solutions of the problems of duplication of the cube and trisection of an angle in which neuseis are applied or points are constructed as intersections of conics, without any evidence to the effect that the writer does not look upon his solution as a correct construction. For the practical performance of neuseis, besides the primitive inserting ruler described above, there were even available the conchoidal compasses of Nicomedes as well. And to what extent, even apart from practical drawing considerations, mentally performed insertion was accepted as a means of construction in proofs is most clearly apparent from the proposition 9.1 discussed above, where, in contrast with the other propositions, the construction of the line segment BE by means of compasses and ruler is possible. That Archimedes nevertheless even in this case has recourse to a neusis, obliges us to reject the view that a neusis was nothing but a substitute for construction in problems where the solution by means of compasses and ruler was not possible.

9.7 As will be seen in the treatise *On Spirals*, the neusis constructions discussed are used by Archimedes in the proofs of indirect passages to a limit. This is based on the circumstance that they are equivalent to certain propositions on limits; this may be understood as follows.

In 9.1 it is proved that B can be so chosen that

$$(BE, \operatorname{arc} B\Gamma) < (R, \text{given circumference of a circle})$$

i.e. that, when ε be any assigned magnitude, B can be so chosen that

$$\frac{BE}{\operatorname{arc} B\Gamma} < \varepsilon\,.$$

This may be written

$$\operatorname*{Lim}_{B \to \Gamma} \frac{BE}{\operatorname{arc} B\Gamma} = 0$$

which is equivalent to

$$\operatorname*{Lim}_{\varphi \to 0} \frac{\sec\varphi - 1}{\varphi} = 0$$

in which $\angle BK\Gamma = \varphi$.

Whereas we nowadays show by calculation that, by the choice of φ, $\dfrac{\sec\varphi - 1}{\varphi}$ can be made less than ε, Archimedes demonstrates geometrically by means of his neusis that the ratio $(BE, \operatorname{arc} B\Gamma)$ may take on any value.

When in 9.2 we call $\angle \Gamma K\Theta = \alpha$ and $EK\Theta = \varphi$, we read for the case where φ approximates to α from below

$$\operatorname*{Lim}_{B \to \Gamma} \frac{BE}{B\Gamma} = \operatorname*{Lim}_{\varphi \to \alpha} \frac{1 - \sec\varphi \cos\alpha}{\alpha - \varphi} = \frac{\Gamma\Theta}{K\Theta} = \operatorname{tg}\alpha\,.$$

In 9.3 it is found in the same way that

$$\operatorname*{Lim}_{B \to \Gamma} \frac{BE}{B\Gamma} = \operatorname{tg}\alpha$$

when φ approximates to α from above.

The propositions 9.4 and 9.5 finally state that

$$\lim_{B \to \Gamma} \frac{BE}{I\Gamma} = \operatorname{tg} \alpha .$$

10. *Elements of Mechanics.*

10. For the treatise *On Spirals* Archimedes requires a couple of lemmas on uniform motion.

10.1. **Spir.** 1. Here it is stated that two distances described by a point in the same uniform motion are proportional to the times of describing them.

Let the motion take place on AB. On this consider two distances $\Gamma\Delta$ and ΔE; represent the times of describing them successively by ZH, $H\Theta$. It has now to be proved that

$$(\Gamma\Delta, \Delta E) = (ZH, H\Theta) .$$

The proof is based directly on the Euclidean definition of proportion. In fact, let $A\Delta$ be a multiple of $\Gamma\Delta$, ΔB a multiple of ΔE, and let $A\Delta > \Delta B$ (this is possible on the ground of the axiom of Eudoxus, which is formulated once more as a lemma in the introduction to *On Spirals*).

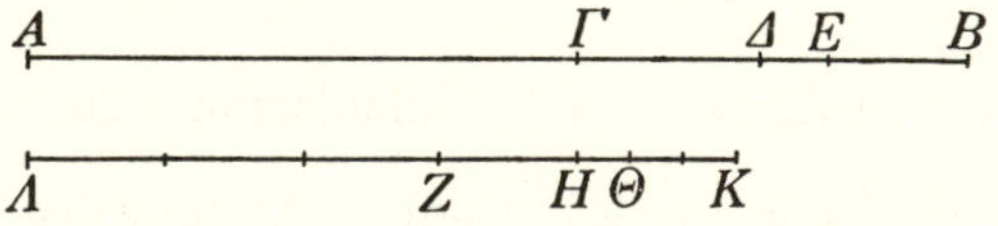

Fig. 50.

Now let ΛH be the same multiple of ZH as $A\Delta$ is of $\Gamma\Delta$, KH the same multiple of ΘH as ΔB of ΔE. Since the moving point needs for all the distances equal to $\Gamma\Delta$ a time equal to ZH, it appears that ΛH stands for the time needed for $A\Delta$, and likewise HK for that required for ΔB. From $A\Delta > \Delta B$ it then follows that $\Lambda H > HK$. In the same way it is realized that from the inequality of any multiples of ZH, $H\Theta$ follows the inequality in the same sense of equal multiples of $\Gamma\Delta$, ΔE, from which the proportion of the proposition is concluded to be true.

As basis of the argument is apparently used a definition (not stated) of uniform motion as motion in which equal distances are described in equal times. It may seem that the proposition follows at once from this, and that Archimedes already makes use of it in the proof. In this way, however, the proposition can only be understood for the case of two distances being in a rational ratio to each other; the proof given serves to demonstrate its validity for irrational ratios of distances as well.

10.2. **Spir.** 2. In this it is shown that two distances described in uniform motion in certain times are proportional to the distances described in a different uniform motion in the same times. This follows at once from 10.1.

CHAPTER IV.

ON THE SPHERE AND CYLINDER

BOOK I.

1. *Introduction.*

The first of the two books into which the treatise *On the Sphere and Cylinder* is divided opens with a letter to Dositheus, in which Archimedes reminds him that on a former occasion he already sent him the proof of the proposition that any segment of an orthotome is four-thirds of the triangle with the same base and equal height[1]). He is now going to demonstrate new propositions:

a) The surface of any sphere is four times its greatest circle[2]).

b) The surface of any segment of a sphere is equal to that of a circle whose radius is equal to the straight line drawn from the vertex of the segment to any point of the circumference of the circle which is the base of the segment[3]).

c) Any cylinder[4]), whose base is equal to the greatest circle of those in the sphere and whose height is equal to the diameter, is

---

[1]) **Q.P.** 17 and 24. *Vide* Chapter X.

[2]) **S.C.** I, 33. The Greeks say "greatest circle" (*μέγιστος κύκλος*), which is more logical than our "great circle".

[3]) **S.C.** I, 42, 43.

[4]) By a cylinder in **S.C.** is meant a right circular cylinder, by a cone a right circular cone.

itself half as large again as the sphere[1]), and its surface is half as large again as the surface of the sphere[2]).

This preliminary and partial summary of the contents of the treatise is followed by a statement the exact purport of which is doubtful, and which we will therefore reproduce in the original, with a specimen translation opposite:

ταῦτα δὲ τὰ συμπτώματα τῇ φύσει προυπῆρχεν περὶ τὰ εἰρημένα σχήματα, ἠγνοεῖτο δὲ ὑπὸ τῶν πρὸ ἡμῶν περὶ γεωμετρίαν ἀνεστραμμένων οὐδενὸς αὐτῶν ἐπινενοηκότος, ὅτι τούτων τῶν σχημάτων ἐστὶν συμμετρία· διόπερ οὐκ ἂν ὀκνήσαιμι ἀντιπαραβαλεῖν αὐτὰ πρός τε τὰ τοῖς ἄλλοις γεωμέτραις τεθεωρημένα καὶ πρὸς τὰ δόξαντα πολὺ ὑπερέχειν τῶν ὑπὸ Εὐδόξου περὶ τὰ στερεὰ θεωρηθέντων, ὅτι πᾶσα πυραμὶς τρίτον ἐστὶ μέρος πρίσματος τοῦ βάσιν ἔχοντος τὴν αὐτήν τῇ πυραμίδι καὶ ὕψος ἴσον, καὶ ὅτι πᾶς κῶνος τρίτον μέρος ἐστὶν τοῦ κυλίνδρου τοῦ βάσιν ἔχοντος τὴν αὐτὴν τῷ κώνῳ καὶ ὕψος ἴσον· καὶ γὰρ τούτων προυπαρχόντων φυσικῶς περὶ ταῦτα τὰ σχήματα, πολλῶν πρὸ Εὐδόξου γεγενημένων

These properties were all along naturally inherent already in the figures referred to, but they were unknown to those who were before our time engaged in the study of geometry, because none of them realized that there exists symmetry between these figures[3]). Therefore I would not hesitate to compare these properties with the insight gained by other geometers and with those of the theorems of Eudoxus on solid figures which to my mind are the most excellent, *viz.* that any pyramid is one-third of the prism which has the same base as the pyramid and equal height, and that any cone is one-third of the cylinder which has the same base as the cone and equal height[4]). For, though these properties also were already naturally inherent in the figures all along, and there were many important

---

1) Greek has no word for volume. This magnitude is denoted by the name of the figure itself. The same is usually done in planimetrical discussions with regard to the area of a figure.

2) S.C. I, 34. Corollary.

3) We are here using the word symmetry in the sense of commensurability, which is to be attributed to it both etymologically and historically.

4) Euclid XII, 7, Porism. XII, 10.

ἀξίων λόγου γεωμετρῶν συνέβαινεν ὑπὸ πάντων ἀγνοεῖσθαι μηδ' ὑφ ἑνὸς κατανοηθῆναι.

geometers before Eudoxus, they have remained unknown to them all and have not been realized by anyone[1]).

If the suggested translation is correct, Archimedes here seems to voice his astonishment that geometrical figures may have remarkable properties inherent in them, *i.e.* without their being stated in the definition we give of them, which properties may long remain unnoticed, in spite of their simplicity. It is the typical mathematician's astonishment at the unsuspected intrinsic wealth of his own definition which is being expressed here.

At the close of his preface the author presents his work to all experts, inviting their verdict; he regrets that it could not be published while Conon was still alive, because the latter would have been eminently capable of grasping its contents and giving his opinion on them. There now follow first the *axiomata* and the *lambanomena* (assumptions) on which the treatise is to be based.

2. *Axiomata.*

The first group of the fundamental propositions has been rightly termed *Axiomata* in so far as they postulate the existence of certain types of curves and surfaces. On the other hand this group also contains two purely nominal definitions.

I. *There are in a plane certain terminated bent lines*[2]) *which either lie wholly on the same side of the straight line joining their extremities or have no part of them on the other side thereof.*

Eutocius comments on this[3]) that the category of the bent lines (*καμπύλαι γραμμαί*) also includes lines which consist wholly or partly

[1]) Elsewhere (introduction to the *Method, Opera* II, 430) Archimedes says that a not inconsiderable part of the merit of these propositions is due to Democritus, who was the first to enunciate them, though without giving any proof (*i.e.* without giving any perfectly exact proof). These two statements are therefore at variance with each other, a fact which is not to be removed by supposing that during his writing of **S.C.** Archimedes did not yet know the things told by him in **Meth.**; for in view of various other reasons the *Method* is assumed to be the earlier of the two treatises.

[2]) By this is meant that the part of the curve which is being considered has two extremities.

[3]) *Opera* III, 4; lines 8 *et seq.*

of straight line segments. Owing to this, among other things, the curves considered by Archimedes may also partially coincide with the straight line determined by the extremities.

II. *I call such a line "concave in the same direction"* (ἐπὶ τὰ αὐτὰ κοίλη) *when it has the property that, if any two points on it are taken, either all the straight lines*[1]) *connecting such points fall on the same side of the line or some fall on one and the same side, while others fall on the line itself, but none on the other side.*

The lines defined in Axiom II therefore belong to the category circumscribed by Axiom I; in most translations this does not emerge sufficiently[2]). In modern terminology it is stated by Axiom II that a line which is concave on the same side includes a convex surface with the line segment joining its extremities.

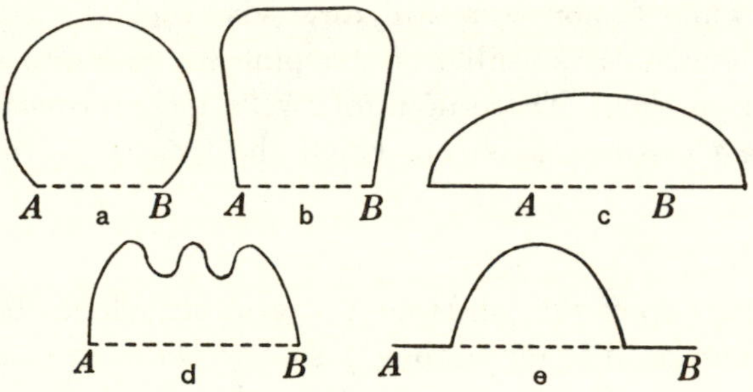

Fig. 51.

Of the lines shown in Fig. 51 with the extremities *A* and *B*, *a*, *b*, and *c* are of the type described, *d* and *e* are not.

III. *Similarly also there are certain terminated surfaces*[3]) *which are not themselves in a plane, but have their boundaries in a plane, and which either lie wholly on the same side of the plane containing the boundaries or have no part of them on the other side thereof.*

Here again the meaning is that the surfaces under consideration may also consist of one or more plane faces.

[1]) The Greeks do not distinguish between a straight line and a straight line segment. In the translations, however, we shall use the latter term wherever this is required for the sake of clarity. In most cases, moreover, the meaning is easily inferred from the context.

[2]) Eutocius states it explicitly (*Opera* III, 4; line 16): ἐκ δὲ τούτων ἦν ἡ ἐπιλογὴ τῶν ἐπὶ τὰ αὐτὰ κοίλων. An erroneous translation is to be found in Czwalina, *Kugel und Zylinder*, p. 8.

[3]) By this is meant that the surface under consideration has a boundary.

IV. *I call such surfaces "concave in the same direction" when they have the property that, if any two points on them are taken, the straight lines connecting such points either all fall on the same side of the surface, or some fall on one and the same side, others upon the surface itself*[1]), *but none on the other side.*

V. *When a sphere is cut by a cone which has its vertex at the centre of the sphere, I use the term solid sector* (τομεύς στερεός) *for the figure comprehended by the surface of the cone and the surface of the sphere included within the cone.*

VI. *When two cones with the same base have their vertices on opposite sides of the plane of the base, so that their axes lie in one straight line, I use the term solid rhombus* (ῥόμβος στερεός) *for the solid figure made up of the two cones.*

3. *Lambanomena.*

By this title is denoted a group of postulates or axioms in the sense of "unproved fundamental propositions on known figures".

I assume the following:

I. *That of the lines which have the same extremities the straight line is the least.*

II. *That of the other lines*[2]), *if, lying in one plane, they have the same extremities, two are unequal whenever both are concave in the same direction and moreover one of them is either wholly included between the other and the straight line which has the same extremities with it, or is partly included by and partly coincides with the other; and that the line which is included is the lesser.*

III. *Similarly, that of the surfaces which have the same boundaries, if the latter are in one plane, the plane surface is the least.*

IV. *That of the other surfaces which have the same boundaries, if the boundaries are in one plane, two are unequal whenever both are concave in the same direction and moreover one of them is either wholly included between the other and the plane which has the same boundaries, or is partly included by and partly coincides with the other; and that the surface which is included is the lesser.*

---

[1]) Here we have to think either of a ruled surface (in this case a cone or a cylinder) or of a surface having one or more plane faces.

[2]) *i.e.* of all lines which are not straight (*i.e.* the καμπύλαι γραμμαί).

V. Ἔτι δὲ τῶν ἀνίσων γραμμῶν καὶ τῶν ἀνίσων ἐπιφανειῶν καὶ τῶν ἀνίσων στερεῶν τὸ μεῖζον τοῦ ἐλάσσονος ὑπερέχειν τοιούτῳ, ὃ συντιθέμενον αὐτὸ ἑαυτῷ δυνατόν ἐστιν ὑπερέχειν παντὸς τοῦ προτεθέντος τῶν πρὸς ἄλληλα λεγομένων.

V. *And also that of unequal lines, unequal surfaces, and unequal solids the greater exceeds the lesser by an amount such that, when added to itself, it may exceed any assigned magnitude of the type of magnitudes compared with one another.*

The "addition to itself" naturally has to be conceived of as repeated any number of times; the meaning therefore is: multiplied by a natural number.

There are great differences between the translations given by various editors of the last words of this postulate *viz.* παντὸς τοῦ προτεθέντος τῶν πρὸς ἄλληλα λεγομένων. Without laying claim to completeness, we mention the following versions:

a) *Editio Princeps*[1]) : *omnem propositam sui generis quantitatem.*
b) *Heiberg*[2]) : *quamvis magnitudinem datam earum, quae cum ea comparari possint.*
c) *Heath*[3]) : *any assigned magnitude among those which are comparable with [it and with] one another.*
d) *Heath*[4]) : *any assigned magnitude of the same kind.*
e) *Ver Eecke*[5]) : *toute grandeur donnée ayant un rapport avec l'une et l'autre des premières.*
f) *Czwalina*[6]) : *jede der beiden gegebenen Grössen.*

When therefore the two unequal lines (surfaces, solids) of which Archimedes speaks are called A and B respectively, and it is assumed that $A-B=C$, the translations a), b), and as far as the addition between brackets is concerned also c) state that there exists a number n such that $n \cdot C > D$ when D is comparable with C or "of the same kind"; in d) also this seems to be suggested. This naturally raises the question when two magnitudes are called of

1) Quoted on p. 41, Note 3; p. 2 of the Latin translation.

2) *Opera* I, 9, and slightly differently in the mention of the *axioma* in **Spir.** (*Opera* II, 13): ...*earum, quae inter se comparari possint.*

3) Heath, *Archimedes*, p. 4.

4) Heath, *Greek Mathematics* II, 35.

5) Ver Eecke, *Archimède*, p. 6.

6) Czwalina, *Kugel und Zylinder*, p. 9.

the same kind or comparable. The answer to this is to be inferred from the Definitions 3 and 4 of Euclid V, which state, when correlated with each other, that two magnitudes are of the same kind (or: have a ratio to one another) when they are capable, upon multiplication, of exceeding one another. In the translations a)–d) it is therefore said that the magnitude C is of the same kind with any magnitude D with which it is of the same kind.

The Greek text, however, contains quite a different statement. It does not refer to *comparable* magnitudes, but to magnitudes *compared* with each other, and by this only the magnitudes A and B can be meant. The postulate therefore states that when C is the difference between two lines (surfaces, solids), the number n may be so chosen that $n \cdot C$ is greater than any line (surface, solid) whatever.

It is thus clear that the translation e), though philologically not exactly in accordance with the text, renders at least the meaning of the statement correctly. In fact, according to the theory of proportions D only then has a ratio to A and to B when D is of the same kind with A and with B.

The translation f), finally, is not at all in agreement with the text.

The view adopted by us above and expressed in the translation is to be found, among others, in the following translations:

Mersenne[1]: *quamcumque dictarum, et inter se collatarum magnitudinum.*

Nizze[2] : *jede gegebene Grösse von der Art der verglichenen.*

We have gone into the matter of the correct translation of the fifth postulate in some detail because it is only after the exact meaning of Archimedes' words has been established that a discussion is possible of the generally neglected, but no less urgent question what motive may have induced him to include this assumption. Indeed, if one of the translations a)–d) is adopted, it is incomprehensible what he may have intended by it. In that case it can only be interpreted as the formulation of Definition 4 of Euclid

[1]) *Universae Geometriae, Mixtaeque Mathematicae Synopsis, Et Bini Refractionum Demonstratarum Tractatus. Studio et Opera* F. M. Mersenne M. Parisiis MDCXLIV.

[2]) Quoted on p. 43, note 2.

V in the form of the postulate which in the literature on the history of mathematics is sometimes termed the postulate of Eudoxus (with regard to two unequal magnitudes a and $b (a < b)$ it is assumed that there exists a number n having the property that $n \cdot a > b$). But on this postulate are based both the whole theory of proportions of Euclid V, which Archimedes makes use of all through his work, and Euclid's lemma (X, 1) about the continued dichotomy of a magnitude, also regularly used by Archimedes. Why then should he have formulated it once more in a separate assumption for a magnitude appearing as the difference between two other magnitudes?

The actual meaning, however, becomes plain when it is considered that in Greek mathematics there existed, side by side with the strict and official method of the indirect passage to a limit, also the less strict, but heuristically more fertile method of indivisibles, and that Archimedes himself diligently used it as a method of investigation[1]). In this method a solid is regarded as the sum of plane sections, a surface as the sum of lines, while it also includes the view of a curve as being generated by juxtaposition of points; this might easily suggest the idea that thus the difference between two solids might be a surface, that between two surfaces a length. We may ignore the question whether this idea was ever applied in practice. But it is in any case understandable that Archimedes, now that he is about to prove his results strictly, with the aid of a method the essential foundation of which is precisely that the difference between two magnitudes of the same kind, however small it may be, satisfies the axiom of Eudoxus with regard to any magnitude of this kind, deems it necessary to banish all unstrict conceptions to which the method of indivisibles might give rise on this point.

From the above it follows that it is desirable to discriminate explicitly between the axiom of Eudoxus and the 5th postulate of the treatise *On the Sphere and Cylinder* of Archimedes. The contents of the latter may be briefly summed up as follows: if two magnitudes satisfy the axiom of Eudoxus in respect of each other, their difference also satisfies this assumption in respect of any magnitude of the same kind homogeneous with both[2]).

[1]) Vide Chapter X.

[2]) J. Hjelmslev, in his treatise *Über Archimedes' Grössenlehre* (Det kgl.

To express it in modern terms, he excludes the existence of actual infinitesimals; the magnitudes he is going to discuss are to form Eudoxian systems.

At the end of the Lambanomena it is mentioned, by way of conclusion from the second, that the perimeter of a polygon inscribed in a circle is less than the circumference of the circle.

4. *Introductory Propositions* (1–6).

In the first Book of *On the Sphere and Cylinder* the ratio form of the compression method (III; 8, 21) is to be repeatedly applied. The group of the propositions 2–6 serves to prepare the way. It is preceded by the proposition 1, in which it is derived from the second assumption that the perimeter of a polygon circumscribed about a circle is greater than the circumference of the circle.

In the following propositions

C denotes a circle, C_n a regular polygon of n sides circumscribed about this circle, I_n a regular polygon of n sides inscribed in this circle. All three symbols at the same time denote the area of the figures they represent. The sides of the polygons are called successively Z_n and z_n.

Danske Videnskabernes Selskab. Matem.-Fysiske Meddelelser XXV, No 15, København 1950, pp. 4, 5), distinguishes the two axioms in question as the axiom of Eudoxus and the lemma of Archimedes. According to him, the object of the lemma is to establish that, when two magnitudes satisfy the axiom of Eudoxus in respect of each other, their difference also satisfies it in respect of all magnitudes of the same kind with a and b. This view is in agreement with the one defended above; it differs from it only in the motivation: the formulation of the new axiom is considered necessary not for the sake of excluding the method of indivisibles, but to give sense to the difference of two homogeneous magnitudes a and b, *e.g.* in the case where a is a circular arc and b a line segment, or a part of the surface of a sphere and b part of a plane. In the theory of proportions of Euclid this axiom, according to the author's view, was not necessary, because $a-b$ always exists as a magnitude of the same kind with a and b. We are not convinced by this argument. Eudoxus (in Euclid V) merely requires of his magnitudes that they shall satisfy his axiom, and does not say at all what magnitudes they are. It cannot be understood why with him a could not be a circular arc and b a line segment.

The axiom of Archimedes is not therefore required because the scope of the geometrical magnitudes under consideration is widened, but it serves to fill up a gap in the theory of proportions of Euclid V (Euclid, for example, tacitly assumes in V, 8 what the axiom of Archimedes explicitly postulates). In fact, through this gap the indivisibles might slip into geometry again.

Proposition 2.

Given two unequal magnitudes, it is possible to find two unequal straight lines such that the greater straight line has to the lesser a ratio less than the greater magnitude has to the lesser.

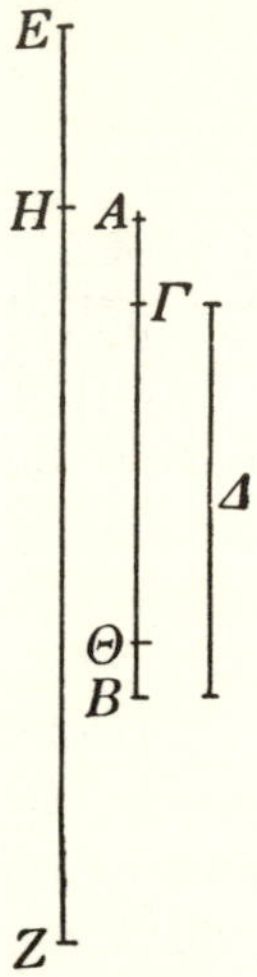

Fig. 52.

Proof: In Fig. 52 let the given magnitudes be AB and Δ ($AB > \Delta$). Measure $B\Gamma = \Delta$. Take any straight line ZH. Measure a multiple $A\Theta$ of $A\Gamma$ such that $A\Theta > \Delta$ (postulate of Eudoxus). Now let HE be the same portion of ZH as $A\Gamma$ is of $A\Theta$. Then

$$(EH, HZ) = (A\Gamma, A\Theta), \text{ whence, because } A\Theta > \Delta\,,$$

$$(EH, HZ) < (A\Gamma, \Delta) = (A\Gamma, \Gamma B)\,.$$

Componendo (III; 0.42):

$$(EZ, HZ) < (AB, \Gamma B) = (AB, \Delta)\,.$$

Upon superficial examination this proof may appear unduly long. In fact, it might be asked why we do not choose any point Θ between A and B, and then conclude that $(A\Theta, \Delta) < (AB, \Delta)$. In that case, however, it would be forgotten that AB and Δ are indeed represented by line segments, in order to fix our ideas, but that in reality they are geometrical magnitudes (μεγέϑη) that are not defined any further, *e.g.* lengths of curves, areas or volumes[1]). EZ and HZ, on the contrary, are real line segments. Although it is not explicitly stated in the proposition, it appears from the given inequality relation and the application of the postulate of Eudoxus that the magnitudes AB and Δ are assumed to be of the same kind.

In modern notation the argument is as follows:

---

[1]) This is also pointed out by Hjelmslev, l.c. p. 7, who observes in this connection that when at the words "Measure $B\Gamma = \Delta$" the text refers to Euclid I, 2 (Heiberg I 12, line 3), in the first place this should read I, 3, but that it has also to be assumed that this is an interpolation, because the given magnitudes AB and Δ are not line segments, but are only represented by line segments. It might be argued against this that, all the same, Euclidean constructions can be applied to these line segments functioning as symbols. For the rest, the above doubt as to the genuineness of the reference is in itself not unjustified. Archimedes never quotes Euclid anywhere else; why should he do it all at once for this extremely elementary question?

If a and b are two unequal magnitudes of the same kind ($a > b$), find a natural number n such that

$$n(a-b) > b.$$

From this it follows that

$$\frac{n+1}{n} < \frac{a}{b}.$$

Proposition 3.

Given two unequal magnitudes and a circle, it is possible to inscribe a polygon in the circle and to circumscribe another about it so that the side of the circumscribed polygon has to the side of the inscribed polygon a ratio less than the greater magnitude has to the less.

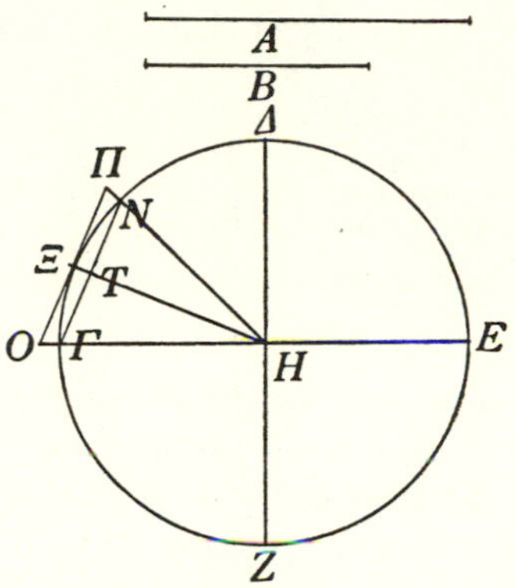

Fig. 53.

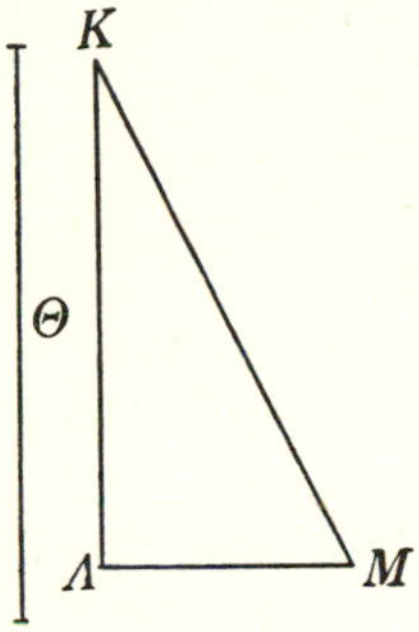

Fig. 54.

Proof (Fig. 53): Assume as given the magnitudes A and B ($A > B$) and the circle H. Find (Prop. 2) two straight lines Θ, $K\Lambda$ ($\Theta > K\Lambda$) such that

$$(\Theta, K\Lambda) < (A, B).$$

Now (Fig. 54) construct a triangle $K\Lambda M$, right-angled in Λ, in which $KM = \Theta$. In the circle H draw two diameters ΓE and ΔZ at right angles. Then apply dichotomy (III; 0.5) to the angle $\Gamma H\Delta$, so that $\angle NH\Gamma < 2 \cdot \angle \Lambda KM$. Now $N\Gamma$ is the side of an inscribed equilateral polygon. Let $H\Xi$ be bisector of $\angle \Gamma HN$, $O\Pi$ tangent to the circle at Ξ, then $O\Pi$ is the side of a circumscribed equilateral polygon. If now $H\Xi$ meet $N\Gamma$ in T, then we have

$$\angle NH\Gamma < 2 \cdot \angle \Lambda KM, \text{ whence } \angle TH\Gamma < \angle \Lambda KM.$$

From this it follows that

$$(\Gamma H, H T) < (KM, K\Lambda)$$

or $$(H\Xi, HT) < (\Theta, K\Lambda),$$

therefore $$(\Pi O, N\Gamma) < (\Theta, K\Lambda) < (A, B).$$

Modern notation: Find a number p such that

$$\frac{p+1}{p} < \frac{A}{B}.$$

Now construct an angle $\varphi(\Lambda KM)$ such that $\cos\varphi = \frac{p}{p+1}$.

By dichotomy find an angle $\alpha = \frac{\frac{1}{2}\pi}{2^m}$, so that $\alpha < 2\varphi$.

Now let α be an angle at the centre of a circumscribed and an inscribed regular polygon of n sides ($n = 2^{m+2}$), then we have for the sides Z_n and z_n thereof

$$\frac{Z_n}{z_n} = \frac{1}{\cos\frac{\alpha}{2}} < \frac{1}{\cos\varphi} = \frac{p+1}{p} < \frac{A}{B}.$$

Proposition 4.

This is similar to Prop. 3, provided one assumes as given, instead of a circle, a sector of a circle, in and about which homologous equilateral segments of a polygon are described. The dichotomy is now applied to the angle at the centre of the sector.

Proposition 5.

Given a circle and two unequal magnitudes, to circumscribe a polygon about the circle and to inscribe another in it, so that the circumscribed polygon may have to the inscribed polygon a ratio less than the greater magnitude has to the less.

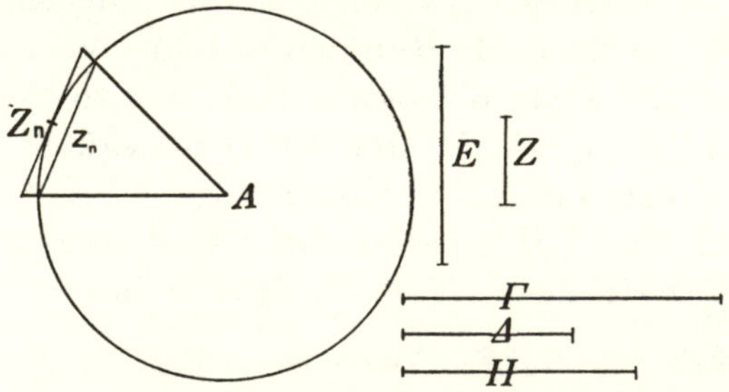

Fig. 55.

Construction: Assume as given (Fig. 55) the circle A and the unequal magnitudes E, Z ($E > Z$). Construct (Prop. 2) two straight lines Γ, Δ ($\Gamma > \Delta$) such that $(\Gamma, \Delta) < (E, Z)$. Find the mean proportional H between Γ and Δ, then $\mathbf{T}(H) = \mathbf{O}(\Gamma, \Delta) < \mathbf{T}(\Gamma)$, whence $H < \Gamma$. Circumscribe (Prop. 3) about A a polygon C_n and inscribe in A a polygon I_n such that the ratio of the sides Z_n and z_n may satisfy the relation

$$(Z_n, z_n) < (\Gamma, H) .$$

Then we also have (III; 0.43)

$$\Delta\Lambda(Z_n, z_n) < \Delta\Lambda(\Gamma, H) = (\Gamma, \Delta) ,$$

therefore
$$(C_n, I_n) < (\Gamma, \Delta) < (E, Z) .$$

Modern notation:

$$\frac{C_n}{I_n} = \frac{Z_n{}^2}{z_n{}^2} < \frac{\Gamma^2}{H^2} = \frac{\Gamma^2}{\Gamma\Delta} = \frac{\Gamma}{\Delta} < \frac{E}{Z}.$$

Proposition 6.

α) In a similar way the corresponding proposition for a sector of a circle may be proved.

β) The reader is reminded of the proposition from the *Elements* (to be found in Euclid XII, 2), which states that upon continued duplication of the number of sides of an inscribed equilateral polygon the sum of the remaining segments of the circle decreases below an assigned area.

γ) Thereafter it is shown that a similar proposition applies fo the sum of the tangent sectors[1]) which lie within a circumscribed equilateral polygon and without the circle.

Proof: Let the assigned area be B and the area of the circle C, then it is possible (Prop. 5) to find a number n such that

$$(C_n, I_n) < (C + B, C), \text{ i.e., because } I_n < C ,$$

$$(C_n, C) < (C + B, C) ,$$

therefore
$$C_n < C + B \quad \text{or} \quad C_n - C < B .$$

[1]) By a tangent sector we understand the figure bounded by the parts of two intersecting tangents to a circle between the point of intersection and the points of contact, and the smaller arc of the circle between the points of contact.

The real meaning of the group of propositions 2–6 becomes clear if we represent the repeatedly recurring ratio of two unequal magnitudes of the same kind by ε (in which therefore $\varepsilon > 1$).

It has then been proved that n can be so chosen that

$$\text{In Prop. 2} \quad 1 < \frac{n+1}{n} < \varepsilon, \quad \text{or}^{1)} \quad \lim_{n\to\infty} \frac{n+1}{n} = 1 .$$

$$\text{In Prop. 3} \quad 1 < \frac{Z_n}{z_n} < \varepsilon, \quad \text{or} \quad \lim_{n\to\infty} \frac{Z_n}{z_n} = 1 .$$

$$\text{In Prop. 5} \quad 1 < \frac{C_n}{I_n} < \varepsilon, \quad \text{or} \quad \lim_{n\to\infty} \frac{C_n}{I_n} = 1 .$$

In Prop. 6 the reader is reminded that with a given number $\delta > 0$, a number n may be chosen such that

$$0 < C - I_n < \delta, \text{ or } \lim_{n\to\infty} (C - I_n) = 0$$

while it is also proved that

$$0 < C_n - C < \delta, \text{ or } \lim_{n\to\infty} (C_n - C) = 0 .$$

5. *Curved Surface of Cylinder and Cone. Propositions* 7–20.

The main theorems of this group are the propositions 13 and 14, in which the curved surfaces of a cylinder and a cone respectively are found. The following propositions serve as an introduction:

Props 7 and 8, in which the lateral surfaces of a regular pyramid inscribed in and of one circumscribed about a cone are found.

Props 9 and 10, in which these lateral surfaces are compared with the curved surface of the cone.

Props 11 and 12, in which the lateral surfaces of regular prisms inscribed in and circumscribed about a cylinder are compared with the curved surface of this cylinder.

Proposition 7.

If in an isosceles cone a pyramid be inscribed having an equilateral base, its surface excluding the base is equal to a triangle having its

[1]) This word implies that the difference between the propositions of Archimedes and the propositions of the theory of variants mentioned behind them is exclusively a difference of notation.

base equal to the perimeter of the base [of the pyramid] and its height equal to the perpendicular drawn from the vertex to one side of the base.

Nowadays this is expressed by saying that the lateral surface of a regular pyramid is equal to half the product of the perimeter of the base and the apothem. This expression, however, is senseless in Greek geometry, because as a rule it will not be possible to represent by numbers the lengths of the line segments multiplied by us. Thus, wherever we use, to denote an area, a product of two factors, the Greek geometer had to introduce a plane figure whose area was equal to that of the figure under consideration. This often makes the argument seem cumbrous to us; it is, however, an essential feature of the Greek point of view which thus becomes manifest.

For the rest, the proof of Prop. 7 is completely identical with the one still commonly used.

In Prop. 8 the corresponding theorem for a circumscribed pyramid is enunciated and proved.

Proposition 9.

If in an isosceles cone a straight line fall within the circle which is the base of the cone, and from its extremities straight lines be drawn to the vertex of the cone, the triangle contained by the chord and the lines drawn to the vertex will be less than the surface of the cone intercepted between the lines drawn to the vertex.

In Fig. 56 let $\Delta . AB\Gamma$ be the given right circular cone, $A\Gamma$ a chord of the circular base. It is required to prove that

$$\triangle\, \Delta A\Gamma < \text{portion of surface of cone } \Delta A\Gamma .$$

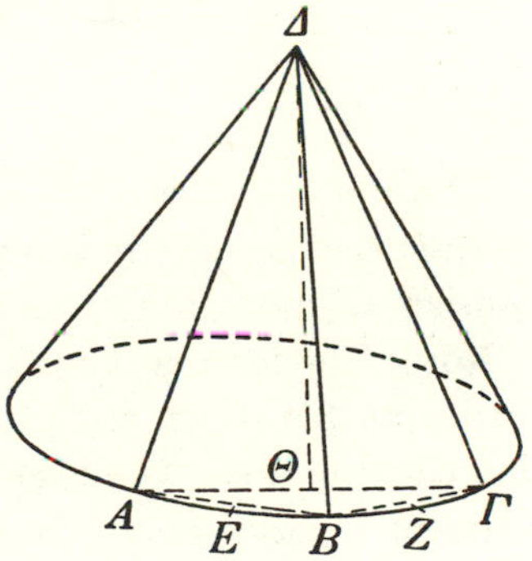

Fig. 56.

Proof: Let B be the middle point of the arc $A\Gamma$, then

$(\alpha)\ \triangle\, \Delta AB + \triangle\, \Delta B\Gamma > \triangle\, \Delta A\Gamma$ (*vide* Note).

$$\text{Suppose } \triangle\, \Delta AB + \triangle\, \Delta B\Gamma - \triangle\, \Delta A\Gamma = \Theta , \tag{1}$$

then either (I) $\Theta \geqq$ segment of circle AB + segment of circle $B\Gamma$

or (II) $\Theta <$ segment of circle AB + segment of circle $B\Gamma$.

Case I. On account of Postulate III we have:

$$\triangle\, \Delta AB < \text{segment of circle } AB + \text{portion of conical surface } \Delta AB$$
$$\triangle\, \Delta B\Gamma < \text{segment of circle } B\Gamma + \text{portion of conical surface } \Delta B\Gamma \quad +$$

$$\triangle\, \Delta AB + \triangle\, \Delta B\Gamma < \text{segment } AB \text{ of circle} + \text{segment } B\Gamma \text{ of circle} + \text{portion of conical surface } \Delta A\Gamma\,,$$

a fortiori:

$$\triangle\, \Delta AB + \triangle\, \Delta B\Gamma < \Theta + \text{portion of conical surface } \Delta A\Gamma\,,$$

from which by (1) it follows that

$$\triangle\, \Delta A\Gamma < \text{portion of conical surface } \Delta A\Gamma.$$

Case II. Apply dichotomy (III; 0.5) to the arcs AB and $B\Gamma$, until the sum of the remaining segments of the circle becomes less than Θ (6β). It is assumed that this result is attained already after one bisection. From Postulate III it now follows as above:

$$\triangle\, \Delta AE + \triangle\, \Delta EB + \triangle\, \Delta BZ + \triangle\, \Delta Z\Gamma < \text{sum of segments of circle } AE, \text{ etc.} + \text{segment of conical surface } \Delta A\Gamma,$$

from which it follows *a fortiori* that

$$\triangle\, \Delta AB + \triangle\, \Delta B\Gamma < \Theta + \text{segment } \Delta A\Gamma\,,$$

whence by (1)

$$\triangle\, \Delta A\Gamma < \text{segment } \Delta A\Gamma\,.$$

The proof can easily be extended to the case where the dichotomy consists of more than one step.

Note: The inequality (α) fundamental for the proof, which Archimedes writes down without accounting for it, in several editions and commentaries is either not proved or proved incorrectly. What Eutocius[1]) says about it is incomprehensible. Heiberg and Ver Eecke are of opinion[2]) that the perpendiculars drawn from Δ to the chords AB, $B\Gamma$, ΓA of the base are equal, and then make use of the inequality $AB + B\Gamma > A\Gamma$.

By Greek methods the proof may be furnished as follows:

---

[1]) *Opera* III, 24.

[2]) *Opera* I, 31. Note 2. Ver Eecke, *Archimède*, p. 17, Note 2.

Because of Euclid XI, 20 we have

$$\angle A\Delta B + \angle B\Delta\Gamma > \angle A\Delta\Gamma$$

whence, if Θ be the middle point of $A\Gamma$, $\angle A\Delta B > \angle A\Delta\Theta$.

We also have

$$\Delta B > \Delta\Theta .$$

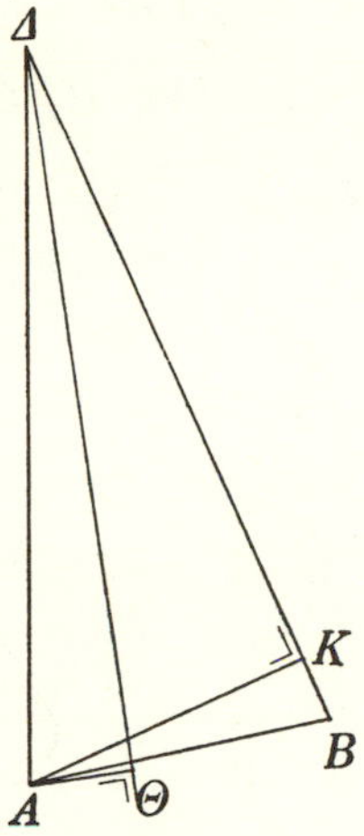

Fig. 57.

With regard to the triangles ΔAB and $\Delta A\Theta$ it is now known that $\Delta A = \Delta A$, $\Delta B > \Delta\Theta$, $\angle A\Delta B > \angle A\Delta\Theta$. If in Fig. 57 the two triangles are drawn in one plane on the same side of $A\Delta$, $\Delta\Theta$ will fall within $\angle A\Delta B$. If $AK \perp \Delta B$, then $AK > A\Theta$, because in the circle on $A\Delta$ as diameter the chord AK subtends a larger arc than does $A\Theta$. Since moreover $\Delta B > \Delta\Theta$, there is no doubt that

$$\triangle \Delta AB > \triangle \Delta A\Theta .$$

We likewise have

$$\triangle \Delta B\Gamma > \triangle \Delta\Theta\Gamma ,$$

from which follows that which was required to be proved.

From the proved proposition it follows that the lateral surface of a pyramid inscribed in a cone is less than the curved surface of the cone. This conclusion is communicated at the end of Prop. 12.

Proposition 10.

If to the circle which is the base of the cone there be drawn tangents which lie in the same plane with the circle and intersect each other, and if from the points of contact and from the point of intersection of the tangents straight lines be drawn to the vertex of the cone, the triangles contained by the tangents and the lines drawn to the vertex are [together] greater than the surface of the cone intercepted by these [lines].

In Fig. 58 let $E \cdot AB\Gamma$ be the given right circular cone, ΔA and $\Delta\Gamma$ tangents to the circle which is the base of the cone.

It is required to be proved that $\triangle EA\Delta + \triangle E\Gamma\Delta >$ segment of conic surface $EA\Gamma$.

Proof: Let B be the middle point of arc $A\Gamma$, $HZ(\parallel A\Gamma)$ the tangent to the circle at B. We now have

$$\Delta\Gamma + \Delta A = \Delta Z + Z\Gamma + \Delta H + HA > \Gamma Z + ZH + HA \, ,$$

whence

$$\triangle E\Delta\Gamma + \triangle E\Delta A > \triangle E\Gamma Z + \triangle EZH + \triangle EHA \, .$$

Assume

$$\triangle E\Delta\Gamma + \triangle E\Delta A - [\triangle E\Gamma Z + \triangle EZH + \triangle EHA] = \Theta \, ,$$

then either (I) $\Theta \geqq$ tangent sector AHB + tangent sector $BZ\Gamma$

or (II) $\Theta <$ tangent sector AHB + tangent sector $BZ\Gamma$.

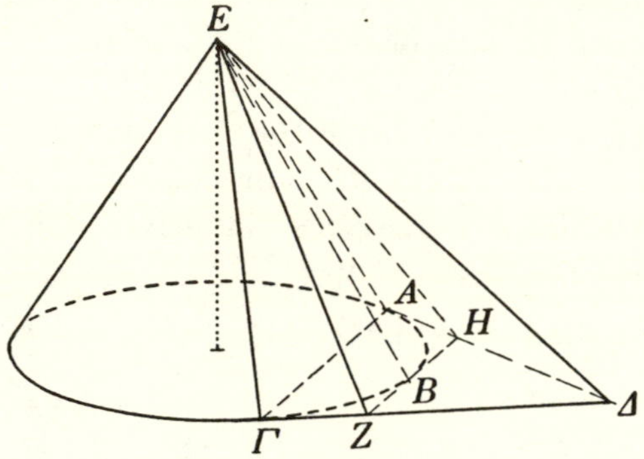

Fig. 58.

Case I. Because of Postulate IV we have:

trapezium $AHZ\Gamma + \triangle EAH + \triangle EHZ + \triangle EZ\Gamma >$ segment of circle $AB\Gamma$ + segment of conical surface $EA\Gamma$.

Therefore:

tangent sector AHB + tangent sector $BZ\Gamma + \triangle EAH + \triangle EHZ + \triangle EZ\Gamma >$ segment $EA\Gamma$.

a fortiori

$$\Theta + \triangle EAH + \triangle EHZ + \triangle EZ\Gamma > \text{segment } EA\Gamma$$

or

$$\triangle E\Delta\Gamma + \triangle E\Delta A > \text{segment } EA\Gamma \, .$$

Case II. Apply dichotomy to the arcs AB and $B\Gamma$ until the sum of the tangent sectors obtained becomes less than Θ (6γ). Thereafter the argument proceeds entirely analogously to that of Prop. 9, Case II.

From the proved proposition it follows that the lateral surface of a circumscribed pyramid is greater than the curved surface of the cone. This conclusion is communicated at the end of Prop. 12.

In Propositions 11 and 12 the corresponding theorems for a cylinder with an inscribed and a circumscribed prism are proved in an analogous manner. From this it follows at the end of Prop. 12 that the curved surface of a cylinder is greater than the lateral surface of an inscribed prism and less than that of a circumscribed prism.

After this, the main propositions of this group can now be proved.

Proposition 13.

The surface of any right circular cylinder excluding its bases is equal to that of a circle whose radius is the mean proportional between the side of the cylinder and the diameter of the base of the cylinder.

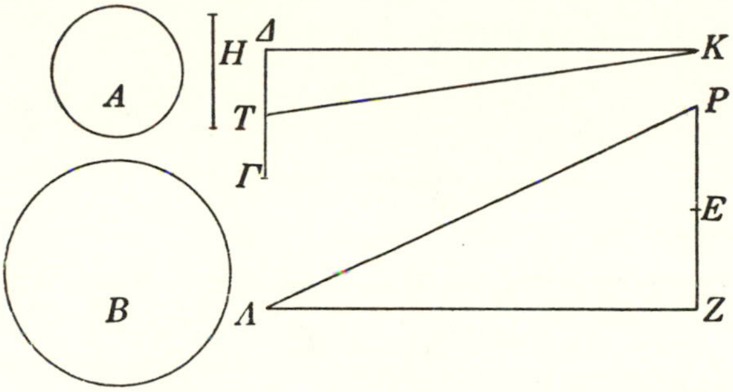

Fig. 59.

In Fig. 59 let the circle A be the base of the cylinder, $\Gamma\Delta$ its diameter, EZ the side (height) of the cylinder, H the mean proportional between $\Gamma\Delta$ and EZ, B the circle with radius H. It has to be proved that the curved surface O of the cylinder is equal to B. If this is not true, then either (I) $B < O$ or (II) $B > O$.

Case I. Because of Prop. 5 it is possible to inscribe in B an equilateral polygon b_n and to circumscribe about B an equilateral polygon B_n, so that

$$(B_n, b_n) < (O, B) . \qquad (\alpha)$$

When n has thus been found, circumscribe about the circle A an equilateral polygon of n sides A_n, and construct the circumscribed prism to the cylinder which has A_n for base. Let the lateral surface thereof be P_n.

Archimedes now proves that

$$B_n = P_n .$$

Since he cannot use any algebraic expressions for these surfaces, they both have to be represented geometrically. Therefore let $K\Delta$ be the perimeter of A_n, and $\Delta T = \frac{1}{2}\Delta\Gamma$, then

$$A_n = \triangle\, T\Delta K .$$

Further let ΔZ be equal to $K\Delta$ and PE to EZ (height of the cylinder), then

$$P_n = \triangle\, PZ\Delta .$$

Now we have

$$(A_n, B_n) = [\mathbf{T}(T\Delta),\ \mathbf{T}(H)] ,$$

in which $$\mathbf{T}(H) = \mathbf{O}(\Delta\Gamma, EZ) = \mathbf{O}(T\Delta, PZ) .$$

From this it follows that

$$(A_n, B_n) = (T\Delta, PZ) = [\triangle\, T\Delta K,\ \triangle\, PZ\Delta] = (A_n, P_n) ,$$

therefore $$B_n = P_n .$$

Modern notation:
Let $\Gamma\Delta$ be equal to d, EZ to h, then the radius of B is:

$$H = \sqrt{dh} .$$

Now we have

$$\frac{A_n}{B_n} = \frac{d^2}{4dh}, \text{ therefore } B_n = \frac{4h}{d} A_n = h \cdot \frac{A_n}{\frac{1}{4}d} = h \cdot \text{perimeter of } A_n = P_n .$$

The inequality (α) now passes into

$$(P_n, b_n) < (O, B)$$

or $$(P_n, O) < (b_n, B) .$$

Since $b_n < B$, $P_n < O$, which is contrary to Prop. 12.
Case II. Now find n such that

$$(B_n, b_n) < (B, O) . \qquad (\beta)$$

Inscribe in the circle A the equilateral polygon of n sides a_n, and construct the prism inscribed in the cylinder, which has a_n

for base. Let the lateral surface thereof be p_n. Since the apothem of a_n is less than ΔT, it is now found that

$$a_n < \triangle\, T\Delta K\,,$$

in which $K\Delta$ represents the perimeter of a_n. Further we have

$$p_n = \triangle\, PZ\Lambda\,.$$

Now, as above,

$$(a_n, b_n) = (\triangle\, T\Delta K,\ \triangle\, PZ\Delta)\,.$$

Since further $a_n < \triangle\, T\Delta K$, also $b_n < \triangle\, PZ\Delta = p_n$ (Euclid V, 14).

From the inequality (β) it now follows that

$$(B_n, p_n) < (B, O)$$

or

$$(B_n, B) < (p_n, O)\,.$$

Now $B_n > B$, therefore $p_n > O$, which is contrary to Prop. 12.

Note the difference between the two proofs: in Case I the surface of the polygon circumscribed *about* B is *equal* to the lateral surface of the prism circumscribed *about* the cylinder; in Case II the surface of the polygon inscribed *in* B is *less* than the lateral surface of the prism inscribed *in* the cylinder.

Proposition 14.

The surface of any isosceles cone excluding the base is equal to the circle whose radius is the mean proportional between the side of the cone and the radius of the circle which is the base of the cone.

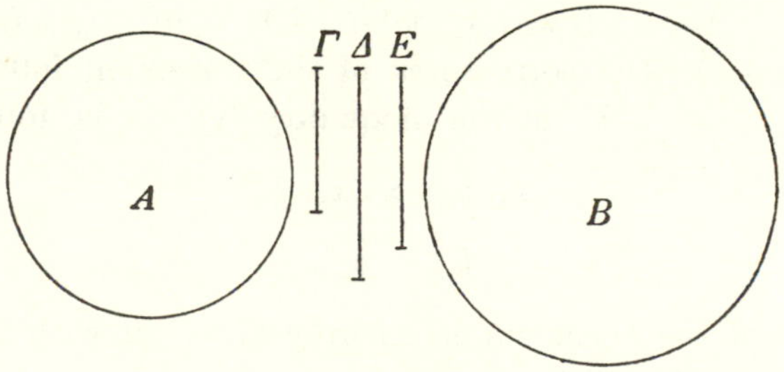

Fig. 60.

In Fig. 60 let the circle A be the base of the cone, and Γ its radius; let Δ be the side (apothem) of the cone, E the mean proportional between Γ and Δ, B the circle with radius E. It has to be proved that the curved surface O of the cone is equal to B.

If this is not true, either (I) $B < O$ or (II) $B > O$.

Case I. In the same way as described in the first part of Prop. 13, the polygons B_n and b_n about and in B, and the polygon A_n about A are found. The latter is the base of a pyramid described about the cone and having a lateral surface P_n. Now we have

$$(A_n, B_n) = [\mathbf{T}(\Gamma), \mathbf{T}(E)] = (\Gamma, \Delta) = (A_n, P_n)$$

therefore
$$B_n = P_n .$$

The rest of the argument is entirely identical with that in Prop. 13.

Case II. Proceeding in the same way as in the second part of Prop. 13, we find a polygon a_n inscribed in A, which is the base of a pyramid inscribed in the cone, whose lateral surface is p_n.

Now again
$$(a_n, b_n) = (\Gamma, \Delta) \qquad (\gamma)$$

Without any proof it is now stated that

$$(\Gamma, \Delta) > (a_n, p_n) .$$

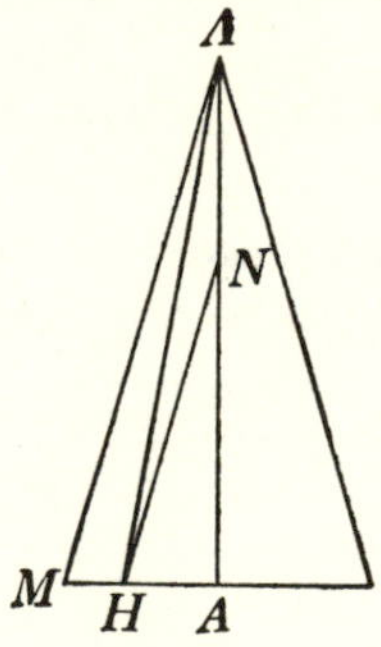

Fig. 61.

The correctness of this becomes apparent when we consider the half meridian section ΛAM of the cone (Fig. 61). If in this, H be the middle point of a side of a_n, then, if $HN \parallel M\Lambda$, we have

$$(\Gamma, \Delta) = (AH, HN) > (AH, H\Lambda) .$$

However, $(AH, H\Lambda) = (a_n, p_n)$, from which the correctness of the statement follows.

From the inequality (γ) it now follows that

$$(a_n, b_n) > (a_n, p_n)$$

therefore
$$b_n < p_n .$$

The rest of the argument is entirely identical with that in the second part of Prop. 13.

The propositions 15–20 contain applications and extensions of the propositions found. In these we will abbreviate "surface of a cone excluding the base" as **E** (Ἐπιφάνεια τοῦ κώνου).

Proposition 15.

The surface of any isosceles cone has the same ratio to its base as the side of the cone has to the radius of the base of the cone.

In fact, if B be the radius of the base of a cone, Γ the side, and E the mean proportional between B and Γ, the curved surface is equal to the circle on E as radius. The ratio of the curved surface to the base therefore is equal to the ratio of $\mathbf{T}(E)$ to $\mathbf{T}(B)$, so that, because $(\Gamma, E) = (E, B)$, it is also equal to the ratio of Γ to B.

Proposition 16.

If an isosceles cone be cut by a plane parallel to the base, the surface of the cone between the parallel planes is equal to a circle whose radius is the mean proportional between the side of the cone between the parallel planes and a straight line which is equal to the sum of the radii of the circles in the parallel planes.

In Fig. 62 let $AB\Gamma$ be a section through the axis of the cone, ΔE the intersection with the cutting plane parallel to the base. Let Θ be a circle, whose radius ϱ is the mean proportional between $A\Delta$ and $(\Delta Z + AH)$. It has to be proved that the surface of the truncated cone is equal to Θ.

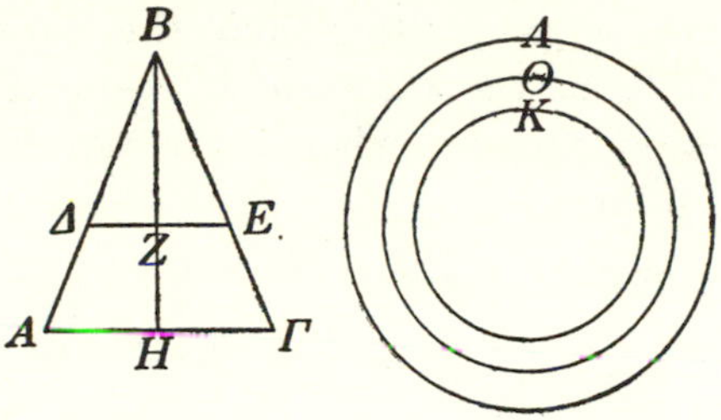

Fig. 62.

Construct

$$\text{a circle } K \text{ on } \varrho_1 \text{ as radius, so that } \mathbf{T}(\varrho_1) = \mathbf{O}(B\Delta, \Delta Z)$$

and $$\text{a circle } \Lambda \text{ on } \varrho_2 \text{ as radius, so that } \mathbf{T}(\varrho_2) = \mathbf{O}(BA, AH)\,,$$

then by prop. 14 $$\Lambda = \mathbf{E}(BA\Gamma)$$

$$K = \mathbf{E}(B\Delta E)\,.$$

Now

$$\mathbf{O}(BA, AH) = \mathbf{O}(B\Delta, \Delta Z) + \mathbf{O}(A\Delta, \Delta Z + AH) \quad (1) \quad (\textit{vide}\ \text{Note})$$

or $$\mathbf{T}(\varrho_2) = \mathbf{T}(\varrho_1) + \mathbf{T}(\varrho)\,.$$

From this it follows that

$$\Lambda = K + \Theta$$

thence, on account of the significance of Λ and K:

$$\Theta = \mathbf{E}(A\Delta E\Gamma) .$$

Note. The equality (1) becomes evident from the consideration of the gnomon figure (Fig. 63), in which BA and AH have been measured as the sides of a rectangle. We now have:

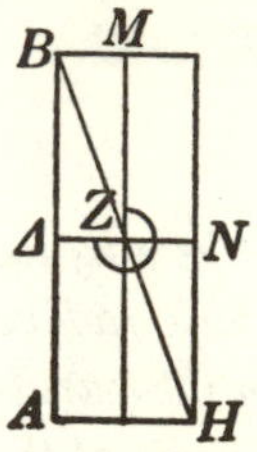

Fig. 63.

$$\mathbf{O}(BA, AH) = \mathbf{O}(B\Delta, \Delta Z) + \mathbf{\Gamma}(Z) ,$$

while $\mathbf{\Gamma}(Z) = \mathbf{O}(A\Delta, AH) + \mathbf{O}(MN)$, whence

(Euclid I, 43)

$$= \mathbf{O}(A\Delta, AH) + \mathbf{O}(AZ)$$

$$= \mathbf{O}(A\Delta, AH) + \mathbf{O}(A\Delta, \Delta Z)$$

$$= \mathbf{O}(A\Delta, \Delta Z + AH) .$$

In a proposition like the present it is quite obvious how cumbrous the argument is frequently rendered by the geometrical method, for the whole derivation can be summed up algebraically as follows:

Suppose $HA = R,\ Z\Delta = r,\ BA = S,\ B\Delta = s$,

then $R:S = r:s$ or $Rs = rS$,

and therefore $\pi RS = \pi rs + \pi(S-s)(R+r)$ [(1) but for the factor π],

from which it follows at once that

$$\Theta = \pi(R+r)(S-s) .$$

The above is followed by a number of lemmas in the form of familiar propositions on the ratios of the volumes of cones and cylinders. They are the propositions Euclid XII, 11, 14, 13, 15, 12 and a corollary of Euclid XII, 10: two cones are in the same ratio as the cylinders having equal bases and heights.

Archimedes now correlates these propositions with his own results as to the surfaces of cones and cylinders.

Proposition 17.

If in two isosceles cones the surface of one cone be equal to the base of the other, while the perpendicular from the centre of the base on the side of the (first) cone be equal to the height (of the second), the cones will be equal.

In Fig. 64 let $AB\Gamma$ and ΔEZ be sections through the axes of the two cones.

Supposition: $\mathbf{K}(B\Gamma) = \mathbf{E}(\Delta EZ)$.

$$\Theta K \perp \Delta Z \cdot AH = \Theta K .$$

What is required to be proved: cone $AB\Gamma$ = cone ΔEZ.

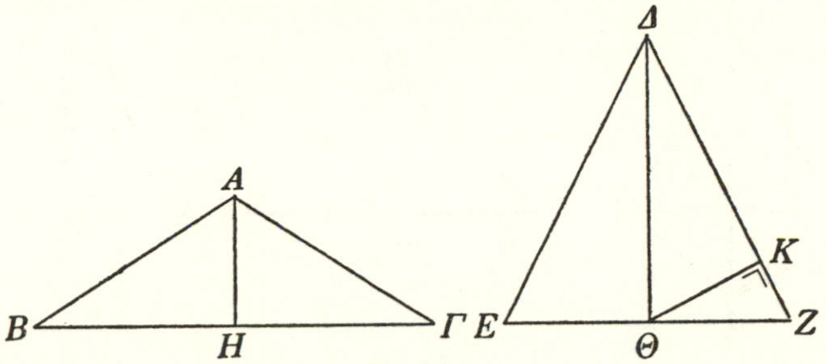

Fig. 64.

Proof: The bases of the cones $AB\Gamma$ and ΔEZ are in the same ratio as the curved surface and the base of ΔEZ, thus (Prop. 15) as AE to ΘE, *i.e.* as $\Delta\Theta$ to ΘK or as $\Delta\Theta$ to AH. The bases therefore are inversely proportional to the heights; consequently the volumes are equal (Lemma 4 preceding Prop. 17 = Euclid XII, 15).

For reasons already set forth it is not possible in Greek geometry to speak, in the strict sense, of the product of an area and a length. Hence this proposition gives an enunciation on ratios, while nowadays it would be formulated by saying that the volume of a right circular cone is equal to one-third of the product of the curved surface and the distance from the centre of the base to a generator.

Proposition 18.

Any solid rhombus consisting of isosceles cones is equal to a cone which has a base equal to the surface of one of the cones composing the rhombus and a height equal to the perpendicular drawn from the vertex of the other cone to one side of the first cone.

In Fig. 65 let the rhombus consist of the cones $AB\Gamma$ and $\Delta B\Gamma$, which have the circle on $B\Gamma$ as diameter for their common base. $\Delta Z \perp AB$. ΘHK is a cone whose base $\mathbf{K}$ (HK) is equal to the curved surface of $AB\Gamma$, and whose height $\Theta\Lambda$ is equal to ΔZ. It has to be proved that cone ΘHK = rhombus A $(B\Gamma)\Delta$.

Proof: Construct a cone $NM\Xi$ such that

$$M\Xi = B\Gamma$$

$$NO = A\Delta .$$

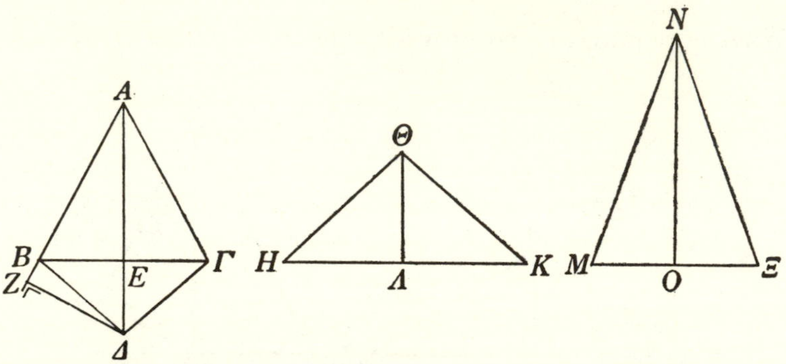

Fig. 65.

Now we have

$$(AB\Gamma, \Delta B\Gamma) = (AE, \Delta E) \qquad \text{(lemma 1 = Euclid XII, 14).}$$

Componendo (III; 0.41)

$$[A(B\Gamma)\Delta, \Delta B\Gamma] = (A\Delta, \Delta E)$$

Also

$$(\Delta B\Gamma, NM\Xi) = (\Delta E, NO) \qquad \text{(lemma 1 = Euclid XII, 14)}$$

Therefore, *ex aequali* (III; 0.45)

$[A(B\Gamma)\Delta, NM\Xi] = (A\Delta, NO)$ and, because $A\Delta = NO$,

$$A(B\Gamma)\Delta = NM\Xi \qquad (\alpha)$$

Further

$$[\mathbf{K}(HK), \mathbf{K}(M\Xi)] = [\mathbf{E}(AB\Gamma), \mathbf{K}(B\Gamma)]$$

$$= (AB, BE) = (A\Delta, \Delta Z) = (NO, \Theta\Lambda) .$$

Consequently (lemma 4 = Euclid XII, 15)

$$\Theta HK = NM\Xi$$

and, because of (α),

$$\Theta HK = A(B\Gamma)\Delta .$$

Algebraically: If $BE = R$, $A\Delta = h$, $AB = s$, $\Delta Z = p$, then:

Volume of rhombus $A(B\Gamma)\Delta = \frac{1}{3}\pi R^2 h = \frac{1}{3}\pi Rsp = \frac{1}{3}p\cdot$curved surface of cone $AB\Gamma$.

If we assume $Rs = \varrho^2$, the volume of the rhombus is found to be equal to that of the right circular cone with ϱ for the radius of the base and p for height.

Proposition 19.

If an isosceles cone be cut by a plane parallel to the base, if on the resulting circle a cone be described which has the centre of the base for its vertex, and if the rhombus so formed be taken away from the whole cone, the part remaining will be equal to a cone which has a base equal to the surface of the cone between the parallel planes, and a height equal to the perpendicular drawn from the centre of the base to one side of the cone.

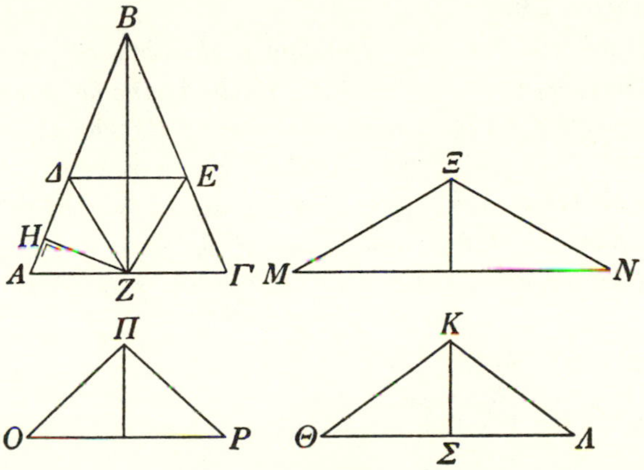

Fig. 66.

In Fig. 66 let $BA\Gamma$ be the section through the axis of the given cone, ΔE the intersection with the cutting plane, $ZH \perp AB$, $K\Theta\Lambda$ the section through the axis of a cone such that

$$\mathbf{K}(\Theta\Lambda) = \mathbf{E}(A\Delta E\Gamma) \quad \text{and} \quad K\Sigma = ZH\,.$$

It has to be proved that

$$\text{Cone } BA\Gamma - \text{Rhombus } B(\Delta E)Z = \text{Cone } K\Theta\Lambda\,.$$

Proof:

Construct a cone ΞMN with $\mathbf{K}(MN) = \mathbf{E}(AB\Gamma)$ and height $= ZH$, and a cone ΠOP with $\mathbf{K}(OP) = \mathbf{E}(B\Delta E)$ and height $= ZH$.

Then (Prop. 17)

$$\Xi MN = BA\Gamma$$

and (Prop. 18)

$$\Pi OP = B(\Delta E)Z .$$

Moreover (lemma 1 = Euclid XII, 11)

$$\Xi MN = K\Theta\Lambda + \Pi OP .$$

Therefore

$$K\Theta\Lambda = \Xi MN - \Pi OP = BA\Gamma - B(\Delta E)Z .$$

Algebraically: If $ZH = p$, then

$$\text{Cone } BA\Gamma - \text{Rhombus } B(\Delta E)Z = \tfrac{1}{3}p[\mathbf{E}(BA\Gamma) - \mathbf{E}(B\Delta E)]$$
$$= \tfrac{1}{3}p . \mathbf{E}(A\Delta E\Gamma) .$$

Proposition 20.

If one of the isosceles cones forming a rhombus be cut by a plane parallel to the base, if on the resulting circle a cone be described which has the same vertex as the other cone, and if the rhombus so formed be taken away from the whole rhombus, the part remaining will be equal to a cone which has a base equal to the surface of the first cone between the parallel planes, and a height equal to the perpendicular drawn from the vertex of the second cone to the side of the first cone.

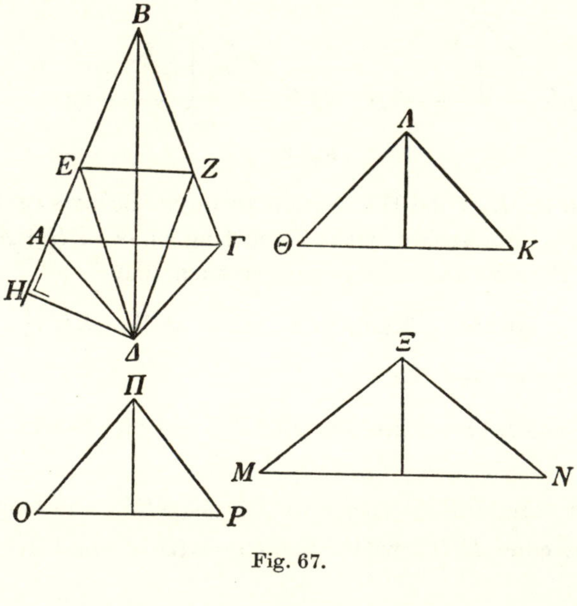

Fig. 67.

In Fig. 67 let $B(A\Gamma)\Delta$ be the section through the axis of the given rhombus, EZ the intersection with the cutting plane, $B(EZ)\Delta$ the section through the axis of the constructed rhombus. $\Delta H \perp BA$. If now $\Lambda\Theta K$ be a cone with base $\mathbf{K}(\Theta K)=\mathbf{E}(AEZ\Gamma)$ and height $=\Delta H$, it has to be proved that

$$B(A\Gamma)\Delta - B(EZ)\Delta = \Lambda\Theta K .$$

Proof: Construct a cone ΞMN with base $\mathbf{K}(MN)=\mathbf{E}(BA\Gamma)$ and height $=\Delta H$,

and a cone ΠOP with base $\mathbf{K}(OP)=\mathbf{E}(BEZ)$ and height $=\Delta H$,

then (Prop. 18)

$$\Xi MN = B(A\Gamma)\Delta$$

$$\Pi OP = B(EZ)\Delta ,$$

while also

$$\Xi MN = \Lambda\Theta K + \Pi OP .$$

From this it follows that

$$\Lambda\Theta K = B(A\Gamma)\Delta - B(EZ)\Delta .$$

Algebraically: If $\Delta H = p$, then

$$\text{Rhombus } B(A\Gamma)\Delta - \text{Rhombus } B(EZ)\Delta$$

$$= \tfrac{1}{3}p[\mathbf{E}(BA\Gamma) - \mathbf{E}(BEZ)] = \tfrac{1}{3}p . \mathbf{E}(AEZ\Gamma) .$$

The propositions 18–20 are together equivalent to the proposition at present in general use in elementary solid geometry, according to which the volume of the solid resulting when a triangle revolves about an axis in its own plane through one of its angular points is equal to one-third of the product of the surface described by the side opposite the angular point on the axis and the perpendicular drawn from that angular point to that side.

6. *Surface and Volume of the Sphere. Propositions* 21–34.

Prop. 21 starts a new group of propositions leading to the derivation of theorems on the surface (Prop. 33) and the volume (Prop.

34) of the sphere. The proper understanding of this group is rendered difficult by its inconvenient arrangement. We shall therefore begin by giving a broad outline of the trend of the argument.

The basic idea consists in that the surface and the volume of the sphere are compared with the surfaces and the volumes respectively of the solids resulting when regular polygons which are inscribed in a great circle of the sphere and the number of whose sides is a multiple of four are caused to revolve about a diagonal which is a diameter of the sphere. As is set forth more in detail in Prop. 23, the solid obtained by the revolution of the inscribed polygon I_n is included between portions of the curved surfaces of cones the bounding circles of which lie in parallel planes on the surface of the sphere. As is stated in Prop. 28, upon revolution of the circumscribed polygon C_n a similar solid is formed, the bounding portions of the conical surfaces, however, in this case touching the sphere in circles lying in parallel planes, while the whole of the solid, in the manner of Prop. 23, is inscribed in a sphere which is concentric with the given sphere and has a greater radius than the latter. Henceforth we will call the surfaces of the solids described by I_n and C_n, respectively, $\mathbf{E}(I_n)$ and $\mathbf{E}(C_n)$, the surface of the given sphere itself **E**, the volumes of the solids described by I_n and C_n —as also these solids themselves—$\mathbf{S}(I_n)$ and $\mathbf{S}(C_n)$, the volume of the sphere and the sphere itself: **S**[1]).

We shall first outline the argument leading up to the finding of the surface of the sphere.

To the conical surfaces described by the sides of I_n upon revolution, the propositions 14 (area of the curved surface of a cone) and 16 (area of the curved surface of a truncated cone) are applicable. With the aid of these propositions, in Prop. 24 an expression for $\mathbf{E}(I_n)$ is derived which, reduced in Prop. 25 on the ground of a planimetrical lemma found in Prop. 21, produces the result that $\mathbf{E}(I_n)$ is less than the surface of a circle **A** which has the diameter of the sphere for radius and thus is four times as great as a great circle of the sphere. Since further $\mathbf{E}(C_n)$ is in turn inscribed in a sphere (with a greater radius), the result obtained may also be applied to the latter. This is done in Prop. 29. In Prop. 30 it is derived from this that $\mathbf{E}(C_n)$ is greater than **A**.

---

1) **E** of ἐπιφάνεια. **S** of στερεόν.

Meanwhile, by application of the fourth postulate it has been realized in Prop. 23 that $\mathbf{E}(I_n)$ is less than the surface $\mathbf{E}$ of the sphere, and also in Prop. 28 that $\mathbf{E}(C_n)$ is greater than $\mathbf{E}$. The results obtained may be summed up as follows:

$$\mathbf{E}(I_n) < \mathbf{A} < \mathbf{E}(C_n)$$

$$\mathbf{E}(I_n) < \mathbf{E} < \mathbf{E}(C_n) .$$

After this, in Prop. 33 the main theorem

$$\mathbf{E} = \mathbf{A}$$

is proved by the compression method (ratio form; III; 8.21) by means of a double *reductio ad absurdum*. The lemma used for this is the result derived in Prop. 32, *viz.* that the ratio of $\mathbf{E}(I_n)$ to $\mathbf{E}(C_n)$ is the duplicate ratio of the sides z_n and Z_n of the polygons I_n and C_n.

In order to find the volume of the sphere, the following line of thought is pursued: In Prop. 26 it is found with the aid of Propositions 18 and 20 that the volume $\mathbf{S}(I_n)$ is equal to that of a cone whose base is equal to $\mathbf{E}(I_n)$ and whose height is equal to the radius R of the sphere. From this it follows in Prop. 27, on the strength of the Prop. 25 already referred to above, that $\mathbf{S}(I_n)$ is less than a cone $\mathbf{X}$ whose base is equal to $\mathbf{E}$ and whose height is equal to R, and in Prop. 31 similarly that $\mathbf{S}(C_n)$ is greater than that cone. Without its being mentioned, it is further assumed that $\mathbf{S}(I_n)$ is less than the sphere itself and $\mathbf{S}(C_n)$ greater than the latter. We therefore have the inequalities

$$\mathbf{S}(I_n) < \mathbf{X} < \mathbf{S}(C_n)$$

$$\mathbf{S}(I_n) < \mathbf{S} < \mathbf{S}(C_n) ,$$

from which in Prop. 34 the main proposition

$$\mathbf{S} = \mathbf{X}$$

is derived by a double *reductio ad absurdum* (cf. III; 8.21). For this, use is made of the result derived in Prop. 32, *viz.* that the ratio of $\mathbf{S}(I_n)$ to $\mathbf{S}(C_n)$ is the triplicate ratio of the sides z_n and Z_n of the polygons I_n and C_n.

We can now, without incurring any risk of obscurity, state the propositions in the order in which Archimedes gives them; it merely has to be noted that Prop. 22 will only be applied in the theory of the sector of a sphere, which begins with Prop. 35.

Proposition 21.

If in a circle a polygon be inscribed with an even number of equal sides and straight lines be drawn which join the angular points of the polygon so that they are parallel to any one of [the straight lines] which subtend two sides of the polygon, all the connecting lines [together] have to the diameter of the circle the same ratio as the straight line subtending half of the sides but one has to the side of the polygon.

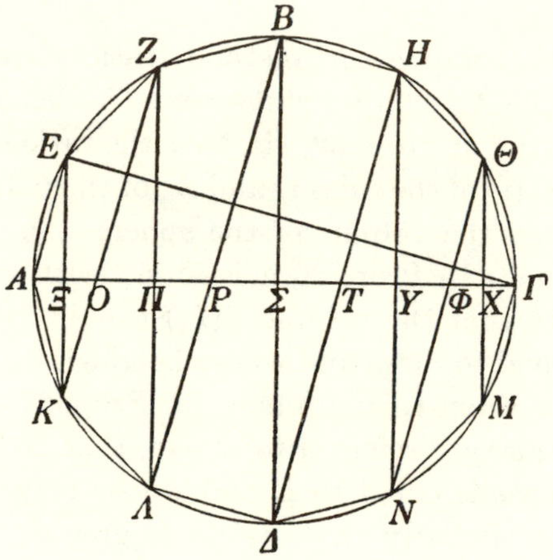

Fig. 68.

If in Fig. 68, $AEZ\ldots K$ be the given regular polygon in the circle, and if EK, $Z\Lambda$, etc. be the parallel chords in question (which meet the diameter $A\Gamma$ successively in Ξ, Π, etc.), it has to be proved that

$$(EK + Z\Lambda + B\Delta \ldots + \Theta M, A\Gamma) = (E\Gamma, AE)\,.$$

Proof: If the parallel chords KZ, ΛB, etc. be drawn, which meet the diameter $A\Gamma$ successively in O, P, etc., then because of similarity of triangles we have

$$(E\Xi, \Xi A) = (K\Xi, \Xi O) = \ldots = (MX, X\Gamma)$$

therefore

$$(E\Xi + K\Xi + \ldots + MX, A\Xi + \Xi O + \ldots + X\Gamma) = (E\Xi, \Xi A) = (E\Gamma, AE)$$

or
$$(EK + Z\Lambda + \ldots + \Theta M, A\Gamma) = (E\Gamma, AE)\,.$$

Goniometrical notation: If the number of sides of the polygon be $2n$, then

$$EK = 2R\sin\frac{\pi}{n},\ Z\Lambda = 2R\sin\frac{2\pi}{n},\ \ldots\Theta M = 2R\sin\frac{(n-1)\pi}{n}$$

while

$$\frac{E\Gamma}{AE} = \cot\frac{\pi}{2n}.$$

The proposition proved is therefore

$$\sin\frac{\pi}{n} + \sin\frac{2\pi}{n} + \ldots + \sin\frac{(n-1)\pi}{n} = \cot\frac{\pi}{2n}.$$

Proposition 22.

If in a segment of a circle a polygon be inscribed such that all its sides excluding the base are equal and their number even, and straight lines be drawn parallel to the base which join the angular points of the polygon, all the straight lines so drawn and half of the base [together] have to the height of the segment the same ratio as the straight line drawn from the diameter of the circle to the side of the polygon has to the side of the polygon.

The formulation of the last part of the proposition is obscure; what is meant is the straight line drawn from the extremity Δ of the diameter $B\Delta$ to the extremity Z of the side BZ.

If in Fig. 69, $AB\Gamma$ be the given segment, it has to be proved that

$$(ZH + E\Theta + A\Xi, B\Xi) = (\Delta Z, ZB).$$

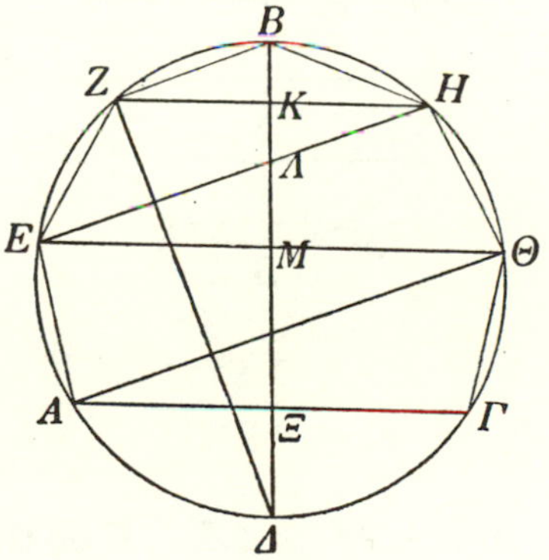

Fig. 69.

The proof proceeds entirely as in Prop. 21. In goniometrical notation the proposition is as follows:

If arc $AB\Gamma$ be equal to 2α and the number of sides of the portion $A\ldots\Gamma$ of the polygon be $2n$, then

$$\frac{2\left[\sin\frac{\alpha}{n} + \sin\frac{2\alpha}{n} + \ldots \sin\frac{n-1}{n}\alpha\right] + \sin\alpha}{1-\cos\alpha} = \cot\frac{\alpha}{2n},$$

which equality in the case of $a = \pi$ passes into that of Prop. 21.

Proposition 23.

In this proposition, whose form differs from the usual type because no definite theorem is formulated, but the desired conclusion is reached in the course of the argument, it is stated (Fig. 68) that the surface of the solid produced by the revolution of the inscribed polygon $A \ldots B \ldots \Gamma \ldots \Delta$ about the diameter $A\Gamma$ is less than the surface of the sphere. This follows at once from the fourth postulate, applied successively to the surfaces of a zone (or segment) of a sphere and a truncated cone (or cone) which have the same circle for their boundary.

Proposition 24.

The surface of the solid inscribed in the sphere is equal to a circle the square on whose radius is equal to the rectangle contained by the side of the (revolving) figure and a straight line equal to [the sum of] all the connecting lines which are parallel to the straight line subtending two sides of the polygon.

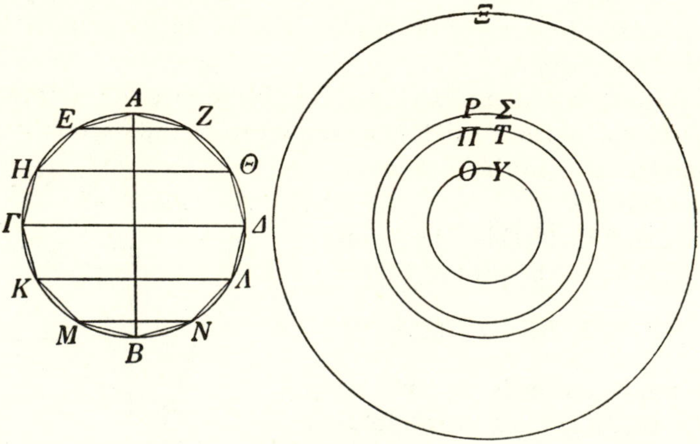

Fig. 70.

In Fig. 70 let $AZ \ldots E$ be the polygon I_n revolving about AB. If Ξ be the circle referred to in the proposition, it has to be proved that

$$\mathbf{E}(I_n) = \Xi .$$

Proof: Construct circles O, Π, R, Σ, T, Y such that the squares described on their radii as sides are successively equal to the rectangles

$$\mathbf{O}(AE, \tfrac{1}{2}EZ),\ \mathbf{O}\left(AE, \frac{EZ+H\Theta}{2}\right)\ldots\mathbf{O}(AE, \tfrac{1}{2}MN)\,,$$

then by Props 14 and 16 these circles represent the surfaces described successively by AE, $EH\ldots MB$. The sum of these squares is equal to the rectangle

$$\mathbf{O}(AE, EZ+H\Theta+\ldots+MN)$$

and consequently to the square described on the radius of Ξ as side. From this it follows that

$$\Xi = O+\Pi+P+\Sigma+T+Y$$

or
$$\mathbf{E}(I_n) = \Xi\,.$$

Algebraically:

$$\mathbf{E}(I_n) = \pi.AE\frac{EZ}{2}+\pi.EH.\frac{EZ+H\Theta}{2}+\ldots+\pi.MB\frac{MN}{2}$$
$$= \pi.AE(EZ+H\Theta+\ldots+MN)\,.$$

Proposition 25.

The surface of the solid inscribed in the sphere, which is contained by the conical surfaces, is less than four times the greatest circle of the sphere.

Proof: In Fig. 68 let P be a circle the square on whose radius is equal to

$$\mathbf{O}(AE, EK+Z\Lambda+\ldots\Theta M)\,,$$

and $\mathbf{K}$ a circle on $A\Gamma$ as diameter,
then by Prop. 24

$$\mathbf{P} = \mathbf{E}(I_n)\,.$$

By Prop. 21 (correlated with Euclid VI, 16), however, we have

$$\mathbf{O}(AE, EK+Z\Lambda+\ldots\Theta M) = \mathbf{O}(A\Gamma, \Gamma E) < \mathbf{T}(A\Gamma)\,,$$

therefore
$$\mathbf{P} < 4.\mathbf{K}\,.$$

Algebraically:

$$\mathbf{E}(I_n) = \pi.AE(EK+Z\Lambda+\ldots+\Theta M)$$
$$= \text{(Prop. 21) } \pi.A\Gamma.\Gamma E < \pi.A\Gamma^2 = 4\pi.\Sigma A^2\,.$$

Proposition 26.

The solid inscribed in the sphere, which is contained by the conical surfaces, is equal to a cone whose base is a circle equal to the surface of the solid inscribed in the sphere and whose height is equal to the perpendicular drawn from the centre of the sphere to one side of the polygon.

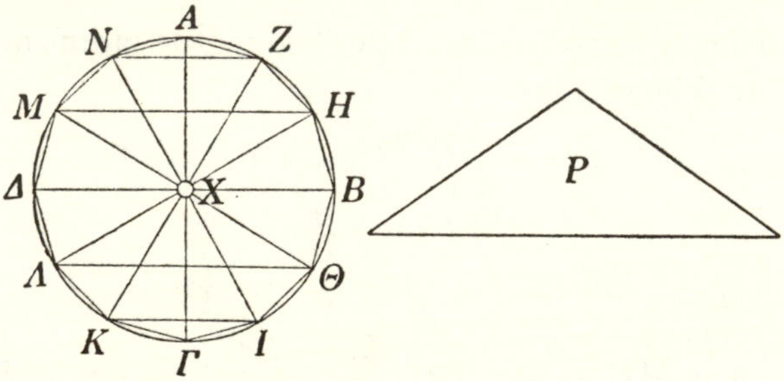

Fig. 71.

Proof (Fig. 71): Let P be the cone whose base is equal to $\mathbf{E}(I_n)$ and whose height is equal to the perpendicular r from X to AN.

The solid is now conceived to be composed of:

1) Rhombus $X(NZ)A$, which by Prop. 18 is equal to a cone with base $\mathbf{E}(AZ)$ and height r.

2) Rest of rhombus with vertex X and base $\mathbf{E}(ZH)$, which by Prop. 20 is equal to a cone with base $\mathbf{E}(MNHZ)$ and height r.

3) Rest of cone with vertex X and base $\mathbf{E}(M\Delta HB)$, which by Prop. 19 is equal to a cone with base $\mathbf{E}(M\Delta HB)$ and height r.

The same applies to the other hemisphere.

By addition we arrive at the proposition.

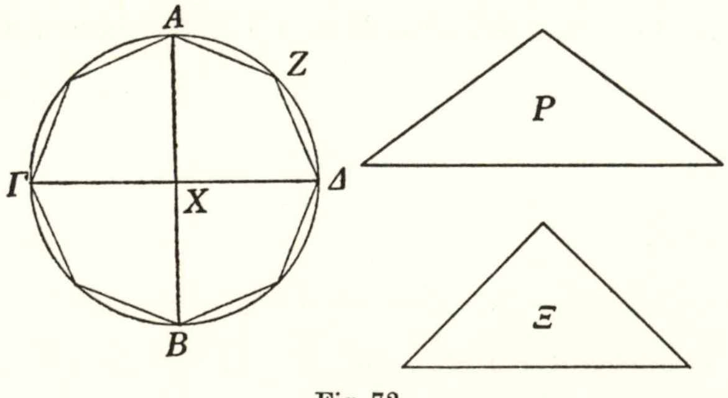

Fig. 72.

Proposition 27.

The solid inscribed in the sphere, which is contained by the conical surfaces, is less than four times the cone whose base is equal to the greatest circle of the sphere and whose height is equal to the radius of the sphere.

Proof (Fig. 72): Let P be the cone with base $\mathbf{E}(I_n)$ and height r, Ξ a cone which has the circle X for base and a height equal to XA.

Since $$\mathbf{E}(I_n) < 4.X \quad \text{and} \quad r < R$$

we have $$P < 4.\Xi,$$

whence $$\mathbf{S}(I_n) < 4.\Xi.$$

Proposition 28.

In this proposition (Fig. 73), which like Prop. 23 is differently formulated, a polygon of $4n$ sides (C_n) is considered, which has been circumscribed about a great circle $AB\Gamma\Delta$ of a sphere and revolves with this circle about EH. C_n generates a solid circumscribed about the sphere. With regard to this solid it is stated that

$$\mathbf{E}(C_n) > \mathbf{E}.$$

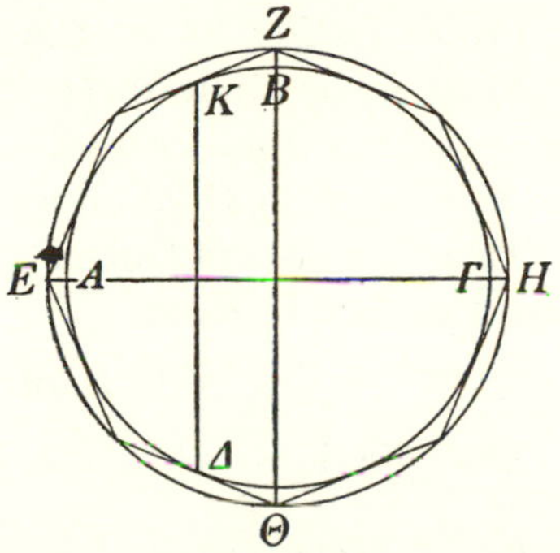

Fig. 73.

This follows at once from the fourth postulate, applied to the parts of the solid and the sphere on either side of the plane of the circle on $K\Delta$ as diameter.

Proposition 29.

The surface of the solid circumscribed about the sphere is equal to a circle the square on whose radius is equal to the rectangle contained by one side of the polygon and a straight line equal to ⟨the sum of⟩ all the lines joining the angular points of the polygon, parallel to one of the lines subtending two sides of the polygon.

Since C_n in turn is inscribed in another circle, this proposition is identical with Prop. 24.

Proposition 30.

The surface of the solid circumscribed about the sphere is greater than four times the greatest circle of the sphere.

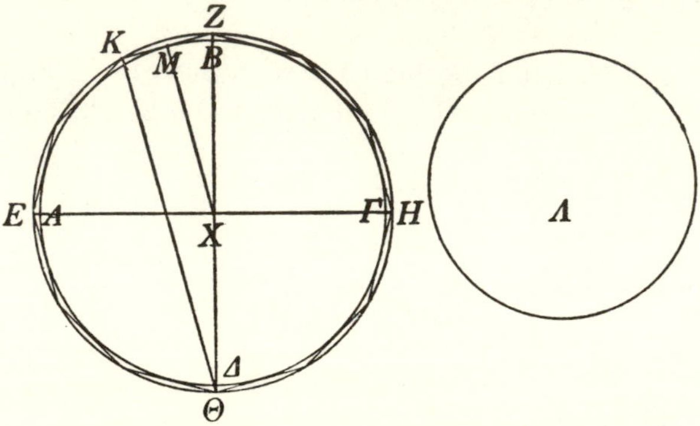

Fig. 74.

Proof: In Fig. 74 let the surface under consideration be equal to that of the circle Λ. In entirely the same way as in Prop. 25 it then follows from Props 21 and 29 that the square on the radius of Λ is equal to $\mathbf{O}(Z\Theta, \Theta K)$.

Therefore the radius of Λ is greater than ΘK, which itself is equal to $B\Delta$. Thence

$$\Lambda > 4.\text{surface of a great circle.}$$

That ΘK is equal to $B\Delta$, follows from $\Theta K = 2XM$.

Proposition 31.

The solid circumscribed about the smallest sphere is equal to a cone whose base is a circle which is equal to the surface of the solid and whose height is equal to the radius of the sphere.

This follows at once from Prop. 26. In a Porism it is concluded, with the aid of Prop. 30, that

$$\mathbf{S}(C_n) > \mathbf{X},$$

in which, by agreement, **X** represents a cone whose base is equal to the surface **E** of the sphere and whose height is equal to the radius.

Proposition 32.

If in a sphere a solid be inscribed and another be circumscribed about it, and the two solids have been generated as above by similar polygons, the surface of the circumscribed solid is to the surface of the inscribed solid in the duplicate ratio of the side of the polygon circumscribed about the greatest circle to the side of the polygon inscribed in the same circle, and the ⟨circumscribed⟩ solid itself is to the ⟨inscribed⟩ solid in the triplicate ratio of the same sides.

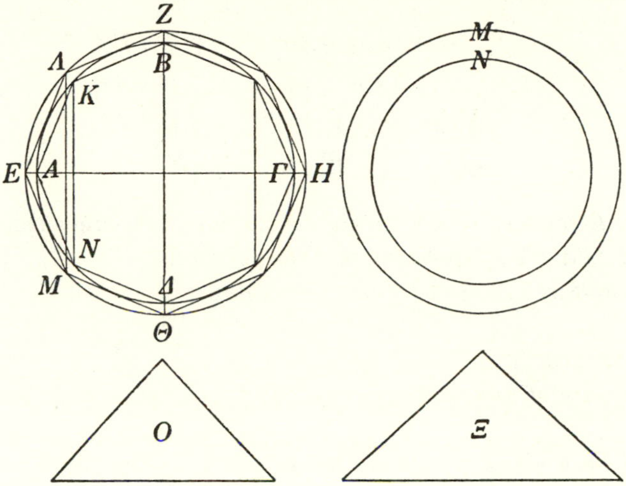

Fig. 75.

Proof (Fig. 75): 1st Part.

Suppose $$M = \mathbf{E}(C_n)$$

$$N = \mathbf{E}(I_n),$$

then

the square on the radius of $M = \mathbf{O}(E\Lambda, \Lambda M + Z\Theta + \ldots)$
the square on the radius of $N = \mathbf{O}(AK, KN + B\Delta + \ldots)$.

Because the polygons are similar, the rectangles **O** are similar as well, and thus, because the ratio (M, N) is equal to that of the squares on their radii, and consequently to that of the said rectangles, the ratio in question according to Euclid VI, 20 is the duplicate ratio of $(E\Lambda, AK)$.

2nd Part. Now let Ξ and O be two cones, whose bases are equal to M and to N respectively, and whose heights are equal respectively to the radius of the sphere and to the perpendicular drawn from the centre to AK; then the circumscribed solid is equal to Ξ

(Prop. 31), the inscribed solid to O (Prop. 26). Since the polygons circumscribed about and inscribed in the circle are similar, the ratio of Z_n to z_n is the same as that of the heights of the cones to the diameters of their bases. According to Euclid XII, 12 the ratio of the cones therefore is the triplicate ratio of $(E\Lambda, AK)$.

Algebraically:

$$\frac{\mathbf{E}(C_n)}{\mathbf{E}(I_n)} = \frac{E\Lambda.[\Lambda M + Z\Theta + \ldots]}{AK.[KN + B\Delta + \ldots]} = \left(\frac{E\Lambda}{AK}\right)^2$$

$$\frac{\mathbf{S}(C_n)}{\mathbf{S}(I_n)} = \frac{R.\mathbf{E}(C_n)}{r.\mathbf{E}(I_n)} = \left(\frac{E\Lambda}{AK}\right)^3,$$

in which R represents the radius of the circle circumscribed about $AKB\ldots$ and r the apothem of this polygon.

The above is followed by the two main propositions:

Propositon 33.

The surface of any sphere is equal to four times that of its greatest circle.

Let $\mathbf{A}$ be a circle whose radius is equal to the diameter of the sphere. It has to be proved that

$$\mathbf{E} = \mathbf{A}.$$

If this is not true, then either (I) $\mathbf{A} < \mathbf{E}$ or (II) $\mathbf{A} > \mathbf{E}$.

Case I. Suppose $\mathbf{A}$ less than $\mathbf{E}$. Construct two line segments B, Γ $(B > \Gamma)$ such that

$$(B, \Gamma) < (\mathbf{E}, \mathbf{A}) \qquad \text{(Prop. 2)}$$

and a line segment Δ such that

$$(B, \Delta) = (\Delta, \Gamma).$$

Find n so that if Z_n and z_n be the sides of C_n and I_n

$$(Z_n, z_n) < (B, \Delta) \qquad \text{(Prop. 3)}$$

Then

$$[\mathbf{E}(C_n), \mathbf{E}(I_n)] < \Delta\Lambda(B, \Delta) = (B, \Gamma) \qquad \text{(Prop. 32);}$$

therefore

$$[\mathbf{E}(C_n), \mathbf{E}(I_n)] < (\mathbf{E}, \mathbf{A}) \qquad (\alpha)$$

However,

$$\mathbf{E}(C_n) > \mathbf{E} \qquad \text{(Prop. 28)}$$

and

$$\mathbf{E}(I_n) < \mathbf{A}, \qquad \text{(Prop. 25)}$$

which is contrary to the inequality (α). In fact, the latter may be written

$$[\mathbf{E}(C_n), \mathbf{E}] < [\mathbf{E}(I_n), \mathbf{A}]$$

and since $\mathbf{E}(I_n)$ is less than $\mathbf{A}$, *a fortiori* $\mathbf{E}(C_n)$ would have to be less than $\mathbf{E}$.

Case II. Suppose $\mathbf{A}$ greater than $\mathbf{E}$.

Proceeding as above, n is found such that

$$[\mathbf{E}(C_n), \mathbf{E}(I_n)] < (\mathbf{A}, \mathbf{E}). \qquad (\beta)$$

However, $\mathbf{E}(C_n) > \mathbf{A}$ (Prop. 30) and $\mathbf{E}(I_n) < \mathbf{E}$ (Prop. 23), from which the impossibility of the inequality (β) becomes evident.

Proposition 34.

Any sphere is equal to four times the cone whose base is equal to the greatest circle of the sphere, and whose height is equal to the radius of the sphere.

Let Ξ be the cone in question, then it has to be proved that

$$\mathbf{S} = 4.\Xi$$

or, according to the notation adopted above,

$$\mathbf{S} = \mathbf{X}.$$

If this is not true, either (I) $\mathbf{X} < \mathbf{S}$ or (II) $\mathbf{X} > \mathbf{S}$.

Case I. Suppose $\mathbf{X}$ less than $\mathbf{S}$.

Construct two line segments K, H ($K > H$) such that

$$(K, H) < (\mathbf{S}, \mathbf{X}) \qquad \text{(Prop. 2)}$$

and then two line segments I, Θ such that K, I, Θ, H form an arithmetical progression. Now find n so that

$$(Z_n, z_n) < (K, I), \qquad \text{(Prop. 3)}$$

then by Prop. 32

$$[\mathbf{S}(C_n), \mathbf{S}(I_n)] = \mathbf{T\Lambda}(Z_n, z_n) < \mathbf{T\Lambda}(K, I) < (K, H)\,[1].$$

[1]) The correctness of the latter inequality is proved as follows: Conceive a descending arithmetical progression K, I, Θ, H, and form a geometrical progression K, I, Λ, M. Then $\Lambda > \Theta$, because the geometric mean of K

Now therefore

$$[\mathbf{S}(C_n),\ \mathbf{S}(I_n)] < (\mathbf{S}, \mathbf{X}). \qquad (\alpha)$$

However,

$$\mathbf{S}(C_n) > \mathbf{S} \qquad \text{(Post. 4)}$$

$$\mathbf{S}(I_n) < \mathbf{X}, \qquad \text{(Prop. 27)}$$

which is contrary to the inequality (α).

Case II. Suppose **X** greater than **S**. As above, we find

$$[\mathbf{S}(C_n),\ \mathbf{S}(I_n)] < (\mathbf{S}, \mathbf{X}) \qquad (\beta)$$

However,

$$\mathbf{S}(C_n) > \mathbf{X} \qquad \text{(Prop. 31)}$$

$$\mathbf{S}(I_n) < \mathbf{S}, \qquad \text{(Post. 4)}$$

which is contrary to the inequality (β).

Algebraically:

$$\mathbf{S} = 4.\tfrac{1}{3}\pi R^2.R\,.$$

In a Porism the property is enunciated which Archimedes wished to have represented on his sepulchral column[1]): the total surface of a cylinder which has a great circle of the sphere for its base and whose height is equal to the diameter is equal to one and a half times the surface of the sphere, and its volume is equal to one and a half times the volume of the sphere.

Algebraically:

a) $\quad 2\pi R.2R + 2\pi R^2 = 6\pi R^2 = \tfrac{3}{2}\mathbf{E}$

b) $\quad \pi R^2.2R = 2\pi R^3 = \tfrac{3}{2}.\tfrac{4}{3}\pi R^3 = \tfrac{3}{2}\mathbf{S}\,.$

7. *Surface of a Segment of a Sphere and Volume of a Sector of a Sphere. Propositions* 35–44.

We are now coming to a group of propositions from Book I of the treatise *On the Sphere and Cylinder*, in which theorems are derived on the surface of a segment of a sphere and on the volume of a sector of a sphere. The sequence of the demonstration is mainly

---

and Λ is equal to the arithmetic mean of K and Θ. Further $I - \Lambda > \Lambda - M$, so that certainly $I - \Theta > \Lambda - M$; therefore $\Theta - H > \Lambda - M$, and, because $\Theta < \Lambda$, certainly $H < M$. Now

$$(K, H) > (K, M) = \mathbf{T\Delta}(K, I)\,.$$

[1]) *Vide* page 32.

parallel to that of the previous group, in which the surface and the volume of the sphere were dealt with. Instead of complete in- and circumscribed polygons, however, it is now parts of polygons which are being considered, inscribed in and circumscribed about the arc of the sector of the circle by whose revolution the sector of the sphere to be dealt with is generated. Since the wording of the propositions now to be discussed differs only slightly from that of the previous group, we shall not mention them *in extenso*, but sum up the whole of the argument in a comprehensive discussion.

On the analogy of Prop. 24 an expression is first derived in Prop. 35 for the surface $\mathbf{E}(I_n)$ of the solid generated by the revolution of the portion of the inscribed polygon I_n (Fig. 76). On the ground of Props 14 and 16 we find that the surfaces described by $\Theta E, E\Gamma, \Gamma A$ are equal successively to circles the squares on whose radii are equal successively to

$$\mathbf{O}(E\Theta, \tfrac{1}{2}EZ),\ \mathbf{O}\left(E\Gamma, \frac{EZ+\Gamma\Lambda}{2}\right),$$

$$\mathbf{O}\left(A\Gamma, \frac{\Gamma\Lambda+AH}{2}\right),$$

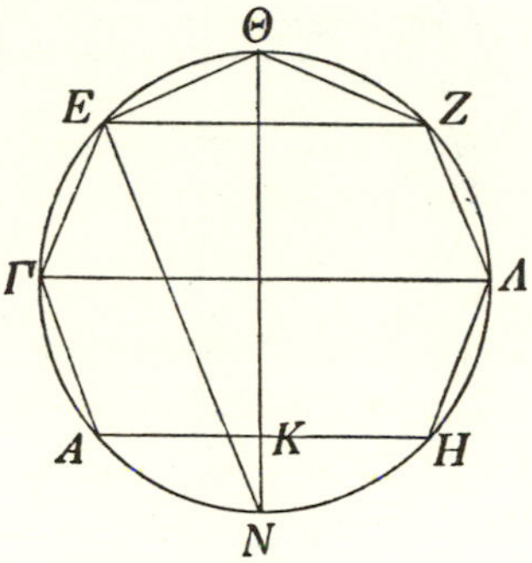

Fig. 76.

from which it follows that $\mathbf{E}(I_n)$ is equal to a circle the square on whose radius is

$$\mathbf{O}(\Theta E, EZ+\Gamma\Lambda+\tfrac{1}{2}AH)\,.$$

Now by Prop. 22

$$(EZ+\Gamma\Lambda+\tfrac{1}{2}AH, \Theta K) = (NE, \Theta E)\,,$$

so that

$$\mathbf{O}(\Theta E, EZ+\Gamma\Lambda+\tfrac{1}{2}AH) = \mathbf{O}(\Theta K, NE)\,.$$

Further

$$\mathbf{O}(\Theta K, NE) < \mathbf{O}(\Theta K, \Theta N) = \mathbf{T}(\Theta A)\,,$$

so that the surface $\mathbf{E}(I_n)$ is found to be less than the circle on $A\Theta$ as radius. This result is reached in Prop. 37. We shall henceforth represent the circle on $A\Theta$ as radius by **K**. Prop. 37 apparently corresponds to Prop. 25. The object is again to compress the surface **E** of the segment between the same boundaries as the circle **K**, to which it will be found equal. To this end it is first recognized in

Prop. 36, on the strength of the fourth postulate, that the surface $\mathbf{E}(I_n)$ is also less than the surface **E** of the segment of the sphere. Then **K** and **E** both have to be compared with the surface $\mathbf{E}(C_n)$ described by the revolution of a portion of a circumscribed polygon. The comparison of $\mathbf{E}(C_n)$ and **E** takes place in Prop. 39. Here (Fig. 77) the portion $Z\Theta \ldots H$ of the polygon is considered, the angular points of which lie on the greater of the two concentric circles Δ, while the sides touch the lesser. AM and BN touch the lesser

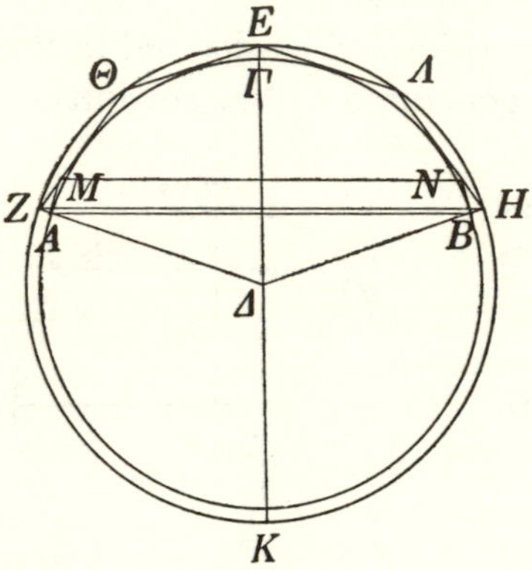

Fig. 77.

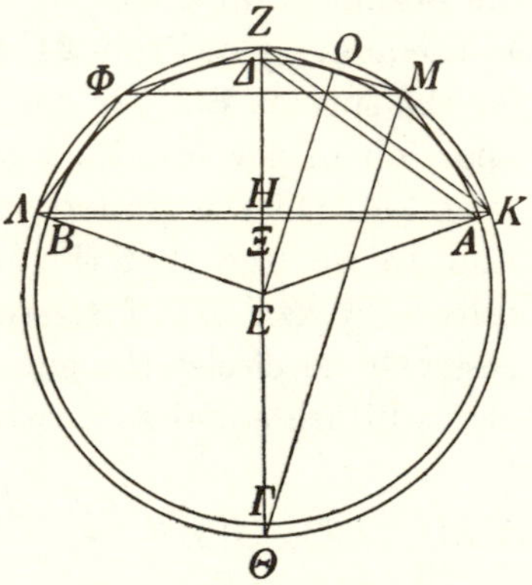

Fig. 78.

circle successively in A and B. On account of the fourth postulate the surface described by $AM\Theta E \ldots NB$ is greater than the surface **E** of the segment generated by the revolution of the arc $A\Gamma B$. *A fortiori*, therefore, the surface described by $ZM \ldots E \ldots NH$ is greater than **E**. In fact, if we compare the conical surfaces described by ZM and AM, these are equal successively to two circles the squares on whose radii are respectively

$$\mathbf{O}\left(ZM, \frac{ZH+MN}{2}\right) \quad \text{and} \quad \mathbf{O}\left(AM, \frac{AB+MN}{2}\right).$$

Since $ZM > AM$ and $ZH > AB$, the former surface is greater than the latter.

In Prop. 40 (Fig. 78), $\mathbf{E}(C_n)$ is compared with **K**. Here the revolving portion of the polygon is $\Lambda\Phi ZMK$, and the surface of the solid of revolution is therefore equal to a circle the square on whose radius is equal to

$$\mathbf{O}(ZM, M\Phi + \tfrac{1}{2}K\Lambda),$$

i.e. by Prop. 22 equal to

$$\mathbf{O}(M\Theta, ZH).$$

Now $M\Theta = 2EO = \Delta\Gamma$ and $ZH > \Delta\Xi$ (because of similarity of the triangles ZKH and $\Delta A\Xi$ in connection with $ZK > \Delta A$).

Therefore $\mathbf{O}(M\Theta, ZH) > \mathbf{O}(\Delta\Gamma, \Delta\Xi) = \mathbf{T}(A\Delta)$.

Consequently

$$\mathbf{E}(C_n) > \mathbf{K}.$$

In order to prove the proposition on the surface of the segment of a sphere we now only require a proposition in which it is proved that the surfaces $\mathbf{E}(C_n)$ and $\mathbf{E}(I_n)$ are to each other in the duplicate ratio of the sides of the revolving portions of the polygons C_n and I_n. This is done in the first part of Prop. 41, which is perfectly analogous to Prop. 32, in which the corresponding result was reached for the argument leading up to the finding of the surface of the sphere.

The way has thus been sufficiently prepared for the proof of the main proposition. It is enunciated in

Proposition 42.

The surface of any segment of a sphere less than a hemisphere is equal to a circle whose radius is equal to the straight line drawn from the vertex of the segment to the circumference of the circle which is the base of the segment of the sphere.

The correctness of this assertion is derived by means of a double *reductio ad absurdum* (III; 8,21) from the two relations obtained, *viz.*

$$\mathbf{E}(I_n) < \mathbf{E} < \mathbf{E}(C_n)$$

and

$$\mathbf{E}(I_n) < \mathbf{K} < \mathbf{E}(C_n)$$

in literally the same way as was done in Prop. 33 for the surface of the sphere.

The restriction that the segment must be less than a hemisphere would be superfluous in our view. That Archimedes does make it, is probably to be understood as follows. The whole chain of reasoning is based on the propositions on the surfaces of conical figures, which have been derived in the first part of Book I; when a segment of a circle which is greater than a semi-circle is considered, one of the sides of I_n and of C_n might be parallel to the axis of revolution and therefore upon revolution describe a cylindrical surface, which according to Greek conceptions cannot be dealt with as conical. In the derivation of the analogous property of the sphere this case could not present itself.

That the proposition, however, is also true of a segment of a sphere which is greater than a hemisphere, is recognized without any difficulty in Prop. 43 by considering the surface of such a segment as the difference of the surfaces of the whole sphere and of the complementary segment.

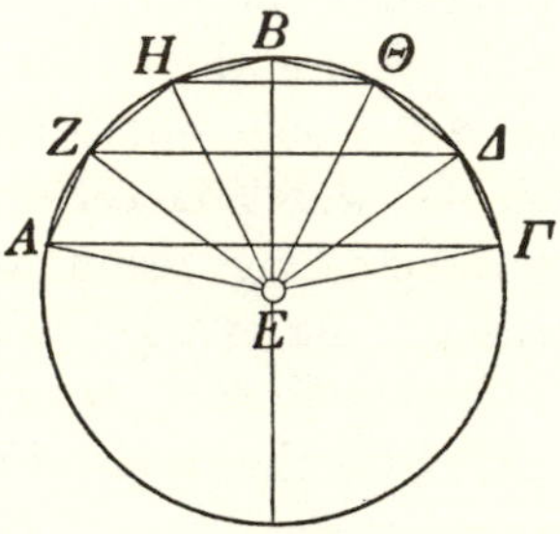

Fig. 79.

The argument leading up to the finding of the volume of a sector of a sphere may be summarized as follows (Fig. 79):

In Prop. 38 the volume $\mathbf{S}(I_n)$ is considered of the solid of revolution which is generated by the revolution of the figure

$$EAZHB\Theta\Delta\Gamma E$$

and which consists of a solid rhombus and further of remainders of rhombi. By the application of Props 18 and 20 we find that $\mathbf{S}(I_n)$ is equal to the volume of the cone which has a base equal to $\mathbf{E}(I_n)$ and a height equal to the apothem of the portion $AZ\ldots\Gamma$ of the polygon. In a Porism it is then stated that the volume of this cone is less than that of a cone X, whose base is equal to a circle on AB as radius and whose height is equal to the radius of the sphere. Indeed, by Prop. 37, $\mathbf{E}(I_n)$ is less than this circle, while the apothem is less than the radius of the sphere. From this it follows that

$$\mathbf{S}(I_n) < \mathbf{X}.$$

In the same way a property of $\mathbf{S}(C_n)$ is derived in the Porisms of Prop. 40. $\mathbf{S}(C_n)$ is found to be equal to a cone whose base is equal to $\mathbf{E}(C_n)$ and whose height is equal to the radius of the sphere, from which it follows that

$$\mathbf{S}(C_n) > \mathbf{X},$$

because, by Prop. 41, $\mathbf{E}(C_n)$ is greater than the base of $\mathbf{X}$. We now have therefore the inequalities

$$\mathbf{S}(I_n) < \mathbf{X} < \mathbf{S}(C_n), \qquad (\alpha)$$

to which there is added in the proof of the main proposition, without its being expressly mentioned,

$$\mathbf{S}(I_n) < \mathbf{S} < \mathbf{S}(C_n), \qquad (\beta)$$

in which **S** represents the volume of the sector of the sphere.

Furthermore it is derived in Prop. 41 that the ratio of $\mathbf{S}(I_n)$ to $\mathbf{S}(C_n)$ is the triplicate ratio of the sides of I_n and C_n. (γ)

The main theorem is then enunciated in

Proposition 44.

The volume of any sector of a sphere is equal to a cone which has a base equal to the surface of the segment of the sphere included in the sector and a height equal to the radius of the sphere.

The derivation of this proposition from the two inequalities (α) and (β) in connection with the proposition (γ) proceeds on lines perfectly analogous to the corresponding argument in Prop. 34.

To be consistent, Archimedes ought first to have enunciated the proposition again only for a sector of a sphere less than a hemisphere.

From our present-day point of view it may appear superfluous that a separate proof is given for the surface and the volume of a sphere, while the surface of a segment of a sphere and the volume of a sector of a sphere are also derived. As has already been remarked above, it is not, however, in accordance with the Greek view to consider the surface of a sphere as a particular case of the surface of a segment of a sphere or to derive its volume from that of a sector of a sphere.

CHAPTER V.

ON THE SPHERE AND CYLINDER

Book II

The second book of the treatise *On the Sphere and Cylinder* opens again with a letter by Archimedes to Dositheus, in which he sums up once more the principal propositions of Book I, announces the solution of some related problems for Book II, and holds out a prospect of sending him soon the promised propositions on spirals and conoids.

The first of the problems dealt with in Book II concerns the construction of a circle whose surface is equal to that of a given sphere; the solution of this problem follows at once from I, 33. This is followed by

Proposition 1.

Given a cone or a cylinder, to find a sphere equal to the cone or the cylinder.

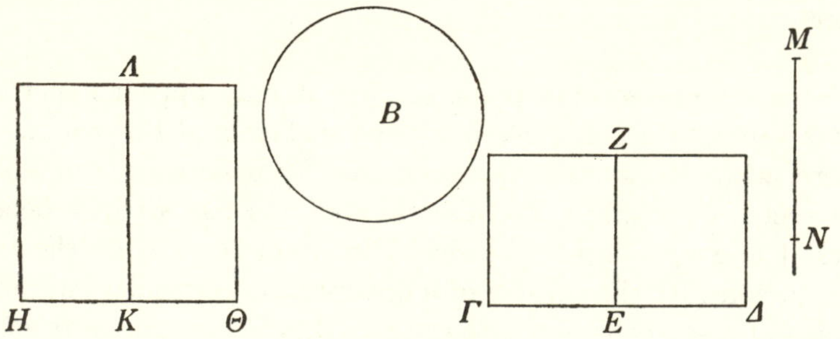

Fig. 80.

The discussion of this is given analytically (Fig. 80). If the sphere B is to be equal to the cone or cylinder A, the cylinder K, of which the diameter of the base $H\Theta$ and the height $K\Lambda$ are both equal to the diameter of B, according to **S.C.** I, 34 will be equal to $\frac{3}{2}B$, *i.e.* equal to a cylinder E which is equal to $\frac{3}{2}A$. If first this cylinder E (diameter of the base $\Gamma\Delta$, height EZ) is constructed, because of Euclid XII, 15 the bases of K and E have to be inversely proportional to the heights, *i.e.*, in connection with Euclid XII, 2,

$$[\mathbf{T}(\Gamma\Delta),\ \mathbf{T}(H\Theta)] = (K\Lambda, EZ)\,. \tag{1}$$

In this, $H\Theta = K\Lambda$. If MN is now constructed, so that

$$\mathbf{T}(H\Theta) = \mathbf{O}(\Gamma\Delta, MN)\,, \tag{2}$$

then (1) passes into

$(\Gamma\Delta, MN) = (H\Theta, EZ)$, *i.e. permutando*, in connection with (2),

$$(\Gamma\Delta, H\Theta) = (H\Theta, MN) = (MN, EZ)\,.$$

The diameter $H\Theta$ of the required sphere B is therefore the first of the two mean proportionals between $\Gamma\Delta$ and EZ[1]. The synthesis is now clear. The construction of the cylinder E is also elucidated by Eutocius; it follows at once from the propositions Euclid XII, 10, 14.

Algebraically: Let the diameter of the given cylinder be d, the height h, the diameter of the required sphere x, then a cylinder with diameter x and height x (which is $\frac{3}{2}$ of the sphere) must have the same volume as a cylinder with diameter d and height $\frac{3}{2}h$.

Consequently we must have

$$\frac{d^2}{x^2} = \frac{x}{\frac{3}{2}h}.$$

If $\dfrac{x^2}{d} = p$, then $\dfrac{d}{p} = \dfrac{x}{\frac{3}{2}h}$, therefore $\dfrac{d}{x} = \dfrac{x}{p} = \dfrac{p}{\frac{3}{2}h}$, so that x is the first of the two mean proportionals between d and $\frac{3}{2}h$.

In the second proposition a theorem on the volume of a segment of a sphere is derived, which is to be applied for the solution of later problems.

Proposition 2.

Any segment of a sphere is equal to a cone which has the same base as the segment and for height a straight line which has to the height of the segment the same ratio as the sum of the radius of the sphere and the height of the remaining segment has to the height of the remaining segment.

---

[1]) The fact that here in the synthesis the two mean proportionals between two given line segments have to be found induces Eutocius to make his invaluable digression on the different solutions given in Greek mathematics of the problem of the two mean proportionals (and consequently of the Delian problem of the duplication of the cube). *Opera* III, 54–106.

In order to elucidate the complicated argument we will reproduce it analytically (Fig. 81).

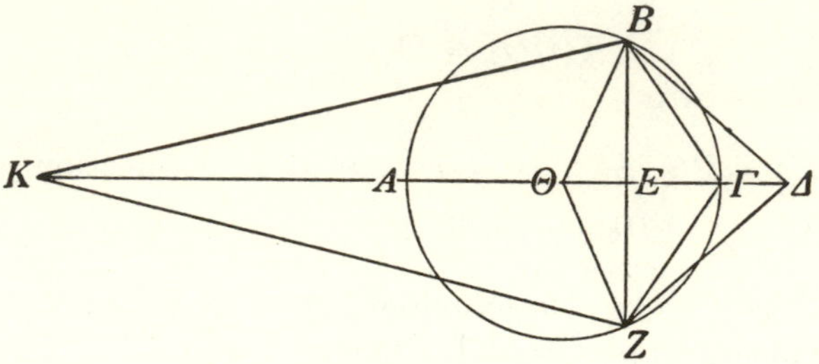

Fig. 81.

To construct a cone which shall be equal to the segment ΓBZ (which is less than a hemisphere), we try to transform the sector $\Theta B\Gamma Z$ of the sphere into a solid rhombus, which has the circle on BZ as diameter for base and one vertex of which lies in Θ. The second cone of the rhombus will then be equal to the segment of the sphere.

Now (I, 42, 44) the sector $\Theta B\Gamma Z$ is equal to a cone the diameter of whose base is $B\Gamma$ and whose height is R. If therefore the other vertex of the rhombus is Δ, we must have (Euclid XII, 15 and 2)

$$[\mathbf{T}(B\Gamma),\ \mathbf{T}(BE)] = (\Theta\Delta, R)$$

or

$$(A\Gamma, AE) = (\Theta\Delta, R)\,,$$

from which *separando* $(\Gamma E, AE) = (\Gamma\Delta, R)\,,$

permutando $(\Gamma E, \Gamma\Delta) = (AE, R)\,,$

componendo $(\Delta E, \Gamma E) = (AE + R, AE)\,.$

If therefore Δ be found such that this relation is satisfied, we find, by reversing the argument,

$$\text{sector } \Theta B\Gamma Z = \text{rhombus } \Theta(BZ)\Delta$$

and therefore also

$$\text{segment } \Gamma BZ = \text{cone } \Delta BZ\,.$$

Algebraically: Suppose $A\Gamma = d$, $\Theta\Gamma = R$, $\Gamma E = h_1$, $AE = h_2$, $BE = r$, then the position of Δ is determined by

$$\tfrac{1}{3}\pi r^2 . \Theta\Delta = \tfrac{1}{3}\pi . B\Gamma^2 . R$$

or

$$\Theta\Delta = R.\frac{B\Gamma^2}{r^2} = R.\frac{d}{h_2}$$

therefore

$$\Delta E = R\frac{d}{h_2} - (R-h_1) = h_1\frac{R+h_2}{h_2}.$$

The expression found for the volume of the segment of the sphere with height h_1 thus becomes

$$I = \tfrac{1}{3}\pi r^2.h_1\frac{R+h_2}{h_2} = \tfrac{1}{3}\pi h_1{}^2(R+h_2) = \tfrac{1}{3}\pi h_1{}^2(3R-h_1)\,.$$

This proof, however, is valid only for segments of spheres which are less than the hemisphere, because only cones with salient vertices are considered. It therefore still has to be proved that, if K is determined by the relation

$$(KE, AE) = (R+\Gamma E, \Gamma E)\,,$$

also segment $ABZ =$ cone KBZ.

To this end it is proved that the whole sphere is equal to the rhombus $K(BZ)\Delta$. Since (**S.C.** I, 34) the sphere is equal to a cone the radius of whose base is $A\Gamma$ and height R,the rhombus, however, to a cone with radius of the base BE and height $K\Delta$, it has to be proved that

$$[\mathbf{T}(A\Gamma),\ \mathbf{T}(BE)] = (K\Delta, R)\,.$$

Now because of

$$\mathbf{T}(BE) = \mathbf{O}(AE, \Gamma E)$$

the first member may also be written

$$[\mathbf{T}(A\Gamma),\ \mathbf{O}(AE, \Gamma E)]\,.$$

In order to give the proof, we now determine separately the ratios

$$(A\Gamma, AE) \quad \text{and} \quad (A\Gamma, \Gamma E)\,.$$

The following relations are given:

$$(\Delta E, \Gamma E) = (R+AE, AE) \qquad (KE, AE) = (R+\Gamma E, \Gamma E)$$

separando

$$(\Gamma\Delta, \Gamma E) = (R, AE) \qquad (KA, AE) = (R, \Gamma E)$$

permutando

$(AE, \Gamma E) = (R, \Gamma \Delta)$ $\qquad\qquad$ $(AE, \Gamma E) = (KA, R)$

$$(R, \Gamma\Delta) = (KA, R)$$

componendo

$$(\Theta\Delta, \Gamma\Delta) = (\Theta K, R)$$

permutando

$$(\Theta\Delta, \Theta K) = (\Gamma\Delta, R)$$

componendo

$$(K\Delta, \Theta K) = (\Theta\Delta, R)$$

or $\mathbf{O}(\Theta K, \Theta\Delta) = \mathbf{O}(K\Delta, R) \quad (\alpha)$

componendo $\qquad\qquad$ *componendo*

$(A\Gamma, AE) = (\Theta\Delta, R)$ $\qquad\qquad$ $(A\Gamma, \Gamma E) = (\Theta K, R)$

$$[\mathbf{T}(A\Gamma),\ \mathbf{O}(AE, \Gamma E)] = [\mathbf{O}(\Theta\Delta, \Theta K),\ \mathbf{T}(R)]$$ [1]

or, because of (α) ,

$$[\mathbf{T}(A\Gamma), \mathbf{O}(AE, \Gamma E)] = [\mathbf{O}(K\Delta, R),\ \mathbf{T}(R)] = (K\Delta, R)\ .$$

q. e. d.

Algebraically: If Δ and K are determined successively by

$$\Delta E = h_1 \frac{R+h_2}{h_2} \quad \text{and} \quad KE = h_2 \frac{R+h_1}{h_1},$$

then it has to be proved that

$$\tfrac{1}{3}\pi r^2 . K\Delta = \tfrac{1}{3}\pi d^2 R$$

or

$$\frac{K\Delta}{R} = \frac{d^2}{r^2} = \frac{d^2}{h_1 h_2}.$$

---

[1]) This passage, which we should nowadays perform by multiplication of the corresponding members of the two proportions derived, is motivated in the Greek theory of proportions by writing

$$(A\Gamma, AE) = [\mathbf{T}(A\Gamma),\ \mathbf{O}(A\Gamma, AE)] = [\mathbf{O}(\Theta\Delta, \Theta K),\ \mathbf{O}(R, \Theta K)]$$
$$(A\Gamma, \Gamma E) = [\mathbf{O}(A\Gamma, AE),\ \mathbf{O}(\Gamma E, AE)] = [\mathbf{O}(R, \Theta K), \mathbf{T}(R)]\ ,$$

from which the required conclusion follows *ex aequali*.

Now it is true that

$$K\Delta = \Delta E + KE = \frac{h_1^2(R+h_2)+h_2^2(R+h_1)}{h_1h_2} = R\frac{(h_1+h_2)^2}{h_1h_2} = R\frac{d^2}{h_1h_2}.$$

Note. The property proved is frequently also used, instead of in the form

$$(\Delta E, \Gamma E) = (R+AE, AE),$$

in the form to be derived *separando, viz.*:

$$(\Delta\Gamma, \Gamma E) = (R, AE).$$

Algebraically:

$$\frac{\Delta\Gamma}{h_1} = \frac{R}{h_2}.$$

Proposition 3.

To cut a given sphere by a plane so that the surfaces of the segments may have to each other a ratio which is the same as a given ratio.

From I, 42, 43 it follows readily that for this the plane has to divide the diameter to which it is perpendicular in the given ratio.

This is followed by one of the great problems of Greek geometry:

Proposition 4.

To cut a given sphere ⟨*by a plane*⟩ *so that the segments of the sphere may have to each other a ratio which is the same as a given ratio*[1]).

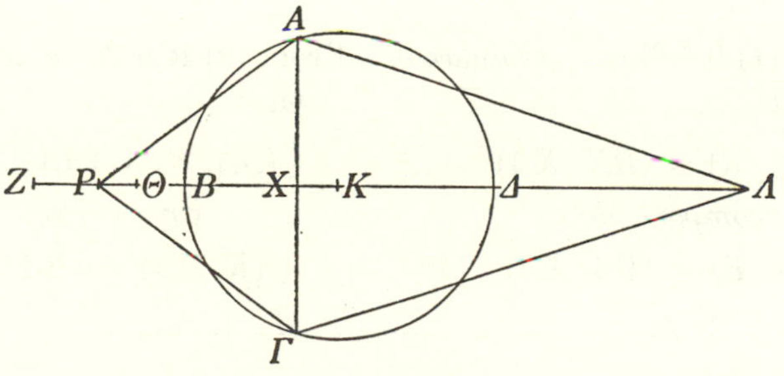

Fig. 82.

[1]) The original formulation of this proposition (we would remind the reader that we only possess an Attic version of *On the Sphere and Cylinder*) is to be read from its mention in the preface to *On Spirals*: τὰν δοθεῖσαν σφαῖραν ἐπιπέδῳ τεμεῖν, ὥστε τὰ τμάματα αὐτᾶς ποτ' ἄλλαλα τὸν ταχθέντα λόγον ἔχειν.

In Fig. 82 let $A\Gamma$ be the intersection of the required plane with the meridian plane of the sphere with centre K perpendicular to it. Apparently we may also require the sphere to have a given ratio to the segment $A\Delta\Gamma$. Let this ratio be equal to

$$(ZB, Z\Theta), \text{ in which } ZB = R\,.$$

Find according to **S.C.** II, 2 the points P and Λ so that

$$\text{Cone } PA\Gamma = \text{segment } BA\Gamma$$

$$\text{Cone } \Lambda A\Gamma = \text{segment } \Delta A\Gamma\,.$$

For this we must have (*vide* Note to II, 2)

$$(PB, BX) = (R, X\Delta) \qquad (1)$$

$$(\Lambda\Delta, \Delta X) = (R, XB)\,. \qquad (2)$$

The problem thus amounts to finding a point X so that

$$(P\Lambda, X\Lambda) = (ZB, Z\Theta)\,, \qquad (3)$$

the positions of the points P and Λ, however, depending according to (1) and (2) on the position of X.

The ratio $(P\Lambda, X\Lambda)$ is compounded of the ratios

$$(P\Lambda, \Delta\Lambda) \quad (\alpha) \quad \text{and} \quad (\Delta\Lambda, X\Lambda) \quad (\beta)\,.$$

These two ratios are first written in a different form:

α. From (1) it follows *permutando* that

$$(PB, R) = (BX, X\Delta)$$

componendo

$$(PK, R) = (B\Delta, X\Delta) \quad (4)$$

From (2) it follows *permutando* that

$$(\Lambda\Delta, R) = (\Delta X, XB)$$

componendo

$$(\Lambda K, \Lambda\Delta) = (B\Delta, X\Delta) \quad (5)$$

$$(PK, R) = (\Lambda K, \Lambda\Delta)$$

permutando

$$(PK, K\Lambda) = (R, \Lambda\Delta)$$

componendo

$$(P\Lambda, K\Lambda) = (K\Lambda, \Lambda\Delta)\,.$$

Apparently therefore $(P\Lambda, \Lambda\Delta)$ is the duplicate ratio of $(K\Lambda, \Lambda\Delta)$ and because of (5) also of $(B\Delta, X\Delta)$. Therefore

$$(P\Lambda, \Lambda\Delta) = [\mathrm{T}(B\Delta), \mathrm{T}(X\Delta)] .$$

β. From (2) it follows *componendo*

$$(\Lambda X, \Lambda\Delta) = (ZX, ZB) \quad \text{or } \textit{invertendo}$$

$$(\Lambda\Delta, \Lambda X) = (ZB, ZX) .$$

The ratio $(P\Lambda, X\Lambda)$ therefore is compounded of the ratios

$$[\mathrm{T}(B\Delta), \mathrm{T}(X\Delta)] \quad \text{and} \quad (ZB, ZX) . \tag{6}$$

The ratio $(ZB, Z\Theta)$ appearing in the 2nd member of (3), however, is compounded of

$$(ZB, ZX) \quad \text{and} \quad (ZX, Z\Theta) . \tag{7}$$

By comparison of the expressions (6) and (7) it follows, because of (3), that

$$(ZX, Z\Theta) = [\mathrm{T}(B\Delta), \mathrm{T}(X\Delta)] . \tag{8}$$

The question therefore amounts to a given line segment $Z\Delta$ $(3R)$ having to be so divided in a point X that the part ZX shall have to a given line segment $Z\Theta$ the same ratio as the square on another given line segment $B\Delta$ $(2R)$ has to the square on the remaining part $X\Delta$.

Archimedes announces that he will deal with this problem analytically and synthetically at the end of the proposition[1]). This, however, is not done. The solution he promises must already have been lost in the days of Diocles (who gives another solution instead). Eutocius, however, thinks he has found it again[2]), in an apparently mutilated form; he presents a reconstruction which we reproduce as follows: Formulated in a general way (*i.e.* without any relation to the notation and the particular ratios of the line

---

[1]) *Opera* I, 192. ἐπὶ τέλει ἀναλυθήσεται καὶ συντεθήσεται.

[2]) *Opera* III, 130–132. Eutocius here says that the solution of Archimedes —recognizable by the Dorian dialect and by the old denomination of the conics—, which did not occur in any codex, was traced by him in a book, but that it was obscure and showed many errors in the figures. He himself in his exposition uses the Apollonian terms parabola and hyperbola, an example which we are following in the reproduction of his argument.

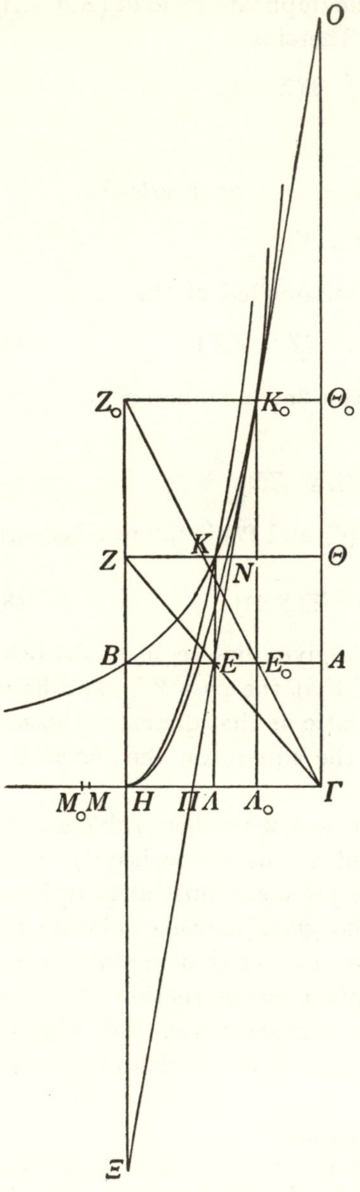

Fig. 83.

segments to each other in II, 4), the problem is as follows: Given (Fig. 83) a line segment AB, a line segment $A\Gamma$, and a surface $\mathbf{\Delta}$, it is required to find on AB a point E such that

$$(AE, A\Gamma) = [\mathbf{\Delta}, \mathbf{T}(BE)] \quad (\gamma)$$

This question is treated analytically: Let E be the required point. From the figure, in which $A\Gamma$ is plotted at right angles to AB and the rectangle $HZ\Theta\Gamma$ is completed, it follows that

$$(AE, A\Gamma) = (H\Gamma, HZ)\,. \quad (\delta)$$

Further suppose that $\mathbf{\Delta} = \mathbf{O}(H\Gamma, HM)$, then, because $BE = ZK$, we must have

$$(H\Gamma, HZ) = [\mathbf{O}(H\Gamma, HM),\ \mathbf{T}(ZK)]$$

or $[\mathbf{O}(H\Gamma, HM),\ \mathbf{O}(HZ, HM)]$
$$= [\mathbf{O}(H\Gamma, HM),\ \mathbf{T}(ZK)]\,,$$

therefore

$$\mathbf{T}(ZK) = \mathbf{O}(HZ, HM)\,.$$

From this it follows (III; 2.0) that K lies on a parabola which has H for vertex, HZ for diameter, and HM for *orthia* (*latus rectum*).

Moreover, because of Euclid I, 43, the rectangle $K\Gamma$ is equal to the rectangle AH, so that K also lies on an orthogonal hyperbola with asymptotes $H\Gamma$ and $\Theta\Gamma$, which passes through B (*Conica* II, 12). K is therefore determined as the point of intersection of two conics.

Before it is possible to proceed to the synthesis, a $\delta\iota o\varrho\iota\sigma\mu\acute{o}\varsigma$[1]) is derived. Indeed, it is evident that for a given value of AB the data $A\Gamma$ and $\mathbf{\Delta}$ are subject to certain limitations, for we have the relation

$$(AE, A\Gamma) = [\mathbf{\Delta}, \mathbf{T}(BE)]$$

which, because of Euclid XI, 34, states that the volume of the parallelepiped with base $\mathbf{T}(BE)$ and height AE is equal to the volume of the parallelepiped whose base has a surface $\mathbf{\Delta}$ and whose height is $A\Gamma$. If we denote the solids in question successively by

$$\Sigma[\mathbf{T}(BE), AE)] \quad \text{and} \quad \Sigma[\mathbf{\Delta}, A\Gamma] \qquad (\Sigma = \sigma\tau\varepsilon\varrho\varepsilon\acute{o}\nu)\,,$$

it has to be found out what is the maximum value which can be taken by the volume of the first solid when E varies; the volume of the second solid is then delimited by this value as its upper limit.

It is now proved that the solid with base $\mathbf{T}(BE)$ and height AE attains its maximum volume when $BE = 2.AE$: to see this we construct the above-mentioned curves once more for the point E_0, which satisfies this condition; the corresponding point K_0 lies on a parabola with diameter HZ_0 and *latus rectum* HM_0 so that

$$\mathbf{T}(Z_0K_0) = \mathbf{O}(Z_0H, HM_0)$$

and also on the orthogonal hyperbola already mentioned above. We now proceed to prove that the parabola HK_0 touches this hyperbola in K_0.

To prove this, on Z_0H produced measure $H\Xi = HZ_0$, then by a familiar property of the tangent to a parabola (III; 2.2), ΞK_0 is tangent to the parabola in K_0. If further ΞK_0 meets the asymptote $\Gamma\Theta$ in O, then, because $BE_0 = 2.AE_0$, apparently $K_0\Pi = K_0O$, so that the line ΠO, in view of a property of the tangent to a hyperbola (III; 5.43), touches this curve in K_0. By this the proposition is proved.

The branch $BK_0\ldots$ of the hyperbola in the figure thus lies completely on one side of the parabola $HK\ldots$ If therefore E be any point of AB (in the figure chosen between B and E_0), the perpendicular through E on AB meets the hyperbola in K, while

[1]) $\Delta\iota o\varrho\iota\sigma\mu\acute{o}\varsigma$ is here to be understood in the sense of "condition which the data of a problem have to satisfy, if it is to permit of solutions". *Elements of Euclid* I, 168.

the line ZK, parallel to AB, meets the parabola HK_0 in N, it is found that

$$ZK < ZN\,.$$

Now K lies on the orthogonal hyperbola; from this follows the equality of the rectangles $K\Gamma$ and $B\Gamma$, and consequently also that of the rectangles KA and $B\Lambda$; in view of the reverse of Euclid I, 43 the points Γ, E, and Z now lie on one straight line.

It had to be proved that

$$\Sigma[\mathbf{T}(BE), AE] < \Sigma[\mathbf{T}(BE_0), AE_0] \qquad (\zeta)$$

or, if we call the particular value of the given surface $\mathbf{\Delta}$, which with $A\Gamma$ given leads to the finding of the point E_0, $\mathbf{\Delta}_0$,

$$\Sigma[\mathbf{\Delta}, A\Gamma] < \Sigma[\mathbf{\Delta}_0, A\Gamma]\,.$$

In this, however,

$$\mathbf{\Delta} = \mathbf{O}(H\Gamma, HM) \quad \text{and} \quad \mathbf{\Delta}_0 = \mathbf{O}(H\Gamma, HM_0)\,,$$

while

$$\mathbf{T}(ZK) = \mathbf{O}(HZ, HM) \quad \text{and} \quad \mathbf{T}(ZN) = \mathbf{O}(HZ, HM_0)\,.$$

Because $ZK < ZN$, we have $HM < HM_0$, therefore $\mathbf{\Delta} < \mathbf{\Delta}_0$, from which the correctness of the proposition becomes evident. For a point E between E_0 and A the proof is analogous.

The *διορισμός* thus found is then applied as follows in the synthesis.

First find E_0 so that $BE_0 = 2.AE_0$. If now

$$\Sigma[\mathbf{\Delta}, A\Gamma] \left.\begin{matrix} > \\ = \\ < \end{matrix}\right\} \Sigma[\mathbf{T}(BE_0), AE_0] \quad \begin{matrix} \text{there is no solution.} \\ E_0 \text{ is the required point.} \\ \text{find the point } M \text{ so that} \end{matrix}$$

$$\mathbf{\Delta} = \mathbf{O}(H\Gamma, HM).$$

Construct the parabola with diameter HZ and *latus rectum* HM, the orthogonal hyperbola through B with asymptotes ΓH and $\Gamma\Theta$, and find the point of intersection K of these two curves, which lies on the same side of E_0K_0 with H. The required point E is then obtained by drawing a perpendicular from K to AB.

The method found may now be applied in the problem posited by Archimedes in Proposition 4. For this, AB from the general discussion is to be replaced by the line segment $Z\Delta = 3R$ of Fig. 82.

$A\Gamma$ is to be replaced by $Z\Theta$ of Fig. 82, and $\mathbf{\Delta}$ is to be taken equal to $\mathbf{T}(B\Delta)$. Then BE_0 becomes equal to $2R$, AE_0 to R. Because $Z\Theta < R$, we now have

$$\Sigma[\mathbf{\Delta}, A\Gamma] = \Sigma[\mathbf{T}(2R), Z\Theta] < \Sigma[\mathbf{T}(2R), R] ,$$

so that the condition for the solubility of the problem, which has been derived in the general discussion, is satisfied.

We have deliberately reproduced the whole argument in a purely classical form. We will now elucidate it by translating it into algebraic symbolism:

If (again in the general problem) we suppose $AB=a$, $BE=x$, $A\Gamma=p$, $\mathbf{\Delta}=q^2$, the equation of the problem (*vide* (γ)) is:

$$\frac{a-x}{p} = \frac{q^2}{x^2} \text{ or } x^2(a-x) = pq^2 \text{ or } x^3-ax^2+pq^2 = 0 . \qquad (\varepsilon)$$

The Greek solution is equivalent to equating each of the two members of

$$\frac{a-x}{p} = \frac{q^2}{x^2}$$

to $\frac{a}{y}$ in conformity with relation (δ), in which $HZ=y$.

We now have the two equations

$$y(a-x) = ap \quad \text{and} \quad x^2 = \frac{q^2}{a} \cdot y ,$$

which represent successively the orthogonal hyperbola and the parabola used by Archimedes. The abscissae of the points of intersection of these two curves represent the solutions of the equation (ε). The found values of x, however, can only be used in the geometrical problem II, 4 of Archimedes when they satisfy the relation $0<x<\frac{2}{3}a$, because x has to be less than $2R$, and a is equal to $3R$.

In order to find out the number of roots, we compute the discriminant

$$\mathbf{D} = -[27pq^2-4a^3]pq^2 ,$$

from which the following becomes apparent:

If pq^2 be less than $\frac{4}{27}a^3$, there are three real roots; since their product is $-pq^2$ and is therefore negative, while the sum a is positive, one of these roots is negative and the other two are positive.

If pq^2 be equal to $\frac{4}{27}a^3$, there are two positive roots, which coincide.

If pq^2 be greater than $\frac{4}{27}a^3$, there is only one real root, and this is negative.

Since in the problem discussed by Archimedes only positive roots can be used, it is a prerequisite that pq^2 be less than $\frac{1}{27}a^3$.

This prerequisite is identical with the condition (ζ) derived by Archimedes. In fact,

$$pq^2 = x^2(a-x)\,,$$

which latter form reaches its maximum, *viz.* $\frac{4}{27}a^3$, when $x = \frac{2}{3}a$.

Of the two positive roots which satisfy the equation in the case of

$$pq^2 < \frac{4}{27}a^3$$

and which correspond to the two points of intersection of hyperbola and parabola, having positive abscissae, in the particular case of II, 4 only one can be used, *viz.* the one which is less than $\frac{2}{3}a$.

That there is always one such root, and only one, is clear upon geometrical consideration: of the two points of intersection of parabola and orthogonal hyperbola, which have positive abscissae, one lies to the left and one to the right of the line K_0E_0.

Algebraically the same is found because the expression

$$x^3 - ax^2 + pq^2$$

appears to have a zero point between $x=0$ and $x=\frac{2}{3}a$, and another between $x=\frac{2}{3}a$, and $x=a$.

Eutocius mentions two more solutions of the proposition **S.C.** II, 4[1]). In the first, which originates from Dionysodorus[2]), the problem is first reduced (by a different method from that of Archimedes) to the form (3). The solution (translated into the notation of II, 4), however, then proceeds as follows:

In Fig. 82 provide for

$$\mathbf{O}(ZX, Z\Theta) = \mathbf{T}(MX) \qquad (\alpha)$$

then by (8) we have:

---

[1]) *Opera* III, 152 *et seq.* and 160 *et seq.*

[2]) Probably Dionysodorus of Caunus, who was a contemporary of Apollonius. Presumably he was the writer of a treatise περὶ τῆς σπείρας (On the tore), quoted by Heron (*Metrica* II, 13, *Heronis Opera* III, 128). In this he derives an expression for the volume of a tore.

$$[\mathbf{T}(MX),\ \mathbf{T}(Z\Theta)] = (ZX, Z\Theta) = [\mathbf{T}(B\Delta),\ \mathbf{T}(X\Delta)]\,,$$

therefore

$$(MX, Z\Theta) = (B\Delta, X\Delta)\,. \qquad (\beta)$$

In the above, (α) represents a parabola with vertex Z, axis $Z\Delta$, and *latus rectum* $Z\Theta$, and (β) an orthogonal hyperbola with centre Δ, the asymptotes of which are ΔZ and the perpendicular erected through Δ on ΔB. X is then determined as the foot of the ordinate of a point of intersection M of these two curves.

Algebraically this means, in the notation adopted above, that for the solution of the equation

$$\frac{a-x}{p} = \frac{q^2}{x^2}$$

we suppose

$$y^2 = p(a-x)\,,$$

so that

$$\frac{y}{p} = \frac{q}{x}.$$

We now have to find the points of intersection of the parabola

$$y^2 = p(a-x)$$

and the orthogonal hyperbola

$$xy = pq\,.$$

In the second solution, which was suggested by Diocles[1]), the problem is generalized even more, as compared with Archimedes. In fact, in Archimedes, when the points B and Δ were given, the points P and Λ were determined by the relations (Fig. 82):

$$(PB, BX) = (R, X\Delta) \quad \text{and} \quad (\Lambda\Delta, \Delta X) = (R, XB)\,,$$

while X then had to satisfy the condition $(PX, \Lambda X) =$ given ratio.

In Diocles, R is replaced in the two first-mentioned relations by a line segment of any length, which is no longer conditioned by the length of $B\Delta$. Formulated once again, the problem now is as follows (Fig. 84):

[1]) Diocles probably lived about 100 B.C. He is known to this day for his cissoid. The solution of **S.C.** II, 4 quoted above has been taken from his treatise περὶ πυρίων (on burning-mirrors).

To divide a given line segment AB at E so that, if the relations

$$(ZA, AE) = (AK, BE)$$

$$(HB, BE) = (BM, AE) \qquad \text{with } AK = BM$$

apply, E satisfies the condition

$$(ZE, HE) = (\Gamma, \Delta)\,,$$

in which Γ and Δ are given line segments.

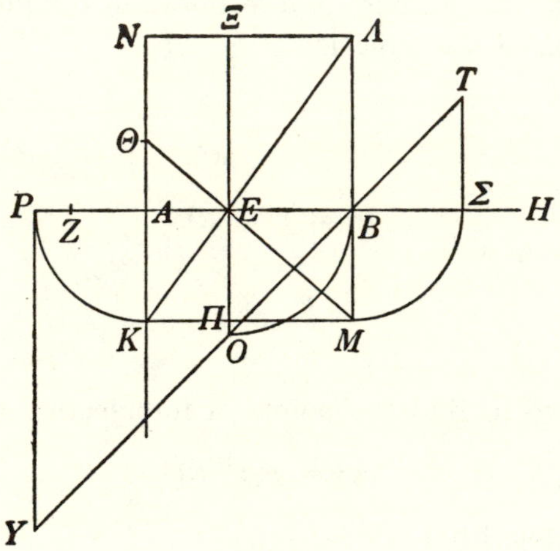

Fig. 84.

This problem is analysed as follows:

Let ME produced meet KA produced in Θ; likewise let KE produced meet MB produced in Λ, then it is proved without difficulty that

$$ZA = \Theta A \quad \text{and} \quad HB = \Lambda B\,.$$

Now

$$(AE, BE) = (\Theta A, MB) = (KA, \Lambda B) = (\Theta A + AE, MB + BE)$$
$$= (KA + AE, \Lambda B + BE)\,;$$

therefore $\mathbf{O}(\Theta A + AE, \Lambda B + BE) = \mathbf{O}(KA + AE, MB + BE)$

or, if $B\Sigma = BM$, $AP = AK$,

$$\mathbf{O}(ZE, HE) = \mathbf{O}(PE, \Sigma E)\,.$$

From this it follows that

$$(\mathit{\Gamma}, \mathit{\Delta}) = (ZE, HE) = [\mathbf{O}(ZE, HE), \mathbf{T}(HE)] = [\mathbf{O}(PE, \mathit{\Sigma}E), \mathbf{T}(HE)] .$$

Now make EO equal to EB and perpendicular to AB, $\mathit{\Sigma}T$ perpendicular to AB with T on OBproduced. From the similarity of the triangles EOB and $\mathit{\Sigma}TB$ it now follows that

$(TB, OB) = (\mathit{\Sigma}B, EB)$, therefore *componendo* $(TO, OB) = (\mathit{\Sigma}E, EB)$.

If the perpendicular to BA through P meet BO produced in Y, then we also have

$$(BO, OY) = (BE, EP) ,$$

from which it follows *ex aequali* that

$$(TO, OY) = (\mathit{\Sigma}E, EP) .$$

Thence

$$[\mathbf{O}(TO, OY), \mathbf{T}(OY)] = [\mathbf{O}(\mathit{\Sigma}E, EP), \mathbf{T}(EP)]$$

or

$$[\mathbf{O}(TO, OY), \mathbf{O}(\mathit{\Sigma}E, EP)] = [\mathbf{T}(OY), \mathbf{T}(EP)] = 2:1 .$$

Therefore

$$\mathbf{O}(TO, OY) = 2\mathbf{O}(\mathit{\Sigma}E, EP) .$$

Above it has been found that

$$[\mathbf{O}(\mathit{\Sigma}E, EP), \mathbf{T}(EH)] = (\mathit{\Gamma}, \mathit{\Delta}) ,$$

therefore, because $EH = \mathit{\Xi}O$,

$$[\mathbf{O}(TO, OY), \mathbf{T}(O\mathit{\Xi})] = (2\mathit{\Gamma}, \mathit{\Delta}) .$$

Now construct a line segment $\mathit{\Phi}$ so that

$$(2\mathit{\Gamma}, \mathit{\Delta}) = (TY, \mathit{\Phi}) ,$$

then it is found that $\mathit{\Xi}$ lies on an ellipse, produced by elliptical application to $\mathit{\Phi}$ with a defect, the ratio of whose sides is $(TY, \mathit{\Phi})$ (III; 1.2). TY is here the diameter conjugated to the direction $\mathit{\Xi}O$.

But because the rectangles KB and $\mathit{\Pi}N$ are equal, $\mathit{\Xi}$ also lies on an orthogonal hyperbola with asymptotes KN and KM, which passes through B. Thus the point $\mathit{\Xi}$ is determined, from which E follows.

Algebraically: If $AB=a$, $AE=x$, $AK=BM=b$, $(\Gamma, \Delta)=m:n$, x has to satisfy the relation

$$\frac{x+\dfrac{b}{a-x}x}{a-x+\dfrac{b}{x}(a-x)} = \frac{m}{n}$$

or

$$\frac{x^2(a-x+b)}{(a-x)^2(x+b)} = \frac{m}{n},$$

which may be reduced to

$$(a-x+b)(x+b) = \frac{m}{n}\left\{a-x+\frac{ab}{x}-b\right\}^2.$$

Now suppose

$$\frac{ab}{x} = y,$$

then the equation reduces to

$$(y+a-x-b)^2 = \frac{n}{m}(a-x+b)(x+b).$$

If we now consider x and y as coordinates in the rectangular system of axes $K(MN)$, we have to intersect the orthogonal hyperbola

$$xy = ab,$$

which has KM and KN for asymptotes and passes through B, by the curve

$$\frac{(y+a-x-b)^2}{(a-x+b)(x+b)} = \frac{n}{m}.$$

Here

$$y+a-x-b = \Xi O,\ a-x+b = \frac{OT}{\sqrt{2}},\ x+b = \frac{OY}{\sqrt{2}}.$$

If we suppose

$$\Xi O = \eta,\ OT = \xi_1,\ OY = \xi_2,$$

we have the equation

$$\frac{\eta^2}{\xi_1\xi_2} = \frac{n}{2m},$$

which stands for an ellipse, written in the two-abscissa form (III;

1.1), which has TY for diameter and the direction of ΞO as conjugate direction.

Proposition 5.

To construct a segment of a sphere similar to a given segment and equal to another given segment of a sphere.

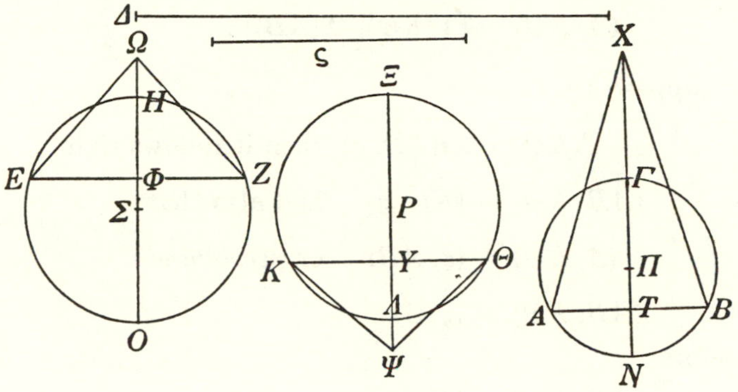

Fig. 85.

In Fig. 85 conceive the required segment $\Lambda K\Theta$ to be equal to the given segment ΓAB and similar to the given segment HEZ[1]). By applying **S.C. II, 2** we replace all three segments by cones. Suppose

$(\Psi Y, \Lambda Y) = (P\Xi + \Xi Y, \Xi Y)$, then segment $\Lambda K\Theta$ = cone $\Psi K\Theta$.

$(XT, \Gamma T) = (\Pi N + NT, NT)$, then segment ΓAB = cone XAB.

$(\Omega\Phi, H\Phi) = (\Sigma O + O\Phi, O\Phi)$, then segment HEZ = cone ΩEZ.

Now it must be true that

1) $$\text{cone } XAB = \text{cone } \Psi K\Theta,$$

whence

$$[\mathbf{T}(K\Theta, \mathbf{T}(AB)] = (XT, \Psi Y) \qquad (\alpha)$$

[1]) Two segments of a sphere are called similar when their heights are in the same ratio as the diameters of their bases, *i.e.* when the meridian sections are similar.

2) $$\text{cone } \Omega EZ \sim \text{cone } \Psi K\Theta$$[1])

whence

$$(\Omega\Phi, EZ) = (\Psi Y, K\Theta)\,.$$

In this, $(\Omega\Phi, EZ)$ is a given ratio; we will equate it to (XT, Δ). We then have *permutando*

$$(XT, \Psi Y) = (\Delta, K\Theta), \text{ i.e. because of } (\alpha)$$

$$(\Delta, K\Theta) = [\mathbf{T}(K\Theta),\ \mathbf{T}(AB)]\,.$$

Now suppose that

$$\mathbf{T}(K\Theta) = \mathbf{O}(AB, \varsigma), \text{ then it follows that}$$

$$(AB, K\Theta) = (K\Theta, \varsigma) \quad \text{and also that}$$

$$(\Delta, K\Theta) = (\varsigma, AB) \quad \text{or } \textit{permutando}$$

$$(AB, K\Theta) = (\varsigma, \Delta)\,.$$

Therefore

$$(AB, K\Theta) = (K\Theta, \varsigma) = (\varsigma, \Delta)\,.$$

From this it appears that $K\Theta$ is the first of the two mean proportionals between the given line segments AB and Δ. The synthesis follows from this[2]).

---

[1]) Indeed, from the similarity of the segments $\Lambda K\Theta$ and HEZ follows the similarity of the cones $\Psi K\Theta$ and ΩEZ. In fact, we have

$$(\Xi Y, \Xi P) = (O\Phi, O\Sigma), \text{ whence}$$

$$(\Xi Y + \Xi P, \Xi Y) = (O\Phi + O\Sigma, O\Phi)\,,$$

from which by **S.C.** II, 2 it follows that

$$(\Psi Y, \Lambda Y) = (\Omega\Phi, H\Phi)\,,$$

and again, in view of the similarity of the segments,

$$(\Psi Y, K\Theta) = (\Omega\Phi, EZ)\,,$$

i.e. the similarity of the cones.

[2]) If the synthesis were furnished by inverting the analysis, we still should have to prove that from the similarity of the cones $\Psi K\Theta$ and ΩEZ the similarity of the segments $\Lambda K\Theta$ and HEZ follows. Archimedes avoids this by making the cone $\Psi K\Theta$, after construction of $K\Theta$, similar to the cone ΩEZ, and the segment $\Lambda K\Theta$ similar to the segment HEZ; he then proves that the segment $\Lambda K\Theta$ is equal to the segment ΓAB.

Proposition 6.

Given two segments of a sphere, either of the same sphere or not, to find a segment of a sphere which shall be similar to one of the given segments and shall have a surface equal to that of the other segment.

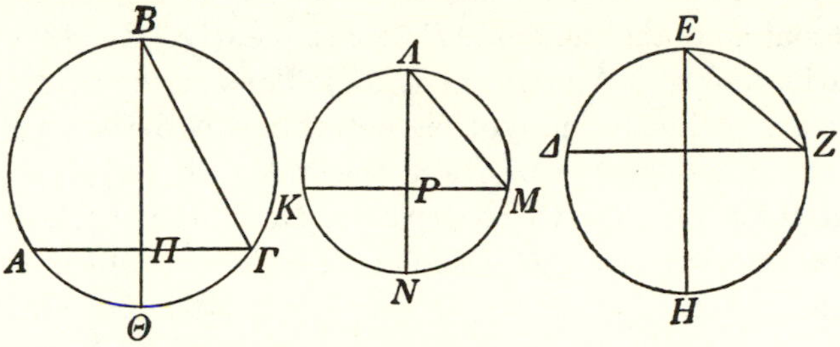

Fig. 86.

If in Fig. 86 the segment ΛKM is to be similar to the segment $BA\Gamma$, it must be true that

$$(\Lambda N, \Lambda M) = (B\Theta, B\Gamma) .$$

But if the surface of the segment ΛKM is to be equal to that of the segment $E\Delta Z$, it must be true that

$$\Lambda M = EZ .$$

Therefore

$$(\Lambda N, EZ) = (B\Theta, B\Gamma) .$$

Thus ΛN is determined, and so is the segment.

Proposition 7.

From a given sphere to cut off a segment by a plane so that the segment shall have a given ratio to the cone which has the same base as the segment and equal height.

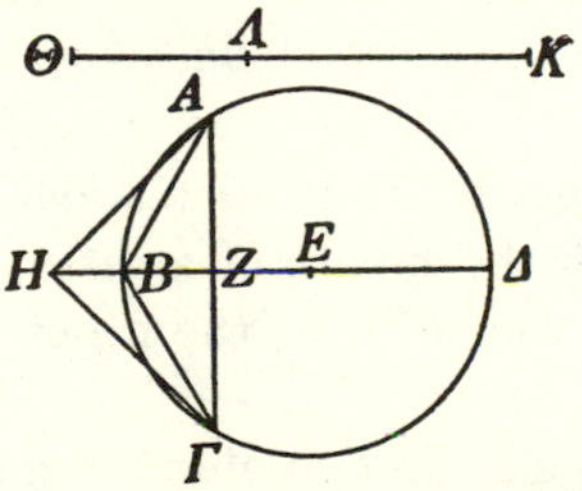

Fig. 87.

If in Fig. 87 $BA\Gamma$ be the required segment, this is equal to the cone $HA\Gamma$, H being determined by

$$(HZ, BZ) = (R+\Delta Z, \Delta Z) .$$

Now the ratio (HZ, BZ) has to be equal to the given ratio of the segment and the cone $BA\Gamma$; consequently $(R+\Delta Z, \Delta Z)$ is a given ratio, and so is therefore $(R, \Delta Z)$. Thus $A\Gamma$ is determined.

Before proceeding to the synthesis, Archimedes derives a διορισμός, *i.e.* he examines what condition the ratio to be given has to satisfy if the solution is to be possible. Now because $\Delta B > \Delta Z$, we also have $(R, \Delta Z) > (R, \Delta B)$ (Euclid V, 8), from which it follows *componendo* that

$$(R+\Delta Z, \Delta Z) > (R+\Delta B, \Delta B) = 3:2 ,$$

so that the given ratio has to be greater than 3:2.

That this condition is also sufficient, becomes apparent from the synthesis. A ratio

$$(\Theta K, \Lambda K) > (3,2)$$

is given, so that

$$(\Theta K, K\Lambda) > (R+\Delta B, \Delta B) .$$

separando

$$(\Theta\Lambda, K\Lambda) > (R, \Delta B) .$$

Now equate $(\Theta\Lambda, K\Lambda)$ to $(R, \Delta Z)$, then ΔZ becomes less than ΔB, so that a point Z is indeed produced which is found to satisfy the condition in question.

Proposition 8 (Fig. 88).

If a sphere be cut by a plane not passing through the centre, the greater segment has to the lesser a ratio less than the duplicate ratio of that which the surface of the greater segment has to the surface of the lesser segment, but greater than the sesquialterate (μείζονα δὲ ἢ ἡμιόλιον).

The conception duplicate ratio (διπλασίων λόγος), which is equivalent to what we call the square of a ratio, is known from Euclid V (*vide* III; 0.31). In view of this, the proof of the first inequality, *viz.*

(Volume segment $BA\Gamma$, Volume segment $\Delta A\Gamma$) <

ΔΛ (surface segment $BA\Gamma$, surface segment $\Delta A\Gamma$)

proceeds fairly simply. Indeed, according to II, 2 the cones $\Theta A\Gamma$ and $HA\Gamma$ are determined, which are equal successively to the spherical segments $BA\Gamma$ and $\Delta A\Gamma$. Owing to this, we have

$$(\Theta Z, BZ) = (R+Z\Delta, Z\Delta) \tag{1}$$

$$(HZ, \Delta Z) = (R+ZB, ZB) \tag{2}$$

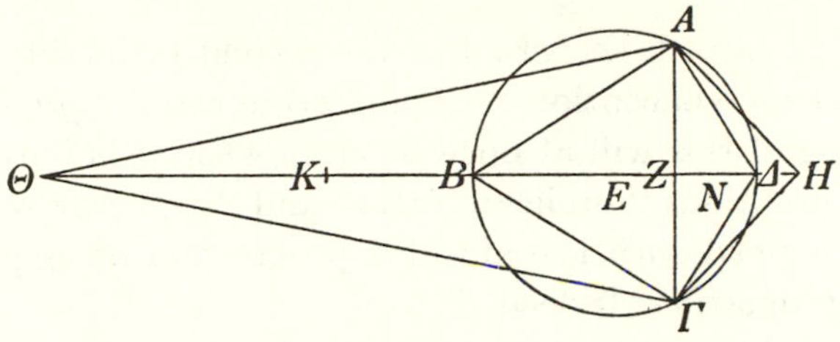

Fig. 88.

The ratio of the volumes of the segments therefore is equal to

$$(Z\Theta, ZH)\,,$$

while the ratio of their surfaces is equal to

$$[\mathbf{T}(AB),\ \mathbf{T}(A\Delta)], \quad i.e. \text{ to } \quad (ZB, Z\Delta)\,.$$

Thus it has to be proved that

$$(Z\Theta, ZH) < \mathbf{\Delta\Lambda}(ZB, Z\Delta) = [\mathbf{T}(ZB),\ \mathbf{T}(Z\Delta)]\,.$$

If $BK = R$, we have

by (1)	by (2)
separando $(B\Theta, ZB) = (BK, Z\Delta)$ (1*a*)	$(ZH, Z\Delta) = [ZK, ZB]$
permutando et invertendo	*permutando*
$(ZB, Z\Delta) = (B\Theta, BK)$	$(ZB, Z\Delta) = (ZK, ZH)$ (3)

$$(B\Theta, BK) = (ZK, ZH)\ldots \tag{4}$$

Because $\Theta Z > \Theta B$, we further have

$$(\Theta Z, \Theta K) > (\Theta B, \Theta K)\,,$$

whence (III; 0.42)

$$(\Theta Z, ZK) < (\Theta B, BK) = (ZK, ZH) \text{ (because of (4))}$$

or

$$\mathbf{O}(\Theta Z, ZH) < \mathbf{T}(ZK) .$$

Therefore

$$[\mathbf{O}(\Theta Z, ZH),\ \mathbf{T}(ZH)] < [\mathbf{T}(ZK),\ \mathbf{T}(ZH)] \text{ or because of (3)}$$

$$(\Theta Z, ZH) < [\mathbf{T}(ZB),\ \mathbf{T}(Z\Delta)] \qquad \text{q.e.d.}$$

In order to furnish the proof of the second inequality, we must first discuss the conception "sesquialterate ratio" (ἡμιόλιος λόγος). The meaning of this will at once be clear when it is borne in mind how the conceptions "duplicate ratio" and "triplicate ratio" could both be explained with the aid of a progression of magnitudes in continued proportion. Indeed, if

$$(A, B) = (B, \Gamma) = (\Gamma, \Delta) ,$$

(A, Δ) is called the triplicate ratio, and (B, Δ) the duplicate ratio of (Γ, Δ). It is now natural that (A, Δ) will be called the sesquialterate ratio of (B, Δ). Symbol: $(A, \Delta) = \mathbf{H\Lambda}(B, \Delta)$. In algebraic symbolism this amounts to

$$\frac{A}{\Delta} = \left(\frac{B}{\Delta}\right)^{\frac{3}{2}},$$

so that, properly speaking, a fractional exponent here occurs for the first time in mathematical literature.

The proof that a ratio (A, Δ) is the sesquialterate ratio of another ratio (B, Δ) can now be furnished by demonstrating the existence of a proportion

$$[\mathbf{T}(A),\ \mathbf{T}(B)] = (B, \Delta) .$$

Indeed, if

$$(B, \Gamma) = (\Gamma, \Delta), \text{ then}$$

$$[\mathbf{T}(A),\ \mathbf{T}(B)] = [\mathbf{T}(B),\ \mathbf{T}(\Gamma)] ,$$

whence

$$(A, B) = (B, \Gamma) = (\Gamma, \Delta) ,$$

whence

$$(A, \Delta) = \mathbf{H\Lambda}(B, \Delta) .$$

In the same way we shall succeed in proving the inequality

$$(A, \Delta) > \mathbf{H\Lambda}(B, \Delta)$$

when we can demonstrate that

$$[\mathbf{T}(\varDelta),\ \mathbf{T}(B)] > (B, \varDelta)\,.$$

Now in the second part of Prop. 8 it has to be proved that

$$(Z\Theta, ZH) > \mathbf{H\Lambda}(ZB, Z\varDelta)$$

and this will be possible according to the scheme just dealt with when a magnitude X is determined such that

$$(ZB, Z\varDelta) = (X, ZH)$$

and when it can then be proved that

$$[\mathbf{T}(Z\Theta),\ \mathbf{T}(X)] > (X, ZH)\,.$$

Now by (3) we had

$$(ZB, Z\varDelta) = (ZK, ZH)\,,$$

so that ZK can be substituted for X. It still has to be proved therefore that

$$[\mathbf{T}(Z\Theta),\ \mathbf{T}(ZK)] > (ZK, ZH)\,. \tag{5}$$

This is achieved as follows by Archimedes: for the ratio (ZK, ZH) there may according to (4) be substituted $(B\Theta, BK)$. In order to transform the latter expression, we remember that in view of Euclid II, 5 the following inequality is true:

$$\mathbf{O}(ZB, Z\varDelta) < \mathbf{O}(EB, E\varDelta)^{1)} \text{ or, because } EB = E\varDelta = BK\,,$$

$$(ZB, BK) < (BK, Z\varDelta), \text{ therefore by (1a)}$$

$$(ZB, BK) < (B\Theta, ZB)$$

or

$$\mathbf{T}(BZ) < \mathbf{O}(B\Theta, BK)\,.$$

Now suppose that

$$\mathbf{O}(B\Theta, BK) = \mathbf{T}(BN)\,,$$

then

$$(B\Theta, BN) = (BN, BK)\,.$$

---

[1]) In fact, $\mathbf{O}(ZB, Z\varDelta) = \mathbf{T}(EB) - \mathbf{T}(EZ) < \mathbf{T}(EB)$.

From this it follows that

$$(B\Theta, BK) = [\mathbf{T}(BN), \mathbf{T}(BK)]$$

componendo

$$(\Theta N, BN) = (KN, BK)$$

permutando

$$(\Theta N, KN) = (BN, BK)$$

$$(B\Theta, BK) = [\mathbf{T}(\Theta N),\ \mathbf{T}(KN)] \text{ and therefore by (4):}$$

$$(ZK, ZH) = [\mathbf{T}(\Theta N),\ \mathbf{T}(KN)]\ .$$

The proof of the desired inequality (5) will therefore have been furnished, if it can also be shown that

$$[\mathbf{T}(Z\Theta),\ \mathbf{T}(ZK)] > [\mathbf{T}(\Theta N),\ \mathbf{T}(KN)]$$

or

$$(Z\Theta, ZK) > (\Theta N, KN)\ .$$

This, however, follows at once from the inequality

$$\Theta Z > KZ\ .$$

In fact, it follows from this (III; 0.45) that

$$(\Theta Z, KZ) > (\Theta Z + ZN, KZ + ZN)$$

or

$$(\Theta Z, KZ) > (\Theta N, KN)\ .$$

Algebraically: Let the radius of the sphere be R, the heights ZB and $Z\Delta$ of the segments $BA\Gamma$ and $\Delta A\Gamma$ successively h_1 and h_2 ($h_1 > h_2$), the heights $Z\Theta$ and ZH of the cones $\Theta A\Gamma$ and $HA\Gamma$, which are equal successively to one of these two segments, H_1 and H_2, then we have for

the ratio of the volumes of the segments	$H_1 : H_2$,
the ratio of the surfaces	$h_1 : h_2$.

It now has to be proved that

$$\left(\frac{h_1}{h_2}\right)^{\frac{3}{2}} < \frac{H_1}{H_2} < \left(\frac{h_1}{h_2}\right)^2,$$

in which H_1 and H_2 successively are determined by the relations

$$\frac{H_1}{h_1} = \frac{R+h_2}{h_2} \quad \text{and} \quad \frac{H_2}{h_2} = \frac{R+h_1}{h_1}.$$

For the ratio $H_1:H_2$ we therefore find

$$\frac{H_1}{H_2} = \frac{R+h_2}{R+h_1} \cdot \left(\frac{h_1}{h_2}\right)^2. \tag{γ}$$

It follows at once from $h_2 < h_1$ that

$$\frac{H_1}{H_2} < \left(\frac{h_1}{h_2}\right)^2.$$

The inequality on the left can be reduced to

$$\sqrt{\frac{h_2}{h_1}} < \frac{R+h_2}{R+h_1}.$$

The correctness of this can be recognized by starting from

$$h_1 h_2 < R^2 .$$

Indeed, from this it follows that

$$\sqrt{h_1 h_2}\left(\sqrt{h_1} - \sqrt{h_2}\right) < R\left(\sqrt{h_1} - \sqrt{h_2}\right) \tag{δ}$$

or

$$(R+h_1)\sqrt{h_2} < (R+h_2)\sqrt{h_1} .$$

This derivation, however, is not the algebraic translation of Archimedes' argument. In the first part of the proof he attempts to arrive at an estimation of $H_1:H_2$ by first estimating $H_1 H_2$, and then dividing this by H_2^2. He finds

$$H_1 H_2 < (R+h_1)^2 ,$$

which we read at once from the expressions for H_1 and H_2, but which he derives from

$$\frac{H_1}{H_1 - h_1 - R} > \frac{H_1 - h_1}{H_1 - h_1 - R}$$

by writing *separando*

$$\frac{H_1}{R+h_1} < \frac{H_1 - h_1}{R} = \frac{h_1}{h_2} = \frac{R+h_1}{H_2}.$$

After this

$$\frac{H_1}{H_2} < \left(\frac{R+h_1}{H_2}\right)^2 = \left(\frac{h_1}{h_2}\right)^2.$$

In the second part he has to prove the inequality (5):

$$\left(\frac{H_1}{R+h_1}\right)^2 > \frac{R+h_1}{H_2}$$

or, which because of

$$\frac{R+h_1}{H_2} = \frac{h_1}{h_2} = \frac{H_1-h_1}{R}$$

comes to the same thing,

$$\left(\frac{H_1}{R+h_1}\right)^2 > \frac{H_1-h_1}{R}.$$

He again starts from the inequality

$$h_1h_2 < R^2$$

or

$$\frac{h_1}{R} < \frac{R}{h_2} = \frac{H_1-h_1}{h_1},$$

from which it follows that

$$h_1^2 < R(H_1-h_1).$$

Now suppose that

$$R(H_1-h_1) = t^2 \ (t = BN), \text{ then } \frac{H-h_1}{t} = \frac{t}{R},$$

therefore

$$\frac{H_1-h_1}{R} = \frac{t^2}{R^2} = \left(\frac{H_1-h_1+t}{R+t}\right)^2.$$

It still has to be proved that

$$\frac{H_1}{R+h_1} > \frac{H_1-h_1+t}{R+t}$$

or

$$\frac{H_1}{h_1-t} > \frac{R+h_1}{h_1-t},$$

which follows from $H_1 > R+h_1$.

In this example we see the cause of the great complexity of the arguments of the Greek theory of proportions: the ancients always wished to operate with ratios of line segments, and therefore could not use expressions such as (γ) or (δ) at all.

It is historically noteworthy in this connection that Archimedes follows up the above treatment of the proposition with another, which deviates from the method usually applied, and can therefore be much shorter (Fig. 89).

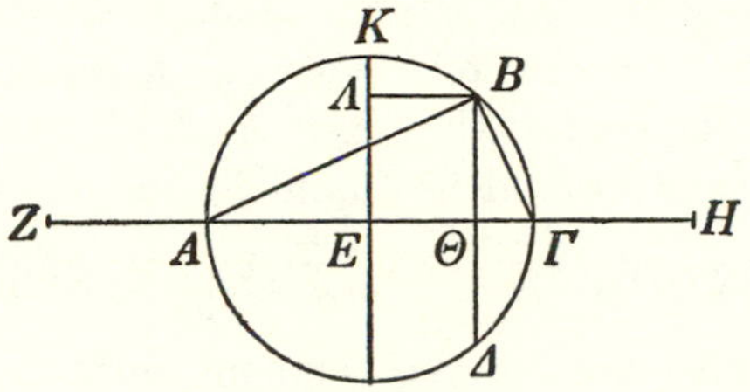

Fig. 89.

It is based on the conception of compound ratio, which is known from Euclid (III; 0.33) and which occupies the same position in the Greek theory of proportions as does the conception of product of two ratios (*i.e.* of two real positive numbers) in later mathematics. In fact, Archimedes considers the ratio of the segments $AB\Delta$ and $\Gamma B\Delta$ as compounded of the ratios

(segment $AB\Delta$, cone $AB\Delta$), (cone $AB\Delta$, cone $\Gamma B\Delta$), (cone $\Gamma B\Delta$, segment $\Gamma B\Delta$).

If $AZ = \Gamma H = EA$, these ratios are equal successively to

$$(H\Theta, \Gamma\Theta) \qquad (A\Theta, \Gamma\Theta) \qquad (A\Theta, Z\Theta)\,.$$

If the first two are compounded according to Euclid VI, 23, the ratio of the segments is found to be compounded of

$$[\mathbf{O}(H\Theta, A\Theta), \mathbf{T}(\Gamma\Theta)] \quad \text{and} \quad (A\Theta, Z\Theta)\,. \qquad (6)$$

Up to this point the argument does not deviate in any respect from the tradition of the Euclidean application of areas. Continuing according to this method, however, one would now have to introduce the fourth proportional X to $\Gamma\Theta$, $A\Theta$, and $H\Theta$, and by this means replace the first of the remaining ratios by $(X, \Gamma\Theta)$, *i.e.* by a ratio of line segments[1]).

One might imagine an extension of this method in the sense that the compound ratio (1) were interpreted as the ratio of two rectangular parallelepipeds the first of which should have the edges $H\Theta$, $A\Theta$, $A\Theta$, the second the edges $\Gamma\Theta$, $\Gamma\Theta$, $Z\Theta$. That would require a stereometrical proposition on the ratio of two such parallelepipeds, analogous to the proposition Euclid VI, 23 on parallelograms; thus an extension of the application of areas to solid geometry would be obtained, which, though it does not occur in Euclid, would fit in completely with the *Elements*. If we now represent a parallelepiped with edges A, B, Γ by $\Sigma(A, B, \Gamma)$, the ratio of the segments would be found to be

$$[\Sigma(H\Theta, A\Theta, A\Theta),\ \Sigma(\Gamma\Theta, \Gamma\Theta, Z\Theta)]\,, \tag{7}$$

while the duplicate ratio of the segment surfaces, *i.e.*

$$[\mathbf{T}(A\Theta),\ \mathbf{T}(\Gamma\Theta)]^{2)}\,,$$

could be written

$$[\Sigma(H\Theta, A\Theta, A\Theta),\ \Sigma(H\Theta, \Gamma\Theta, \Gamma\Theta)]\,. \tag{8}$$

Since $Z\Theta$ is greater than $H\Theta$, what was required to be proved follows at once from a comparison of (7) and (8).

What Archimedes actually does, deviates only terminologically from the procedure described above; he does not speak about the ratio of parallelepipeds, but treats the terms occurring in the ratios (6) (in agreement with what was later to become common practice, but contrary to the Euclidean tradition) as if they were numbers which can be multiplied one by the other. He therefore says, for example, that the ratio compounded of that of the

---

1) Indeed, one would then obtain

$$\mathbf{O}(H\Theta, A\Theta) = \mathbf{O}(\Gamma\Theta, X)\,,$$

whence

$$[\mathbf{O}(H\Theta, A\Theta), \mathbf{T}(\Gamma\Theta)] = [\mathbf{O}(\Gamma\Theta, X),\ \mathbf{T}(\Gamma\Theta)] = (X, \Gamma\Theta)\,.$$

2) For the ratio of the surfaces is

$$[\mathbf{T}(AB),\ \mathbf{T}(B\Gamma)] = (A\Theta, \Gamma\Theta)\,,$$

therefore their duplicate ratio is

$$[\mathbf{T}(A\Theta),\ \mathbf{T}(\Gamma\Theta)]\,.$$

rectangle on $H\Theta$, $A\Theta$ and the square on $\Gamma\Theta$ with that of $A\Theta$ to $Z\Theta$ is the same as the ratio of the rectangle on $H\Theta$, $A\Theta$ *multiplied by* (ἐπὶ) $A\Theta$ to the square on $\Gamma\Theta$ *multiplied by* $Z\Theta$. He thus takes the first step on the road to arithmetization of the geometrical argument, which is not to be fully developed until the 17th century.

It is now also easier to furnish the proof of the second part of II, 8. For the sesquialterate ratio of the surfaces of the segments, *i.e.* of $[\mathbf{T}(AB), \mathbf{T}(\Gamma B)]$, the ratio of the cubes with edges AB and ΓB successively is substituted[1]). These cubes have the same ratio as the cubes with edges $A\Theta$ and $B\Theta$ successively, the ratio of which is compounded of

$$[\mathbf{T}(A\Theta),\ \mathbf{T}(B\Theta)] \quad \text{and} \quad (A\Theta, B\Theta)\,.$$

Now

$$(A\Theta, B\Theta) = [\mathbf{T}(B\Theta),\ \mathbf{O}(B\Theta, \Gamma\Theta)]\,,$$

so that, because of the presence of the common term $\mathbf{T}(B\Theta)$, the ratio of the cubes is equal to

$$[\mathbf{T}(A\Theta),\ \mathbf{O}(B\Theta, \Gamma\Theta)]\,.$$

If we now consider these two surfaces as bases of parallelepipeds with common height $H\Theta$[2]), the sesquialterate ratio of the segment surfaces becomes equal to

$$[\Sigma(A\Theta, A\Theta, H\Theta),\ \Sigma(B\Theta, \Gamma\Theta, H\Theta)]$$

and from a comparison with (7) it appears that it still has to be proved that

$$\Sigma(\Gamma\Theta, \Gamma\Theta, Z\Theta) < \Sigma(B\Theta, \Gamma\Theta, H\Theta)\,.$$

This amounts to

$$[\mathbf{T}(\Gamma\Theta),\ \mathbf{O}(B\Theta, \Gamma\Theta)] < (H\Theta, Z\Theta)$$

or

$$(\Gamma\Theta, B\Theta) < (H\Theta, Z\Theta)$$

or

$$(\Gamma\Theta, B\Theta) < (H\Theta, A\Theta + EK)\,. \tag{9}$$

---

[1]) This shows that Archimedes constantly argues stereometrically, and that the ἐπί locution in the style of our "base multiplied by height" serves to denote volumes of solids.

[2]) Expressed by ἐπί in Archimedes.

For this it is sufficient that

$$(\Gamma H, A\Theta + \Lambda K) > (\Gamma\Theta, B\Theta)\,. \tag{10}$$

Indeed, when this has been proved, we have *permutando*

$$(\Gamma H, \Gamma\Theta) > (A\Theta + \Lambda K, B\Theta)$$

or *componendo*

$$(\Theta H, \Gamma\Theta) > (A\Theta + KE, B\Theta)$$

or *permutando*

$$(\Theta H, A\Theta + KE) > (\Gamma\Theta, B\Theta),\ \textit{i.e.}\ (9)\,.$$

In order to prove (10), it may also be shown that

$$(\Gamma H, A\Theta + \Lambda K) > (B\Theta, A\Theta) = (\Lambda E, A\Theta)$$

or *permutando*

$$(KE, \Lambda E) > (A\Theta + K\Lambda, A\Theta)$$

or *separando*

$$(K\Lambda, \Lambda E) > (K\Lambda, A\Theta)$$

or

$$\Lambda E < A\Theta\,,$$

which is indeed true.

Formulated algebraically, this argument is as follows:

The ratio of the volumes is

$$\frac{I_1}{I_2} = \frac{H_1}{H_2} = \frac{H_1}{h_1}\cdot\frac{h_1}{h_2}\cdot\frac{h_2}{H_2} = \frac{(R+h_2)h_1^2}{(R+h_1)h_2^2},$$

that of the surfaces

$$\frac{O_1}{O_2} = \frac{h_1}{h_2}.$$

It follows at once from $h_1 > h_2$ that $\dfrac{I_1}{I_2} < \left(\dfrac{O_1}{O_2}\right)^2$.

Archimedes therefore appears to apply here the argument which we gave above as the most obvious one.

In the second part it is written (if $\Theta B = r$):

$$\left(\frac{O_1}{O_2}\right)^{\frac{3}{2}} = \left(\frac{AB}{\Gamma B}\right)^3 = \left(\frac{h_1}{r}\right)^3 = \frac{h_1^2}{rh_2} = \frac{h_1^2(R+h_2)}{rh_2(R+h_2)}.$$

From a comparison with $\dfrac{I_1}{I_2}$ it appears that it is still to be proved that

$$h_2(R+h_1) < r(R+h_2)$$

or

$$\frac{h_2}{r} < \frac{R+h_2}{R+h_1} \quad \text{or} \quad \frac{r}{h_1} < \frac{R+h_2}{R+h_1}.$$

Now $h_1 > r$, whence

$$\frac{h_1}{R-r} > \frac{r}{R-r},$$

whence

$$\frac{h_1}{R-r+h_1} > \frac{r}{R},$$

whence

$$\frac{r}{h_1} < \frac{R}{R-r+h_1},$$

whence *a fortiori*

$$\frac{r}{h_1} < \frac{R+h_2}{R+h_1}.$$

Proposition 9.

Of all segments of spheres which have equal surfaces the hemisphere is the greatest.

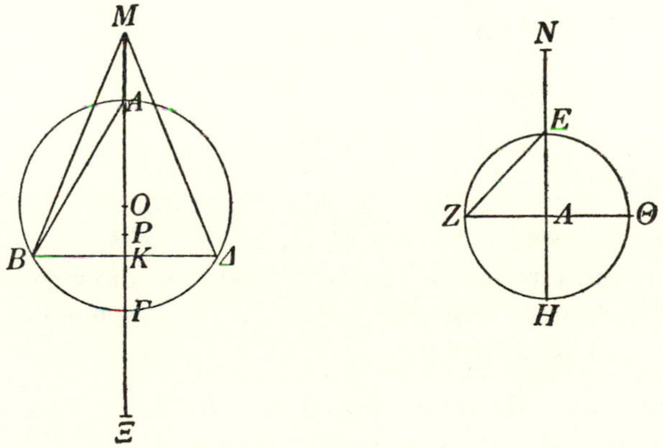

Fig. 90.

In Fig. 90 let $AB\Delta$ and $EZ\Theta$ be two segments of spheres, of which only the latter is a hemisphere. It is given that the surfaces are equal, therefore

$$AB = EZ.$$

It has to be proved that the volume of $EZ\Theta$ is greater than that of $AB\Delta$.

In order to prove this, the cones are constructed which are successively equal to the segments under consideration. To this end suppose that $\Gamma\Xi = \Gamma O$, the radius of the sphere O of which $AB\Delta$ forms a part, $NE = E\Lambda$, the radius of the sphere Λ of which $EZ\Theta$ is one half. Then we have

$$\text{segment } AB\Delta = \text{cone } MB\Delta, \text{ when } (MK, AK) = (K\Xi, K\Gamma)$$

or

$$\mathbf{O}(MK, K\Gamma) = \mathbf{O}(AK, K\Xi) \tag{1}$$

Moreover the hemisphere $EZ\Theta$ is equal to the cone $NZ\Theta$.

The proof will have been furnished, if it has been shown that

$$(N\Lambda, MK) > [\mathbf{T}(BK), \mathbf{T}(Z\Lambda)]$$

or

$$(E\Lambda, MK) > [\mathbf{T}(BK), 2\mathbf{T}(E\Lambda)] = [\mathbf{T}(BK), \mathbf{T}(EZ)] .$$

Now measure $AP = E\Lambda$, then, because $AB = EZ$, it still has to be proved that

$$(AP, MK) > [\mathbf{T}(BK), \mathbf{T}(AB)] = (K\Gamma, A\Gamma)$$

or

$$\mathbf{O}(AP, A\Gamma) > \mathbf{O}(MK, K\Gamma) .$$

Because of (1) this is equivalent to

$$\mathbf{O}(AP, A\Gamma) > \mathbf{O}(AK, K\Xi) ,$$

an inequality which will have to be derived from the position of P on $A\Gamma$, as compared with that of K.

Now the nucleus of the proof consists in showing that P is nearer to the centre O than K. Indeed, when this has been proved, then by Euclid II, 5 we have:

$$\mathbf{O}(AP, P\Gamma) > \mathbf{O}(AK, K\Gamma)\text{[1]}$$

Add to either member

$$\mathbf{T}(AP) = \mathbf{T}(E\Lambda) = \tfrac{1}{2}\mathbf{T}(EZ) = \tfrac{1}{2}\mathbf{T}(AB) = \mathbf{O}(AK, \Gamma\Xi) ,$$

---

[1]) Indeed:

$$\mathbf{O}(AP, P\Gamma) + \mathbf{T}(PO) = \mathbf{O}(AK, K\Gamma) + \mathbf{T}(KO) = \mathbf{T}(AO) .$$

Because

$PO < KO$, we therefore have $\mathbf{O}(AP, P\Gamma) > \mathbf{O}(AK, K\Gamma)$.

then we have

$$\mathbf{O}(AP, A\Gamma) > \mathbf{O}(AK, K\Xi) \qquad \text{q.e.d.}$$

In order to prove that P lies between O and K, we have to distinguish the cases when $AB\Delta$ is greater or less than the half of the sphere O. Archimedes only deals with the first case. In Fig. 90 we have

$$2\mathbf{T}(AP) = \mathbf{T}(AB) \begin{cases} > 2\mathbf{T}(AO) \\ = \mathbf{O}(AK, A\Gamma) < 2\mathbf{T}(AK), \end{cases}$$

whence $AO < AP < AK$; therefore P lies between O and K.

In the second case (Fig. 91), however, we have

$$2\mathbf{T}(AP) = \mathbf{T}(AB) \begin{cases} < 2\mathbf{T}(AO) \\ = \mathbf{O}(AK, A\Gamma) > 2\mathbf{T}(AK), \end{cases}$$

whence $AO > AP > AK$; therefore again P lies between O and K.

Algebraically: In sphere O let the radius be R, $AK = h_1$, $MK = H$, $AB = k$, $BK = r$, $K\Gamma = h_2$, and in sphere Λ let the radius be ϱ, then it is given that

$$2\varrho^2 = k^2 \quad \text{or} \quad \varrho^2 = Rh_1$$

and it has to be proved that

$$\tfrac{2}{3}\pi\varrho^3 > \tfrac{1}{3}\pi r^2 H$$

or

$$2\varrho^3 > r^2 H.$$

Archimedes proves

$$|R - \varrho| < |R - h_1|$$

by stating that $2\varrho^2 = k^2 = 2Rh_1$ always lies between $2R^2$ and $2h_1^2$.

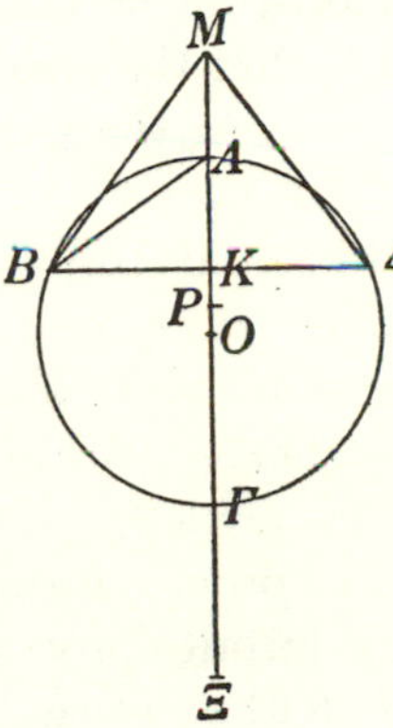

Fig. 91.

Now

$$\varrho(2R - \varrho) > h_1 h_2,$$

from which it follows, when adding $\varrho^2 = Rh_1$ to either member and multiplying either member by h_1, that

$$\varrho \cdot 2Rh_1 > h_1^2(h_2 + R)$$

or

$$\varrho k^2 > h_1 h_2 H$$

or

$$2\varrho^3 > r^2 H.$$

CHAPTER VI.

MEASUREMENT OF THE CIRCLE

1. The work on the measurement of the circle, in which Archimedes derives the ratio between the circumference and the diameter of a circle, which was to become one of the most popular results of his mathematical investigations, is a very short treatise comprising only three propositions. As appears both from its language, from which all traces of the Siculo-Dorian dialect have vanished, and from the argumentation, which is scrappy and rather careless, it has not come down to us in its original form. It is quite possible that the fragment we possess formed part of a longer work, which is quoted by Pappus[1]) under the title *On the Circumference of the Circle* (*περὶ τῆς τοῦ κύκλου περιφερείας*), and that the latter also dealt with the more general question as to the ratio between the length of an arc of a circle and that of its chord[2]).

The first proposition has for its object to reduce the arithmetical quadrature of the circle to the arithmetical rectification, *i.e.* to show that the area of a circle with a given radius can be calculated as soon as the circumference is known.

Proposition 1.

The area of any circle is equal to a right-angled triangle in which one of the sides about the right angle is equal to the radius, and the base [*i.e. the other of the sides about the right angle*] *to the circumference* [*of the circle*][3]).

The proof is furnished by means of the indirect method for dealing with infinite processes (difference form of the compression method III; 8.21). Let the area of the circle be K, that of the right-angled triangle in question Δ, that of the polygons of n sides inscribed in and circumscribed about the circle respectively I_n and C_n.

---

[1]) Pappus, *Collectio* V, 2; 312, line 20. Unless the reader is inclined to assume with Hultsch (l. c. 313, Note 1) that the writer is here quoting the title of Archimedes' treatise in a free form.

[2]) This is supposed by A. Favaro, *Archimede*; Roma 1923, p. 53.

[3]) Free translation of the dubious Greek text: *πᾶς κύκλος ἴσος ἐστὶ τριγώνῳ ὀρθογωνίῳ, οὗ ἡ μὲν ἐκ τοῦ κέντρου ἴση μιᾷ τῶν περὶ τὴν ὀρθήν, ἡ δὲ περίμετρος τῇ βάσει.*

Now suppose $K > \Delta$. According to the argument of Euclid XII, 2[1]) (*i.e.* on the strength of the postulate of Eudoxus), n can then be so determined that

$$K - I_n < K - \Delta ,$$

whence

$$I_n > \Delta .$$

I_n, however, is equal to the area of a right-angled triangle in which the sides about the right angle are equal successively to the apothem and the perimeter of the regular polygon of n sides inscribed in the circle; these are less successively than the radius and the circumference of the circle, *i.e.* less than the corresponding sides of the triangle having an area Δ. From this it follows that

$$I_n < \Delta ,$$

which is contrary to the supposition.

Secondly, suppose $K < \Delta$. It is now possible (*vide* S.C. I,6) to determine a number n such that

$$C_n - K < \Delta - K ,$$

whence

$$C_n < \Delta .$$

In this, C_n is equal to the area of a right-angled triangle in which the sides about the right angle are equal successively to the radius of the circle and the perimeter of the regular circumscribed polygon of n sides, *i.e.* successively equal to and greater than the corresponding sides of the triangle having an area Δ. From this it follows that

$$C_n > \Delta ,$$

which is contrary to the supposition. The only possibility is therefore that

$$K = \Delta .$$

The next proposition, *viz.* Prop. 2, is based on Prop. 3; we wi therefore mention it after the latter.

Proposition 3.

The circumference of any circle is three times the diameter and exceeds it by less than one-seventh of the diameter and by more than ten-seventyoneths.

---

1) *Elements of Euclid* II, 225. Cf. S.C. I, 6.

Therefore

$$\left(3+\tfrac{10}{71}\right) \text{Diameter} < \text{Circumference} < \left(3+\tfrac{1}{7}\right) \text{Diameter} .$$

The upper limit is the familiar Archimedean approximation $\pi \sim \frac{22}{7}$.

To enable the reader to follow Archimedes' argument as closely as possible, we will first give a broad outline of the proof and then render the results of the calculation in Greek symbols, elucidated by an Indo-Arabian transcription.

In the first part of the proposition an upper limit for the required ratio is derived.

In Fig. 92 let $A\Gamma$ be the diameter of a circle with centre E, ΓZ the tangent at Γ, $\angle\, ZE\Gamma$ one-third of a right angle. Then the ratio

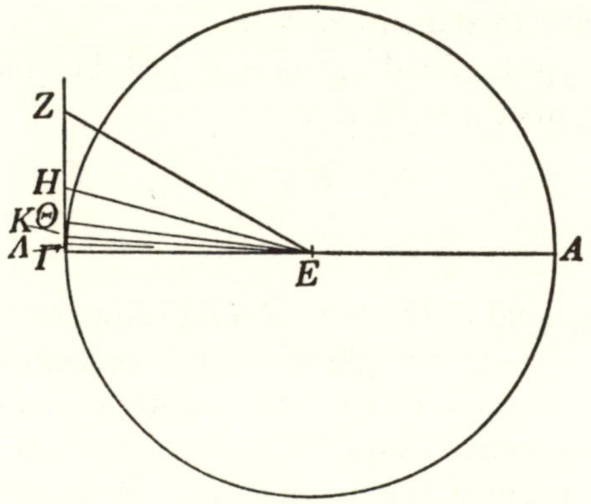

Fig. 92.

$(EZ, \Gamma Z)$ is known, *viz.* $2:1$, while for $(E\Gamma, \Gamma Z)$, *i.e.* in our notation $\sqrt{3}:1$, a rational approximation on the low side $(\frac{265}{153})$ appears to be known. If the bisector of $\angle\, ZE\Gamma$ meets the line ΓZ in H, we have

$$(ZE, \Gamma E) = (ZH, \Gamma H)\,,$$

from which it follows *componendo* and *permutando* that

$$(ZE+E\Gamma, \Gamma Z) = (\Gamma E, \Gamma H)\,.$$

Thus an approximation value for $(\Gamma E, \Gamma H)$ is known, and consequently for $[\mathbf{T}(\Gamma E),\ \mathbf{T}(\Gamma H)]$, therefore *componendo* for

$$[\mathbf{T}(HE),\ \mathbf{T}(\Gamma H)]\,,$$

and from this an approximate value for the ratio $(HE, \Gamma H)$ is found. It is obvious that through the knowledge of the ratios $(EH, \Gamma H)$ and $(E\Gamma, \Gamma H)$ the same standpoint has been reached as when originally the ratios $(EZ, \Gamma Z)$ and $(E\Gamma, \Gamma Z)$ were known.

If Z_n denotes the side of the circumscribed regular polygon of n sides, we have

$$(E\Gamma, Z\Gamma) = (R, \tfrac{1}{2}Z_6)$$

$$(E\Gamma, H\Gamma) = (R, \tfrac{1}{2}Z_{12})$$

and therefore, proceeding in this way, finally an approximation is bound to be reached for the ratio $(R, \frac{1}{2}z_{96})$, and thus also for the ratio between the perimeter of the regular circumscribed polygon of 96 sides and R; from this, however, an upper limit for the ratio between the circumference and the diameter of the circle then follows.

In modern symbols the method is equivalent to a repeated application of the formula

$$Z_{2n} = \frac{RZ_n}{R + \sqrt{R^2 + \frac{1}{4}Z_n{}^2}}.$$

Let us now see how Archimedes performs the calculation. Without any comment he puts

$$(E\Gamma, \Gamma Z) = (\overline{\sigma\xi\varepsilon}, \overline{\varrho\nu\gamma})\ ^{1)} \qquad (265:153)$$

and in this connection writes

$$(EZ, \Gamma Z) = (\overline{\tau\varsigma}, \overline{\varrho\nu\gamma}) \qquad (306:153).$$

We therefore get

$$(\Gamma E, \Gamma H) > (\overline{\varphi o\alpha}, \overline{\varrho\nu\gamma}) \qquad (571:153)$$

$$[\mathrm{T}(\Gamma E), \mathrm{T}(\Gamma H)] = (\overset{\lambda\beta}{M}\ \overline{,\varsigma\mu\alpha}, \overset{\beta}{M}\ \overline{,\gamma\upsilon\vartheta})\ ^{2)} \qquad (326041:23409)$$

$$[\mathrm{T}(EH), \mathrm{T}(\Gamma H)] = (\overset{\lambda\delta}{M}\ \overline{,\vartheta\upsilon\nu}, \overset{\beta}{M}\ \overline{,\gamma\upsilon\vartheta}) \qquad (349450:23409)$$

$$(EH, \Gamma H) = (\overline{\varphi\text{ϙ}\alpha}\ \eta', \overline{\varrho\nu\gamma}) \qquad (591\tfrac{1}{8}:153).$$

In his terminology Archimedes does not always distinguish be-

---

[1]) The meaning seems to be $(E\Gamma, \Gamma Z) > (\overline{\sigma\xi\varepsilon}, \overline{\varrho\nu\gamma})$. For the Greek numerical system compare III; 0.6.

[2]) This line does not occur in the text.

tween exact and approximate equality. The argument, however, only has conclusive force when, just as $(E\Gamma, \Gamma Z) > (\overline{\sigma\xi\varepsilon}, \overline{\varrho\nu\gamma})$ was started from, the final result is also $(EH, \Gamma H) > (\overline{\varphi\text{ϙ}\alpha}\ \eta', \overline{\varrho\nu\gamma})$.

As will readily be seen, the > sign has remained in force through all the preceding transformations of the proportions; this will also be the case in the extraction of the square root finally carried out, since $\overline{\varphi\text{ϙ}\alpha}\ \eta'$ is an approximation on the low side.

For the moment we shall ignore the question how Archimedes arrives at this approximation, and we shall first follow the rest of the calculation.

By repeated construction of bisectors, after H the points Θ, K, Λ successively are found. Now it was already found above that

$$(EH, \Gamma H) = (\overline{\varphi\text{ϙ}\alpha}\ \eta', \overline{\varrho\nu\eta}), \text{ in which } \Gamma H = \tfrac{1}{2}Z_{12}.$$

Archimedes further finds that

$$(E\Gamma, \Gamma\Theta) > (\overline{,\alpha\varrho\xi\beta}\ \eta', \overline{\varrho\nu\gamma}), \quad i.e. \quad \frac{R}{\frac{1}{2}Z_{24}} > \frac{1162\frac{1}{8}}{153}$$

$$(E\Gamma, \Gamma K) > (\overline{,\beta\tau\lambda\delta}\ \delta', \overline{\varrho\nu\gamma}), \quad i.e. \quad \frac{R}{\frac{1}{2}Z_{48}} > \frac{2334\frac{1}{4}}{153}$$

$$(E\Gamma, \Gamma\Lambda) > (\overline{,\delta\chi o\gamma}\ \text{L}', \overline{\varrho\nu\gamma}), \quad i.e. \quad \frac{R}{\frac{1}{2}Z_{96}} > \frac{4673\frac{1}{2}}{153}.$$

Now it therefore appears that

(Diameter, Perimeter of the circumscribed polygon of 96 sides)

$$> (\overline{,\delta\chi o\gamma}\ \text{L}', \overset{\alpha}{M}\ \overline{,\delta\chi\pi\eta})$$

in which the last number has been found as the product of $\overline{\text{ϙ}\varsigma}$ and $\overline{\varrho\nu\gamma}$.

Now $\overset{\alpha}{M}\ \overline{,\delta\chi\pi\eta}$ comprises three times $\overline{,\delta\chi o\gamma}\ \text{L}'$ with a remainder $\overline{\chi\xi\zeta}\ \text{L}'$, which is less than one-seventh of $\overline{,\delta\chi o\gamma}$. Thus *a fortiori* the circumference of the circle is less than three times the diameter plus one-seventh of it.

In Indo-Arabian symbols the last part of the calculation proceeds as follows:

$$\frac{d}{Z_{96}} > \frac{4673\frac{1}{2}}{153},$$

whence

$$96 . Z_{96} < \frac{96 . 153}{4673\frac{1}{2}} d < \frac{22}{7} d ,$$

whence *a fortiori*

$$\text{Circumference of the circle} < \tfrac{22}{7} d .$$

In the second part of the proposition a lower limit for the ratio between the circumference and the diameter is derived by considering inscribed regular polygons of an increasing number of sides. In Fig. 93 let ΓB be the side of the inscribed regular hexagon in

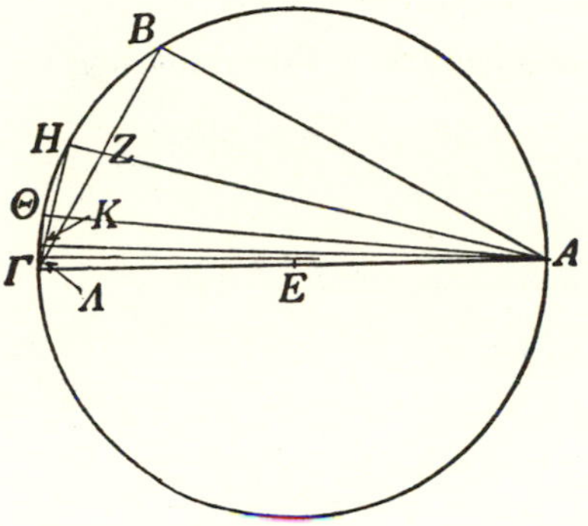

Fig. 93.

the circle with diameter $A\Gamma$ and centre E. A rational approximation to $(AB, B\Gamma)$, *i.e.* to $\sqrt{3}:1$, on the high side $(\frac{1351}{780})$ is now started from. Further $(A\Gamma, \Gamma B)$ is known. If now the bisector of $\angle BA\Gamma$ meets the side $B\Gamma$ in Z and the arc of the circle in H, it follows from the similarity of the triangles $AH\Gamma$ and ΓHZ that

$$(AH, \Gamma H) = (A\Gamma, \Gamma Z) = (AB, BZ) = (A\Gamma + AB, B\Gamma) ,$$

so that $(AH, \Gamma H)$ is known. From this follows $[\mathbf{T}(AH), \mathbf{T}(\Gamma H)]$, thence *componendo* $[\mathbf{T}(A\Gamma), \mathbf{T}(\Gamma H)]$, and thence $(A\Gamma, \Gamma H)$. Since $(A\Gamma, \Gamma H)$ and $(AH, \Gamma H)$ are known, with respect to $H\Gamma$ the same point has been reached as originally with respect to $B\Gamma$.

Indeed, if z_n denotes the side of the inscribed regular polygon of n sides, and d the diameter, we have

$$(A\Gamma, \Gamma B) = (d, z_6)$$

$$(A\Gamma, \Gamma H) = (d, z_{12}) .$$

Proceeding in this way, we finally arrive at an approximation for (d, z_{96}), and consequently for the ratio between the perimeter of the inscribed regular polygon of 96 sides and the diameter; from this, a lower limit for the ratio between the circumference and the diameter of the circle follows.

When the complementary chord of z_n is denoted by k_n $({k_n}^2 = d^2 - {z_n}^2)$, the method is equivalent to a repeated application of the formula

$$\frac{k_{2n}}{z_{2n}} = \frac{d+k_n}{z_n},$$

from which after squaring it follows *componendo* that

$$\frac{d^2}{{z_{2n}}^2} = \frac{2d^2+2dk_n}{{z_n}^2}.\ ^{1)}$$

We now mention the successive results. If the points obtained after H are Θ, K, Λ, successively, we first had

$$(AB, B\Gamma) < (\overline{,\alpha\tau\nu\alpha}, \overline{\psi\pi}) \qquad (1351{:}780)$$

and

$$(A\Gamma, B\Gamma) = (\overline{,\alpha\varphi\xi}, \overline{\psi\pi}) \qquad (1560{:}780)$$

---

[1]) The two methods used by Archimedes might be replaced by a single one by proving the truth of the following relations:

$$Z_{2n} = \frac{z_n Z_n}{z_n + Z_n} \qquad Z_{2n} = \frac{2z_{2n}^2}{z_n}$$

If the perimeters of the circum- and inscribed polygons of n sides are denoted by P_n and p_n respectively, it follows from this that

$$P_{2n} = \frac{2p_n P_n}{p_n + P_n} \qquad p_{2n} = \sqrt{p_n P_{2n}}.$$

These formulae express that in the progression

$$P_n, p_n, P_{2n}, p_{2n}, P_{4n}, p_{4n}, \text{ etc.},$$

starting from the third term, every term P_{2i} is the harmonic mean and every term p_{2i} the geometric mean of the preceding two. If, starting from P_6 and p_6, we now compute successively the terms of this progression, we arrive at the values of P_{96} and p_{96} required by Archimedes. As will be seen, this method differs appreciably from the one actually used by Archimedes. It is therefore decidedly wrong to say, as has been done by H. Dörrie (*Triumph der Mathematik* (Breslau 1933), pp. 183–187), that Archimedes calculated his approximations to π in this way.

whence $(AH, H\Gamma) = (\overline{,\beta \text{ϡ}\iota\alpha}, \overline{\psi\pi})$ (2911:780).

It is then found that

$$(A\Gamma, \Gamma H) < (\overline{,\gamma\iota\gamma}\ \text{L}'\delta', \overline{\psi\pi}), \quad i.e.\ \frac{d}{z_{12}} < \frac{3013\frac{3}{4}}{780}$$

$$(A\Gamma, \Gamma\Theta) < (\overline{,\alpha\omega\lambda\eta}\ \overline{\vartheta}\ \iota\alpha', \overline{\sigma\mu}), \quad i.e.\ \frac{d}{z_{24}} < \frac{1838\frac{9}{11}}{240}$$

$$(A\Gamma, \Gamma K) < (\overline{,\alpha\vartheta}\ \varsigma', \overline{\xi\varsigma}), \quad i.e.\ \frac{d}{z_{48}} < \frac{1009\frac{1}{6}}{66}$$

$$(A\Gamma, \Gamma\Lambda) < (\overline{,\beta\iota\zeta}\ \delta', \overline{\xi\varsigma}), \quad i.e.\ \frac{d}{z_{96}} < \frac{2017\frac{1}{4}}{66}$$

therefore $(\Gamma\Lambda, A\Gamma) > (\overline{\xi\varsigma}, \overline{,\beta\iota\zeta}\ \delta')$ or $\frac{z_{96}}{d} > \frac{66}{2017\frac{1}{4}}$.

From this it follows that

(Perimeter of the inscribed regular polygon of 96 sides, Diameter)

$$> (\overline{,\varsigma\tau\lambda\varsigma}, \overline{,\beta\iota\zeta}\ \delta')$$

in which $\overline{,\varsigma\tau\lambda\varsigma}$ has been obtained as the product of $\overline{\text{ϙ}\varsigma}$ and $\overline{\xi\varsigma}$.

Now $\overline{,\varsigma\tau\lambda\varsigma}$ comprises more than three times $\overline{,\beta\iota\zeta}\ \delta'$ plus ten-seventyoneths of this. From this it follows that the circumference of the circle is greater than three times its diameter plus ten-seventyoneths of it.

The last part of the calculation can be rendered as follows:

$$\frac{z_{96}}{d} > \frac{66}{2017\frac{1}{4}},$$

whence

$$\frac{96 z_{96}}{d} > \frac{96.66}{2017\frac{1}{4}} = \frac{6336}{2017\frac{1}{4}} > 3 + \frac{10}{71},$$

therefore *a fortiori*

$$\text{Circumference of the circle} > 3\tfrac{10}{71}\ \text{diameter}.$$

2. Now the question remains to be discussed how Archimedes can in general have proceeded for the rational approximation to square roots, and how in particular he can have found the two rational limits for $\sqrt{3}$ which formed the starting points for the two

parts of the calculation. We here touch upon a cluster of problems which, more than any other point in Greek mathematics, has given rise to historical investigation and mathematical reconstruction. The literature on this subject is so vast that we have to refrain from a complete critical discussion[1]); we therefore confine ourselves to an exposition of some solutions of the problem which are particularly trustworthy because of their close affiliation with genuinely Greek calculating methods.

As their common basis may be considered a passage in Heron's *Metrica*[2]), where an approximation to the square root of 720 is found as follows: the root from the square number nearest to 720 is 27; now divide 720 by 27; the quotient is $26+\frac{2}{3}$; a better approximation to $\sqrt{720}$ than 27 itself is now

$$\tfrac{1}{2}(27+26+\tfrac{2}{3}) = 26+\tfrac{1}{2}+\tfrac{1}{3}\,.$$

The acceptability of this procedure is obvious: if 27 is too great, the quotient 720:27 is too small; it is to be expected that the arithmetic mean of the two values will be an improved approximation.

Formulated in a general way, the method therefore consists in that, if α_1 is an approximation to $\sqrt{d}$, another (usually better[3])) approximation is obtained by calculating $\beta_1=\dfrac{d}{\alpha_1}$, and then forming

$$\alpha_2 = \frac{\alpha_1+\beta_1}{2}. \tag{1}$$

---

[1]) The reader who is interested in this subject will find a critical survey of the older literature in T. L. Heath, *Archimedes. Introduction* lxxxiv–xcix.
More up to date are the discussions by Jos. E. Hofmann:

a. Erklärungsversuche für Archimed's Berechnung von $\sqrt{3}$. Archiv f. Gesch. Math. Nat. Techn. 12 (1930) 386–408.

b. Ueber die Annäherung von Quadratwurzeln bei Archimedes und Heron. Jahresbericht D.M.V. 43 (1934) 187–210.

[2]) *Heronis Opera* III, 18. The manuscript of the *Metrica* was not traced until 1896; most of the older, partly complicated, attempts at a reconstruction of the Archimedean method have lost much of their value through the discovery of this unique classical testimony.

[3]) In order to find out when α_2 is a better approximation than α_1, we write

$$|\sqrt{d}-\alpha_2| = \frac{(\sqrt{d}-\alpha_1)^2}{2\alpha_1}\,.$$

Continuing this procedure with α_2, we have to determine

$$\beta_2 = \frac{d}{\alpha_2} = \frac{2\alpha_1\beta_1}{\alpha_1+\beta_1},$$

so that β_2 appears to be the harmonic mean between the original values α_1 and β_1.[1])

The method applied by Heron therefore consists in that, starting from two numbers having $\sqrt{d}$ for their geometric mean, the arithmetic and the harmonic mean of these numbers successively are formed. These again have $\sqrt{d}$ for their geometric mean, so that the operation may be started over again.

We also observe that the approximation obtained, *viz.* α_i $(i \geqq 2)$, will always be on the high side of $\sqrt{d}$, because the arithmetic mean of two unequal numbers is greater than the geometric mean.

The same result as the relation (1) is arrived at by means of another formula, the occurrence of which in Greek mathematics is borne out by several references[2]), and the application of which has even been shown in Babylonian mathematics of about 2000 B.C.[3]). In modern notation it is as follows:

A necessary and sufficient condition for α_2 to be better than α_1 is then

$$\frac{(\sqrt{d}-\alpha_1)^2}{2\alpha_1} < |\sqrt{d}-\alpha_1| \quad \text{or} \quad |\sqrt{d}-\alpha_1| < 2\alpha_1 \quad \text{or} \quad \sqrt{d} < 3\alpha_1 .$$

If α_1 was a reasonably good approximation, this condition will naturally be fulfilled.

[1]) As appears from certain passages in Plato and Archytas, the formation of these means is among the problems of Greek mathematics which were already dealt with at an early date. It is quite probable, though it cannot be directly proved, that it was used in connection with the approximation to square roots even before Archimedes. The passages in question from Plato and Archytas are to be found side by side in the treatise by O. Toeplitz to be quoted in Note 2 on p. 237.

[2]) *Stereometrica. Heronis Opera,* V. (Leipzig 1914). We give the following examples:

$$\text{page 150} \quad \sqrt{1125} = 33+\tfrac{1}{2}+\tfrac{1}{22}\,.$$
$$\text{page 152} \quad \sqrt{1081} = 32+\tfrac{1}{2}+\tfrac{1}{4}+\tfrac{1}{8}+\tfrac{1}{64}\,.$$
$$\sqrt{108} = 10+\tfrac{1}{3}+\tfrac{1}{15}\,.$$

Cf. further: P. Tannery, *L'Arithmétique des Grecs dans Héron d'Alexandrie.* Mém. Scientif. I (Paris 1912), 204 *et seq.*

[3]) O. Neugebauer, *Ueber die Approximation irrationaler Quadratwurzeln in der babylonischen Mathematik.* Archiv für Orientforschung VII (1931), Pp. 90 *et seq.*

Let p be an approximation to $\sqrt{d}$ on the low side, so that $d = p^2 + r$. A better approximation is then

$$q = p + \frac{r}{2p}. \tag{2}$$

The truth of this may have been recognized with the aid of the rule for the squaring of a sum which we render algebraically by $(a+b)^2 = a^2 + 2ab + b^2$, and which is expressed geometrically in the Greek application of areas (Euclid II, 4). Indeed, the squaring of the second member gives d, if we neglect the square of $\frac{r}{2p}$, which will be small if the approximation p is good enough.

It is also conceivable that the rule in question was an application of the one occurring in Heron. In fact, with the approximation to $\sqrt{d}$: $\alpha_1 = p$ is to be associated as β_1 the value $p + \frac{r}{p}$; the arithmetic mean of the two approximations is then $p + \frac{r}{2p}$.

If $d = p^2 - r$, in the same way the following approximation is found

$$\sqrt{d} \backsim p - \frac{r}{2p}.$$

Instances of this are also known in Greek mathematics[1]).

Henceforth we shall consider the two above formulae as a single rule, and refer to it as the Babylonian rule.

It now appears that the vast majority of all the approximations to square roots occurring in Archimedes and Heron can be accounted for quite naturally by the application of this Babylonian rule[2]). Generally, however, the rule is not applied in perfect strictness, for, as already appeared above, it was sufficient for Archimedes in his calculations in the *Measurement of the Circle* to use as approximations mixed numbers, the proper fraction part of which in the majority of cases had a denominator 2, 4 or 8, while 6 and 11 each appeared once as denominator. Now the exact application of the rule usually yields much more complicated fractions; Archimedes, however, succeeds in avoiding these by rounding off the value of $\frac{d}{\alpha_1}$ (or, which

[1]) Heron, *Stereometrica* (*vide* Note 2 to p. 231), p. 34: $\sqrt{63} = 8 - \frac{1}{16}$.

[2]) Hofmann, b. (Note 1 to p. 230).

comes to the same thing, of $\frac{r}{\alpha_1}$ when $d=\alpha_1^2+r$) in a suitable way, so that the rest of the calculation is considerably simplified. This naturally has the result that the approximations found by the formation of the arithmetic mean will not always lie on the high side of $\sqrt{d}$. Besides, such approximations would not even be suitable in the first part of the calculation, where an upper limit for the circumference of the circle is to be found. In the choice of the approximations this has naturally been taken into account; probably it was ascertained after each operation whether the desired object had really been attained.

We are giving a few examples to illustrate this point[1]):

In the first part of the proposition the following had to be calculated:

$$\sqrt{571^2+153^2} \quad \text{or} \quad \sqrt{349450}\,.$$

Now put $\alpha_1=571$, then $\beta_1=\frac{d}{\alpha_1}=571+\frac{153^2}{571}$.

In this we have $\frac{153^2}{571}\backsim 40$ (41 is nearer, but 40 leads to a whole number); the new approximation therefore becomes

$$\alpha_2 = \frac{\alpha_1+\beta_1}{2} = 591\,.$$

Now $\beta_2=\frac{d}{\alpha_2}=591+\frac{169}{591}$, in which $\frac{169}{591}>\frac{150}{600}=\frac{1}{4}$. Archimedes now takes for β_2 the value $591+\frac{1}{4}$, and thus finds the new approximation

$$\alpha_3 = 591+\tfrac{1}{8}\,.$$

Since $\frac{1}{4}.591<169$, $\alpha_3<\sqrt{d}$.

In the second part of the proposition the following had to be calculated:

$$\sqrt{2911^2+780^2} \quad \text{or} \quad \sqrt{9082321}\,.$$

From $\alpha_1=2911$ it follows that $\beta_1=2911+\frac{780^2}{2911}\backsim 2911+209$ (remainder 1).

[1]) The calculations are found in Hofmann, b.

Therefore

$$\alpha_2 = 2911 + 104 + \tfrac{1}{2} = 3015 + \tfrac{1}{2}\,.$$

From the small amount of the remainder, *viz.* 1, it may be inferred that α_2 will be greater than $\sqrt{d}$, which is confirmed by calculation.

Now 9082321 has to be divided by $(3015+\frac{1}{2})$; in doing so, the result has to be rounded off in such a way that the quotient becomes slightly too great, while it has to be as simple as possible. The quotient found is 3012, so that the value $3013+\frac{1}{2}+\frac{1}{4}$ is found for the approximation to the root, this being the mean of $3015+\frac{1}{2}$ and 3012.

3. We are now coming to the more special question how Archimedes can have arrived at the two rational approximations $\frac{265}{153}$ and $\frac{1351}{780}$ for $\sqrt{3}$. It is natural to find out whether they can be obtained by means of the Babylonian rule. This is quite easy for the value $\frac{1351}{780}$, if we start from $\frac{5}{3}$ as the first approximation.

The following values are then successively found:

$$\alpha_1 = \tfrac{5}{3},\ \beta_1 = \tfrac{9}{5} \qquad \alpha_2 = \tfrac{1}{2}\left(\tfrac{5}{3}+\tfrac{9}{5}\right) = \tfrac{26}{15}$$

$$\alpha_2 = \tfrac{26}{15},\ \beta_2 = \tfrac{45}{26} \qquad \alpha_3 = \tfrac{1}{2}\left(\tfrac{26}{15}+\tfrac{45}{26}\right) = \tfrac{1351}{780}\,.$$

The starting point $\frac{5}{3}$, however, cannot be found by this method, but it may be imagined as determined on the strength of the discovery that 27, whose square root is three times $\sqrt{3}$, is close to the square of 5. From $27 \backsim 5^2$ it then follows that $\sqrt{3} \backsim \frac{5}{3}$.

Another possibility is that we have to do here with the application of a subsidiary form of the Babylonian rule[1]), in which for $\sqrt{p^2-r}$ we do not write the too high value $p-\dfrac{r}{2p}$, but the too low value

$$p-\frac{r}{2p-1}\,. \tag{3}$$

We then get

$$\sqrt{3} = \sqrt{2^2-1} \backsim 2-\tfrac{1}{3} = \tfrac{5}{3}\,.$$

A direct proof of the use of $\frac{5}{3}$ as approximation to $\sqrt{3}$, indeed,

---

[1]) F. Hultsch, *Die Näherungswerte irrationaler Quadratwurzeln bei Archimedes.* Göttinger Nachrichten 1893. 367–428.

can no more be given than that of the familiarity of the general rule from which it is derived. The assumption that this rule was known, however, explains a great many authentic approximations in such a simple manner that it seems quite warranted to make this assumption.

In the first place the limit $\frac{265}{153}$ can be derived from it[1]) by starting from the approximation $\frac{26}{15}$ found above and then assuming that Archimedes (or whoever of his predecessors may have determined the limits, which perhaps had long been known already) noted that

$$26^2 = 3.15^2+1 .$$

Now

$$\sqrt{3} = \tfrac{1}{15}\sqrt{26^2-1} \backsim \tfrac{1}{15}[26-\tfrac{1}{51}] = \tfrac{265}{153} .$$

The modified Babylonian rule (3) subsequently leads up in a simple way to an explanation of a method of calculation[2]) found by Hofmann in the analysis of approximations to square roots in Heron's works, which may be rendered algebraically as follows:

$$\sqrt{x^2-1} \backsim x-\frac{1}{2x-1}+\frac{1}{(2x-1)(2x+1)} \tag{4}$$

The truth of this can be recognized algebraically as follows: Let the following approximation be known:

$$\alpha_1 = x-\frac{1}{2x-1} ,$$

then we have

$$\beta_1 = \frac{(x^2-1)(2x-1)}{2x^2-x-1} = \frac{(x+1)(2x-1)}{2x+1} = x-\frac{1}{2x+1},$$

so that according to the Babylonian rule the next approximation is:

$$\frac{1}{2}\left[2x-\frac{1}{2x-1}-\frac{1}{2x+1}\right] = x-\frac{1}{2x-1}+\frac{1}{(2x-1)(2x+1)}.$$

As an authentic example in which this rule (4) can have been applied, we mention the reduction occurring in Heron[3])

[1]) F. Hultsch, l.c. p. 401.

[2]) Hofmann, b (Note 1 to p. 230).

[3]) *Geometrica. Heronis Opera* IV, 322.

$$\sqrt{216} = 14+\tfrac{2}{3}+\tfrac{1}{33},$$

which can be accounted for as

$$\sqrt{216} = 3\sqrt{5^2-1} = 3\left[5-\tfrac{1}{9}+\tfrac{1}{9\,.\,11}\right] = 14+\tfrac{2}{3}+\tfrac{1}{33}.$$

By means of the rule (4) the approximation $\frac{1351}{780}$ for $\sqrt{3}$ appears to be obtainable in one step[1]). To achieve this, we write:

$$\sqrt{3} = \tfrac{1}{4}\sqrt{7^2-1} = \tfrac{1}{4}\left[7-\tfrac{1}{13}+\tfrac{1}{13\,.\,15}\right] = \tfrac{1351}{780}.$$

As will be seen, the limits used by Archimedes can be arrived at by many different routes. Of the numerous other methods suggested we mention that of K. Vogel[2]), who in the Babylonian rule

$$\sqrt{\alpha_1{}^2+r} \backsim \alpha_1+\frac{r}{2\alpha_1} = \alpha_2$$

substitutes for $2\alpha_1$ in the denominator $\alpha_1+\alpha_2$, and thus gets a new approximation

$$\alpha_2{}' = \alpha_1+\frac{r}{\alpha_1+\alpha_2},$$

which for

$$\alpha_1 = \tfrac{5}{3} \qquad \alpha_2 = \tfrac{26}{15} \qquad r = \tfrac{2}{9}$$

gives

$$\alpha_2{}' = \tfrac{5}{3}+\frac{\frac{2}{9}}{\frac{5}{3}+\frac{26}{15}} = \tfrac{265}{153}.$$

One may observe that such a procedure after all has no demonstrative force; from a historical point of view this can hardly be called an objection: it is precisely in the field of arithmetic that people were long content (certainly up to the 17th century) to suggest rules and state that they give good results. In this connection it is significant that in his commentary on the *Measurement of the Circle*[3]) Eutocius does not elucidate the passages in which the approximations to $\sqrt{3}$ are mentioned in any other way but by a verifying calculation, from which it appears that 1351^2 differs little from $3\,.\,780^2$, and likewise 265^2 differs little from $3\,.\,153^2$.

---

[1]) Hofmann, b (Note 1 to p. 230).

[2]) Kurt Vogel, *Die Näherungswerte des Archimedes für* $\sqrt{3}$. Jahresbericht D.M.V. XLI (1932), p. 155.

[3]) *Opera* III, 234; 246.

Another objection that might be raised to the above reconstructions is that they do not use the same system at the lower and the upper limit; this again is no conclusive historical argument, since it is not known for certain whether the Greek logisticians actually possessed a fixed system for approximations to square roots. Mathematically, however, it may be felt as an aesthetic objection; those who are of this opinion will perhaps prefer another divination of Archimedes' train of thought, which was suggested by C. Müller[1]) and generalized by O. Toeplitz[2]) and can be rendered algebraically as follows:

Let a and b be two approximations to $\sqrt{d}$, then we have

$$c = \frac{ab+d}{a+b}, \tag{5}$$

a new approximation, which lies on the low or the high side of $\sqrt{d}$, according as a and b lie on opposite sides or on the same side of it. The truth of this assertion can be recognized at once by writing

$$c = \frac{ab+d}{a+b} = \sqrt{d} + \frac{(a-\sqrt{d})(b-\sqrt{d})}{a+b}.$$

In order to examine on what condition c is a better approximation than a, we write

$$\left|\frac{c-\sqrt{d}}{a-\sqrt{d}}\right| = \frac{|b-\sqrt{d}|}{a+b}.$$

In order that

$$|c-\sqrt{d}| < |a-\sqrt{d}|$$

it is necessary and sufficient that

$$\left|\frac{b-\sqrt{d}}{a+b}\right| < 1 \quad ,$$

[1]) C. Müller, *Wie fand Archimedes die von ihm gegebenen Näherungswerte von* $\sqrt{3}$? Quellen and Studien z. Gesch. d. Math. Astr. u. Phys. Abt. B. II (1932) 281–285.

[2]) O. Toeplitz, *Bemerkungen zu der Arbeit von Conrad Müller "Wie fand Archimedes die von ihm gegebenen Näherungswerte von* $\sqrt{3}$*?" im gleichen Heft.* ibid. 286–290.

from which it is derived that

$$2b + a > \sqrt{d}\,.$$

In the same way c is a better approximation than b only when

$$2a + b > \sqrt{d}\,.$$

According to this rule we now get as successive approximations to $\sqrt{3}$:

$$\text{from } a = 1,\ \ b = 2 \qquad c = \tfrac{5}{3} \ < \sqrt{3}$$

$$\text{from } a = \tfrac{5}{3},\ \ b = \tfrac{5}{3} \qquad c = \tfrac{26}{15} \ > \sqrt{3}$$

$$\text{from } a = \tfrac{5}{3},\ \ b = \tfrac{26}{15} \qquad c = \tfrac{265}{153} \ < \sqrt{3}$$

$$\text{from } a = \tfrac{26}{15},\ \ b = \tfrac{26}{15} \qquad c = \tfrac{1351}{780} \ > \sqrt{3}\,.$$

Thus the two limits used by Archimedes are arrived at in few and systematic steps.

If the rule is applied for the case that a is an approximation α to $\sqrt{d}$, and if then we put $b = a$, we find

$$c = \frac{\alpha^2 + d}{2\alpha} = \frac{1}{2}\left[\alpha + \frac{d}{\alpha}\right] = \frac{1}{2}[\alpha + \beta] \text{ when } \beta = \frac{d}{\alpha}.$$

If we put $b = \beta$, we find

$$c = \frac{2\alpha\beta}{\alpha + \beta}.$$

We thus get precisely the two means (arithmetic and harmonic) between α and β which were found upon repeated application of the Babylonian rule.

It is doubtful whether, as Toeplitz thinks, this may be considered to constitute a historical argument showing that the general rule (5), in which there is no connection between a and b, was known. Through the existence of the Babylonian rule the rule (5) is authenticated only for those cases where $b = a$ and $ab = d$. In the above application for determining the limits for $\sqrt{3}$ (the value of which is sought especially in the fact that it yields both limits in the course of one and the same calculation), however, in half of the steps neither the one nor the other is the case.

4. Finally we mention

Proposition 2.

The area of the circle is to the square on its diameter as 11 *to* 14.

This follows at once from the propositions 1 and 3. Indeed, if in Fig. 94 ΓH be the square circumscribed about the circle on AB as diameter, while further $\Delta E = 2AB$, $EZ = \frac{1}{7}AB$, we find

$$(A\Gamma Z, A\Gamma\Delta) = (22,7) .$$

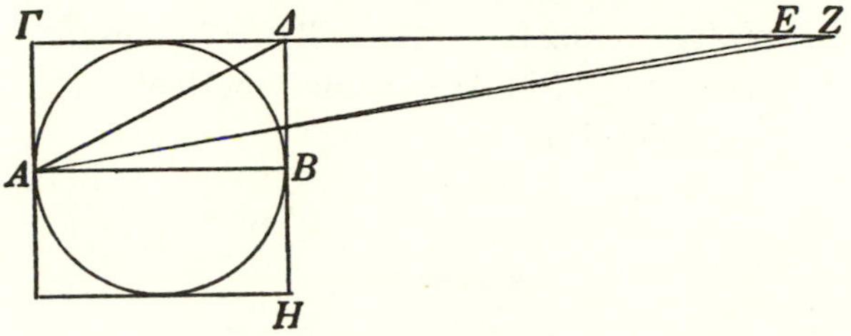

Fig. 94.

However, $\Gamma H = 4A\Gamma\Delta$ and $A\Gamma Z =$ Circle, consequently

$$(\text{Circle, Square on the Diameter}) = (11,14) .$$

5. According to a statement by Heron[1]) Archimedes is credited with a more accurate approximation to the ratio between the circumference and the diameter of the circle in the treatise (now lost) *On Plinthides*[2]) *and Cylinders* (περὶ πλινθίδων καὶ κυλίνδρων). Apparently, however, the values given in the Greek text have not come down correctly; in fact, if the two limits mentioned, each of which has been given as a ratio of large numbers, are reduced to the decimal notation, the lower limit is found to be greater than the true value of π and the upper limit greater than the approximation $\frac{22}{7}$ already known. Various attemps[3]) have been made to reconstruct the original result of Archimedes by the application of small corrections in the numbers occurring in Heron; in this way more accurate limits for π than those in the *Measurement of the Circle* are indeed arrived at.

---

[1]) Heron, *Metrica* (Note 2 to p. 230) I, 26. p. 66.

[2]) According to Heron, *Definitiones* (*Heronis Opera* IV; Note 2 to p. 231) Def. 113 (p. 70) a *plinthis* is a rectangular parallelepiped whose length is less than its breadth and its depth.

[3]) On this see T. L. Heath, *Greek Mathematics* I, 232. E. Hoppe, *Die*

From a mathematical point of view the finding of such closer limits according to the method of the *Measurement of the Circle* naturally is not very important; it is obvious that if we continue in the direction once taken, we are bound to approach the true value of π more and more closely. The great value of the result attained in the *Measurement of the Circle*, however, resides in the remarkably good approximation to π that can be expressed by means of such simple numbers as 22 and 7. A real advance as compared with this was only to be expected as a result of the elaboration of new approximation methods, which would call for less extensive calculations than those involved in Archimedes' method; Chr. Huygens[1]) was the first to achieve this.

CHAPTER VII.

ON CONOIDS AND SPHEROIDS

This treatise deals with theorems on the volumes of segments of conoids and spheroids, *i.e.* of the solids comprehended by a plane and the surface of either a right-angled conoid (*i.e.* a paraboloid of revolution) or an obtuse-angled conoid (*i.e.* a sheet of a hyperboloid of revolution of two sheets), or an oblong or flat spheroid (*i.e.* an ellipsoid of revolution). Each time Archimedes first discusses the case that the segment is right, *i.e.* that the cutting plane is at right angles to the axis of revolution, and subsequently devotes a new proposition to the oblique segment, with the cutting plane having any given position relative to the axis of revolution. In order to simplify matters, we shall confine ourselves to a discussion

zweite Methode des Archimedes zur Berechnung von π. Archiv f. Gesch. d. Naturw. u. d. Techn. IX (1922) 104–107. Heron's statement has again been discussed in the articles:

E. M. Bruins, *Mathematici en Physici*; Euclides 20 (1943–44), 12–16.

Hans Freudenthal, *Hoe hebben de Ouden gerekend?* Euclides 24 (1948–49), 12–34.

E. M. Bruins, *Hoe hebben de Ouden gerekend? (Antwoord aan Prof. Dr H. Freudenthal)*. Euclides 24 (1948–49), 169–185.

E. M. Bruins, *Square Roots in Babylonian and Greek Mathematics*. Proc. Kon. Ned. Acad. v. Wet. LI (1943) No. 3. 121–130.

[1]) Chr. Huygens, *De circuli magnitudine inventa*. Oeuvres Complètes XII (The Hague 1910) 91–181.

of the general case of the oblique segment, because everything else can be deduced from this by specialization. For the terminology (vertex, base, axis, axis produced, diameter of the segment, etc.) Chapter III; 6.21–6.23 should be consulted.

The determinations of the volumes are all based on the compression method (III; 8.2); in and about the solids to be discussed there are constructed figures consisting of piled-up frusta of cylinders, the volumes of which can be found on the strength of theorems on the cylinder and its parts. For the application of the method of the indirect passage to the limit it then has to be proved in each case that the difference between the volumes of the circumscribed solid (C_n) and of the inscribed solid (I_n) can be made less than any given volume by the choice of the number of frusta. The possibility of this is stated for right segments in Prop. 19, for oblique segments in Prop. 20; the latter is worded as follows:

Proposition 20.

Given a segment cut off by a plane not perpendicular to the axis from one of the two conoids or from one of the two spheroids (so that it is not larger than half the spheroid), it is possible to inscribe in the segment a solid figure consisting of frusta of cylinders of equal height, and to circumscribe about the segment another similar solid figure, so that the circumscribed figure exceeds the inscribed figure by a volume less than any given solid magnitude (Fig. 95).

Proof: Suppose (also in the following propositions) the plane of the paper to be the plane through the axis of revolution λ at right angles to the cutting plane, and the intersection of the latter with the plane of the paper $A\Gamma$. The intersection of the cutting plane with the solid is an oxytome with diameter $A\Gamma$ (III; 6.31; 6.41; 6.51–52). Let B be the vertex of the segment determined by the straight line $A\Gamma$ and the meridian section, *i.e.* the point where the tangent to the section is parallel to $A\Gamma$, Δ being the middle point of $A\Gamma$; for the case of the right-angled conoid $B\Delta$ is then parallel to the diameter, while for the obtuse-angled conoid and the spheroids $B\Delta$ passes through the centre. There now exists (III; 3.4) an oblique circular cylinder with axis $B\Delta$, of which the oxytome with diameter $A\Gamma$ forms the base and the whole surface of which falls without the segment of a conoid or spheroid here considered (III; 6.32; 6.42; 6.551). Now apply dichotomy (III; 0.5) to the axis of the segment $B\Delta$, and through the points of division thus obtained construct planes parallel to the plane of the base of the segment. The intersections of these planes with the segments are similar oxytomes (III; 6.61), which in each case are the base of a circum-

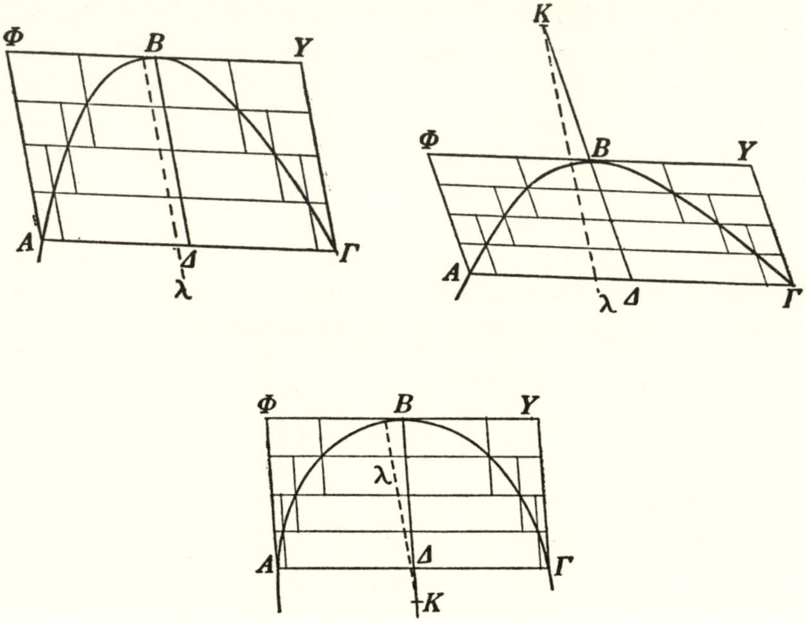

Fig. 95.

scribed and of an inscribed frustum of a cylinder, whose axis lies on $B\Delta$. All the frusta obtained have the same height. If therefore we subtract from the sum C_n of all the circumscribed frusta the sum I_n of all the inscribed frusta, the remainder is the circumscribed frustum having the oxytome with diameter $A\Gamma$ for its base. Since the height of the latter has been obtained by continued dichotomy from the height of the given segment, it can, if the dichotomy is continued far enough, be made less than any given length, so that the volume under consideration can likewise be made less than any given solid magnitude. It has thus been proved that for any given solid magnitude δ it is possible to find a number m such that for $n = 2^m$ the inequality

$$C_n - I_n < \delta$$

is true.

The same statement is nowadays written in the abbreviated form

$$\lim_{n\to\infty}(C_n - I_n) = 0 .$$

After this general introductory proposition, Prop. 21 deals with a right, and Prop. 22 with an oblique segment of a right-angled conoid.

Proposition 22 (Fig. 96).

If by a plane not perpendicular to the axis a segment be cut off from a right-angled conoid, this segment will be half as large again as the segment of a cone which has the same base as the segment and the same axis.

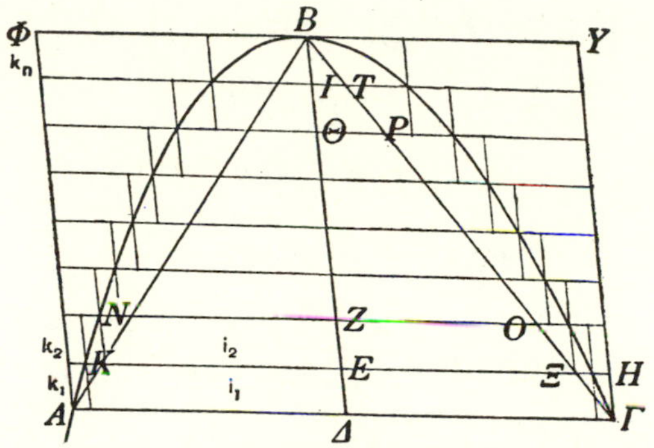

Fig. 96.

Let the plane of the paper intersect the right-angled conoid in the orthotome $AB\Gamma$. Let Ψ be a cone which is half as large again as the segment of a cone $BA\Gamma$, **K** the frustum of a cylinder $A\Gamma Y\Phi$, then we have (III; 6.11)

$$\mathbf{K} = 3 . \text{segment of a cone } BA\Gamma = 2\Psi .$$

It is required to be proved that $\Psi =$ segment of conoid $BA\Gamma$. Suppose this is not true, then

$$\text{either } BA\Gamma > \Psi \text{ (I)} \quad \text{or} \quad BA\Gamma < \Psi \text{ (II)}$$

Case I. Suppose $BA\Gamma > \Psi$. Now continue the dichotomy of $B\Delta$ (Prop. 20) until

$$C_n - I_n < BA\Gamma - \Psi ,$$

then *a fortiori*

$$BA\Gamma - I_n < BA\Gamma - \Psi .$$

Therefore $$I_n > \Psi .$$

Now compare each time the inscribed frusta i_1, i_2, etc. (starting from Δ) with the frusta k_1, k_2, etc. of the frustum of a cylinder **K** comprehended between the same planes.

The ratio of the frusta i_1 and k_1 having the same height is equal to the ratio of their bases. Since these bases are similar oxytomes, whose diameters are EK and ΔA respectively, we have

$$(i_1, k_1) = [\mathbf{T}(EK), \mathbf{T}(\Delta A)] = (BE, B\Delta) = (E\Xi, \Delta\Gamma) .$$

Also

$$(i_2, k_2) = [\mathbf{T}(ZN), \mathbf{T}(EH)] = [\mathbf{T}(ZN), \mathbf{T}(\Delta\Gamma)] = (BZ, B\Delta)$$
$$= (ZO, \Delta\Gamma)$$

etc., up to

$$(i_{n-1}, k_{n-1}) = (IT, \Delta\Gamma) .$$

Archimedes concludes from this that

$$(i_1 + i_2 + \ldots + i_{n-1}, k_1 + k_2 + \ldots + k_{n-1})$$
$$= (E\Xi + ZO + \ldots + IT, (n-1)\Delta\Gamma) . \qquad (1)$$

Algebraically this conclusion is evident; in fact, $k_1 = k_2$ etc.; by adding together all the proportions with a first member (i_m, k_m) we obtain

$$\frac{i_1 + \ldots i_{n-1}}{k_1} = \frac{E\Xi + \ldots IT}{\Delta\Gamma},$$

consequently also

$$\frac{i_1 + \ldots i_{n-1}}{k_1 + \ldots k_{n-1}} = \frac{E\Xi + \ldots IT}{(n-1)\Delta\Gamma}.$$

In order to reach this conclusion, Archimedes must have had recourse to **C.S.** 1 (III; 7.21), by considering therein

as series III: $k_1, k_2 \ldots k_{n-1}$

as series I: $i_1, i_2 \ldots i_{n-1}$

as series II: $E\Xi, ZO \ldots IT$

as series IV: $\Delta\Gamma, \Delta\Gamma \ldots \Delta\Gamma$

From this the proportion (1) actually follows, for we have

$$(i_1, k_1) = (E\Xi, \Delta\Gamma) \text{ etc., and also } (i_1, i_2) = (E\Xi, ZO) \text{ etc.}$$

Archimedes now further notes that the line segments IT, $\Theta P \ldots E\Xi$, $\Delta\Gamma$ form an arithmetical progression, the common difference of which is equal to the least term, and from this he concludes that

$$(n-1)\Delta\Gamma > 2(E\Xi + ZO + \ldots + IT). \tag{2}$$

This conclusion apparently is wrong: the two members are equal. It was no doubt the intention to apply the arithmetical lemma III; 7.1, but this leads to

$$n.\Delta\Gamma > 2(E\Xi + \ldots + IT) \tag{3}$$

Archimedes now concludes from (2) in relation with (1) that

$$k_1 + \ldots k_{n-1} > 2(i_1 + \ldots i_{n-1}),$$

and thence *a fortiori* that

$$\text{Frustum of cylinder } \mathbf{K} > 2I_n,$$

whence $\Psi > I_n$, which is contrary to the supposition.

This result is correct, for because $k_1 = k_2$ etc. we may also write for (1):

$$(i_1 + i_2 + \ldots + i_{n-1}, k_1 + k_2 + \ldots + k_n) = (E\Xi + \ldots IT, n.\Delta\Gamma)$$

and then it follows indeed from (3) that

$$\mathbf{K} > 2I_n \text{[1]}.$$

Case II. Secondly suppose $BA\Gamma < \Psi$. Now continue the dichotomy of $B\Delta$ until

$$C_n - I_n < \Psi - BA\Gamma,$$

then *a fortiori*

$$C_n - BA\Gamma < \Psi - BA\Gamma,$$

whence $$C_n < \Psi.$$

[1] An interesting and most plausible hypothesis as to the origin of the incorrect relation (2) was put forward by S. Heller, *Ein Fehler in einer Archimedes-Ausgabe, seine Entstehung und seine Folgen*. Abh. Bayer. Akad. d. Wiss. Math.-Nw. Klasse. N. F. Heft 63. München 1954.

If we now compare the circumscribed frusta $c_1, c_2 \ldots$ with the frusta $k_1, k_2, \ldots$, it appears that

$$c_1 = k_1, \text{ whence } (c_1, k_1) = (\Delta\Gamma, \Delta\Gamma)$$

$$(c_2, k_2) = [\mathbf{T}(EK), \mathbf{T}(EH)] = [\mathbf{T}(EK), \mathbf{T}(\Delta\Gamma)] = (BE, B\Delta)$$

$$= (E\Xi, \Delta\Gamma) \text{ etc., up to } (c_n, k_n) = (IT, \Delta\Gamma)\,.$$

Now again, because of **C.S.** 1 (III; 7.21), we have

$$(c_1 + c_2 + \ldots + c_n, k_1 + k_2 \ldots + k_n) = (IT \ldots + \Delta\Gamma, n.\Delta\Gamma)$$

and by the lemma (III; 7.1)

$$2(IT + \ldots + \Delta\Gamma) > n.\Delta\Gamma\,,$$

whence

$$2(c_1 + c_2 \ldots + c_n) > k_1 + k_2 \ldots + k_n = \mathbf{K}\,.$$

Therefore $$c_1 + \ldots c_n = C_n > \Psi\,,$$

which is contrary to the supposition.

Algebraic formulation:

If I is the volume of the segment of the right-angled conoid under consideration, Prop. 20 is equivalent to

$$I = \lim_{n\to\infty} I_n\,.$$

In Prop. 22 suppose the equation of the orthotome relative to the system of axes $B(Y\Delta)$ to be: $x^2 = py$.

Suppose $B\Delta = y_0$, $\Gamma\Delta = x_0$. Let $B\Delta$ be divided into n equal parts (with Archimedes $n = 8$). Let the volume of the mth inscribed frustum (starting from B) be i_m, then we have

$$\frac{i_m}{\frac{1}{n}\mathbf{K}} = \frac{x^2}{x_0^2} = \frac{y}{y_0} = \frac{m}{n}\,,$$

whence

$$i_m = \frac{m}{n^2}\mathbf{K}\,.$$

Therefore $$I_n = \sum_1^{n-1} i_m = \mathbf{K}\frac{1 + 2 + \ldots (n-1)}{n^2},$$

whence

$$I = \mathbf{K}.\lim_{n\to\infty} \frac{\frac{1}{2}n(n-1)}{n^2} = \tfrac{1}{2}\mathbf{K}.$$

The propositions 23 and 24 contain corollaries to the propositions on the volume of a segment of a right-angled conoid.

Proposition 23.

If from a right-angled conoid two segments be cut off by planes one of which is perpendicular to the axis and the other is not, and the axes of the segments are equal, the segments will also be equal.

Proof: In the plane of the paper (Fig. 97) through the two axes of the segments let $B\Gamma E$ and ΛZA be the intersections with the

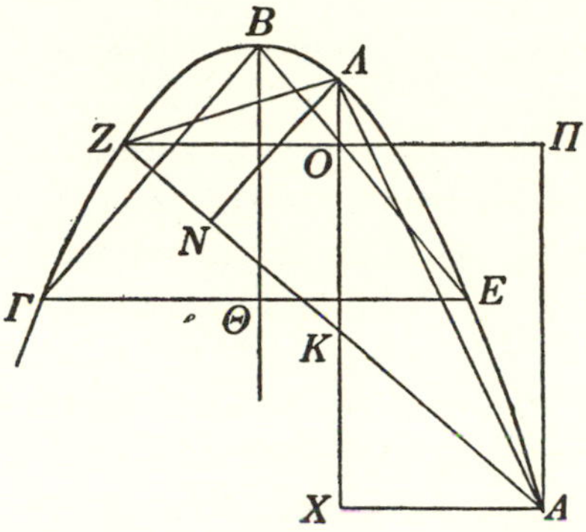

Fig. 97.

segments, $B\Theta$ and ΛK the equal axes. Because of Prop. 22 the proposition will have been proved when the equality of the cone $B\Gamma E$ and the segment of a cone ΛZA has been proved. The base of the cone is a circle on $E\Gamma$ as diameter, that of the segment is an oxytome with major axis AZ and minor axis ΠZ (III; 6.31); the height of the first solid is $B\Theta$, that of the second $\Lambda N \perp AZ$. The surfaces of the bases are to one another as $\mathbf{T}(\Gamma E)$ to $\mathbf{O}(AZ, \Pi Z)$ (III; 3.11), which is the same ratio as

$$[\mathbf{T}(\Theta E), \mathbf{O}(AK, AX)].$$

Now if M be the *orthia* of the diameter $B\Theta$, the *orthia* μ for the oblique conjugation of the orthotome to the straight line ΛK is determined by

$$(\mu, M) = [\mathbf{T}(AK), \mathbf{T}(AX)] \qquad \text{(III; 2.321)}.$$

Thence, because $B\Theta = \Lambda K$,

$$[\mathbf{T}(AK),\ \mathbf{T}(E\Theta)] = (\mu, M) = [\mathbf{T}(AK),\ \mathbf{T}(AX)]\ ,$$

whence $$E\Theta = AX\ .$$

The ratio between the bases of the cone and the segment of a cone is therefore (AX, AK), which is the same ratio as $(\Lambda N, \Lambda K)$ or $(\Lambda N, B\Theta)$. The bases are therefore inversely proportional to the heights, from which follows the equality of the volumes (III; 6.01).

Proposition 24.

If from a right-angled conoid two segments be cut off by planes drawn in any manner, the segments will be to one another as the squares on their axes.

Proof: Construct the right segments successively having the same heights as the given segments. By the preceding proposition it will be sufficient if the proposition is proved for two such segments. In that respect, however, it follows immediately from Prop. 22, because, as will be readily recognized, the volumes of the inscribed cones are to one another as the squares on the heights.

The propositions 25 and 26 relate successively to the volumes of a right and of an oblique segment of an obtuse-angled conoid. We can again confine ourselves to the discussion for the oblique segment in

Proposition 26.

If by a plane not perpendicular to the axis a segment be cut off from an obtuse-angled conoid, this segment will be to the segment of a cone having the same base as the segment and the same axis as the sum of the axis of the segment and triple the axis produced to the sum of the axis and double the axis produced.

Let the plane of the paper cut the obtuse-angled conoid in the amblytome $AB\Gamma$ (Fig. 98).

Let Θ be the vertex of the enveloping cone of the conoid, *i.e.* the centre of the amblytome $AB\Gamma$; further let $B\Theta = \Theta Z = ZH$. Let Ψ be a cone whose volume is to the volume of the segment of a cone $BA\Gamma$ as the two lengths referred to in the proposition, *i.e.* as $H\Delta$ to $Z\Delta$.

It has then to be proved that the segment of the conoid is equal to Ψ. If **K** be the frustum of a cylinder $A\Gamma Y\Phi$ and if $BP = \frac{1}{3}\Delta B$, whence $H\Delta = 3\Theta P$, we have

$$(\mathbf{K}, \text{segment of cone } BA\Gamma) = (H\Delta, \Theta P)$$

$$(\text{segment}, \Psi) = (Z\Delta, H\Delta)\,,$$

therefore, *ex aequali*: $(\mathbf{K}, \Psi) = (Z\Delta, \Theta P)$.

If the segment of the conoid is not equal to Ψ, it is either greater or less than Ψ.

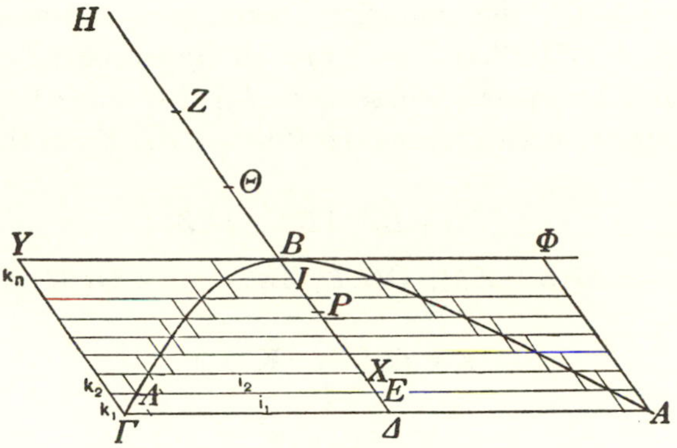

Fig. 98.

Case I. First suppose $BA\Gamma > \Psi$. Now continue the dichotomy of $B\Delta$ until (Prop. 20)

$$C_n - I_n < BA\Gamma - \Psi, \text{ then } a\ fortiori$$

$$BA\Gamma - I_n < BA\Gamma - \Psi, \text{ whence } I_n > \Psi\,.$$

Now compare each time the inscribed frusta of a cylinder $i_1, i_2 \ldots i_{n-1}$ with the frusta $k_1, k_2 \ldots$ of the frustum **K** comprehended between the same planes.

The ratio between the frusta i_1 and k_1, which have the same height, is equal to the ratio between their bases, *i.e.* between the similar oxytomes with homologous diameters $E\Delta$ and $\Delta\Gamma$. Therefore

$$(i_1, k_1) = [\mathbf{T}(E\Delta, \mathbf{T}(\Delta\Gamma)] = [\mathbf{O}(BE, ZE), \mathbf{O}(B\Delta, Z\Delta)]\,.$$

Also

$$(i_2, k_2) = [\mathbf{O}(BX, ZX), \mathbf{O}(B\Delta, Z\Delta)]$$

etc., up to

$$(i_{n-1}, k_{n-1}) = [\mathbf{O}(BI, ZI),\ \mathbf{O}(B\Delta, Z\Delta)]\,.$$

From this it follows, as in Prop. 22, by the application of **C.S. 1** (III; 7.21) that

$$(i_1+i_2\ldots i_{n-1}, n.k_1) = (I_n, \mathbf{K})$$
$$= [\mathbf{O}(BE, ZE)+\ldots+\mathbf{O}(BI, ZI), n.\mathbf{O}(B\Delta, Z\Delta)]\,.$$

The series of the rectangles

$$\mathbf{O}(B\Delta, Z\Delta),\ \mathbf{O}(BE, ZE)\ldots\mathbf{O}(BI, ZI)$$

apparently satisfies the conditions mentioned in the arithmetical lemma **C.S. 2** (III; 7.4). They are all hyperbolically applied to BZ, exceeding by squares whose sides $B\Delta$, $BE\ldots BI$ form an arithmetical progression with common difference BI. From this it follows that

$$(B\Delta+BZ, \tfrac{1}{3}B\Delta+\tfrac{1}{2}BZ)$$
$$< [(n.\mathbf{O}(B\Delta, Z\Delta),\ \mathbf{O}(BE, ZE)+\ldots+\mathbf{O}(BI, ZI)]$$

or

$$(Z\Delta, \Theta P) < (\mathbf{K}, I_n)\,,$$

whence

$$(\mathbf{K}, \Psi) < (\mathbf{K}, I_n)\,,$$

whence $I_n < \Psi$, which is contrary to the supposition.

Case II. Secondly suppose $BA\Gamma < \Psi$. Now continue the dichotomy of $B\Delta$ until

$$C_n - I_n < \Psi - BA\Gamma\,,$$

then *a fortiori*

$$C_n - BA\Gamma < \Psi - BA\Gamma\,,$$

whence

$$C_n < \Psi\,.$$

If we again compare the circumscribed frusta $c_1, c_2\ldots$ with the frusta $k_1, k_2\ldots$, it appears that

$$c_1 = k_1, \text{ whence } (c_1, k_1) = [\mathbf{O}(B\Delta, Z\Delta),\ \mathbf{O}(B\Delta, Z\Delta)]$$
$$(c_2, k_2) = [\mathbf{O}(BE, ZE),\ \mathbf{O}(B\Delta, Z\Delta)]$$
$$\vdots \qquad\qquad \vdots$$
$$(c_n, k_n) = [\mathbf{O}(BI, ZI),\ \mathbf{O}(B\Delta, Z\Delta)]\,,$$

whence

$$(c_1+\ldots c_n, n.k_1) = [\mathbf{O}(B\Delta, Z\Delta)+\ldots+\mathbf{O}(BI, ZI), n.\mathbf{O}(B\Delta, Z\Delta)].$$

Now by the application of **C.S.** 2 (III; 7.4), as above, we have

$$[n.\mathbf{O}(B\Delta, Z\Delta),\ \mathbf{O}(B\Delta, Z\Delta)+\ldots+\mathbf{O}(BI, ZI)] < (Z\Delta, \Theta P),$$

whence

$$(\mathbf{K}, C_n) < (Z\Delta, \Theta P) = (\mathbf{K}, \Psi),$$

whence $C_n > \Psi$, which is contrary to the supposition.

Algebraically: Let the equation of the hyperbola in the two-abscissa form relative to the system of axes $B(\Phi\Delta)$ be

$$\frac{x^2}{y(y+2a)} = C,$$

in which $a = B\Theta$. Further let $\Delta A = x_0$, $B\Delta = y_0$, and let $B\Delta$ be divided into n equal parts. If then the volume of the mth inscribed frustum of a cylinder (starting from B) be i_m, we have

$$\frac{i_m}{\frac{1}{n}\mathbf{K}} = \frac{x^2}{x_0^2} = \frac{y(y+2a)}{y_0(y_0+2a)} = \frac{\frac{m}{n}y_0\left[\frac{m}{n}y_0+2a\right]}{y_0(y_0+2a)}$$

or

$$i_m = \mathbf{K}\frac{m(my_0+2na)}{n^3(y_0+2a)}.$$

Consequently

$$I_n = \mathbf{K}\frac{y_0[1^2+2^2+\ldots(n-1)^2]+2na[1+2+\ldots(n-1)]}{n^3(y_0+2a)}$$

Now $I = \lim_{n\to\infty} I_n$. Nowadays we should write

$$I_n = \mathbf{K}\frac{\frac{1}{6}(n-1)n(2n-1)y_0+n^2(n-1)a}{n^3(y_0+2a)},$$

from which it follows that

$$I = \mathbf{K}\frac{\frac{1}{3}y_0+a}{y_0+2a} = \tfrac{1}{3}\mathbf{K}\frac{y_0+3a}{y_0+2a}.$$

The remaining part of the treatise *On Conoids and Spheroids* deals with the determination of the volume of a segment of a spheroid.

Archimedes first devotes two separate propositions to the special case where the plane by which the segment is cut off passes through the centre of the spheroid, in which case, as appears from **C.S.** 18 (III; 6.56), it divides the whole solid into two equal parts. The determination of the volume is carried out in Prop. 27 for a cutting plane perpendicular to the axis, in Prop. 28 for any cutting plane passing through the centre. We shall discuss again the latter proposition:

Proposition 28.

If a spheroid be cut by a plane through the centre not perpendicular to the axis, half the spheroid is equal to twice the segment of a cone having the same base as the segment and the same axis.

Let the plane of the paper, through the axis of revolution at right angles to the cutting plane, cut the spheroid in the oxytome $AB\Gamma\Delta$ with centre Θ, and the cutting plane in $A\Gamma$ (Fig. 99).

Let Ψ be a cone, whose volume is equal to twice the volume of the segment of a cone $BA\Gamma$, **K** the frustum of a cylinder $AK\Lambda\Gamma$, then we apparently have

$$\mathbf{K} = \tfrac{3}{2}\Psi .$$

If half the spheroid $BA\Gamma$ is not equal to Ψ, it is either greater or less than Ψ.

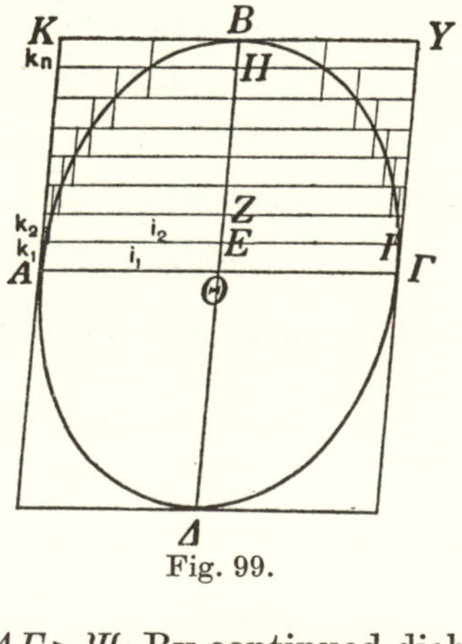

Fig. 99.

Case I. Suppose $BA\Gamma > \Psi$. By continued dichotomy of $B\Theta$ in the manner elucidated in Propositions 22 and 26 it is now possible to obtain an inscribed solid figure for which

$$I_n > \Psi .$$

A comparison of the inscribed frusta $i_1, i_2 \ldots$ with the corresponding frusta $k_1, k_2 \ldots$ of **K** now leads to

$$(i_1, k_1) = [\mathbf{T}(EI),\ \mathbf{T}(\Theta A)] = [\mathbf{O}(BE, \varDelta E),\ \mathbf{T}(B\Theta)] \quad \text{(III; 3.0)}$$
$$(i_2, k_2) = [\mathbf{O}(BZ, \varDelta Z),\ \mathbf{T}(B\Theta)]$$
$$(i_{n-1}, k_{n-1}) = [\mathbf{O}(BH, \varDelta H),\ \mathbf{T}(B\Theta)],$$

from which it follows by **C.S.** 1 (III; 7.21):

$$(i_1 + i_2 \ldots + i_{n-1},\ n.k_1)$$
$$= [\mathbf{O}(BE, \varDelta E) + \ldots + \mathbf{O}(BH, \varDelta H),\ n.\mathbf{T}(B\Theta)]. \tag{1}$$

Now by Euclid II, 5 we have:

$$\mathbf{O}(BE, \varDelta E) = \mathbf{T}(B\Theta) - \mathbf{T}(\Theta E).$$

The successive rectangles **O** may therefore be represented (Fig. 100) as gnomons formed by taking away from $\mathbf{T}(B\Theta)$ successively

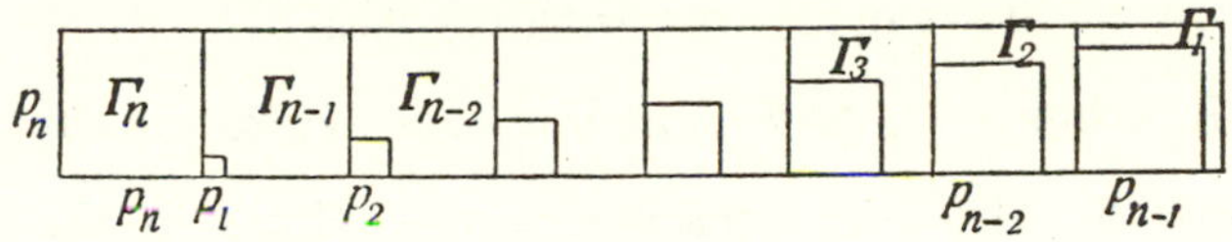

Fig. 100.

$\mathbf{T}(\Theta E), \mathbf{T}(\Theta Z) \ldots \mathbf{T}(\Theta H)$. If we denote the sides of these squares by $p_1, p_2, \ldots p_{n-1}$ respectively, the corresponding gnomons by Γ_{n-1}, $\Gamma_{n-2}, \ldots \Gamma_1$, while $\mathbf{T}(B\Theta)$ may be referred to as Γ_n and $B\Theta$ as p_n, the proved equality (1) may be written

$$(I_n, \mathbf{K}) = [\Gamma_1 + \Gamma_2 + \ldots + \Gamma_{n-1},\ n.\mathbf{T}(B\Theta)]. \tag{2}$$

Now by III; 7.31 we have

$$n.\mathbf{T}(p_n) < 3[\mathbf{T}(p_1) + \ldots + \mathbf{T}(p_n)],$$

or

$$3n.\mathbf{T}(p_n) - 2n.\mathbf{T}(p_n) < 3[\mathbf{T}(p_1) + \ldots + \mathbf{T}(p_n)],$$

whence

$$3[\mathbf{T}(p_n) - \mathbf{T}(p_1) + \mathbf{T}(p_n) - \mathbf{T}(p_2) + \ldots + \mathbf{T}(p_n) - \mathbf{T}(p_{n-1})] < 2n.\mathbf{T}(p_n)$$

or

$$3[\Gamma_1 + \Gamma_2 + \ldots + \Gamma_{n-1}] < 2n.\mathbf{T}(B\Theta).$$

From this, however, it follows in relation with (2) that $3I_n < 2\mathbf{K}$, whence $I_n < \Psi$, which is contrary to the supposition.

Case II. Now suppose $BA\Gamma < \Psi$. It is now possible to obtain a circumscribed solid figure such that

$$C_n < \Psi,$$

upon which a comparison of corresponding frusta of this circumscribed figure and of $\mathbf{K}$ leads to the relation

$$(c_1 + c_2 \ldots + c_n, n.k_1)$$
$$= [\mathbf{O}(B\Theta, \Delta\Theta) + \mathbf{O}(BE, \Delta E) + \ldots + \mathbf{O}(BH, \Delta H), n.\mathbf{T}(B\Theta)]. \quad (3)$$

Representing the rectangles by gnomons again, we draw, as above, from

$$3n.\mathbf{T}(p_n) > 3[\mathbf{T}(p_1) + \ldots + \mathbf{T}(p_{n-1})]$$

the conclusion that

$$3[\Gamma_1 + \Gamma_2 + \ldots + \Gamma_n] > 2n.\mathbf{T}(B\Theta),$$

from which it follows in relation with (3) that

$3C_n > 2\mathbf{K}$, whence $C_n > \Psi$, which is contrary to the supposition.

Algebraically: Let the equation of the ellipse in the two-abscissa form relative to the system of axes $B(Y\Delta)$ be

$$\frac{x^2}{y(2a-y)} = C,$$

in which $B\Delta = 2a$. Let $\Theta\Gamma$ be equal to b, and let $B\Delta$ be divided into n equal parts. Then we have for the volume of the mth inscribed frustum of a cylinder (starting from B):

$$\frac{i_m}{\frac{1}{n}\mathbf{K}} = \frac{x^2}{b^2} = \frac{y(2a-y)}{a^2}.$$

Archimedes now transforms $y(2a-y)$ geometrically in a way which is equivalent algebraically to $y(2a-y) = a^2 - (a-y)^2$, and thus obtains

$$\frac{i_m}{\frac{1}{n}\mathbf{K}} = \frac{a^2 - \left(a - \frac{m}{n}a\right)^2}{a^2}.$$

From this it follows that

$$I_n = \mathbf{K}\frac{(n-1)a^2 - \left\{\left(\frac{a}{n}\right)^2 + \left(\frac{2a}{n}\right)^2 + \ldots + \left[\frac{(n-1)a}{n}\right]^2\right\}}{na^2}.$$

Because $I = \lim\limits_{n\to\infty} I_n$, it now follows from this that

$$I = \mathbf{K}\left[1 - \lim_{n\to\infty}\frac{1^2+2^2+\ldots(n-1)^2}{n^3}\right] = \tfrac{2}{3}\mathbf{K}.$$

In the following pair of propositions the volume is derived of the lesser of the two segments into which a plane not passing through the centre divides a spheroid. The more general proposition is

Proposition 30.

If a spheroid be cut by a plane not perpendicular to the axis and not passing through the centre, the lesser of its segments will be to the segment of a cone having the same base as the segment and the same

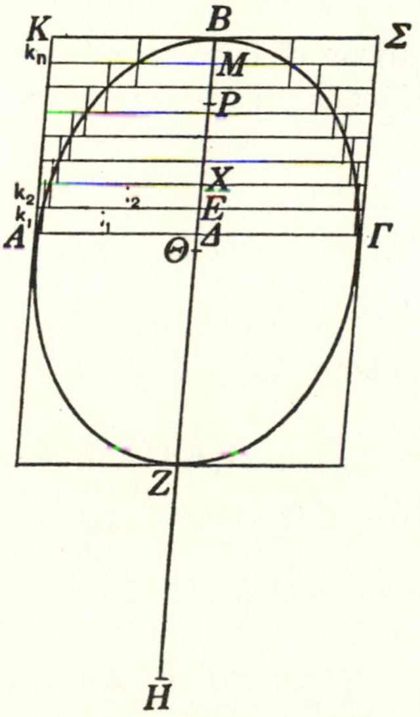

Fig. 101.

axis as the sum of half the line connecting the vertices of the segments and the axis of the greater segment to the axis of the greater segment.

Let the plane of the paper cut the spheroid in the oxytome $AB\Gamma Z$

with centre Θ (Fig. 101); further let ZH be equal to $Z\Theta$. Let Ψ be a cone, whose volume is to the volume of the segment of a cone $BA\Gamma$ as the two lengths referred to in the proposition, *i.e.* as $H\Delta$ to $Z\Delta$. It has now to be proved that the segment $BA\Gamma$ is equal to Ψ. If **K** be the frustum of a cylinder $A\Pi\Sigma\Gamma$ and if $BP = \frac{1}{3}B\Delta$, whence $H\Delta = 3\Theta P$, we have

$$(\mathbf{K}, \text{segment of a cone } BA\Gamma) = (H\Delta, \Theta P)$$

$$(\text{segment of a cone } BA\Gamma, \Psi) = (Z\Delta, H\Delta),$$

whence

$$(\mathbf{K}, \Psi) = (Z\Delta, \Theta P).$$

If the segment of the conoid is not equal to Ψ, it is either greater or less than Ψ.

Case I. Suppose $BA\Gamma > \Psi$. In the manner already repeatedly elucidated there is now constructed an inscribed solid figure such that

$$I_n > \Psi,$$

upon which a comparison of corresponding frusta of the inscribed solid figure and of **K** leads, in the same way as in Prop. 28, to the relation:

$$(i_1 + i_2 + \ldots + i_{n-1}, n.k_1)$$
$$= [\mathbf{O}(BE, ZE) + \ldots + \mathbf{O}(BM, ZM), n.\mathbf{O}(B\Delta, Z\Delta)]. \qquad (1)$$

The rectangles **O** cannot now be represented as gnomons; the method of Prop. 28, however, may now be extended as follows (Fig. 102):

$$\mathbf{O}(BE, ZE) = \mathbf{O}(B\Delta - \Delta E, Z\Delta + \Delta E)$$
$$= \mathbf{O}(B\Delta, Z\Delta) - \mathbf{O}(\Delta E, \Delta E + Z\Delta - B\Delta).$$

Now suppose $Z\Delta = N\Xi$ and $B\Delta = O\Xi$.

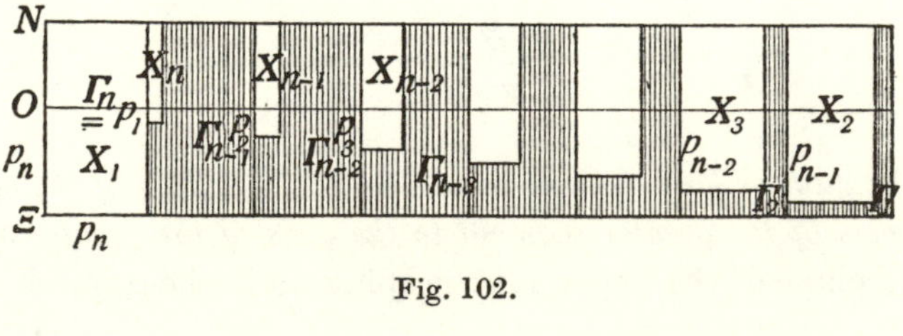

Fig. 102.

$\mathbf{O}(B\Delta, Z\Delta)$, henceforth to be denoted by Γ_n, can be applied hyperbolically to the line segment $NO = Z\Delta - B\Delta$, exceeding by a square $\mathbf{T}(B\Delta)$. If $p_1 = \Delta E$, we find $\mathbf{O}(BE, ZE)$ by subtracting from Γ_n $\mathbf{T}(p_1)$ and $\mathbf{O}(p_1, NO)$, so that the shaded area Γ_{n-1} is left. In the same way we find the following rectangles as $\Gamma_{n-2} \ldots \Gamma_1$, while therefore $p_2 = \Delta X, \ldots p_{n-1} = \Delta M$. The non-shaded rectangles X_2, $X_3 \ldots X_n$ that have been taken away now satisfy, together with Γ_n as X_1, the conditions of **C.S.** 2 (III; 7.4). From this it follows that

$$(n.X_1, X_1 + X_2 + \ldots X_n) < (NO + p_n, \tfrac{1}{3}p_n + \tfrac{1}{2}NO),$$

in which

$$NO = Z\Delta - B\Delta = 2\Theta\Delta, p_n = B\Delta, \tfrac{1}{3}p_n = BP.$$

Consequently we have

$$(n.X_1, X_1 + \ldots X_n) < (Z\Delta, BP + \Theta\Delta),$$

whence *convertendo*

$$[n.X_1, \Gamma_1 + \ldots \Gamma_{n-1}] > (Z\Delta, \Theta P) = (K, \Psi),$$

whence in relation with (1) (*invertendo*)

$$(\mathrm{K}, I_n) > (\mathrm{K}, \Psi), \text{ whence } I_n < \Psi,$$

which is contrary to the supposition.

Case II. Secondly suppose $BA\Gamma < \Psi$. Now construct a circumscribed solid figure such that

$$C_n < \Psi.$$

A comparison of corresponding frusta of a cylinder now gives

$$(c_1 + c_2 + \ldots c_n, n.k_1)$$
$$= [\mathbf{O}(B\Delta, Z\Delta) + \mathbf{O}(BE, ZE) + \ldots + \mathbf{O}(BM, ZM), n.\mathbf{O}(B\Delta, Z\Delta)] \quad (2)$$

We now represent the successive rectangles in the second member, in the same way as above, by the surfaces $\Gamma_n, \ldots \Gamma_1$. By **C.S.** 2 (III; 7.4) we now have

$$(n.X_1, X_2 + X_3 \ldots X_n) > (Z\Delta, BP + \Theta\Delta),$$

whence *convertendo*

$$(n.X_1, \Gamma_1 + \ldots \Gamma_n) < (Z\Delta, \Theta P) = (\mathrm{K}, \Psi),$$

whence in relation with (2) (*invertendo*)

$$(\mathbf{K}, C_n) < (\mathbf{K}, \Psi) \text{ or } C_n > \Psi,$$

which is contrary to the supposition.

Algebraically: Suppose (with the same notations as above) that $BZ = 2a$, $B\Delta = y_0$, $\Delta\Gamma = x_0$, then it has to be proved that

$$I = \tfrac{1}{3}\mathbf{K}\frac{3a - y_0}{2a - y_0}.$$

Now we have

$$\frac{i_m}{\frac{1}{n}\mathbf{K}} = \frac{x^2}{x_0^2} = \frac{y(2a-y)}{y_0(2a-y_0)}.$$

Archimedes now writes

$$y(2a-y) = y_0(2a-y_0) - (y_0-y)[(y_0-y)+2(a-y_0)].$$

If we now denote $2(a-y_0)$ by t,
we have

$$\frac{i_m}{\frac{1}{n}\mathbf{K}} = \frac{y_0(y_0+t)-(y_0-y)(y_0-y+t)}{y_0(y_0+t)},$$

whence, because

$$y = \frac{m}{n}y_0,$$

$$\frac{i_m}{\frac{1}{n}\mathbf{K}} = \frac{y_0(y_0+t) - \frac{n-m}{n}y_0\left[\frac{n-m}{n}y_0+t\right]}{y_0(y_0+t)},$$

whence

$$I_n = \mathbf{K}\frac{(n-1)(y_0+t) - y_0\left[\left(\frac{1}{n}\right)^2 + \ldots\left(\frac{n-1}{n}\right)^2\right] - t\frac{1+2+\ \ldots(n-1)}{n}}{n(y_0+t)},$$

whence

$$I = \lim_{n\to\infty} I_n = \mathbf{K}\frac{y_0+t-\frac{1}{3}y_0-\frac{1}{2}t}{y_0+t}$$

$$= \tfrac{1}{3}\mathbf{K}\frac{2y_0+\frac{3}{2}t}{y_0+t} = \tfrac{1}{3}\mathbf{K}\frac{3a-y_0}{2a-y_0}.$$

Finally the volume is also determined of the greater of the two segments into which a plane not passing through the centre divides a spheroid. The more general proposition is Prop. 32, which is perfectly identical with Prop. 30, provided the words "greater" and "lesser" be interchanged. The proof, however, differs considerably from that of Prop. 30. The latter was based on Prop. 20, in which (with a view to the theorem mentioned in III; 6.551) it had to be assumed that the segment which is cut off by the cutting plane of the spheroid under consideration is less than half the spheroid. This is not the case here, and consequently the method of Prop. 30 cannot now be applied. The highly complicated proof is as follows (Fig. 103):

Let the plane of the paper, taken as above, cut the spheroid in the oxytome $AB\Gamma\Delta$, the base of the segment in AB, the plane through the centre Θ parallel to the base in $K\Lambda$. Further suppose $\Delta H = BZ = \frac{1}{2}B\Delta$.

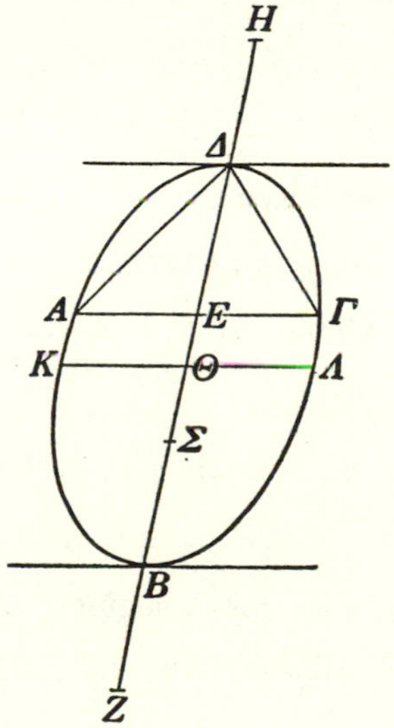

Fig. 103.

It has now to be proved that:

(segment of spheroid $BA\Gamma$, segment of cone $BA\Gamma$) = $(HE, \Delta E)$.

Proof: By **C.S.** 30 we have

(segment of spheroid $\Delta A\Gamma$, segment of cone $\Delta A\Gamma$) = (ZE, BE).

Further the ratio

(segment of cone $\Delta K\Lambda$, segment of cone $\Delta A\Gamma$) is compounded of

$$[\mathbf{T}(\Theta K),\ \mathbf{T}(EA)] \quad \text{and} \quad (\Delta\Theta, \Delta E), \quad \text{thence of}$$

$$[\mathbf{O}(\Delta\Theta, B\Theta),\ \mathbf{O}(\Delta E, BE)] \quad \text{and} \quad (\Delta\Theta, \Delta E)\,.$$

Now construct a point Σ such that

$$(\Delta E, \Delta\Theta) = (\Delta\Theta, \Delta\Sigma), \text{ consequently}$$

$$(\Delta\Theta, \Delta E) = [\mathbf{O}(B\Theta, \Delta\Sigma),\ \mathbf{O}(B\Theta, \Delta\Theta)]\,.$$

From this it follows that

(segment of cone $\Delta K\Lambda$, segment of cone $\Delta A\Gamma$)

$$= [\mathbf{O}(B\Theta, \Delta\Sigma),\ \mathbf{O}(BE, \Delta E)]\,.$$

We also have

(segment of cone $\Delta A\Gamma$, segment of spheroid $\Delta A\Gamma$) $= (BE, ZE)$

$= [\mathbf{O}(BE, \Delta E),\ \mathbf{O}(ZE, \Delta E)]$, so that *ex aequali*

(segment of cone $\Delta K\Lambda$, segment of spheroid $\Delta A\Gamma$)

$$= [\mathbf{O}(B\Theta, \Delta\Sigma),\ \mathbf{O}(ZE, \Delta E)]\,.$$

By **C.S.** 28 we further have

(spheroid, segment of cone $\Delta K\Lambda$) $= (4, 1) = (HZ, B\Theta)$

$$= [\mathbf{O}(HZ, \Delta\Sigma),\ \mathbf{O}(B\Theta, \Delta\Sigma)]\,,$$

whence again *ex aequali*

(spheroid, segment of spheroid $\Delta A\Gamma$) $= [\mathbf{O}(HZ, \Delta\Sigma),\ \mathbf{O}(ZE, \Delta E)]$,

whence

(segment of spheroid $BA\Gamma$, segment of spheroid $\Delta A\Gamma$)

$$= [\mathbf{O}(HZ, \Delta\Sigma) - \mathbf{O}(ZE, \Delta E),\ \mathbf{O}(ZE, \Delta E)]\,.$$

In this

$$\mathbf{O}(HZ, \Delta\Sigma) - \mathbf{O}(ZE, \Delta E) = \mathbf{O}(HE, \Delta\Sigma) + \mathbf{O}(ZE, \Delta\Sigma) - \mathbf{O}(ZE, \Delta E)$$

$$= \mathbf{O}(HE, \Delta\Sigma) + \mathbf{O}(ZE, E\Sigma)\,.$$

Consequently

(segment of spheroid $BA\Gamma$, segment of spheroid $\Delta A\Gamma$)

$$= [\mathbf{O}(HE, \Delta\Sigma) + \mathbf{O}(ZE, E\Sigma),\ \mathbf{O}(ZE, \Delta E)]$$

$$(\text{segment of spheroid } \Delta A\Gamma, \text{ segment of cone } \Delta A\Gamma)$$
$$= (ZE, BE) = [\mathbf{O}(ZE, \Delta E),\ \mathbf{O}(BE, \Delta E)]$$
$$(\text{segment of cone } \Delta A\Gamma, \text{ segment of cone } BA\Gamma) = (\Delta E, BE)$$
$$= [\mathbf{O}(BE, \Delta E),\ \mathbf{T}(BE)]\,,$$

whence *ex aequali*

$$(\text{segment of spheroid } BA\Gamma, \text{ segment of cone } BA\Gamma)$$
$$= [\mathbf{O}(HE, \Delta\Sigma) + \mathbf{O}(ZE, E\Sigma),\ \mathbf{T}(BE)]\,.$$

In this

$$\mathbf{T}(BE) = \mathbf{T}(B\Theta + \Theta E) = \mathbf{T}(B\Theta) + \mathbf{T}(\Theta E) + 2.\mathbf{O}(B\Theta, \Theta E)$$
$$= \mathbf{T}(B\Theta) + \mathbf{O}(\Theta E, ZE) = \mathbf{O}(\Delta E, \Delta\Sigma) + \mathbf{O}(\Theta E, ZE)\,,$$

so that

$$(\text{segment of spheroid } BA\Gamma, \text{ segment of cone } BA\Gamma)$$
$$= [\mathbf{O}(HE, \Delta\Sigma) + \mathbf{O}(ZE, E\Sigma),\ \mathbf{O}(\Delta E, \Delta\Sigma) + \mathbf{O}(\Theta E, ZE)]\,.$$

The ratio in the second member can now be reduced with the aid of the relation by which Σ is determined:

$(\Delta E, \Delta\Theta) = (\Delta\Theta, \Delta\Sigma)$. In fact, it follows from this that

$(\Theta E, \Delta E) = (\Theta\Sigma, \Delta\Theta)$, whence

$(\Theta E, \Theta\Sigma) = (\Delta E, \Delta\Theta)$ or

$(\Sigma E, \Theta E) = (HE, \Delta E)$.

Consequently

$$[\mathbf{O}(HE, \Delta\Sigma),\ \mathbf{O}(\Delta E, \Delta\Sigma)]$$
$$= [\mathbf{O}(ZE, E\Sigma),\ \mathbf{O}(\Theta E, ZE)] = (HE, \Delta E)\,.$$

Therefore

$$(\text{segment of spheroid } BA\Gamma, \text{ segment of cone } BA\Gamma) = (HE, \Delta E)\,.$$

In this proof, several peculiar features of Greek mathematics (the inconvenient arrangement resulting from the low development of symbolic notation, the limitation to terms at most of the second degree, but also the great ingenuity in carrying out algebraic reductions by geometrical means) are so clearly manifested that it is worth while considering it somewhat more in detail.

Let us denote the volumes of the spheroid itself and of the two segments of the spheroid $\Delta A\Gamma$ and $BA\Gamma$ by I, I_1, and I_2 respectively, those of the segments of the cone $\Delta K\Lambda$, $\Delta A\Gamma$, and $BA\Gamma$ by K, K_1, and K_2 respectively; further suppose $\Delta B = 2a$, $\Delta E = h_1$, $BE = h_2$, $\Theta K = b$, $EA = p$.

The proof now consists in deducing from the ratios

$$\frac{I_1}{K_1}, \frac{K_1}{K}, \frac{K}{I}$$

by multiplication the ratio $\frac{I_1}{I}$.

From this we then find $\frac{I_2}{I}$, and subsequently, when

$$\frac{I}{K} \text{ and } \frac{K}{K_2} \text{ are known,}$$

by multiplication again $\frac{I_2}{K_2}$.

Formulated algebraically, this would nowadays be as follows:

$$\frac{I_1}{K_1} = \frac{a+h_2}{h_2} \qquad \text{C.S. 30}$$

$$\frac{K_1}{K} = \frac{p^2 h_1}{b^2 a} = \frac{h_1{}^2 h_2}{a^3}$$

$$\frac{K}{I} = \frac{1}{4} \qquad \text{C.S. 28}$$

Therefore

$$\frac{I_1}{I} = \frac{(a+h_2)h_1{}^2}{4a^3}$$

Therefore

$$\frac{I_2}{I} = \frac{4a^3 - (a+h_2)h_1{}^2}{4a^3}$$

$$\frac{I}{K} = 4 .$$

$$\frac{K}{K_2} = \frac{b^2 a}{p^2 h_2} = \frac{a^3}{h_1 h_2{}^2}.$$

From this it follows that

$$\frac{I_2}{K_2} = \frac{4a^3-(a+h_2)h_1^2}{h_1h_2^2},$$

from which we find after a certain amount of calculation, with the aid of the relation

$$2a = h_1+h_2,$$

that

$$\frac{I_2}{K_2} = \frac{a+h_1}{h_1}.$$

Translated into algebraic formulation, the argument of Archimedes is slightly different: he introduces a magnitude t (with him $\Delta\Sigma$) by means of the relation

$$a^2 = h_1 t$$

and thus finds

$$\frac{K_1}{K} = \frac{h_1h_2}{at} \quad\text{and}\quad \frac{K}{K_2} = \frac{at}{h_2^2}.$$

Thus he obtains

$$\frac{I_2}{K_2} = \frac{4at-h_1(a+h_2)}{h_2^2}.$$

It is now particularly curious to see how he reduces this expression. He writes for the numerator:

$$(a+h_1)t+(a+h_2)t-h_1(a+h_2) = (a+h_1)t+(a+h_2)(t-h_1),$$

for the denominator:

$$[a+(a-h_1)]^2 = a^2+(a-h_1)(a+h_2),$$

so that

$$\frac{I_2}{K_2} = \frac{(a+h_1)t+(a+h_2)(t-h_1)}{h_1t+(a+h_2)(a-h_1)}.$$

Now

$$\frac{h_1}{a} = \frac{a}{t} = \frac{a-h_1}{t-a}$$

or

$$\frac{a-h_1}{h_1} = \frac{t-a}{a} = \frac{t-h_1}{a+h_1},$$

whence

$$\frac{(a+h_1)t}{h_1t} = \frac{(t-h_1)(a+h_2)}{(a-h_1)(a+h_2)} = \frac{I_2}{K_2};$$

therefore

$$\frac{I_2}{K_2} = \frac{a+h_1}{h_1}.$$

CHAPTER VIII.

ON SPIRALS

1. The discussion of the so-called Archimedean spiral, to which the whole of the treatise *On Spirals* is devoted, after eleven introductory propositions, which have already been dealt with in Chapter III (10.1–2; 9; 7.30; 7.50), opens with a number of definitions, the first of which is rendered here *in extenso.*

If in a plane a straight line be drawn and, while one of its extremities remains fixed, after performing any number of revolutions at a uniform rate return again to the position from which it started, while at the same time a point moves at a uniform rate along the straight line, starting from the fixed extremity, the point will describe a spiral (ἕλιξ).

The fixed extremity is called the *origin of the spiral* (ἀρχὰ τᾶς ἕλικος; invariably to be represented by A); the starting position of the revolving line, *origin* of the revolution (ἀρχὰ τᾶς περιφορᾶς; to be rendered by *initial line*); the line segments of length L traversed during the first, second, etc. complete revolution by the moving point on the revolving radius vector, *first, second, etc. line segment* (εὐθεῖα πρώτα, δευτέρα etc.) the areas traversed by the segment of the radius vector between A and the moving point during the first, second, etc. revolution, *first, second, etc. area* (χωρίον πρῶτον, δεύτερον etc.). Of the straight line joining the origin with a point P of the curve we call *forward part* (τὰ προαγούμενα) the half of the straight line with extremity P lying on the side towards which the revolution takes place, *i.e.* the half of the straight line on which the origin does not lie; *backward part* (τὰ ἑπόμενα) the other half of the straight line. The circles with A as centre and L, $2L$, etc. respectively as radii are referred to as *first, second, etc. circle* (κύκλος πρῶτος, δεύτερος etc.).

The propositions 12–17 contain introductory geometrical theorems on radii vectores and tangents to the spiral.

Proposition 12.

If towards the turn described in any revolution any number of straight lines be drawn from the origin, which make equal angles with each other, they will exceed each other by an equal magnitude (Fig. 104).

This follows at once from the definition of the curve.

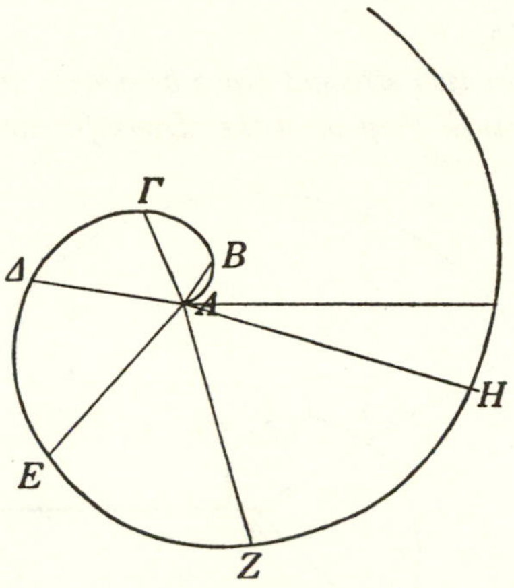

Fig. 104.

Proposition 13.

If a straight line touch the spiral, it will touch it in one point only.

For a discussion of the Greek tangent concept, see III; 1.6. What is meant here is that, if a straight line meet the spiral in a given point Γ (Fig. 105), while none of its points lies within the turn to which Γ belongs, this straight line cannot meet the same turn again in another point H. In other words, it has to be proved that a turn of a spiral has no double tangents.

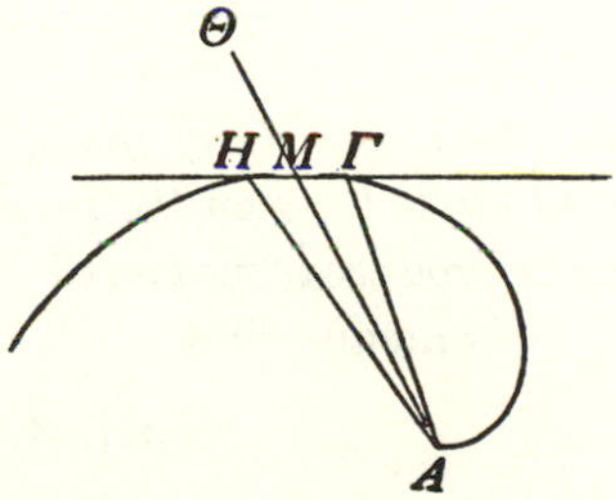

Fig. 105.

Now suppose that the straight line met the spiral in the manner described above (*i.e.* touched it) in Γ and in H. Now draw the bisector AM of $\triangle A\Gamma H$, then by a familiar planimetrical theorem we have

$$A\Gamma + AH > 2.AM .$$

On AM lies a point Θ of the spiral which is determined by

$$A\Gamma + AH = 2.A\Theta .$$

From this it follows that $AM < A\Theta$, so that Θ lies outside $\triangle A\Gamma H$, and M therefore within the turn of the spiral, which is contrary to the supposition that ΓH does not have any point within the turn of the spiral.

Proposition 14.

If from the origin two straight lines be drawn to the first turn and they are produced until they meet the circumference of the first circle,

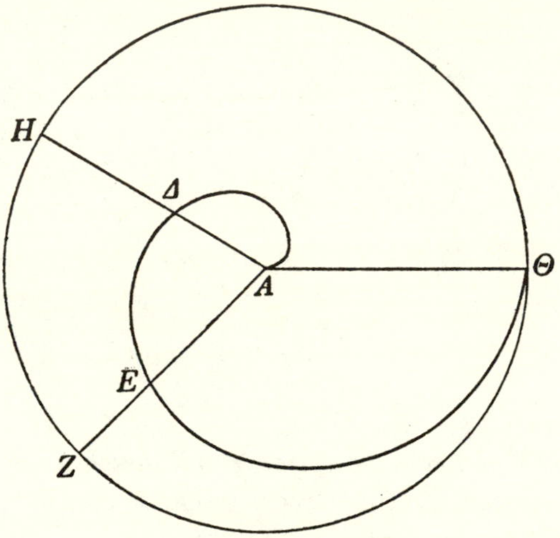

Fig. 106.

the radii vectores of the spiral will be to each other as the arcs of the first circle between the initial line and the extremities of the radii vectores produced, measured in the direction of revolution (Fig. 106).

This means that

$$(A\Delta, AE) = (\text{arc}\,\Theta H, \text{arc}\,\Theta Z)\ .$$

Proposition 15.

Radii vectores of the second turn are to each other in the same ratio as the arcs mentioned (in Prop. 14), *each augmented with the entire circumference of the first circle.*

Both these propositions follow at once from the definition of the curve.

Proposition 16.

If a straight line touch the first turn and the point of contact be joined with the origin, the angles which the tangent makes with the radius vector will be unequal, the forward angle being obtuse, the backward angle acute.

In Fig. 107 let EZ touch the first turn at Δ[1]. Let the circle with A as centre through Θ be the first circle. Draw the circle with A

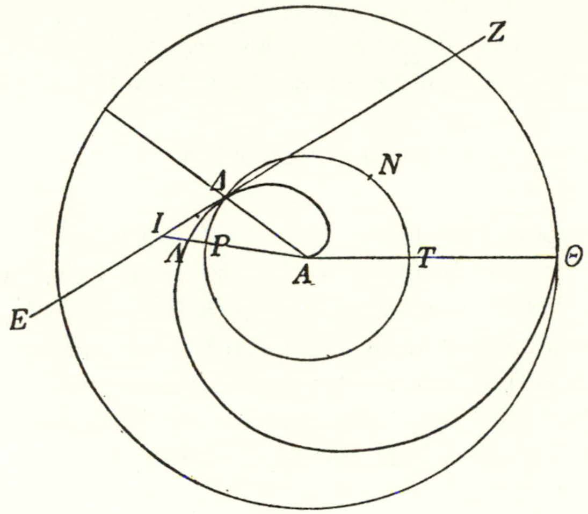

Fig. 107.

as centre through Δ, then arc $TN\Delta$ of this circle will lie outside the spiral and arc ΔPT within it; consequently $\angle A\Delta E$ is greater than the angle of a semi-circle, which itself is greater than any acute angle[2]. Therefore $\angle A\Delta E$ is not acute. It still has to be proved that it is not right either, *i.e.* that EZ does not touch the circle $TN\Delta$ at Δ. Suppose this to be the case, then it would be possible (by III; 9.1) to draw through A a straight line meeting the circle in P, the spiral in Λ, the straight line EZ in I, so that

$$(PI, AP) < (\text{arc}\,\Delta P, \text{arc}\,\Delta NT),$$

---

[1]) By the forward angle between the tangent and the radius vector is meant $\angle E\Delta A$, by the backward angle $\angle Z\Delta A$.

[2]) By "angle of a semi-circle" is meant the so-called horn-like angle which a circle makes with its diameter (not therefore the right angle between a diameter and the tangent to the circle at one of its extremities). The theorem applied here is Euclid III, 16, 3rd paragraph. See *Elements of Euclid* II, 34, 37. In view of the use made of it by Archimedes, it cannot be maintained, as there asserted, that Euclid included the concept of the hornlike angle for historical reasons only.

whence *componendo*

$$(AI, AP) < (\text{arc}\,TN\Delta P, \text{arc}\,\Delta NT) = (A\Lambda, A\Delta)\,.$$

However, $AP = A\Delta$, whence $AI < A\Lambda$, which is contrary to the supposition that EZ is the tangent to the spiral.

In Prop. 17 the same theorem is set forth with regard to tangents to other turns and to tangents at the extremities of turns.

2. *Application of the Archimedean Spiral to the Rectification of Circular Arcs.*

In the propositions 18–20 the spiral is turned to account to rectify circular arcs. In Prop. 18 it is stated that the circumference of the first circle is equal to the segment of the perpendicular erected on the initial line in the origin between the origin and the point of intersection with the tangent to the spiral at the extremity of the first turn; in Prop. 19 the same theorem is stated with regard to the circumferences of the second, the third circle, etc. The proofs are entirely identical with that of Prop. 20, of which the two preceding propositions are particular cases. We shall therefore confine ourselves to discussing

Proposition 20.

If a straight line touch the first turn of the spiral at a point other than the extremity[1]), *if from the point of contact to the origin a straight line be drawn, if then a circle be described with the origin as centre and the straight line as radius and from the origin a straight line be drawn perpendicular to the radius vector of the point of contact, this perpendicular will meet the tangent, and the segment between the point of intersection and the origin will be equal to the arc of the circle between the point of intersection with the initial line and the point of contact, measured in the direction of revolution of the radius vector.*

In Fig. 108 let the tangent at the point Δ to the spiral meet the perpendicular erected in A on $A\Delta$ in Z. That the point of intersection Z exists, follows from the result found in Prop. 16, *viz.* that the backward angle between the tangent and the radius vector is acute. Let the circle through Δ with A as centre intersect the

[1]) Neither of the two restrictions, *viz.* "first turn" and "at a point other than the extremity", is essential.

initial line in K. ΔZ does not touch the circle, and will therefore meet it once more in N; let M be the middle point of the chord ΔN. It has to be proved that AZ is equal to the circular arc $KN\Delta$, which we represent by Π. If this is not true, either $AZ > \Pi$ or $AZ < \Pi$.

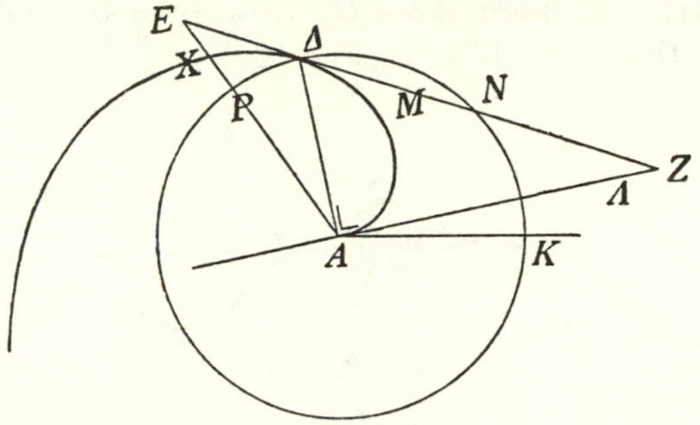

Fig. 108.

Case I. Suppose $AZ > \Pi$. Then there exists between A and Z a point Λ such that also $A\Lambda > \Pi$[1]).

Now in the constructed circle apply the Prop. III; 9.3 to the chord ΔN in order to find a point E on $N\Delta$ produced so that if AE meets the circle in P and the spiral in X, the ratio $(PE, P\Delta)$ shall have an assigned value which has to be greater than $(\Delta M, AM)$, and consequently also greater than $(A\Delta, AZ)$. E can therefore be so determined that

$$(PE, P\Delta) = (A\Delta, A\Lambda),$$

whence *permutando*

$$(PE, A\Delta) = (P\Delta, A\Lambda),$$

whence, because $A\Delta = AP$,

$$(PE, AP) < (\text{arc}\, P\Delta, \Pi)$$

or *componendo*

$$(AE, AP) < (\Pi + \text{arc}\, P\Delta, \Pi),$$

1) The existence of the point Λ follows from Prop. 4 of the treatise: *Given two unequal lines, viz. a straight line and a circular arc, it is possible to find a straight line less than the greater of the given lines and greater than the less.* Proof: Let the circular arc be A, the straight line B; find n such that n times the difference between A and B is greater than B; now divide B into n equal parts, then the required line segment is the sum of one-nth of B and the lesser of the magnitudes A and B; it is not stated how a straight line segment equal to A is made.

which latter ratio, because of the symptom of the spiral, is equal to $(AX, A\Delta)$. The inequality

$$(AE, AP) < (AX, A\Delta)\,,$$

however, is impossible, because $AE > AX$, while $AP = A\Delta$.

Case II. Now suppose $AZ < \Pi$. Now take on AZ produced a point I such that also $AI < \Pi$ (Fig. 109).

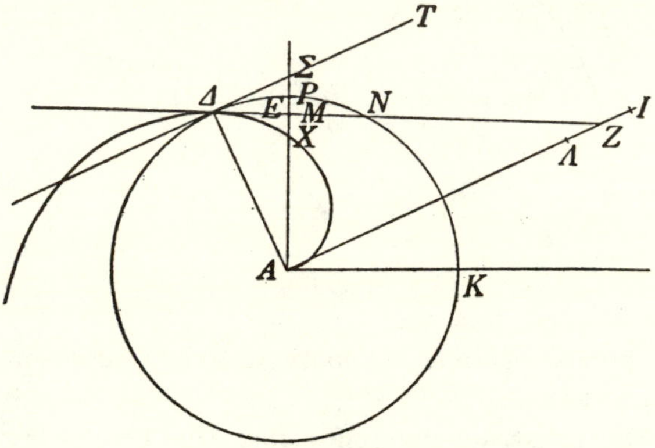

Fig. 109.

Now in the constructed circle apply to the chord ΔN with middle point M and the line ΔT, which touches the circle at Δ, the Proposition III; 9.4 in order to find a point E between Δ and N such that, if AE meets the spiral in X, the circle in P, and the tangent ΔT in Σ, the ratio $(EP, \Delta\Sigma)$ will have an assigned value which must be less than $(\Delta M, AM)$, and consequently less than $(A\Delta, AZ)$. E can therefore be so determined that

$$(PE, \Delta\Sigma) = (A\Delta, AI)\,,$$

whence *permutando*

$$(PE, A\Delta) = (\Delta\Sigma, AI)$$

or

$$(PE, AP) > (\text{arc}\,\Delta P, \Pi)\,,$$

whence *separando*

$$(AE, AP) < (\text{arc}\,KNP, \Pi)\,,$$

whence (symptom of the spiral)

$$(AE, AP) < (AX, A\Delta)\,.$$

This inequality, however, is impossible, because $AE > AX$, while $AP = A\Delta$.

The truth of the proposition proved above is very easily recognized with present-day means.

Let the equation of the spiral in polar coordinates with origin O and initial line on the initial line of the curve be

$$\varrho = \lambda\varphi\,,$$

then the acute backward angle Θ, made by the radius vector of a point Δ (ϱ, φ) with the tangent, is determined by

$$\operatorname{tg}\Theta = \frac{\varrho}{\varrho'} = \frac{\lambda\varphi}{\lambda} = \varphi$$

and we therefore find for the polar subtangent

$$AZ = \varrho\,.\operatorname{tg}\Theta = \varrho\varphi\,.$$

i.e. the length of the arc with length φ of a circle with radius φ.

The reasoning by which Archimedes may have arrived at this curious proposition may perhaps be elucidated further as follows:

In Prop. 16 it was recognized that the angle between a tangent and the radius vector of the point of contact is obtuse in the forward direction, while by studying the figure we are soon led to suspect that this angle will constantly increase upon continued revolution of the radius vector. Now Archimedes may have desired to make a quantitative statement about the extent of this increase, and may therefore have noted that when in Fig. 108 X lies on the curve close to Δ, the angle $AE\Delta$ may serve as approximation to the angle $A\Delta N$. Because of the symptom of the curve we now have

$$(AX, A\Delta) = (\text{arc}\,PK, \text{arc}\,\Delta K)$$

or *separando*

$$(PX, A\Delta) = (\text{arc}\,P\Delta, \text{arc}\,\Delta K)\,.$$

If in this we replace by approximation PX by PE and the arc $P\Delta$ by the chord $P\Delta$, then we have

$$(PE, A\Delta) \backsim (P\Delta, \text{arc}\,\Delta K)$$

or *permutando*

$$(PE, P\Delta) \backsim (A\Delta, \text{arc}\,\Delta K)\,.$$

From this relation we learn how the form of $\triangle\, PE\Delta$ alters upon revolution of the radius vector. It also shows, however, that the arc ΔK can be approximately constructed as the side of a triangle with one side $A\Delta$ and similar to $\triangle PE\Delta$. When X approximates to Δ along the curve, $\angle EP\Delta$ will approximate to a right angle and $\angle PE\Delta$ to $\angle A\Delta N$. We may therefore suspect that the arc ΔK will be equal to the side AZ of a triangle $A\Delta Z$ which is right-angled in A.

The correctness of this supposition is proved after the custom of Greek mathematics by means of a double *reductio ad absurdum*; for this, use is made of the theorems proved in III; 9 with regard to the limit to which the ratio $(PE, P\Delta)$ tends when P tends to Δ along the circle (III; 9.7). The first part of the proof can be rendered as follows with the aid of the terminology of limits:

By III; 9.3 we have

$$\lim_{P \to \Delta} \frac{PE}{P\Delta} = \frac{\Delta M}{A\Delta} = \frac{A\Delta}{AZ}.$$

In view of the fact that ΔE is a tangent and therefore deviates little from the curve in the neighbourhood of Δ, we may write

$$\lim_{P \to \Delta} \frac{PE}{P\Delta} = \lim_{X \to \Delta} \frac{XE}{P\Delta} = \lim_{X \to \Delta} \frac{XE}{\text{arc } P\Delta}.$$

Because of the symptom of the curve, however, the latter fraction is equal to $\dfrac{A\Delta}{\text{arc } \Delta K}$, which fraction is constant when X tends to Δ.

From this it follows that

$$\frac{A\Delta}{AZ} = \frac{A\Delta}{\text{arc } \Delta K}, \text{ whence } AZ = \text{arc}\,\Delta K\,.$$

Archimedes formulates all this by means of inequalities. He knows that $(PE, P\Delta)$ tends to $(A\Delta, AZ)$ on the high side, and that $(PE, P\Delta)$ can therefore be made less than any magnitude exceeding $(A\Delta, AZ)$. If $AZ > \Pi$, it is possible to make

$$(PE, P\Delta) < (A\Delta, \Pi)\,,$$

which is contrary to

$$(PE, P\Delta) > (PX, \text{arc}\,P\Delta) = (A\Delta, \Pi)\,.$$

It is now at once obvious why the second part of the proof cannot be given entirely analogously to the first part, *i.e.* by the application of III; 9.2 instead of III; 9.3., In fact if (in Fig. 109) E lies between Δ and N, $PE < PX$ and, since also $P\Delta < \text{arc}\,P\Delta$, it is not known whether $(PE, P\Delta)$ is greater or less than $(PX, \text{arc}\,P\Delta)$.

Archimedes therefore now makes use of III; 9.4. According to this theorem $(PE, \Delta\Sigma)$ tends to $(A\Delta, AZ)$ on the low side, and can therefore be made greater than any magnitude less than $(A\Delta, AZ)$. If $AZ < \Pi$, it is thus possible to make

$$(PE, \Delta\Sigma) > (A\Delta, \Pi),$$

which is contrary to

$$(PE, \Delta\Sigma) < (XP, \text{arc}\,\Delta P) = (A\Delta, \Pi).$$

The whole argument therefore appears to be closely akin to that followed in differential geometry. If the equation of the curve is

$$\varrho = \lambda\varphi$$

and $\angle A\Delta N = \Theta$, we have

$$\text{tg}\,\Theta = \lim_{X\to\Delta}\frac{P\Delta}{XP} = \lim_{\Delta\varphi\to 0}\frac{\varrho\Delta\varphi}{\Delta\varrho} = \lim_{\Delta\varphi\to 0}\frac{\varrho\Delta\varphi}{\lambda\Delta\varphi} = \frac{\varrho}{\lambda} = \varphi,$$

which is therefore equivalent to the result of a differentiation.

From the proved property we at once derive the familiar property of the Archimedean spiral that the polar subnormal is constant, *viz.*

$$\varrho\cot\Theta = \frac{\varrho}{\varphi} = \lambda.$$

Archimedes does not mention this theorem. In order to prove it, he would have had to consider two points Δ and Δ_1 and describe the circular arcs ΔK and $\Delta_1 K_1$ with A as centre. If the normals of Δ and Δ_1 meet the lines ZA and Z_1A_1 produced respectively in B and B_1, we have in view of the proved theorem

$$(AB, A\Delta) = (A\Delta, \text{arc}\,\Delta K) \tag{1}$$

$$(AB_1, A\Delta_1) = (A\Delta_1, \text{arc}\,\Delta_1 K_1) \tag{2}$$

and when $A\Delta_1$ meets the circle $K\Delta$ in H (symptom of the spiral):

$$(A\Delta, A\Delta_1) = (\text{arc}\,\Delta K, \text{arc}\,HK).$$

From (1) it follows in view of this that

$$(AB, A\Delta) = (A\Delta_1, \text{arc}\,HK),$$

while also

$$(A\Delta, A\Delta_1) = (\text{arc}\,HK, \text{arc}\,K_1\Delta_1)\,.$$

Ex aequali we now have

$$(AB, A\Delta_1) = (A\Delta_1, \text{arc}\,K_1\Delta_1)\,,$$

whence, because of (2),

$$(AB, A\Delta_1) = (AB_1, A\Delta_1)\,,$$

therefore

$$AB = AB_1\,.$$

3. *Quadrature of the Spiral.*

The remaining part of the treatise is devoted to the discussion of the quadrature of the Archimedean spiral. Here the difference form of the compression method (III; 8.2) is applied again, so that it first has to be proved that figures can be circumscribed about and inscribed in the figure the area of which is required to be found, for which figures the difference between the areas can be made less than any assigned small magnitude. This is done in Prop. 21 for the first area, in Prop. 22 for the second and subsequent, in Prop. 23 for a sector bounded by any part of a turn and the radii vectores of the extremities thereof. The method is elucidated sufficiently by the discussion of the last-mentioned case.

Proposition 23.

When the area is taken which is bounded by an arc of the spiral which is less than a turn and which does not have the origin for extremity and by the radii vectores of the extremities, it is possible to circumscribe about this area a plane figure consisting of similar sectors, and to inscribe in it another such that the circumscribed figure exceeds the inscribed figure by less than any assigned area.

In Fig. 110 let the spiral arc $B\ldots Z$ be given. Describe a circle through Z with A as centre; on this the radii vectores of B and Z determine an arc HZ. Apply dichotomy to this arc until there be formed a sector AKH, less than the assigned area ε. Now draw circular arcs with A as centre respectively through $B\ldots Z$, each bounded by two successive radii vectores as obtained by the dichotomy. A circumscribed and an inscribed figure are thus obtained, whose areas may be referred to as C_n and I_n respectively;

both these figures consist of sectors of circles which are all similar. Now the drawing shows at once that

$$C_n - I_n = \text{sector } A\Lambda Z - \text{sector } ABM < \text{sector } AKH < \varepsilon .$$

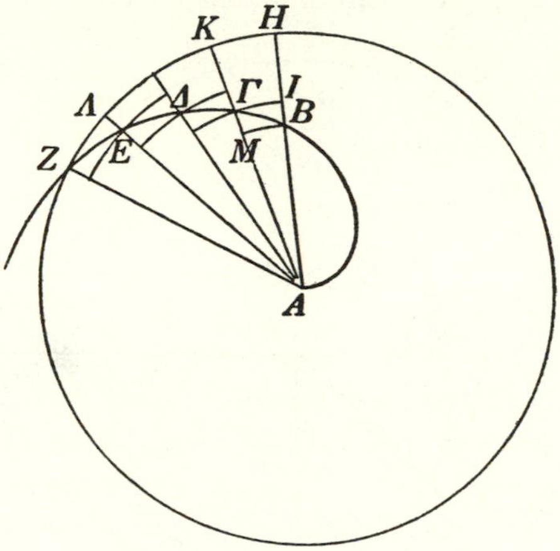

Fig. 110.

If the area of the figure bounded by the spiral arc BZ and the radii vectores AB and AZ is equal to O, we have *a fortiori*

$$O - I_n < \varepsilon \qquad\qquad C_n - O < \varepsilon .$$

After this the quadrature proper starts with

Proposition 24.

The area bounded by the first turn of the spiral and the first line segment is equal to one-third of the first circle.

Let X_1 be the area in question, K_1 the first circle. If it is not true that $X_1 = \frac{1}{3}K_1$, either $X_1 < \frac{1}{3}K_1$ or $X_1 > \frac{1}{3}K_1$.

Case I. Suppose $X_1 < \frac{1}{3}K_1$. It is now possible according to Prop. 23 to determine n such that

$$C_n - X_1 < \tfrac{1}{3}K_1 - X_1 ,$$

so that

$$C_n < \tfrac{1}{3}K_1 . \qquad (1)$$

The circumscribed figure (Fig. 111) consists of similar sectors of circles $S_1, S_2 \ldots, S_n$ respectively with radii $a_1, a_2 \ldots, a_n$, which form an arithmetical progression, the common difference of which is equal to the least term. Now we have by III; 7.33:

$$n.S_n < 3(S_1+S_2+\ldots+S_n),$$

in which

$$n.S_n = K_1.$$

Therefore $\qquad C_n > \frac{1}{3}K_1$, which is contrary to (1).

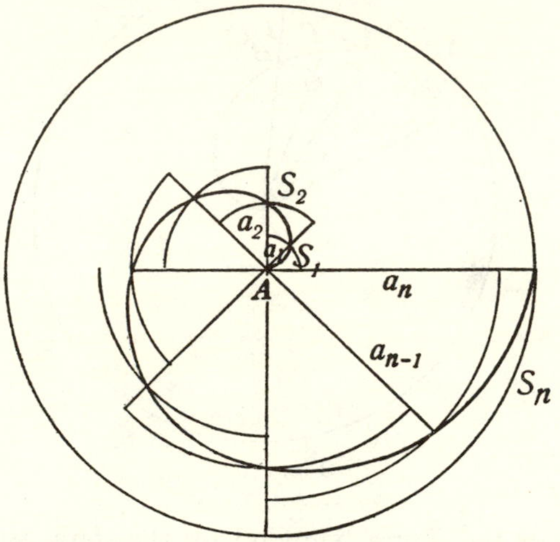

Fig. 111.

Case II. Secondly suppose $X_1 > \frac{1}{3}K_1$. It is now possible according to Prop. 23 to determine n such that

$$X_1 - I_n < X_1 - \frac{1}{3}K_1,$$

so that

$$I_n > \frac{1}{3}K_1. \tag{2}$$

The inscribed figure consists of similar sectors of circles S_1 $S_2 \ldots, S_{n-1}$ respectively, with the radii $a_1, a_2 \ldots, a_{n-1}$ already mentioned. Now we have by III; 7.33:

$$3(S_1+S_2+\ldots S_{n-1}) < n.S_n,$$

whence

$$I_n < \frac{1}{3}K_1, \text{ which is contrary to (2)}.$$

Algebraically: The radii of the successive inscribed sectors of circles are

$$\frac{L}{n}, \frac{2L}{n} \ldots \frac{(n-1)L}{n}.$$

We therefore find

$$I_n = \sum_1^{n-1} i \frac{1}{2}\left(\frac{iL}{n}\right)^2 \cdot \frac{2\pi}{n} = \pi L^2 \frac{1^2+2^2+\ldots(n-1)^2}{n^3}$$

$$= \pi L^2 \frac{\frac{1}{6}(n-1)n(2n-1)}{n^3},$$

whence

$$I = \lim_{n\to\infty} I_n = \tfrac{1}{3}\pi L^2 .$$

The operation is equivalent to

$$I = \int_0^{2\pi} \tfrac{1}{2}\varrho^2 d\varphi, \text{ in which } \varrho = \frac{\varphi}{2\pi} L .$$

Here, too, we find $I = \frac{1}{3}\pi L^2$.

Proposition 25.

The area bounded by the second turn of the spiral and the second of the line segments on the initial line is to the second circle in the ratio of 7 *to* 12, *being the same ratio which the sum of the rectangle comprehended by the radii of the second and of the first circle and one-third of the square on the difference by which the radius of the second circle exceeds that of the first has to the square on the radius of the second circle*[1]).

In Fig. 112 let $E\Delta\ldots\Theta$ be the second turn of the spiral, and the circle through Θ with A as centre the second circle K_2; and let the second area be called X_2. In order to elucidate Archimedes' result, we give the discussion—which is indirectly synthetical in his treatise—in the analytical form in which to all likelihood he himself will also have given it when he found the theorem.

Imagine therefore a circumscribed and an inscribed figure, each consisting of similar sectors, the number of these sectors being $n = 2^k$ and the radii forming an arithmetical progression. These sectors are:

[1]) It is to be noted that the figure under consideration is *not* bounded by the first turn, the second turn, and the second line segment, but only by the second turn and the second line segment. In the drawing the first turn should therefore be thought away.

for $C_n : S_2$ with radius $a_2 = A\Delta$, S_3 with radius $a_3 = A\Lambda$, ..., S_{n+1} with radius $a_{n+1} = A\Theta$.

for $I_n : S_1$ with radius $a_1 = AE$, S_2 with radius $a_2 = A\Delta$, ..., S_n with radius $a_n = AH$.

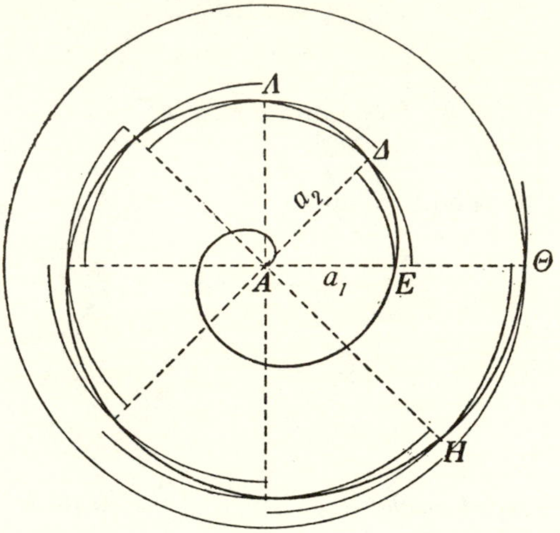

Fig. 112.

Both figures may be compared with the second circle K_2, whose area is equal to $n . S_{n+1}$.

To the said sectors of circles applies the proposition III; 7.52 (**Spir.** 11), provided it is freed of the restrictive condition that the common difference of the arithmetical progression of the radii should be equal to the least term. This leads to the following result:

$$(n . S_{n+1}, S_{n+1} + \ldots + S_2) < [\mathbf{T}(a_{n+1}), \mathbf{O}(a_{n+1}, a_1) + \tfrac{1}{3}\mathbf{T}(a_{n+1} - a_1)]$$
$$< (n . S_{n+1}, S_n + \ldots + S_1)$$

$$(K_2, C_n) < [\mathbf{T}(2a_1), 2\mathbf{T}(a_1) + \tfrac{1}{3}\mathbf{T}(a_1)] < (K_2, I_n)$$
$$(K_2, C_n) < (12,7) < (K_2, I_n) .$$

Since this relation holds good for any value of $n = 2^k$ and the difference between C_n and I_n can decrease below any assigned value owing to the choice of k, Archimedes could already suspect the value 12 : 7 for the ratio (K_2, X_2), which nowadays we should have to derive through a limiting process. The proof of the correctness of this supposition (which he gives at once synthetically) consists in that he constructs a circle M, the square on whose radius is equal to

$$\mathbf{O}(A\Theta, AE) + \tfrac{1}{3}\mathbf{T}(\Theta E) ,$$

the area of which circle (because $A\Theta = 2.AE$) is to that of K_2 as 7 to 12. It now has to be proved that $X_2 = M$. If this is not true, either $X_2 < M$ or $X_2 > M$.

Case I. Suppose $X_2 < M$, then there exists a circumscribed figure such that also

$$C_n < M . \tag{1}$$

When we apply the proposition III; 7.52, we find as above

$$(K_2, C_n) < (12,7) = (K_2, M), \text{ whence}$$

$$C_n > M, \text{ which is contrary to (1).}$$

Case II. Suppose $X_2 > M$, then there exists an inscribed figure such that also

$$I_n > M . \tag{2}$$

Application of the prop. III; 7.52 now gives

$$(K_2, I_n) > (12,7) = (K_2, M) ,$$

whence

$$I_n < M, \text{ which is contrary to (2).}$$

Algebraically: We now have to determine

$$I = \lim_{n\to\infty} \pi L^2 \frac{n^2 + (n+1)^2 + \ldots (2n-1)^2}{n^3} ,$$

for which we find

$$\tfrac{7}{3}\pi L^2 = \tfrac{7}{12}\pi(2L)^2 .$$

The equivalent from the integral calculus is here

$$I = \int_{2\pi}^{4\pi} \tfrac{1}{2}\varrho^2 d\varphi, \text{ in which } \varrho = \frac{\varphi}{2\pi} L .$$

We find $I = \tfrac{7}{3}\pi L^2$.

In a Corollary the result here found is extended as follows[1]):

[1]) Additions between brackets for greater clarity.

The area (X_i) *bounded by a given turn of the spiral and the line segment of the initial line called by the same number (i.e. described by the radius vector of a point traversing a given turn of the curve) has to the circle* (K_i) *with the same number the same ratio as the sum of the rectangle comprehended by the radius* (iL) *of that circle and the radius* $[(i-1)L]$ *of the circle with a number one lower, plus one-third of the square on the difference between those two radii, has to the square on the radius of the greater of the said circles.*

This can be recognized by the same method as was applied in Prop. 25. The whole proof remains unchanged, provided we omit to introduce the relation $a_{n+1}=2a_1$. The ratio in question appears to be

$$(X_i, K_i) = [i(i-1)+\tfrac{1}{3}, i^2]\,, \tag{1}$$

which is equivalent to

$$X_i = \pi L^2[i(i-1)+\tfrac{1}{3}]\,.$$

For $i=1$ the value $\frac{1}{3}$ of Prop. 24 follows from (1); for $i=2$, the value 7:12 of Prop. 25.

Proposition 26.

The argument of Prop. 25 can be used unchanged for the case (Fig. 113) of an arc of a spiral $E\Theta$ which is a genuine part of a turn. The area O of the sector $AE\Theta$ is now compared with that of the sector of a circle $S(AZ\Theta)$ with radius $A\Theta$. We find

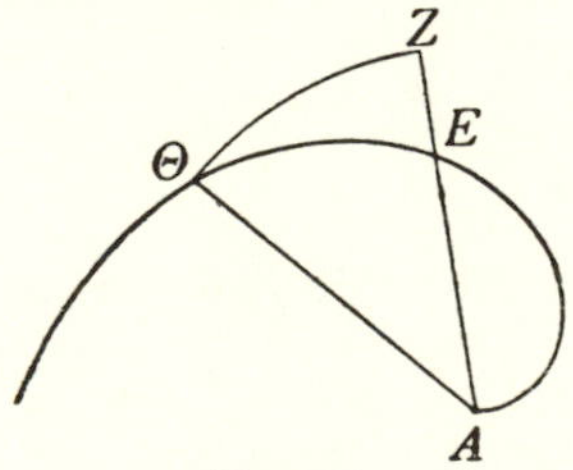

Fig. 113.

$$(O, S) = [\mathbf{O}(AE, A\Theta)+\tfrac{1}{3}\mathbf{T}(A\Theta-AE),\ \mathbf{T}(A\Theta)]\,.$$

This result implies that the sector of the spiral bounded by the radii vectores ϱ_1 and $\varrho_2 (\varrho_2>\varrho_1)$, belonging successively to the arguments φ_1 and φ_2, has an area

$$O = \frac{\varrho_1\varrho_2+\frac{1}{3}(\varrho_2-\varrho_1)^2}{\varrho_2{}^2}\cdot\tfrac{1}{2}\varrho_2{}^2(\varphi_2-\varphi_1)$$

or because

$$\varphi = \frac{\varrho}{L}2\pi\,,$$

$$O = \frac{\pi}{L}(\varrho_2 - \varrho_1)[\varrho_1\varrho_2 + \tfrac{1}{3}(\varrho_2 - \varrho_1)^2] = \pi\frac{\varrho_2^3 - \varrho_1^3}{3L}.$$

In the preceding propositions reference has repeatedly been made to the area of the sectors of a spiral described by the radius vector from A when an arc of the spiral is traversed. From this a theorem is now derived about the areas of the various spiral rings R_i which are bounded by the $(i-1)$th line segment on the initial line, the $(i-1)$th and the ith turn, and the ith line segment. Apparently

$$R_1 = X_1, \ldots R_i = X_i - X_{i-1} (i > 1).$$

This theorem is as follows:

Proposition 27.

Of the (ring) areas bounded by the turns of the spiral and the line segments on the initial line, the third is the double of the second, the fourth the triple, the fifth the quadruple, and thus each subsequent ring area will be the subsequent multiple of the second; the first, however, is one-sixth of the second (Fig. 114).

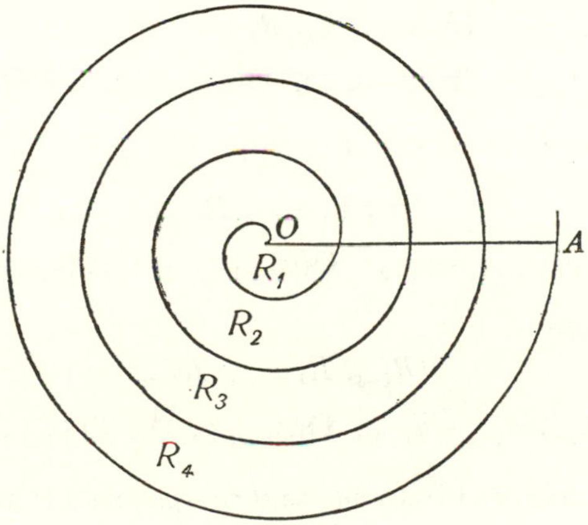

Fig. 114.

The last result follows at once from Prop. 26.
In fact, according to this proposition we have

$$(X_2, K_2) = (7,12).$$

Also

$$(K_2, K_1) = (4,1) = (12,3).$$

Thence

$$(X_2, K_1) = (7,3) .$$

We also have (**Spir.** 25)

$$(K_1, X_1) = (3,1) .$$

Thence

$$(X_2, X_1) = (7,1) \text{ or } (R_2, R_1) = (6,1) .$$

After this it is first proved that $R_3 = 2R_2$, while then the proof is furnished for any number. This last argument is as follows in the notation introduced above:

By **Spir.** 25, Cor. we have

$$(R_1 + \ldots R_i, K_i) = [\mathbf{O}(a_i, a_{i-1}) + \tfrac{1}{3}\mathbf{T}(a_i - a_{i-1}), \mathbf{T}(a_i)] .$$

Also $$(K_i, K_{i-1}) = [\mathbf{T}(a_i), \mathbf{T}(a_{i-1})] ,$$

and, again by **Spir.** 25, Cor.,

$$(K_{i-1}, R_1 + \ldots R_{i-1}) = [\mathbf{T}(a_{i-1}), \mathbf{O}(a_{i-1}, a_{i-2}) + \tfrac{1}{3}\mathbf{T}(a_{i-1} - a_{i-2})] ,$$

whence *ex aequali*

$$(R_1 + \ldots R_i, R_1 + \ldots R_{i-1})$$
$$= [\mathbf{O}(a_i, a_{i-1}) + \tfrac{1}{3}\mathbf{T}(a_i - a_{i-1}), \mathbf{O}(a_{i-1}, a_{i-2}) + \tfrac{1}{3}\mathbf{T}(a_{i-1} - a_{i-2})]$$

and *separando*, because $a_i - a_{i-1} = a_{i-1} - a_{i-2}$,

$$(R_i, R_1 + \ldots R_{i-1})$$
$$= [\mathbf{O}(a_{i-1}, a_i - a_{i-2}), \mathbf{O}(a_{i-1}, a_{i-2}) + \tfrac{1}{3}\mathbf{T}(a_{i-1} - a_{i-2})] . \qquad (1)$$

We also have

$$(R_{i-1}, R_1 + \ldots R_{i-2})$$
$$= [\mathbf{O}(a_{i-2}, a_{i-1} - a_{i-3}), \mathbf{O}(a_{i-2}, a_{i-3}) + \tfrac{1}{3}\mathbf{T}(a_{i-2} - a_{i-3})] ,$$

from which it follows *invertendo* and *componendo* that

$$(R_1 + \ldots R_{i-1}, R_{i-1})$$
$$[\mathbf{O}(a_{i-2}, a_{i-1}) + \tfrac{1}{3}\mathbf{T}(a_{i-2} - a_{i-3}), \mathbf{O}(a_{i-2}, a_{i-1} - a_{i-3})] ,$$

thus *ex aequali* with (1) in connection with

$$a_i - a_{i-2} = a_{i-1} - a_{i-3}$$

$$(R_i, R_{i-1}) = [\mathbf{O}(a_{i-1}, a_i - a_{i-2}), \mathbf{O}(a_{i-2}, a_{i-1} - a_{i-3})] = (a_{i-1}, a_{i-2}).$$

From this it follows that

$$(R_{i-1}, R_{i-2}) = (a_{i-2}, a_{i-3}) \text{ etc.}$$

and thence

$$(R_i, R_2) = (a_{i-1}, a_1) = (i-1).$$

Algebraically:

When the result of Prop. 25, Cor. is transformed as follows:

$$\frac{X_i}{K_i} = \frac{i(i-1)+\frac{1}{3}}{i^2},$$

whence

$$X_i = \frac{i(i-1)+\frac{1}{3}}{i^2}\pi(iL)^2 = [i(i-1)+\tfrac{1}{3}]K_1,$$

we find at once

$$R_i = X_i - X_{i-1} = 2(i-1)K_1 = (i-1)\,R_2\;(i > 1).$$

The method followed by Archimedes is as follows algebraically:

From

$$\frac{X_i}{K_i}, \frac{K_i}{K_{i-1}}, \text{ and } \frac{K_{i-1}}{X_{i-1}}$$

he finds

$$\frac{X_i}{X_{i-1}} = \frac{i(i-1)+\frac{1}{3}}{(i-1)(i-2)+\frac{1}{3}},$$

whence, because $R_i = X_i - X_{i-1}$,

$$\frac{R_i}{X_{i-1}} = \frac{2(i-1)}{(i-1)(i-2)+\frac{1}{3}}.$$

He therefore knows

$$\frac{R_{i-1}}{X_{i-2}}$$

and consequently also $\dfrac{R_{i-1}}{X_{i-1}}$.

Thus he finds

$$\frac{R_i}{R_{i-1}} = \frac{i-1}{i-2},$$

whence

$$\frac{R_i}{R_2} = i-1.$$

The treatise is concluded by

Proposition 28.

If on any turn of a spiral two points be taken which are not its extremities, and circles be described which have the origin for centre and the radii vectores of those points for radii, the area bounded by the larger circular arc between the two radii vectores, the spiral arc between the two, and the radius vector produced has to the area bounded by the smaller circular arc, the same spiral arc, and the line joining their extremities the same ratio as the radius of the smaller circle plus two-thirds of the difference between the radii of the two circles has to the radius of the smaller circle plus one-third of the same difference (Fig. 115).

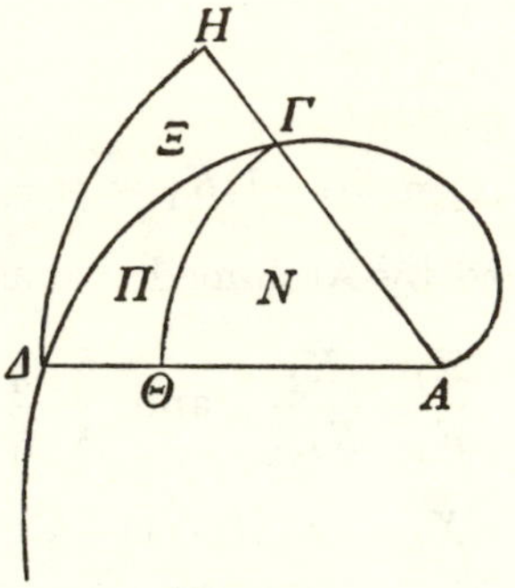

Fig. 115.

If Γ and Δ are the points under consideration, Ξ and Π the areas referred to, N the sector of a circle $A\Gamma\Theta$, it has therefore to be proved that

$$(\Xi, \Pi) = (A\Theta + \tfrac{2}{3}\Delta\Theta, A\Theta + \tfrac{1}{3}\Delta\Theta) .$$

It is known (**Spir.** 26) that

$$(N+\Pi, N+\Pi+\Xi) = [\mathbf{O}(A\Delta, A\Theta) + \tfrac{1}{3}\mathbf{T}(\Delta\Theta),\ \mathbf{T}(A\Delta)] \qquad (1)$$

Also

$$(N+\Pi+\Xi, N) = [\mathbf{T}(A\Delta),\ \mathbf{T}(A\Theta)] ,$$

whence *ex aequali*

$$(N+\Pi, N) = [\mathbf{O}(A\Delta, A\Theta) + \tfrac{1}{3}\mathbf{T}(\Delta\Theta),\ \mathbf{T}(A\Theta)] ,$$

from which it follows *convertendo* that

$$(N+\Pi, \Pi) = [\mathbf{O}(A\Delta, A\Theta) + \tfrac{1}{3}\mathbf{T}(\Delta\Theta),\ \mathbf{O}(A\Theta, \Delta\Theta) + \tfrac{1}{3}\mathbf{T}(\Delta\Theta)] . \quad (2)$$

From (1) it follows, because of

$$\mathbf{T}(A\Delta)-\mathbf{O}(A\Delta, A\Theta)-\tfrac{1}{3}\mathbf{T}(\Delta\Theta) = \mathbf{O}(A\Delta, \Delta\Theta)-\tfrac{1}{3}\mathbf{T}(\Delta\Theta)$$
$$= \mathbf{O}(A\Theta, \Delta\Theta)+\mathbf{T}(\Delta\Theta)-\tfrac{1}{3}\mathbf{T}(\Delta\Theta),$$

invertendo and *separando*

$$(\Xi, N+\Pi) = [\mathbf{O}(A\Theta, \Delta\Theta)+\tfrac{2}{3}\mathbf{T}(\Delta\Theta),\ \mathbf{O}(A\Delta, A\Theta)+\tfrac{1}{3}\mathbf{T}(\Delta\Theta)],$$

whence *ex aequali* with (2)

$$(\Xi, \Pi) = [\mathbf{O}(A\Theta, \Delta\Theta)+\tfrac{2}{3}\mathbf{T}(\Delta\Theta),\ \mathbf{O}(A\Theta, \Delta\Theta)+\tfrac{1}{3}\mathbf{T}(\Delta\Theta)]$$
$$= [A\Theta+\tfrac{2}{3}\Delta\Theta, A\Theta+\tfrac{1}{3}\Delta\Theta].$$

Algebraically: Let $A\Theta$ be equal to ϱ, $A\Delta$ to $\varrho+\nu$, then we have

$$\frac{N+\Pi}{N+\Pi+\Xi} = \frac{\varrho(\varrho+\nu)+\frac{1}{3}\nu^2}{(\varrho+\nu)^2}. \tag{3}$$

We then find with the aid of

$$\frac{N+\Pi+\Xi}{N} = \frac{(\varrho+\nu)^2}{\varrho^2}$$

$$\frac{N+\Pi}{N} = \frac{\varrho(\varrho+\nu)+\frac{1}{3}\nu^2}{\varrho^2},$$

whence

$$\frac{N+\Pi}{\Pi} = \frac{\varrho(\varrho+\nu)+\frac{1}{3}\nu^2}{\varrho\nu+\frac{1}{3}\nu^2},$$

From (3) it follows that

$$\frac{\Xi}{N+\Pi} = \frac{\varrho\nu+\frac{2}{3}\nu^2}{\varrho(\varrho+\nu)+\frac{1}{3}\nu^2},$$

so that

$$\frac{\Xi}{\Pi} = \frac{\varrho+\frac{2}{3}\nu}{\varrho+\frac{1}{3}\nu}.$$

CHAPTER IX.

ON THE EQUILIBRIUM OF PLANES
OR
CENTRES OF GRAVITY OF PLANES

Book I.

1. The treatise on the equilibrium of planes occupies a place apart in the work of Archimedes. In fact, whereas in all his mathematical treatises he builds on foundations long ago established, in this work he concerns himself with an investigation into the very foundations; moreover he leaves the domain of pure mathematics for that of natural science considered from the mathematical point of view: he sets forth certain postulates on which he bases a chapter from the theory of equilibrium, and he is thus the first to establish the close interrelation between mathematics and mechanics, which was to become of such far-reaching significance for physics as well as mathematics.

Through this extension of the field of his activity Archimedes comes into contact directly with the fundamental difficulties inherent in the foundation of mechanics on postulates. The way in which he solves these difficulties, in so far as they are manifested in his subject, even to our own day gives rise to differences of opinion between his commentators and critics: the same argument (in Prop. 6) which is fundamental for the further development, not only of the work itself, but also of two of the purely mathematical treatises still to be discussed, is rejected by some as a paralogism and accepted as correct by others. In the following pages we shall have to define our standpoint with regard to this matter; since, however, the discussion concerns not only the course of the disputed argument itself, but also the structure of the underlying system of postulates, we shall first submit to the reader's attention without any interruption all the postulates and those propositions which precede the theorem in question.

Postulates.

I. *We postulate that equal weights at equal distances are in equilibrium, and that equal weights at unequal distances are not in equilibrium, but incline towards the weight which is at the greater distance.*

II. *that if, when weights at certain distances are in equilibrium, something be added to one of the weights, they are not in equilibrium, but incline towards that weight to which something has been added.*

III. *similarly that, if anything be taken away from one of the weights, they are not in equilibrium, but incline towards that weight from which nothing has been taken away.*

IV. *when equal and similar figures are made to coincide, their centres of gravity*[1]) *likewise coincide.*

V. *in figures which are unequal, but similar, the centres of gravity will be similarly situated*[2]).

VI. *if magnitudes at certain distances be in equilibrium, other [magnitudes] equal to them will also be in equilibrium at the same distances.*

VII. *in any figure whose perimeter is concave in the same direction the centre of gravity must be within the figure.*

This having been postulated:

Proposition 1.

Weights which are in equilibrium at equal distances are equal.

For, if they were unequal, by taking away from the greater the weight by which it exceeds the lesser, we should disturb the equilibrium on account of postulate III, whereas because of postulate I there would precisely have to be equilibrium in the new position.

Proposition 2.

Unequal weights at equal distances are not in equilibrium, but will incline towards the greater weight.

If from the greater weight is taken away the weight by which it exceeds the lesser, there will be equilibrium (postulate I). If then the original position is restored, the truth of the proposition becomes apparent with the aid of postulate II.

---

[1]) We shall revert presently to the question as to the meaning of this term.

[2]) To this it is added: *we say that points are similarly situated in relation to similar figures if straight lines drawn from these points to the equal angles make equal angles with the homologous sides.*

Proposition 3.

Unequal weights can [only] be in equilibrium at unequal distances, the greater [weight] being at the lesser [distance][1]).

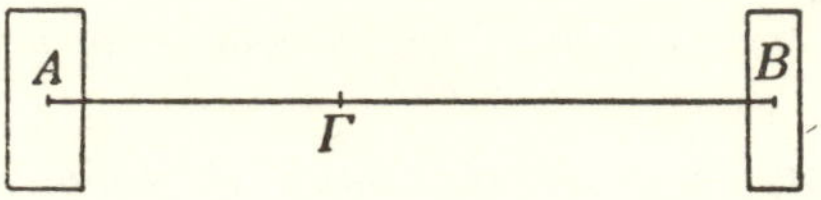

Fig. 116.

In Fig. 116 let the weights A and B ($A > B$) be in equilibrium at the distances $A\Gamma$ and ΓB. It has to be proved that

$$A\Gamma < \Gamma B.$$

Proof: If the weight $A - B$ is taken away from A, the weights have to incline towards B (postulate III). This, however, is impossible, if $A\Gamma = \Gamma B$ (postulate I) and also when $A\Gamma > \Gamma B$ (postulate I). Consequently $A\Gamma$ has to be less than ΓB.

It is obvious that also weights which are in equilibrium at unequal distances are unequal, the greater [weight] being at the lesser [distance].

This is not elucidated any further. Apparently this enunciation contains nothing but a contraposition of the second part of postulate I[2]).

Proposition 4.

If two equal magnitudes have not the same centre of gravity, the centre of gravity of the magnitude composed of the two magnitudes will be the middle point of the straight line joining the centres of gravity of the magnitudes.

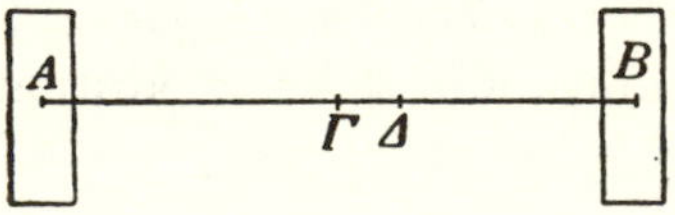

Fig. 117.

In Fig. 117 let A and B be successively the centres of gravity of the magnitudes A and B, Γ the middle point of the line segment AB. If now Γ is not the centre of gravity of the magnitude composed of A and B, let it be some other point Δ of AB. A and B therefore are in equilibrium when Δ is held. This, however, is impossible because of postulate I. Consequently Γ is the centre of gravity.

---

[1]) This proposition is formulated very elliptically: *τὰ ἄνισα βάρεα ἀπὸ τῶν ἀνίσων μακέων ἰσορροπησοῦντι, καὶ τὸ μεῖζον ἀπὸ τοῦ ἐλάσσονος.*

[2]) Postulate I, 2nd part says: if weight $A_1 =$ weight A_2 and distance $l_1 \neq$ distance l_1, there is no equilibrium. Contraposition: if there is equilibrium and also $l_1 \neq l_2$, then $A_1 \neq A_2$. It then follows from the first part of Prop. 3 that the weight at the lesser distance must be the greater of the two.

Proposition 5.

If of three magnitudes the centres of gravity are on a straight line and the magnitudes have equal weight, and if also the straight lines between the centres are equal, the centre of gravity of the magnitude composed of all the magnitudes will be the point which is also the centre of gravity of the middle magnitude.

This is recognized by noting that according to Prop. 4 the centre of gravity of the system of the extreme magnitudes coincides with that of the middle magnitude.

In Porism I the theorem is extended to any odd number of magnitudes the centres of gravity of which are on a straight line at equal distances, while any two magnitudes which are equidistant from the middle point of the line segment joining the two extreme centres are of equal weight.

In Porism II it is stated that if from the system described above the middle magnitude be omitted, the centre of gravity of the system will be the same point as before.

This is followed by the famous and much disputed proposition in which the so-called lever principle is enunciated.

Proposition 6.

Commensurable magnitudes are in equilibrium at distances reciprocally proportional to the weights.

Let the commensurable magnitudes be A and B, of which A and B are the centres, and let $E\Delta$ be a given distance, and let the distance $\Delta\Gamma$ be to the distance ΓE as A to B. It has to be proved that the centre of gravity of the magnitude composed of A and B is Γ (Fig. 118).

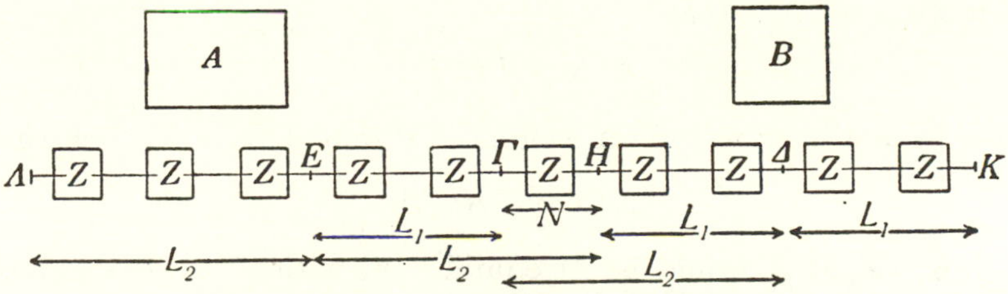

Fig. 118.

Proof: Since A and B are commensurable, so are $\Delta\Gamma$ and $E\Gamma$. Let N be a common measure of these two distances. Make $\Delta H =$

$\Delta K = E\Gamma$ and $E\Lambda = \Delta\Gamma$. Apparently EH is also equal to $\Delta\Gamma$. Since $H\Lambda = 2.\Delta\Gamma$ and $HK = 2.E\Gamma$, we also have

$$A:B = \Lambda H:HK .$$

Now let the magnitude Z be contained as many times in A as the distance N in ΛH, whence also as many times in B as N in HK. Divide ΛH and HK each into equal parts N, A and B each into equal parts Z. Place on each of the line segments N a magnitude Z, so that in each case the centre of gravity of Z is the middle point of N, then the centre of gravity of all the magnitudes Z placed on the parts of ΛH will be the point E (Prop. 5, Porism II), while in the same way the centre of gravity of all the magnitudes Z placed on the parts of HK will be the point Δ. *Now therefore A will be at E and B at Δ. There will now be equal magnitudes on a straight line, the centres of gravity of which are equidistant from one another and the number of which is even. It is now obvious that of the magnitude composed of all the magnitudes the middle point of the straight line bounded by the centres of the middle magnitudes will be the centre of gravity. So that the centre of gravity of the magnitude composed of all the magnitudes is the point Γ. If therefore A is at E and B at Δ, they will be in equilibrium about Γ.*

It has thus been proved that the inverse proportionality of force and arm is a sufficient condition for the equilibrium of a lever supported in its centre of gravity under the influence of two weights on either side of the fulcrum. It is not proved by Archimedes that this condition is also necessary; the proof of this might have been given as follows:

Let weights A and B be in equilibrium at distances ΓA and ΓB. If it were not true that

$$A:B = \Gamma B:\Gamma A ,$$

there would have to exist a weight Δ, different from B, such that

$$A:\Delta = \Gamma B:\Gamma A .$$

This weight Δ would be in equilibrium with A by the proved Proposition 6. If then the original situation is restored, the equilibrium will be disturbed (on account of the postulates II and III), which is contrary to the supposition that A and B are in equilibrium.

It is to be considered a flaw in the treatise on equilibrium that Archimedes omits to give this proof; indeed, later on (see Chapter XII) he repeatedly applies the condition of inverse proportionality as a necessary condition of equilibrium.

2. Let us now study the proof of Proposition 6 somewhat more closely. Its real nucleus appears to consist in the manner in which the magnitudes A and B are placed at the points E and Δ; this is done by dividing each of them into equal parts Z, which are suspended on the segments of the straight line ΛK in such a way that the centres of gravity of the two systems come to lie successively at E and Δ. On the lever are thus suspended, not the originally given magnitudes A and B, but two systems of magnitudes which are successively of the same weight as A and B and whose centres of gravity are at the points E and Δ, which are looked upon successively as positions of A and B. This implies that the influence exerted on the equilibrium by a body suspended on a lever is judged exclusively by the gravity of the body and the place of its centre of gravity, and that the shape is immaterial.

We now have to ask ourselves: was Archimedes aware of the fact that the conclusion he draws is based on this premiss? If so, has he expressly formulated this premiss as such? And if so again, does he account for it any further?

This threefold question (which is often wrongly not split up into three members) has been answered in various ways, which testify to great differences in appreciation of the argument of Archimedes.

Let us listen in the first place to the opinion of a well-known writer on the history of Mechanics, Ernst Mach[1]), whose views enjoy great authority. We are giving his words *in extenso*:

"So überraschend uns nun auf den ersten Blick die Leistung von Archimedes ... erscheint, so steigen uns bei genauer Betrachtung doch Zweifel an der Richtigkeit derselben auf. Aus der bloszen Annahme des Gleichgewichts gleicher Gewichte in gleichen Abständen wird die verkehrte Proportion zwischen Gewicht und Hebelarm abgeleitet! Wie ist das möglich?

Wenn wir schon die blosze Abhängigkeit des Gleichgewichts vom Gewicht und Abstand überhaupt nicht aus uns herausphilosophieren konnten, sondern aus der Erfahrung holen muszten, um

[1]) Ernst Mach, *Die Mechanik in ihrer Entwicklung historisch-kritisch dargestellt.* 7e Auflage (Leipzig 1912). p. 14.

wieviel weniger werden wir die Form dieser Abhängigkeit, die Proportionalität, auf spekulativem Wege finden können.

Wirklich wird von Archimedes und allen Nachfolgern die Voraussetzung dasz die (gleichgewichtstörende) Wirkung eines Gewichts P im Abstand L von der Achse durch das Produkt $P.\ L$ (das sogenannte statische Moment) gemessen sei, mehr oder weniger versteckt oder stillschweigend eingeführt. Zunächst ist klar, dasz bei vollkommen symmetrischer Anordnung das Gleichgewicht unter Voraussetzung irgendeiner beliebigen Abhängigkeit des gleichgewichtstörenden Moments von L, also $P.f(L)$, besteht; demnach kann aus diesem Gleichgewicht unmöglich die bestimmte Form $P.L$ abgeleitet werden. Der Fehler der Ableitung musz also in der vorgenommenen Transformation liegen, und liegt hier auch. Archimedes setzt die Wirkung zweier gleicher Gewichte unter allen Umständen gleich der Wirkung des doppelten Gewichts mit dem Angriffspunkte in der Mitte. Da er aber einen Einflusz der Entfernung vom Drehpunkt kennt und voraussetzt, so darf dies nicht von vornherein angenommen werden, wenn die beiden Gewichte ungleiche Entfernung vom Drehpunkt haben. Wenn nun ein Gewicht, das seitwärts vom Drehpunkt liegt, in zwei gleiche Teile geteilt wird, welche symmetrisch zu dem ursprünglichen Angriffspunkt verschoben werden so nähert sich das eine Gewicht dem Drehpunkt so viel, als sich das andere von dem selben entfernt. Nimmt man nun an, dasz die Wirkung hierbei dieselbe bleibt, so ist hiermit schon über die Form der Abhängigkeit des Moments von L entschieden, denn dies ist nur möglich bei der Form $P.\ L$, bei Proportionalität zu L. Dann ist aber jede weitere Ableitung überflüssig. Die ganze Ableitung enthält den zu beweisenden Satz, wenn auch nicht ausdrücklich ausgesprochen und in anderer Form, schon als Voraussetzung."

To Mach's view, therefore, Archimedes in the first place was not aware that his argument is based on the premiss that a lever equilibrium is not disturbed, if one of the suspended magnitudes is made to change its shape while preserving its weight and centre of gravity, and in the second place the whole proof becomes superfluous as soon as this premiss is consciously adopted, because it already implies the theorem to be proved (*viz.* that inverse proportionality of two weights to the lengths of the arms on which they are suspended constitutes a sufficient condition for the equilibrium of a lever). The second part of the objection may also be worded as follows in present-day formulation: once we have accepted the fact that the function $P.f(L)$ satisfies the functional equation[1])

$$\tfrac{1}{2}P.f(L+h)+\tfrac{1}{2}P.f(L-h) = P.f(L)\,,$$

it is superfluous to prove that $P.f(L)$ has the form $P.L$, for the functional equation in question exists only if $f(L)$ has the form L.

Neither of Mach's objections here summarized seems to us to be wholly valid. In fact, we found Archimedes explicitly declaring that owing to the position of the parts Z on the line segments N the common centre of gravity of the parts of A lies at E and that of the parts of B at Δ, and that *consequently* the magnitude A is placed at E and the magnitude B at Δ, whereas actually neither A nor B are present in their original form. From this it is already clearly apparent that he judges the influence of a magnitude solely by the weight and the position of the centre of gravity. But in addition to this he has expressed in postulate VI that when magnitudes at given distances are in equilibrium, magnitudes equal to the former will be in equilibrium at the same distances. Upon first consideration this may seem to be a perfectly superfluous tautology. As soon as we assume, however, that by "magnitudes at the same distances" he understands "magnitudes the centres of gravity of which lie at the same distances from the fulcrum", we have conferred a reasonable meaning on this postulate and at the same time have found the explicit formulation of the premiss applied in Prop. 6, which Mach deems to be lacking. The first of his objections may thus be considered to have been refuted[2]).

And as regards the second: its untenability is evident as soon as we pay attention to the mathematical formulation given of it above.

---

[1]) In fact, this equation expresses that the influence of the equilibrium of a weight P at a distance l is equivalent to the combined influences of weights $\tfrac{1}{2}P$ at points which are symmetrically situated in relation to the point where P was first suspended. It is naturally also based on the supposition that the influences in question can be combined additively.

[2]) The conception of the postulate VI here described is due to O. Toeplitz, whom I thank for the oral and written explanation of his point of view. The ideas of Toeplitz have been elaborated in a very careful study by W. Stein, *Der Begriff des Schwerpunktes bei Archimedes*. Quellen und Studien zur Geschichte der Mathematik, Physik und Astronomie. Abt. B: Quellen. I (1930), 221–244.

Indeed, if it is not necessary to solve the functional equation in this case, when is there any sense in doing so? And if in general the drawing of a conclusion Q from a group of premisses P may be called superfluous when P applies only if Q is true, one may just as well reject as superfluous all mathematical proofs, for if Q is not true, it is certain that P will not hold!

Now we also found Mach having recourse to the general argument that from the existence of equilibrium with symmetrical loads one can never draw a conclusion about the nature of the function $f(L)$, because this equilibrium would exist with any form of this function. To this we can only reply that Archimedes does not draw his conclusion from the mere existence of this equilibrium, but that this is only one among the premisses on which his argument is based. This argument can be judged only by considering the whole system of postulates, definitions, and propositions already proved which are used in the course of it. It is the really weak spot in Mach's view of the question that he has neglected this consideration.

The same defect in a somewhat different form is also found in the author of the German Archimedes translation, A. Czwalina[1]). The latter remarks that if the lever principle were to state the inverse proportionality of the force and the square of the arm as a condition of equilibrium, the propositions 1–5 would stand unchanged, whereas Prop. 6 would be incorrect, from which he concludes that Prop. 6 does not therefore result from Props 1–5, and that consequently the derivation of Prop. 6 is not correct.

By this argumentation the proof of Euclid I, 29 (equality of alternating interior angles as a necessary condition of parallelism) and the proof of I, 32 based on this (the sum of the angles of a triangle is 180°) would not be correct either; in fact, if the sum of the angles of a triangle were to be less than 180°, the propositions 1–27 would stand unchanged, whereas 29 and 32 would not be true. The analogy between the two cases is perfect. It is true that the propositions 29 and 32 do not result from 1–27, but from the latter supplemented with the fifth postulate; in the same way the sixth proposition of the treatise on the equilibrium of planes does not result from the propositons 1–5, but from the latter, combined with the postulate VI that has meanwhile been introduced. As soon as

---

[1]) A. Czwalina, *Ueber das Gleichgewicht ebener Flächen* (see p. 45, Note 6).

we accept this postulate in the above interpretation, we can no longer find any fault with the proof of Prop. 6.

A slightly different point of view in relation to the question under consideration is taken by O. Hölder[1]). The latter agrees with Mach in considering the proof of Prop. 6 insufficient, because he, too, is of opinion that Archimedes neglects to postulate or prove the admissibility of combining or splitting up weights while maintaining the position of the centre of gravity. In his view, however, the proof can be corrected (and will then acquire value) if only we succeed in supplementing the gap presumed to have been left by Archimedes. It would seem to us, however, that the way in which he attempts to do this (by superposition of positions of equilibrium, involving, *inter alia*, the concept of the reactive force exerted by a fulcrum) fits in very little with the character of Greek mechanics, while moreover there is not the slightest evidence for it in the text.

We have now got to the point where we can answer the first two members of the above question with some certainty: Archimedes was fully aware of the fact that the proof of the sixth proposition is based on the premiss that the influence of a weight suspended on a lever on the equilibrium depends on the gravity of the body and the position of its centre of gravity; and he has explicitly formulated this premiss in the sixth postulate, in which he postulates, apparently tautologically, that the lever equilibrium is not disturbed if the weights suspended on it are replaced by other weights, which are equal to the first and are suspended in the same places.

We are now left with the question whether he accounts for this statement or not, a question which is related with another, *viz.* what he understands by the centre of gravity of a body and how he conceives this concept to be introduced. In fact, it is striking that he always refers to the centre of gravity as if it were a perfectly familiar thing: the term is used, starting with postulate IV, without any explicit definition, and in the proofs of the propositions 4 and 6 it is identified in the case of a loaded lever, without any

---

[1]) O. Hölder, *Die Mathematische Methode. Logisch-erkenntnistheoretische Untersuchungen im Gebiete der Mathematik, Mechanik und Physik* (Berlin, 1924) § 12. *Der Hebelbeweis des Archimedes*, p. 39 *et seq.*

motivation, with the point in which this lever has to be supported in order to be in (indifferent[1])) equilibrium.

With regard to the questions thus raised we may take either of the following two standpoints:

a) it is possible that Archimedes, when writing the treatise on the equilibrium of planes, could assume the theory of the centre of gravity to be familiar to a certain extent, because this theory had already been developed either by earlier students of mechanics or by himself in a treatise now lost[2]).

b) it is possible that the work on the equilibrium of planes is an entirely autonomous treatise, and that the definition of the concept of centre of gravity is to be conceived of as being implied in the postulates on which this work is built.

Both these points of view have been defended: the former by G. Vailati[3]) and the latter by Toeplitz and Stein[4]). We shall first discuss the latter view. According to this, all the terms to be found in the postulates of the work, in so far as they relate to statics (such as *βάρος*, weight; *ἰσορροπεῖν*, to be in equilibrium; *κέντρον τοῦ βάρεος*, centre of gravity), are to be considered so many unknowns, for the finding of which the postulates serve as equations, while other similar equations can be found by explicitly formulat-

---

[1]) It is possible to ask with W. Stein (*l. c.*, p. 228) whether in his considerations of equilibrium Archimedes is really thinking exclusively of indifferent equilibrium or whether he also admits the possibility of stable equilibrium. This amounts to asking whether the condition that the lever is supported in its centre of gravity is considered as sufficient only or as necessary as well. If this question is to be answered, the view of Stein that in the latter case the Prop. 4 is only a trivial consequence of Postulate I, from which he concludes that Archimedes does not postulate it as a necessary condition of equilibrium that the lever should be supported in the centre of gravity, seems to us to carry little conviction: in Greek mathematics there are numerous instances of trivial consequences which nevertheless constitute the subject of a separate proposition. It would seem more important that in **Q.P.** Archimedes invariably considers really stable equilibria, so that it does not after all seem to be his intention to identify the centre of gravity and the fulcrum.

[2]) This might then have been one of the works quoted by Pappus and Simplicius, which we mentioned in Chapter II in the enumeration of the lost treatises sub 3).

[3]) G. Vailati, *Del concetto di centro di gravità nella Statica d'Archimede.* Atti della R. Accad. d. Scienze di Torino. 32 (1896–97), 742–758.

[4]) See Note 2 of pag. 293.

ing the assumptions which are tacitly made in the proofs of the propositions. The system of postulates thus completed then comprises implicitly the definitions of all the terms in question, and it is then in particular no longer necessary to ask for an explicit definition of the meaning of the term "centre of gravity".

The investigation defined above has been made very carefully by W. Stein, and undoubtedly has considerably clarified our insight into the statics of Archimedes. It would, however, seem doubtful whether this method will enable us actually to follow his line of thought. Not because the idea of a definition being implicit in a system of postulates was entirely alien to the Greeks: when in the *Elements* Euclid uses the concept of a "straight line", he never refers to the property of length without breadth, which is given as the characteristic of this concept in the definitions, but exclusively to the properties of its being determined by two points and of the unlimited extensibility of any line segment mentioned as characteristics in the postulates. To that extent it may be said that he uses an implicit definition of the concept of a straight line, if not *ex confesso*, at least *de facto*. But this is practically the only instance of this method of definition in the *Elements*: it is not used for other terms, where it might have been applied, such as the area of a figure or the volume of a solid; the meaning of these terms is apparently considered to be intuitively known, and the postulates given about them (congruent figures have equal areas, the whole is greater than the part), though they do constitute a step in the direction of the implicit definition (which would be given by a complete system of postulates), do not yet testify to the conscious desire to define the concepts in question entirely in this way. This cannot be altered by the fact that it is possible afterwards, by formulating so-called tacit assumptions (which are not yet assumptions or conscious suppositions from the point of view of the author), to complete the underlaying system of postulates in such a way that it becomes sufficient for implicit definition.

Now we ask ourselves whether it is likely that, whereas in the elementary geometry of the Greeks the conscious implicit definition by means of a system of postulates was as yet so little used, this method of definition would have been carried so far in Greek mechanics that not only one term, as in the geometrical examples, but seven terms at a time were defined in this way. Is it really to be believed that Archimedes forced himself, when using the words "to be in equilibrium", "incline", "weight", to think of nothing but the relations established between them in the postulates? Is it not much more probable that he constantly had in mind an idealized lever, which he mentally saw inclining or remaining in equilibrium under the influence of weights (in the form of planimetric figures) suspended on it, that the postulates contain nothing but the formulation of the results of the simplest observations he was able to make, and that the clarity of the images thus obtained completely prevented the desire for the abstract definition of the words used from arising?

If this is true, the term "centre of gravity", which is introduced just as unemphatically as the other terms mentioned, must have had an intuitive meaning that was plain to all. When we read the postulates and the propositons with an unbiassed mind, we do get the impression that the author is thinking of a lever supported in its fulcrum (the lever being idealized into a straight line), on which thin plates (idealized into planimetric figures) are attached in their centres of gravity (as is clear from the frequent use of the same letter to refer to the suspended figure, the centre of gravity of that figure, and the point of the lever indicating the position of that figure). Since, however, the term "centre of gravity" cannot possibly have had a meaning intuitively as clear as "be in equilibrium" or "incline", there is every reason to assume with Vailati that this term could be supposed to be familiar to the readers of the work in view of the advance knowledge they might be expected to have; thus we arrive at the first of the possibilities referred to above, and we therefore have to ascertain what data are available to us in the history of Greek natural sciences about an elementary theory of statics in which the term "centre of gravity" may have been introduced.

It is mainly a group of remarks in Heron's *Mechanica*[1]) and a coherent exposition devoted to the subject by Pappus in the *Collectio*[2]) which may be considered for this purpose. Both Heron and

[1]) Of this work a complete text has only been preserved in Arabic, while in Greek there are fragments. Both are to be found with a German translation in: Heronis *Opera* II, 1. The passages in question are: Book I, Cap. 24; Book II, Cap. 35 *et seq.* (II, 35 also in Greek).

[2]) Pappus, *Collectio* VIII, 5; 1030 *et seq.*

Pappus are authors who came long after Archimedes, and it may therefore at first view seem a little illogical to quote them in this connection. The former, however, writes in such an elementary way and the latter is so encyclopaedic that it does not appear very risky to consult them on those points which may have remained beneath the threshold of exposition in the work of their great predecessor, whose level neither of them attains; moreover, Greek mechanics was arrested so much in the stage of elementary principles that in authors of the first and third centuries A.D. we must not by any means expect any further development of the subject than may already have been reached in the third century B.C.

Apparently Heron and Pappus drew from the same source in their treatment of the subject "centre of gravity"; since Heron, however, writes about it sketchily and indistinctly, we shall mainly give Pappus' words, only occasionally quoting Heron for confirmation.

In the eighth book of the *Collectio*, in which Pappus speaks about mechanics, the theory of the centre of gravity (*κέντρον τοῦ βάρους*) is called the starting point and element of the barycentric theory (*ἀρχὴ καὶ στοιχεῖον τῆς κεντροβαρικῆς πραγματείας*) because after the exposition of this the other parts of the theory automatically become clear. This is followed by an explicit definition: *We say that the centre of gravity of any body is a point within that body which is such that, if the body be conceived to be suspended from that point, the weight carried thereby remains at rest and preserves its original position*[1]).

Heron expresses himself in approximately the same way when he defines a point which is rendered in the German translation of the available Arabic text by *Aufhängepunkt*, which point, according to him, was distinguished by Archimedes and his adherents from the centre of gravity[2]). For the centre of gravity, however, he quotes a definition of the Stoic Poseidonius, who again says approximately the same thing in a somewhat careless manner[3]), so that it does not

---

1) *ibidem* line 11: *λέγομεν δὲ κέντρον βάρους ἑκάστου σώματος εἶναι σημεῖόν τι κείμενον ἐντός, ἀφ' οὗ κατ' ἐπίνοιαν ἀρτηθὲν τὸ (βάρος) ἠρεμεῖ φερόμενον καὶ φυλάσσει τὴν ἐξ ἀρχῆς θέσιν.* Here *βάρος* is to be considered synonymous with *σῶμα.*

2) *Heronis Opera* II, 1. p. 64.

3) *ibidem*, p. 63: *der Schwerpunkt ist ein solcher Punkt, dasz wenn die*

become quite clear to what distinction he is actually referring. It is, however, significant that he repeatedly uses a term which is rendered by *Schwer- oder Neigungspunkt* in the German translation, and which will probably have been *κέντρον τοῦ βάρους ἢ τῆς φορᾶς* in Greek. In this term *φορά* simply refers to the natural motion of a heavy body; the idea therefore seems to be that, since the gravity is the cause of this motion, the downward tendency may be conceived to be concentrated in the centre of gravity, so that this point can also be styled falling centre (a concept which is analogous to the later *centrum oscillationis* or *centrum percussionis*).

With regard to this centre of gravity or falling centre Heron goes on to say[1]) that it is a point through which pass all the verticals of the points of suspension; here point of suspension refers, in contrast with the *Aufhängepunkt* used above, to any point in which the body may be suspended, and the verticals in question are those lines of the body which in the position of equilibrium coincide with the verticals of the points of suspension. Heron also observes that the centre of gravity or falling centre may also be situated outside the substance of the body, for example with rings or wheels.

In order to determine the centre of gravity Pappus now imagines a vertical plane *αβγδ*, on whose horizontal upper edge *αβ* the body is so placed as to be in equilibrium. Now the plane *αβγδ*, when extended, will divide the body into two parts balancing each other about this plane as plane of support[2]), *i.e.* into parts of equal apparent weight. The body is now placed again on *αβ* in a different position, so that there is once more equilibrium. It is again divided into two parts of equal apparent weight by the plane. The two intersections which the plane *αβγδ* has determined in the body in the two positions successively taken up by the body will have to

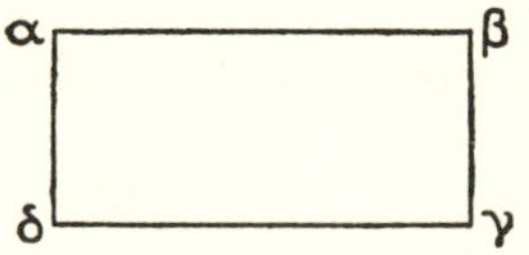

Fig. 119.

Last in demselben aufgehängt wird, sie in zwei gleiche Teile geteilt wird; where "gleich" apparently was to have been understood as "keeping each other in equilibrium", though, as will appear, it is often "of equal weight" which is meant.

[1]) *ibidem*, p. 36.

[2]) Pappus, *Collectio* VIII, 5; 1030, line 26. *τεμεῖ τὸ ἐπικείμενον σῶμα εἰς ἰσόρροπα δύο μέρη, οἷον περὶ ἄρτημα τὸ ἐπίπεδον ἰσορροποῦντα.*

intersect each other. In fact, if these two intersections were parallel, the same parts would at the same time be of equal apparent weight and not of equal apparent weight, which is absurd[1]).

The body will also be in equilibrium when it rests on the straight line, which the two intersections have in common, as on a support[2]). If we now find another straight line, which may also serve as support, this line, when produced, will have to meet the one first mentioned. Indeed, if this were not the case, it would be possible to draw through the two straight lines two parallel planes, both of which would divide the body into parts which would be of equal apparent weight when considered in one way, and not of equal apparent weight when considered in another way[3]).

From this it follows that all the supporting lines obtained in the way described above pass through one point, *viz.* the centre of gravity as defined above[4]). In fact, any plane through this point divides the body into two parts balancing each other when supported in said plane[5]).

According to Pappus, this is the most essential part (*τὸ μάλιστα σύνεχον*) of the barycentric theory. For the elements of what can be proved with the aid of this he refers to the work of Archimedes on the equilibrium of planes and to Heron's mechanics[6]). He himself discusses, as an application, a planimetric theorem which does

---

[1]) Pappus, *Collectio* VIII, 5; 1032, line 2. *εἰ γὰρ μὴ τεμεῖ, τὰ αὐτὰ μέρη καὶ ἰσόρροπα καὶ ἀνισόρροπα γενήσεται ἀλλήλοις, ὅπερ ἄτοπον.*

[2]) It appears to be meant that the body is supported in the lowest point of intersection of its surface with the said straight line. The body can therefore be kept in equilibrium both by supporting it along a horizontal straight line (or in two points of the latter) and by supporting it in a point. This is what is probably meant by Heron (II, 1; p. 64) when he quotes from Archimedes: *Lasten neigen sich nicht auf einer Linie und auf einem Punkte.* Here *sich neigen* = falling. The passage thus implies: the body can be prevented from falling by being supported along a straight line or in a point.

[3]) This conclusion is based on the assumption that any plane through a supporting line obtained as above divides the body into two parts of equal apparent weight.

[4]) Pappus *Collectio* VIII, 5; 1032, line 26.

[5]) Ibidem, line 30.

[6]) Ibidem 1034; lines 1–4. Pappus therefore does not in any case consider the work of Archimedes as an entirely independent treatise, but as an application of the theory of the centre of gravity.

not interest us here as such[1]), but which is significant on account of the way in which he is found to determine the centre of gravity of a triangle on the basis of the above considerations. Indeed, he observes (Fig. 120) that if the triangle $\alpha\beta\gamma$ is placed (in the horizontal position) with the median $\alpha\delta$ on the upper edge of a vertical supporting plane, the figure will be in equilibrium, because the areas of the triangles $\alpha\beta\delta$ and $\alpha\gamma\delta$ are equal. The same applies to the median $\beta\varepsilon$, and the point of intersection ζ of $\alpha\delta$ and $\beta\varepsilon$ is therefore the centre of gravity of the triangle.

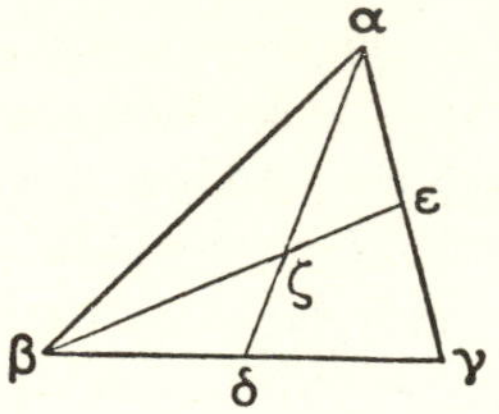

Fig. 120.

The fragment from Pappus which we have rendered here is of interest for the history of mechanics in more than one respect; in fact, the characteristic feature of the evolution of this branch of physics consists in the very early tendency of dealing with the theory of motion and equilibrium in an axiomatic manner. With regard to the last-mentioned subject, that of statics, this tendency is already very plain: on the basis of a number of physical experiences concerning the equilibrium of bodies which are supported by a narrow horizontal beam, a deductive treatment of the theory of equilibrium is given very soon, in which the conditions are idealized to the same extent as was the case with the deductive treatment of geometry. The supporting beam becomes a horizontal straight line; for the supported body is taken a planimetric figure; no attention is paid to the degree of stability of the equilibrium; it is not considered what procedure would have to be followed to support any body in its centre of gravity; probably we should not imagine the experience gained as greatly varied either; it is sufficient to think of rectangular blocks or of thin plates of a rectangular or triangular form.

It seems moreover that the process of idealization here involved has indeed been considered theoretically more in detail, to wit by Archimedes himself. At least, Heron observes[2]) that it is naturally

[1]) The theorem is as follows: if on the sides $\alpha\beta$, $\beta\gamma$, $\gamma\alpha$ of a triangle $\alpha\beta\gamma$ be situated the points η, ϑ, $\varkappa$ successively, in such a way that

$$\alpha\eta : \eta\beta = \beta\vartheta : \vartheta\gamma = \gamma\varkappa : \varkappa\alpha ,$$

the triangles $\alpha\beta\gamma$ and $\eta\vartheta\varkappa$ have the same centre of gravity.

[2]) *Heronis Opera* II, 1. p. 62.

only possible to speak of gravity and natural motion in the case of physical bodies, but that Archimedes has made it sufficiently plain in what sense a centre of gravity may also be assigned to solid or plane mathematical figures.

Experience will undoubtedly further have taught that a beam or a thin plate, when supported successively along two parallel straight lines of the same plane boundary surface, could not be in equilibrium in both cases. This fact, however (as appears from the indirect reasoning of Pappus on the subject), is at once invested with the character of logical evidence. The relation in which the two parts determined in the body by the supporting plane, when extended, are to each other is looked upon as a relation of equality between two magnitudes, and the parallelism of two such supporting planes is felt as absurd, because, when one plane yields the parts A and B, and the other the parts A_1 and B_1, it is not possible that simultaneously with

$$A > A_1, \quad B < B_1 \tag{1}$$

it is also true that

$$A = B, \quad A_1 = B_1 . \tag{2}$$

This logical evidence is of course only apparent. Pappus does not see that the equilibrium-disturbing effect exerted by each of the parts of the body is not determined by the weight of that part alone, and that therefore from the absurdity of the simultaneous existence of the relations (1) and (2) for the weights of the separate parts no logical conclusion can be drawn as to the impossibility of the influence on the equilibrium. That he is, however, thinking exclusively of the amounts of the weights, is quite clear from his derivation of the centre of gravity of a triangle. The two parts into which a median divides a triangle, in his view, balance when supported along this median because their areas are equal (the weights being conceived to be proportional to the areas). According to this argument there would also have to be equilibrium if the triangle were supported along any other straight line dividing the area into two equal parts; such a straight line, however, only passes through the centre of gravity, if it also contains an angular point. And moreover, since it is assumed as self-evident that any vertical plane through the centre of gravity divides the body into parts of equal apparent weight, any line through the centre of gravity of a triangle would have to divide the area into two equal parts, which is not true either.

It is this which throws an unexpectedly clear light on the object which Archimedes may have aimed at with his work on the equilibrium of planes, the subsidiary title of which refers to centres of gravity of plane figures: he recognized the erroneous nature of the method for the determination of a centre of gravity just described, which apparently dates further back and was preserved by Pappus with curious thoughtlessness; he understood that parts balancing each other in general do not have equal weight, but that the position of their respective centres of gravity, too, has to be taken into account. He thus came to consider a lever on whose arms the parts of the body were attached in their centres of gravity, and in this way the theory of lever furnished him with the means for determining centres of gravity, because he only had to ask in what point the lever was to be supported in order to obtain equilibrium. This theory in itself, however, was based on the barycentric theory—probably studied long before his day—, which we saw exposed by Pappus and to which, in spite of palpable logical defects, a certain physically convincing effect cannot be denied. Thus the methodically somewhat complex situation arose that the lever principle could be on the one hand an application and on the other hand the basis of the theory of the centre of gravity.

It has thus likewise been elucidated how Archimedes came to adopt, without any further motivation, the principle—formulated none too clearly in postulate VI—which protects the proof of the sixth proposition, as we now have come to realize, against the objections raised to it by Mach: it is in the centre of gravity or falling centre that from the very beginning the whole downward tendency constituting the essence of gravity has been considered concentrated. Was it not bound to seem evident that the influence which a body could exert on a lever as a result of this tendency did not change as long as its intensity and the place where it resides remained unchanged?

3. The above discussion may be considered to have sufficiently elucidated the questions to which the first part of the treatise *On the Equilibrium of Planes* gave rise, so that we can now continue the discussion of the work itself.

Proposition 7.

However, even if the magnitudes are incommensurable, they will be in equilibrium at distances reciprocally proportional to the magnitudes.

The necessity of a separate discussion of the case that the weights suspended on the lever have no common measure follows from the essential significance of the common measure Z of the magnitudes A and B in the proof of Prop. 6.

The somewhat cursorily written proof may be rendered as follows:

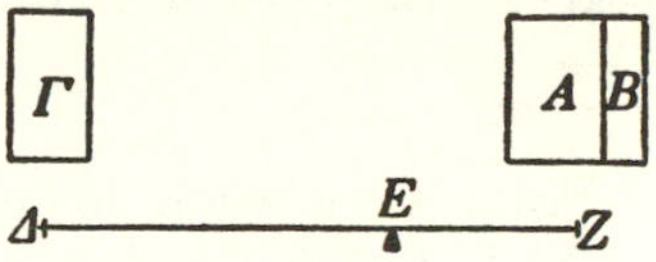

Fig. 121.

In Fig. 121 let the incommensurable magnitudes be $A+B$ and Γ, the arms of the lever on which they are suspended, EZ and $E\Delta$ respectively. The supposition is then

$$(E\Delta, EZ) = (A+B, \Gamma). \tag{1}$$

It has to be proved that E is the centre of gravity of $(A+B)$ in Z and Γ in Δ (*i.e.* that the centres of gravity of the magnitudes lie in these points successively), or in other words that the lever, when supported in E, is in equilibrium.

Suppose this is not true, then $(A+B)$ will be either too great or too small for equilibrium. Let $(A+B)$ be too great. Then take away from it such an amount that the remainder is still too great for equilibrium, but commensurable with Γ.

Let this remainder be A. Now we have

$$(A, \Gamma) < (E\Delta, EZ), \tag{2}$$

so that the lever will incline towards Δ, which is contrary to the supposition that A alone is still too great for equilibrium. In the same way the case that $(A+B)$ is too small can be dealt with.

The proof apparently contains considerable gaps. The possibility of determining the remainder A in such a way that it is too great for equilibrium and at the same time commensurable with Γ is not based either on a postulate or on a theorem. That it follows from the inequality (2) that the lever will incline towards Δ is indeed physically plausible, but logically not justified. It is naturally derived from the consideration that there would be equilibrium if A were replaced by $A' > A$, so that

$$(A', \Gamma) = (E\Delta, EZ), \tag{3}$$

i.e. by Prop. 6, followed by the application of postulate III. But it appears from (1) that $E\Delta$ and EZ are incommensurable, and it is therefore only possible to conclude from (3) that there is equilibrium, if Prop. 7 has first been proved. In the proof of Prop. 7 it is not permissible to make use of this conclusion.

The proof might be improved, though not quite saved, if the line segment ΔZ were divided in a point H in such a way that

$$(A, \Gamma) = (H\Delta, HZ)\,.$$

From this it would follow by Prop. 6 that the lever, when loaded with A at Z and with Γ at Δ, is in equilibrium. If upon this it is to be concluded that there can be no equilibrium when the lever is supported in E, it would have to be postulated that a system of bodies has only one centre of gravity, or the consideration used by Pappus, *viz.* that upon the lever being successively supported along two parallel straight lines there cannot be equilibrium in both cases, would have to be adopted as a postulate. For the recognition that if (2) is true, the lever will incline towards Δ, however, another postulate would be required, *viz.* that upon the support being shifted to one side a lever originally in equilibrium will incline towards the other side.

Proposition 8.

If from a magnitude another magnitude be taken away which does not have the same centre as the whole, when the straight line joining the centres of gravity of the whole magnitude and the magnitude taken away be produced towards the side where the centre of the whole magnitude is situated, and when from the produced part of the line joining the said centres a segment be cut off such that it has to the segment between the centres the same ratio as the weight of the magnitude taken away has to the remaining magnitude, the extremity of the segment cut off will be the centre of gravity of the remaining magnitude.

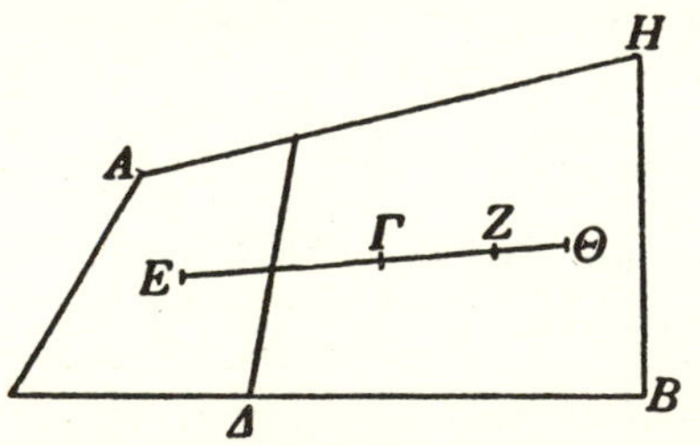

Fig. 122.

In Fig. 122 let the segment $A\Delta$ with centre of gravity E be taken away from the magnitude AB with centre of gravity Γ. On $E\Gamma$ produced Z has been determined so that

$$(Z\Gamma, E\Gamma) = (A\Delta, H\Delta). \tag{1}$$

It has to be proved that Z is the centre of gravity of ΔH.

Proof: Suppose another point, Θ, to be the centre of gravity of ΔH. Since AB is composed of $A\Delta$ and $H\Delta$, the centre of gravity of AB must be a point O on $E\Theta$, determined by

$$(\Theta O, EO) = (A\Delta, H\Delta).$$

Therefore Γ is not the centre of gravity of AB, which is contrary to the supposition.

It might be asked why Θ is collinear with E and Γ; in fact, if this is not the case, no contradiction will arise. The proof might therefore be given more effectively as follows:

Since Γ is the centre of gravity of AB, Γ has to lie on $E\Theta$, so that

$$(\Theta\Gamma, E\Gamma) = (A\Delta, H\Delta).$$

From a comparison with (1) it now appears at once that Θ coincides with Z.

Proposition 9.

The centre of gravity of any parallelogram lies on the straight line joining the middle points of opposite sides of the parallelogram.

Let the parallelogram $AB\Gamma\Delta$ (Fig. 123) be given, in which E and Z are successively the middle points of AB and $\Gamma\Delta$. It has to be proved that the centre of gravity of $AB\Gamma\Delta$ lies on EZ.

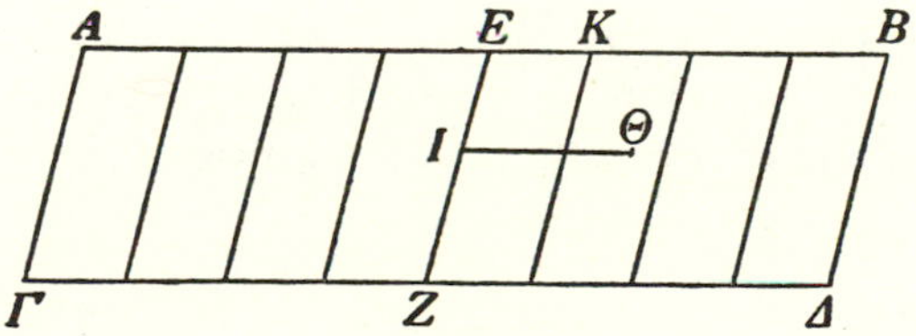

Fig. 123.

Suppose this is not true, but the centre of gravity is a point Θ outside EZ. Then let the straight line through Θ parallel to AB meet the straight line EZ in I. Now apply dichotomy (III; 0.5) to EB until the parts obtained (each equal to EK) are less than ΘI, and through the points of division thus obtained draw lines parallel to EZ. Proceeding in the same way on the other side of EZ, $AB\Gamma\Delta$

is divided into an even number of parallelograms, which are all congruent with KZ. When all these parallelograms are successively applied to KZ, the centres of gravity coincide with that of KZ (postulate IV). All these centres therefore lie on a straight line parallel to AB. By the application of Prop. 5, Corollary 2 we now recognize that the centre of gravity of AB must lie on the line segment having the centres of the central parallelograms for extremities. It cannot therefore be Θ; for $EK < I\Theta$.

Proposition 10.

The centre of gravity of any parallelogram is the point of intersection of its diagonals.

According to Prop. 9 the centre of gravity lies on each of the two straight lines joining the middle points of opposite sides; it is, however, also through the point of intersection of these straight lines that the diagonals pass.

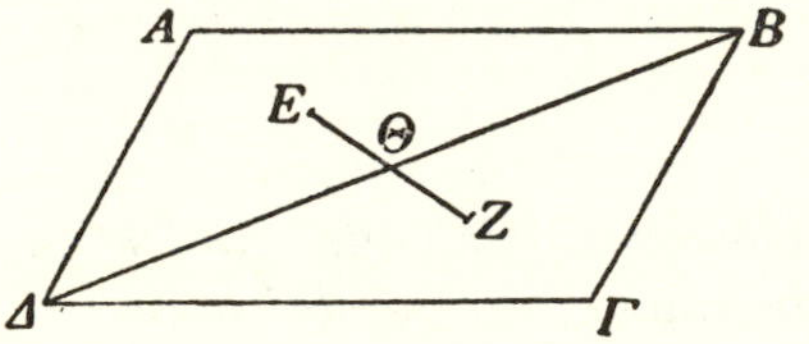

Fig. 124.

A second proof (Fig. 124) is given by considering the triangles into which the diagonal $B\Delta$ divides the parallelogram $AB\Gamma\Delta$. Since these triangles are congruent, the centres of gravity will coincide when the triangles are applied to each other (postulate IV). Now let E be the centre of gravity of $\triangle\ AB\Delta$, Θ the middle point of ΔB, Z a point on E produced, so that $E\Theta = \Theta Z$. When $\triangle\ AB\Delta$ is applied to $\triangle\ \Gamma\Delta B$, E will fall on Z, therefore Z is the centre of gravity of $\triangle\ \Gamma\Delta B$, and consequently, by Prop. 4, Θ is the centre of gravity of $AB\Gamma\Delta$.

Proposition 11.

If two triangles be similar to each other and within these triangles two points be similarly situated with respect to the triangles, and one point be the centre of gravity of the triangle in which it is situated, the other point will also be the centre of gravity of the triangle in which it is situated.

The meaning of the term "similarly situated" has been explained in the discussion of postulate V. The proof is a *reductio ad absurdum*, it being assumed that another point were the centre of gravity, upon which postulate V is applied; it is thus based on the unambiguousness of the relation of similar situation.

Proposition 12.

If two triangles be similar and the centre of gravity of one triangle lie on the straight line drawn from an angular point to the middle point of the base, the centre of gravity of the other triangle will lie on the straight line similarly drawn.

The proof is based on Prop. 11 in relation to the planimetric theorem that points which divide homologous medians of similar triangles into homologous proportional parts are similarly situated with respect to those triangles.

It appears that this proposition is not applied anywhere.

After these introductory theorems the situation of the centre of gravity of a triangle is found. The chief work is done in

Proposition 13.

In any triangle the centre of gravity lies on the straight line joining any vertex to the middle point of the base.

In Fig. 125 let Δ be the middle point of the base $B\Gamma$ of $\triangle\ AB\Gamma$. Suppose the centre of gravity Θ of $\triangle\ AB\Gamma$ not to lie on $A\Delta$. We

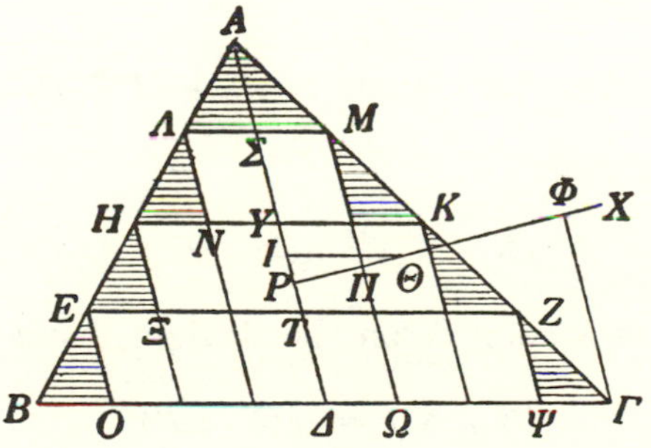

Fig. 125.

then know only that Θ lies within $\triangle\ AB\Gamma$ (postulate VII). Let the straight line through Θ parallel to $B\Gamma$ meet the straight line $A\Delta$ in I. Now apply dichotomy (III; 0.5) to $B\Gamma$ until the parts thus obtained (each equal to $\Delta\Omega$) are less than ΘI; through the points of division draw straight lines parallel to $A\Delta$, and divide the triangle in the manner indicated in Fig. 125 into parallelograms (MN,

$K\Xi$, etc.) and triangles ($A\Sigma M$, $Z\Psi\Gamma$, etc.). By Prop. 9 the centres of gravity of the parallelograms all lie on $A\Delta$, consequently the centre of gravity P of the figure $\mathbf{X}_\Pi$, consisting of all the parallelograms, also lies on $A\Delta$ (Prop. 6). Join P and Θ; let this straight line meet the straight line ΩM in Π and the straight line drawn through Γ parallel to $A\Delta$ in Φ. Π then lies between P and Θ, and Φ lies outside the triangle.

It is then ascertained how the centre of gravity of the figure $\mathbf{X}_\Delta$, consisting of all the shaded triangles, must be situated with respect to the points P, Θ, and Φ.

For this we first compare the area of $\triangle\ AB\Gamma$ with the sum of the areas of the shaded triangles. By similarity we have:

$$(A\Delta\Gamma, A\Sigma M) = [\mathbf{T}(A\Gamma),\ \mathbf{T}(AM)],\ \text{etc.},$$

whence

$$(A\Delta\Gamma, A\Sigma M + \ldots + Z\Psi\Gamma) = [\mathbf{T}(A\Gamma),\ \mathbf{T}(AM) + \ldots + \mathbf{T}(Z\Gamma)]$$

$$= [\mathbf{T}(A\Gamma),\ \mathbf{O}(AM, A\Gamma)] = (A\Gamma, AM)\,.$$

Likewise

$$(A\Delta B, A\Sigma\Lambda + \ldots + EOB) = (AB, A\Lambda) = (A\Gamma, AM)\,.$$

Therefore

$$(AB\Gamma, \mathbf{X}_\Delta) = (A\Gamma, AM) = (\Delta\Gamma, \Delta\Omega) = (P\Phi, P\Pi) > (P\Phi, P\Theta)$$

and thence *separando*

$$(\mathbf{X}_\Pi, \mathbf{X}_\Delta) > (\Theta\Phi, \Theta P)\,.$$

Now determine a point X on the straight line $P\Phi$ such that

$$(\mathbf{X}_\Pi, \mathbf{X}_\Delta) = (\Theta X, \Theta P)\,,$$

then $\Theta X > \Theta\Phi$, therefore X lies on $P\Phi$ produced.

By Prop. 8, X is now the centre of gravity of the figure $\mathbf{X}_\Delta$, consisting of the shaded triangles, which is impossible, because all these triangles are on the opposite side of the straight line drawn through Γ parallel to $A\Delta$ from X.

The latter conclusion is not, as Heiberg thinks[1]), based on postu-

---

[1]) *Opera* II, 155, Note 3.

late VII, for the perimeter of the figure X_Δ is not "concave in the same direction". It should rather be imagined to have been made in view of the consideration that, if the centre of gravity is found of a figure whose component parts all lie on the same side of a straight line by combining with the aid of Prop. 6 two parts, combining their combination with a third part, etc., the centre of gravity of the whole figure must lie on the same side of the straight line on which all the parts lie.

In a second proof of Prop. 12 (Fig. 126) Archimedes joins the supposed centre of gravity Θ to A, B, and Γ, and through the

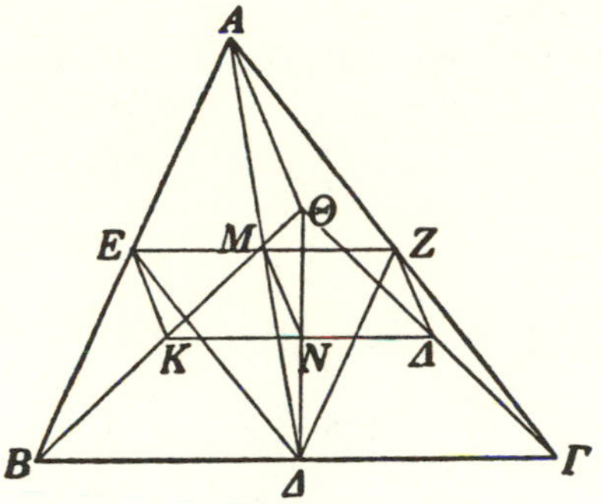

Fig. 126.

middle points E of AB and Z of $A\Gamma$ draws straight lines parallel to $A\Theta$, which meet $B\Theta$ and $\Gamma\Theta$ successively in K and Λ. In the similar triangles $AB\Gamma$ and $EB\Delta$, Θ and K are similarly situated points, therefore (Prop. 11) K is the centre of gravity of $\triangle\ EB\Delta$, and likewise Λ is the centre of gravity of $\triangle\ Z\Gamma\Delta$. Since further the areas of the triangles $EB\Delta$ and $Z\Gamma\Delta$ are equal, the centre of gravity of their combination lies at the middle point of $K\Lambda$, *i.e.* the point of intersection N with $\Theta\Delta$. In the parallelogram $AE\Delta Z$, M is the centre of gravity; the centre of gravity of $\triangle\ AB\Gamma$ therefore lies on MN, and cannot thus be Θ. Therefore Θ cannot lie outside $A\Delta$.

Proposition 14.

In any triangle the centre of gravity is the point in which the straight lines of the triangle joining the vertices to the middle points of the sides meet.

This follows at once from Prop. 13.

To conclude Book I, the centre of gravity of a trapezium is determined.

Proposition 15.

In any trapezium having two parallel sides[1]) *the centre of gravity lies on the straight line joining the middle points of the parallel sides, in such a way that the segment of it having the middle point of the smaller of the paralllel sides for extremity is to the remaining segment as the sum of double the greater plus the smaller is to the sum of double the smaller plus the greater of the parallel sides.*

In Fig. 127, in the trapezium $AB\Gamma\Delta$ ($A\Delta \parallel B\Gamma$) let the non-parallel sides produced meet in H; Z and E are successively the

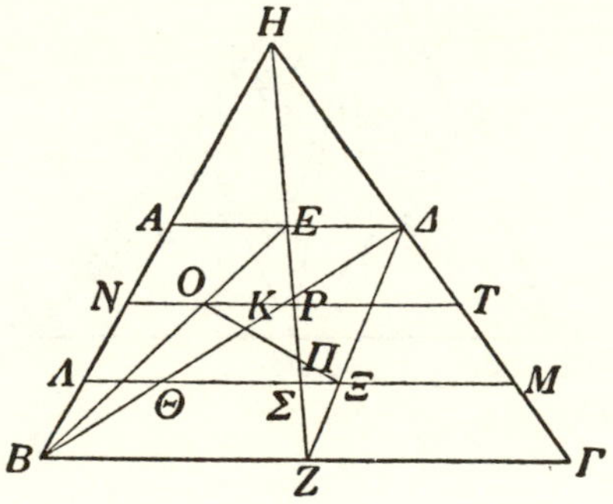

Fig. 127.

middle points of $B\Gamma$ and $A\Delta$. The centres of gravity of the triangles $HB\Gamma$ and $HA\Delta$ by Prop. 13 lie on HZ, therefore (Prop. 8 and postulate VII) the required centre of gravity of $AB\Gamma\Delta$ lies on the line segment EZ.

Now divide $B\Delta$ into three equal parts by means of the points K and Θ, and through these points draw NT and ΛM parallel to $B\Gamma$. Now the point of intersection Ξ of ΛM and ΔZ is the centre of gravity of $\triangle\ \Delta B\Gamma$, and likewise the point of intersection O of $N\Gamma$ and BE is the centre of gravity of $\triangle\ BA\Delta$. The centre of gravity of the trapezium therefore is the point of intersection of EZ and $O\Xi$. Now we have by Prop. 6 or 7:

$$(\Delta B\Gamma, BA\Delta) = (O\Pi, \Xi\Pi) = (P\Pi, \Sigma\Pi)\,,$$

[1]) Trapezium without any further indication refers to a quadrilateral.

and consequently also

$$(B\Gamma, A\Delta) = (P\Pi, \Sigma\Pi),$$

whence

$$(B\Gamma, P\Pi) = (A\Delta, \Sigma\Pi) = (2B\Gamma + A\Delta, 2P\Pi + \Sigma\Pi)$$
$$= (B\Gamma + 2A\Delta, P\Pi + 2\Sigma\Pi).$$

From this it follows that

$$(2B\Gamma + A\Delta, \Pi E) = (B\Gamma + 2A\Delta, Z\Pi),$$

which is equivalent to that which was to be proved.

CHAPTER X

THE METHOD OF MECHANICAL THEOREMS

1. We shall now, deviating from the order in which Archimedes' works appear in Heiberg's edition of the text, first discuss the treatise *The Method of Mechanical Theorems, for Eratosthenes*, to be briefly designated as *The Method*. In fact, when we know this work, it is easier to understand the *Quadrature of the Parabola*, the contents of which in turn are assumed as known in Book II of *On the Equilibrium of Planes*.

The discovery and decipherment of the manuscript of the *Method* has already been discussed in Chapter II. The object of the work becomes clear from the introductory letter to Eratosthenes, which we first give in translation:

Archimedes to Eratosthenes greeting!

On an earlier occasion I sent you some of the theorems found by me, the propositions of which I had written down, urging you to find the proofs which I did not yet communicate at the time. The propositions of the theorems I sent were the following:

firstly: if in a right prism having a square[1]) *for its base a cylinder be inscribed which has its bases in the squares facing each other and its sides in the other faces of the prism*[2]), *and a plane be drawn through*

[1]) The word used is παραλληλόγραμμον, but it is clear from the context that a square is meant.

[2]) The meaning is that the curved surface of the cylinder touches the vertical faces.

the centre of the circle which is the base of the cylinder and one side of the square in the opposite base, this plane will cut off from the cylinder a portion which is bounded by two planes and the surface of the cylinder, viz. the plane drawn, the plane in which lies the base of the cylinder, and the cylinder surface between the said planes; and the portion cut off from the cylinder is one-sixth of the whole prism[1]).

The proposition of the other theorem is as follows: if in a cube a cylinder be inscribed which has its bases in opposite squares and the surface of which touches the four other faces, and if in the same cube another cylinder be inscribed which has its bases in other squares and the surface of which touches the four other faces, the solid bounded by the surfaces of the cylinders, which is enclosed by the two cylinders, is two-thirds of the whole cube ...[2])

I will send you the proofs of these theorems in this book.

Since, as I said, I know that you are diligent, an excellent teacher of philosophy, and greatly interested in any mathematical investigations that may come your way, I thought it might be appropriate to write down and set forth for you in this same book a certain special method, by means of which you will be enabled to recognize certain mathematical questions with the aid of mechanics. I am convinced that this is no less useful for finding the proofs of these same theorems. For some things, which first became clear to me by the mechanical method, were afterwards proved geometrically, because their investigation by the said method does not furnish an actual demonstration. For it is easier to supply the proof when we have previously acquired, by the method, some knowledge of the questions than it is to find it without any previous knowledge.

That is the reason why, in the case of the theorems the proofs of which Eudoxus was the first to discover, viz. on the cone and the pyramid, that the cone is one-third of the cylinder and the pyramid one-third of the prism having the same base and equal height, no small share of the credit should be given to Democritus, who was the first to state the fact about the said figure[3]), *though without proof.*

---

[1]) The derivation of this result is found in Propositions 12–15.

[2]) The derivation of this result is not to be found in the part of the *Method* that has been preserved.

[3]) It has struck students that Archimedes uses the singular here, whereas he first referred to theorems on the cone and the pyramid. It does not, however, seem likely that any inferences can be made from this.

My own experience is also that I discovered the theorem now published[1]), *in the same way as the earlier ones*[2]).

I now wish to describe the method in writing, partly, because I have already spoken about it before, that I may not impress some people as having uttered idle talk[3]), *partly because I am convinced that it will prove very useful for mathematics; in fact, I presume there will be some among the present as well as future generations who by means of the method here explained will be enabled to find other theorems which have not yet fallen to our share.*

We will now first write down what first became clear to us by the mechanical method, viz. that any segment of an orthotome is larger by one-third than the triangle which has the same base and equal height, and thereafter all the things that have become clear in this way. At the end of the book we will give the geometrical proofs of the theorems the propositions of which we sent you on an earlier occasion[4]).

In this exceptionally interesting document Archimedes therefore vouchsafes us a much more intimate glimpse of his mathematical workshop than was ever granted by any other Greek mathematician. In fact, Greek mathematics is characterized—and in this respect, too, it founded a tradition which was to last down to our own time—by a care of the form of the mathematical argument which, superficially viewed, seems almost exaggerated. It demands the inexorably proceeding, irrefutably persuading sequence of logical conclusions constituting the synthetic method of demonstration, but to this it sacrifices the reader's wish to gain also an insight into the method by which the result was first discovered. It is this wish, however, which Archimedes meets in his *Method*: he will reveal how he himself, long before he knew how to prove his theorems, became convinced of their truth.

2. The work opens with a number of lemmas on centres of gravity,

[1]) Here again Archimedes is evidently referring to the two above mentioned propositions.

[2]) This refers to a number of theorems from **S.C.**, **C.S.**, and **Q.P.**, as will become clear in the subsequent discussion of the *Method*.

[3]) In the preface to **Q.P.** (*Opera* II, 262, lines 11-13) Archimedes says that the theorem on the area of a segment of an orthotome first became clear to him by the mechanical method, and that he then proved it geometrically.

[4]) These are apparently the two theorems mentioned at the beginning of the letter.

some of which we have already encountered as postulates or propositions in *On the Equilibrium of Planes*. They state that, if from a magnitude a portion be taken away, the centre of gravity of the remainder coincides with that of the whole if the removed portion had the same centre of gravity, and that otherwise it is found by Prop. I, 8 of the work in question; that the centre of gravity of a system of magnitudes whose centres of gravity lie on one and the same straight line lies on the same line; that the centre of gravity of a straight line (*i.e.* a homogeneous straight line segment) is its middle point, that of a triangle the point of intersection of the medians, that of a parallelogram the point of intersection of the diagonals, that of a circle the centre, that of a cylinder the middle point of the axis, that of a prism the same[1]), and that of a cone the point which divides the axis in such a way that the segment towards the vertex is three times the remaining segment. Moreover the proposition **C.S.** 1 (III; 7.20) is mentioned as the last lemma.

This part is followed by the propositions. We shall first give the first application of the method to be described, partly in a literal translation and without any comment:

Proposition 1 (Fig. 128).

Let the segment $\alpha\beta\gamma$ be given, comprehended by the straight line $\alpha\gamma$ and the orthotome $\alpha\beta\gamma$; let $\alpha\gamma$ be bisected in δ, let $\delta\beta\varepsilon$ be drawn parallel to the diameter, and let $\beta\alpha$ and $\beta\gamma$ be joined.

I say that the segment $\alpha\beta\gamma$ is larger by one-third (ἐπίτριτον) than the triangle $\alpha\beta\gamma$[2]).

Proof: Let the straight line drawn through α parallel to $\delta\beta$ meet the tangent to the curve at γ in ζ and the straight line $\gamma\beta$ in $\varkappa$; ε is the point of intersection of $\gamma\zeta$ and $\beta\delta$. Through a variable point

[1]) As appears from the application in Prop. 13, *axis* here refers to the line segment joining the centres of gravity of the two bases. Archimedes here deviates from his usual terminology, in which ἄξων invariably means the axis of revolution.

[2]) The formulation of the proposition deviates from the usual type; in his more official works Archimedes always avoids any reference to the figure and any designation of points and lines by letters. From this again it is clear that the *Method* is to be considered as a private communication to Eratosthenes rather than a treatise meant for publication.

ξ of $\alpha\gamma$ a straight line is drawn parallel to $\delta\beta$, which meets the curve in o, $\gamma\beta$ in ν, $\gamma\zeta$ in μ.

Further make $\varkappa\vartheta$ equal to $\varkappa\gamma$.

Now it is known that

$$\delta\beta = \beta\varepsilon \text{ (III; 2.2)},$$

whence

$$\alpha\varkappa = \varkappa\zeta \text{ and } \xi\nu = \nu\mu .$$

Further we have

$$(o\xi, o\mu) = (\alpha\xi, \xi\gamma) \text{ (III; 2.7)},$$

whence

$$(o\xi, \xi\mu) = (\alpha\xi, \alpha\gamma)$$
$$= (\varkappa\nu, \varkappa\gamma) = (\varkappa\nu, \varkappa\vartheta) .$$

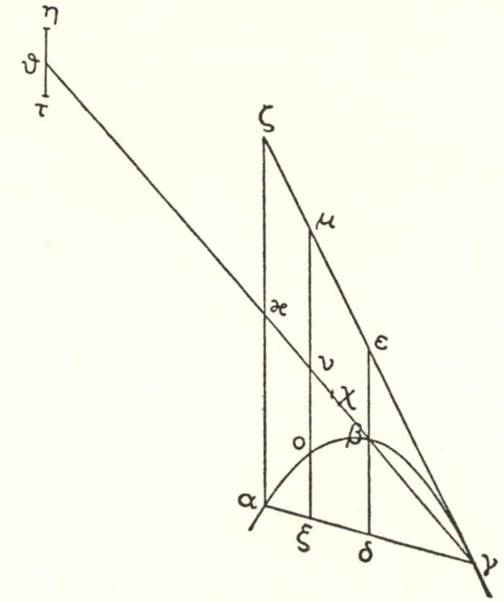

Fig. 128.

And[1]) *since ν is the centre of gravity of the straight line $\mu\xi$, if we take $\tau\eta$ equal to ξo and its centre of gravity ϑ, so that $\tau\vartheta = \vartheta\eta$, $\tau\vartheta\eta$ will balance $\mu\xi$, when remaining in its place, because $\vartheta\nu$ is divided into segments which are inversely proportional to the gravities $\tau\eta$ and $\mu\xi$, namely in such a way that $\mu\xi$ is to $\eta\tau$ as $\vartheta\varkappa$ to $\varkappa\nu$, so that $\varkappa$ is the centre of gravity of the weight made up of both. In the same way, however many straight lines there be drawn in the triangle $\zeta\alpha\gamma$ parallel to $\varepsilon\delta$, when they remain in their places, they will balance the portions cut off from them by the curve, when transferred to ϑ...*

And since the triangle $\gamma\zeta\alpha$ is made up of all the parallel lines in the triangle $\gamma\zeta\alpha$, and the segment $\alpha\beta\gamma$ is made up of all the parallel lines drawn inside the curve in the manner of ξo, the triangle $\zeta\alpha\gamma$, when remaining in its place, will balance about the point $\varkappa$ the segment of the curve placed about ϑ as centre of gravity, so that $\varkappa$ is their common centre of gravity.

Now let χ be the centre of gravity of $\triangle\ \alpha\zeta\gamma$. χ lies on $\gamma\varkappa$ in such a way that

$$\gamma\varkappa = \varkappa\vartheta = 3\varkappa\chi .$$

Since there is equilibrium between $\triangle\ \alpha\zeta\gamma$, remaining in its place,

[1]) *Opera* II, 436, lines 10–30.

and the segment $\alpha\beta\gamma$ about ϑ as centre of gravity, we have (reverse of **Pl.Ae.** I, 6, 7)

$$(\triangle\ \alpha\zeta\gamma,\ \text{segment}\ \alpha\beta\gamma) = (\varkappa\vartheta, \varkappa\chi) = (3,1)\ ,$$

whence

$$\triangle\ \alpha\zeta\gamma = 3.\text{segment}\ \alpha\beta\gamma\ .$$

Since

$$\triangle\ \alpha\zeta\gamma = 2.\triangle\ \alpha\varepsilon\gamma = 4.\triangle\ \alpha\beta\gamma\ ,$$

it is found that

$$\text{segment}\ \alpha\beta\gamma = \tfrac{4}{3}.\triangle\ \alpha\beta\gamma\ .$$

This has not therefore been proved by the above, but a certain impression has been created that the conclusion is true. Since we thus see that the conclusion has not been proved, but we suppose it is true, we shall mention the previously published geometrical proof, which we ourselves have found for it, in its appointed place[1]).

It would seem here that Archimedes intended to collect at the end of the *Method* the exact mathematical proofs of all the theorems found by the method described, even if, as is the case with the theorem in *Quadrature of the Parabola* which has just been discussed, they had already been published previously.

3. The method which Archimedes wishes to explain emerges so clearly from the proposition dealt with that we can proceed to discuss it already here.

We may note first of all that it is characterized by the application of two different principles: in the first place it makes use of considerations taken from mechanics in that it conceives geometrical figures to be attached to a lever in such a way that the latter remains in equilibrium, and then draws up conditions for such equilibrium; and it is further based on the view that the area of a plane figure is to be looked upon as the sum of the lengths of all the line segments drawn therein in a given direction and of which the figure is imagined to be made up; this view will be extended to space in the following propositions in the sense that a solid, too, is conceived to be made up of all the intersections determined therein by a plane of fixed inclination that is displaced, and that subsequently also the volume of the solid is looked upon as the sum of the areas of those intersections. We shall designate these two me-

[1]) *Opera* II, 438, lines 16–21.

thodic principles by the references: "barycentric method" and "method of indivisibles".

We further saw that Archimedes is not prepared to recognize the results obtained with this twofold method as actually proved conclusions. It might now be asked where in his view resides the lack of exactness, in the barycentric character of the arguments, in the application of indivisibles, or in both.

The answer to this question may be given without much doubt: the mathematical deficiency is exclusively a consequence of the use of indivisibles; there is not the least objection from the mathematical point of view against properly founded barycentric considerations, such as we already found applied in Prop. 1.

That this is actually the view of Archimedes is particularly evident from the fact that in his treatise *Quadrature of the Parabola*, which constitutes an official publication satisfying all requirements of exactness, he proves the insight gained in Prop. 1 on the area of any segment of an orthotome once more by means of statical considerations, but this time without indivisibles[1]). Moreover, the way in which in *On the Equilibrium of Planes* he bases the theory of the lever on postulates strongly creates the impression that he does not see any essential difference between his Elements of Statics and the systematization of planimetry as Euclid had given it[2]).

Although Archimedes could therefore make use unconcernedly of the barycentric method for dealing with mathematical problems, which had probably been introduced by himself, he was bound to experience a great deal of doubt and uncertainty with regard to the application of the method of indivisibles, for here he touched upon a question which in the centuries preceding his own had given rise

---

[1]) It is true that he also furnishes the proof once more by purely geometrical means; there is, however, not the slightest cause to assume that he did not consider the two proofs equivalent.

[2]) The view that the application of mechanical methods was the very thing which gave occasion to consider the proofs from the *Method* inexact is taken by H. de Vries, *Historische Studiën* (Groningen 1926), p. 139. It is true that the remark in the introductory letter that the consideration according to the method there mentioned has no demonstrative force, together with the fact that the method is termed mechanical, is an argument in favour of this view. On account of the above argumentation, however, we cannot share it.

to violent controversy more than any other questions in Greek mathematics[1]). It was the profound question of atomism or continuity, on which, though originating from physics, opinions were divided also in mathematics, and which finds its clearest expression in the aporia that worried Democritus: if the circular sections that can be made in a cone parallel to the base are congruent, how can the cone differ from a cylinder; and if they grow smaller towards the vertex, is not then the curved surface, which should be smooth, scalariform?

The great influence which this question exerted on the history of Greek mathematics cannot indeed be reconstructed in detail, but its essential features can be recognized readily enough. The spectacular intervention of Zeno of Elea in the evolution of mathematical thought seems to have been largely caused by the embarrassment into which the human intellect had been thrown by the mathematical continuum; it was perhaps the most powerful source of the famous crisis of principles, which disturbed the gradual growth of mathematics about 400 B.C.; the reconstruction, with which the name of Eudoxus is associated, was brought about not in the last place by the conquest of this intellectual problem.

This marked the end of the unconcern with which the infinite had always been referred to as if the word connoted nothing but something very large, but finite; the method of the indirect limiting process tied down the application of infinite processes to the rigorous forms which so far we have found Archimedes observing in all his works; and if one did not know that in mathematics a discovery is one thing and a proof quite another, and that the method by which the reader is convinced of the truth of a theorem in many cases is quite different from the way in which it was first found, one might believe that the method of indivisibles had disappeared definitely from Greek mathematics after Eudoxus.

The *Method* has revealed to us—and it is this which constitutes the eminent importance of its discovery—that the indivisibles had only been banished from the published treatises, but that in the workshop of the producing mathematician they held undiminished sway, as they were to do so frequently in later periods, *e.g.* in Cavalieri, Galilei, Huygens, Leibniz. Unconcerned about the ra-

[1]) For a more detailed discussion of this matter reference is made to *Elements of Euclid* I, p. 41 *et seq.*

tional untenability of the view, and as a remarkable instance of the fertility which may be inherent to irrational modes of thought even in the most rational of all sciences, Archimedes, as long as he is seeking to find new results, considers a segment of a parabola as the sum of all its ordinates, or a solid as the sum of all its intersections in parallel planes; and when he is speaking about a figure, we find him already glibly using the expressions "all the lines" or "all the circles" which are to become current coin in the 17th century and which denote the set of all parallel indivisibles which "fill up" the figure, as it is technically called.

This is the important new insight which the publication of the *Method* has furnished us. Moreover, as will be shown more in detail with examples, it enables us to follow in various cases the development of a proposition from its non-rigorous, intuitively convincing discovery to its impeccable, abstractly persuasive exposition in a published work.

Archimedes always continues to distinguish sharply between these two phases of the process: Democritus has found the theorem that any pyramid is one-third of the prism having the same base and height (perhaps also by means of indivisibles), but only Eudoxus has proved it. That any segment of a parabola is greater by one-third than the triangle having the same base and vertex is to be expected in view of Prop. 1 of the *Method*; but it is only in the long chain of propositions contained in the treatise *Quadrature of the Parabola* that this assertion is to be raised to the rank of a proved assertion.

4. Before proceeding to the discussion of the other propositions of the *Method*, we still have to give an explanation of the mechanical part proper of the method applied. It is found that each time one of the straight lines of the figure is considered as an immaterial balance (ζυγός) supported in its centre of gravity, that on this balance in one or more points plane or solid figures are so attached that the point of attachment is at the same time the centre of gravity of the figure, and that of other figures "all the lines" or "all the intersections" are transferred to one of the ends of the lever (in Prop. 1 all the ordinates $o\xi$ of the segment of the orthotome being transferred to ϑ). The idea is then that, after this transfer, from all these lines or intersections the figure from which they are taken is built up again, but in such a way that the end in which

each of them is placed with its centre of gravity is also the centre of gravity of the reconstructed figure again. The fact that the established equilibrium of the lever is not disturbed by this is guaranteed by postulate VI of *On the Equilibrium of Planes*: the reconstruction leaves the weight and the place of the figure unchanged, and the equilibrium once established is thus maintained. The fundamental importance which this postulate is thus found to possess for the *Method* convinces us once again of the correctness of the identical interpretation which we gave of it, in accordance with Toeplitz, in *On the Equilibrium of Planes*.

5. We will now continue our discussion of the propositions.

Proposition 2.

That the volume of any sphere is four times that of the cone which has its base equal to the greatest circle of the sphere and its height equal to the radius of the sphere, and that the volume of the cylinder which has its base equal to the greatest circle of the sphere and its height equal to the diameter of the sphere is one and a half times that of the sphere, is recognized according to this method as follows.

In Fig. 129 let $\alpha\beta\gamma\delta$ be a greatest circle of the sphere, $\alpha\gamma$ and $\beta\delta$ two diameters of it at right angles to each other. Consider the cone with vertex α, whose base is the greatest circle in the plane through $\beta\delta$ at right angles to $\alpha\gamma$. The extended surface of this cone intersects the plane through γ at right angles to $\alpha\gamma$ in a circle on $\varepsilon\zeta$ as diameter. This circle is the base of a cylinder $\varepsilon\zeta\eta\lambda$ with height $\alpha\gamma$.

Make $\alpha\vartheta = \alpha\gamma$, and consider $\gamma\vartheta$ as a balance with fulcrum α. A variable plane $\mu\nu$ at right angles to $\alpha\gamma$ intersects cone, sphere, and cylinder in circles whose diameters are successively $\pi\varrho$[1]), ξo, and $\mu\nu$, and which we designate by $\mathbf{K}(\pi\varrho)$, $\mathbf{K}(\xi o)$, $\mathbf{K}(\mu\nu)$.

Now because

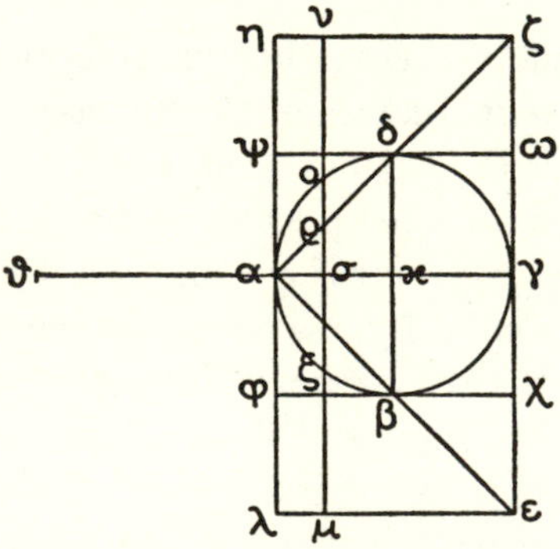

Fig. 129.

$$\varrho\sigma = \alpha\sigma,$$

[1]) At the point of intersection of the straight lines $\alpha\varepsilon$ and $\mu\nu$ the letter π has been dropped.

we have

$$\mathbf{T}(\varrho\sigma)+\mathbf{T}(o\sigma) = \mathbf{T}(\alpha o) = \mathbf{O}(\alpha\sigma, \alpha\gamma) ,$$

whence because

$$\nu\sigma = \alpha\gamma ,$$

we have

$$[\mathbf{T}(\nu\sigma), \ \mathbf{T}(\varrho\sigma)+\mathbf{T}(o\sigma)] = [\mathbf{T}(\alpha\gamma), \ \mathbf{O}(\alpha\sigma, \alpha\gamma)] =$$

$$(\alpha\gamma, \alpha\sigma) = (\alpha\vartheta, \alpha\sigma) \tag{1}$$

or

$$[\mathbf{K}(\mu\nu), \ \mathbf{K}(\pi\varrho)+\mathbf{K}(\xi o)] = (\alpha\vartheta, \alpha\sigma) .$$

The circle $\mathbf{K}(\mu\nu)$ *suo loco*[1]) can therefore balance the circles $\mathbf{K}(\pi\varrho)$ and $\mathbf{K}(\xi o)$, both placed in ϑ (*i.e.* so placed that their centres of gravity fall in ϑ). Consequently there is also equilibrium between the cylinder $\varepsilon\zeta\eta\lambda$ *suo loco* and the combination in ϑ of the sphere $\alpha\beta\gamma\delta$ and the cone $\alpha\varepsilon\zeta$ (for the three solids are filled up by the above-mentioned circles if the plane $\mu\nu$ moves from $\eta\lambda$ to $\zeta\varepsilon$).

Since $\varkappa$ is the centre of gravity of the cylinder, the following relation holds:

$$(\text{Sphere}+\text{Cone, Cylinder}) = (\alpha\varkappa, \alpha\vartheta) ,$$

whence

$$\text{Cylinder} = 2\,(\text{Sphere}+\text{Cone}) .$$

The cylinder is three times the cone, so that it follows from the last-mentioned relation that

$$\text{Cone } \alpha\varepsilon\zeta = 2 . \text{Sphere } \alpha\beta\gamma\delta ,$$

whence

$$\text{Sphere } \alpha\beta\gamma\delta = 4 . \text{Cone } \alpha\beta\delta .$$

The theorem on the ratio between the sphere and the cylinder, too, is now immediately obvious.

When this had been recognized, the suspicion arose that the surface of any sphere is four times that of a greatest circle on the sphere. In fact, it was assumed that as any circle is equal to a triangle which has

---

[1]) By this Latin translation of the Greek expression αὐτοῦ μένων we indicate that the figure under consideration remains where it is.

the circumference of the circle for its base and whose height is equal to the radius of the circle, so any sphere is equal to a cone which has the surface of the sphere for its base and whose height is equal to the radius of the sphere.

A truly striking sidelight is here thrown on the way in which Archimedes found his two famous theorems 33 and 34 of *On the Sphere and Cylinder*, for it is revealed that the theorem on the volume was the first to be found, and that subsequently the suspected analogy with the relation between the surface and the circumference of the circle led to the theorem on the surface.

Proposition 3.

By this method it is also recognized that the cylinder which has a base equal to the greatest circle of a spheroid and a height equal to the axis of the spheroid is one and a half times the spheroid; when this has been recognized, it is obvious that if a spheroid be cut by a plane through the centre at right angles to the axis, half the spheroid will be the double of the cone which has the same base as the segment and the same axis.

As regards the mechanical argument, the proof is identical with that of Prop. 2. The proportion (1), however, requires a longer derivation, because in Fig. 129 $\alpha\beta\gamma\delta$ now has to be conceived as an oxytome, which generates the spheroid (ellipsoid of revolution) by revolution about one of its axes, $\alpha\gamma$.

By the symptom of the oxytome (III; 3.0) we have

$$[\mathbf{T}(o\sigma),\ \mathbf{O}(\alpha\sigma,\gamma\sigma)] = [\mathbf{T}(\delta\varkappa),\ \mathbf{T}(\alpha\varkappa)] = [\mathbf{T}(\varrho\sigma),\ \mathbf{T}(\alpha\sigma)]\,,$$

whence

$$[\mathbf{T}(o\sigma),\ \mathbf{T}(\varrho\sigma)] = (\gamma\sigma,\alpha\sigma) = (\nu\zeta,\alpha\sigma) = (\nu\varrho,\varrho\sigma)$$
$$= [\mathbf{O}(\nu\varrho,\varrho\sigma),\ \mathbf{T}(\varrho\sigma)]\,,$$

whence

$$\mathbf{T}(o\sigma) = \mathbf{O}(\varrho\sigma,\nu\varrho)\,.$$

From this it follows that

$$\mathbf{T}(\varrho\sigma)+\mathbf{T}(o\sigma) = \mathbf{O}(\varrho\sigma,\varrho\sigma+\nu\varrho) = \mathbf{O}(\varrho\sigma,\nu\sigma)\,.$$

Now we have

$$[\mathbf{T}(\nu\sigma),\ \mathbf{T}(\varrho\sigma)+\mathbf{T}(o\sigma)] = (\nu\sigma,\varrho\sigma) = (\zeta\gamma,\varrho\sigma) = (\alpha\gamma,\alpha\sigma) = (\alpha\vartheta,\alpha\sigma)\,.$$

Further the proof proceeds on the same lines as in Prop. 2.

The exact proof of the theorem on half the spheroid has been furnished in **C.S.** 27.

Proposition 4.

That any segment cut off from an orthoconoid by a plane at right angles to the axis is one and a half times the cone which has the same base as the segment and the same axis, is recognized by this method as follows.

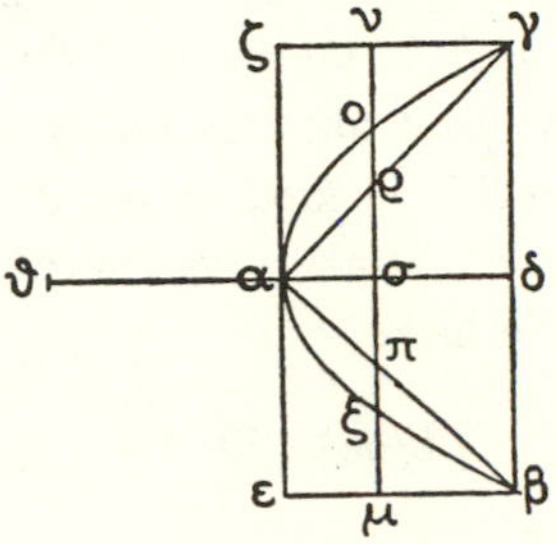

Fig. 130.

In Fig. 130 let the orthotome $\alpha\beta\gamma$ generate, by revolution about the diameter $\alpha\delta$, an orthoconoid, from which a segment is cut off by a plane through δ at right angles to $\alpha\delta$. A variable plane $\mu\nu$ at right angles to $\alpha\delta$ intersects the orthoconoid in a circle on $o\xi$ as diameter, and the cylinder which has its base and height in common with the segment in a circle on $\mu\nu$ as diameter. Again make $\alpha\vartheta = \alpha\delta$, and consider $\delta\vartheta$ as a balance with fulcrum α.

We now have by the symptom of the orthotome (III; 2.0):

$$[\mathbf{T}(\nu\sigma),\ \mathbf{T}(o\sigma)] = [\mathbf{T}(\gamma\delta),\ \mathbf{T}(o\sigma)] = (\alpha\delta,\ \alpha\sigma) = (\alpha\vartheta,\ \alpha\sigma)\ .$$

From this it appears that $\mathbf{K}(\mu\nu)$ *suo loco* balances $\mathbf{K}(o\xi)$ in ϑ, so that also the cylinder *suo loco* balances the segment in ϑ. Since the centre of gravity of the cylinder is the middle point $\varkappa$ of $\alpha\delta$, it follows from this that

$$\text{Cylinder } \beta\gamma\zeta\varepsilon = 2.\text{Segment } \alpha\beta\gamma\ ,$$

whence

$$\text{Segment } \alpha\beta\gamma = \tfrac{3}{2}.\text{Cone } \alpha\beta\gamma\ .$$

The exact proof is to be found in **C.S.** 21.

6. In the propositions hitherto discussed it was possible to find, from the consideration of the equilibrium of the balance, the volume of the solid which, resolved into its intersections, had been transferred to one of the ends, because with regard to the solid which was attached to the balance *suo loco* both the volume and the position of the centre of gravity were known. If, however, the

volume of the transferred solid is known, by the same method the position of the centre of gravity of the non-displaced solid can be found. This is done in Prop. 5 for a right segment of an orthoconoid, in Prop. 6 for a hemisphere, in Prop. 9 for any segment of a sphere.

Proposition 5.

That the centre of gravity of any segment cut off from an orthoconoid by a plane at right angles to the axis lies on the straight line which is the axis of the segment, which line is so divided by it that the part towards the vertex is twice the remaining part, is recognized by this method as follows.

We shall now compare in Fig. 130 the circles $\mathbf{K}(o\xi)$ and $\mathbf{K}(\pi\varrho)$. We have

$$[\mathbf{T}(o\sigma),\ \mathbf{T}(\gamma\delta)] = (\alpha\sigma,\ \alpha\delta) = (\varrho\sigma,\ \gamma\delta) = [\mathbf{T}(\varrho\sigma),\ \mathbf{O}(\varrho\sigma,\ \gamma\delta)],$$

whence

$$[\mathbf{T}(o\sigma),\ \mathbf{T}(\varrho\sigma)] = (\gamma\delta,\ \varrho\sigma) = (\alpha\delta,\ \alpha\sigma) = (\alpha\vartheta,\ \alpha\sigma)\ .$$

From this it follows that $\mathbf{K}(o\xi)$ *suo loco* balances $\mathbf{K}(\pi\varrho)$ in ϑ, so that the segment of the conoid *suo loco* balances the cone $\alpha\beta\gamma$ in ϑ. Since the volume of the segment is one and a half times that of the cone, the distance from α to the centre of gravity is therefore $\frac{2}{3}\alpha\delta$.

Proposition 6.

The centre of gravity of any hemisphere lies on the straight line which is its axis, which is so divided by it that the part adjacent to the surface of the hemisphere is to the other part in the ratio of five to three.

The hemisphere is generated (Fig. 131) by revolution of the semicircle $\alpha\delta\beta$ about $\alpha\gamma$. We now have as above:

$$[\mathbf{T}(\varrho\varepsilon),\ \mathbf{T}(\varrho\varepsilon)+\mathbf{T}(o\varepsilon)] = [\mathbf{T}(\alpha\varepsilon),\ \mathbf{T}(\alpha o)] = (\alpha\varepsilon,\ \alpha\gamma) = (\alpha\varepsilon,\ \alpha\vartheta)\ .$$

From this it follows that $\mathbf{K}(\pi\varrho)$ and $\mathbf{K}(o\xi)$ *suis locis* balance together $\mathbf{K}(\pi\varrho)$ in ϑ, so that the hemisphere $\alpha\delta\beta$ and the cone $\alpha\delta\beta$ *suis locis* balance together the cone alone in ϑ. Now place in ϑ a cylinder M, which balances the cone *suo loco*, and a cylinder N, which balances the hemisphere *suo loco*. (The idea is that ϑ shall be the centre of gravity of both cylinders.)

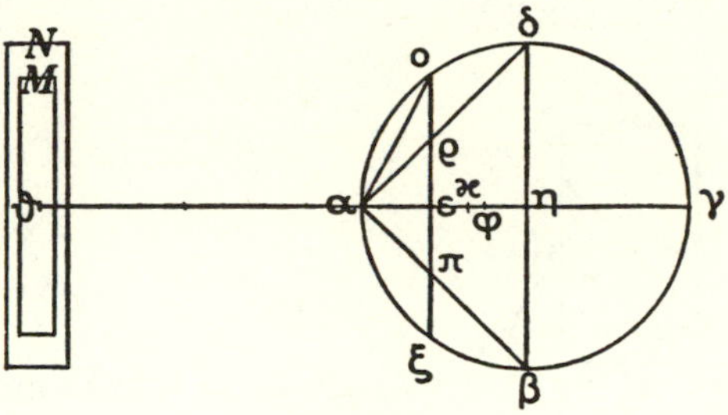

Fig. 131.

If now φ be the centre of gravity of the cone, and $\varkappa$ that of the hemisphere, we have

$$\alpha\varphi = \tfrac{3}{4}\alpha\eta = \tfrac{3}{8}\alpha\vartheta\,,$$

whence

$$M = \tfrac{3}{8}.\text{Cone } \alpha\delta\beta\,.$$

Therefore we have

$$N = \tfrac{5}{8}.\text{Cone } \alpha\delta\beta = \tfrac{5}{16}.\text{Hemisphere } \alpha\delta\beta\,,$$

and since N balances in ϑ the hemisphere *suo loco*, we also have

$$\alpha\varkappa = \tfrac{5}{16}\alpha\vartheta = \tfrac{5}{8}\alpha\eta\,,$$

from which follows that which was required to be proved.

Proposition 7.

By this method it is also recognized that any segment of a sphere is to the cone which has the same base as the segment and the same axis in the same ratio as the sum of the radius of the sphere and the height of the remaining segment is to the height of the remaining segment.

The segment is generated (Fig. 132) by revolution of the segment of the circle $\alpha\beta\delta$ about $\alpha\gamma$. $\upsilon\tau$ is the diameter at right angles to $\alpha\gamma$. Just as above, $\alpha\vartheta = \alpha\gamma$, and $\vartheta\gamma$ is considered as a balance with fulcrum α. Further $\eta\varkappa = \eta\lambda = \alpha\gamma$. $\varkappa\lambda$ is the diameter of the base of a cylinder with height $\alpha\eta$. Just as in Prop. 2, it is recognized that this cylinder *suo loco* balances the combination of the cone $\alpha\varepsilon\zeta$ and the segment of the sphere, both in ϑ. If now χ be the centre of gravity of the cylinder, so that $\alpha\chi = \frac{1}{2}\alpha\eta$, we have:

$$(\text{Cone } \alpha\varepsilon\zeta + \text{Segment, Cylinder}) = (\alpha\chi, \alpha\vartheta) = [\mathbf{O}(\alpha\vartheta, \alpha\chi), \mathbf{T}(\alpha\vartheta)]. \quad (1)$$

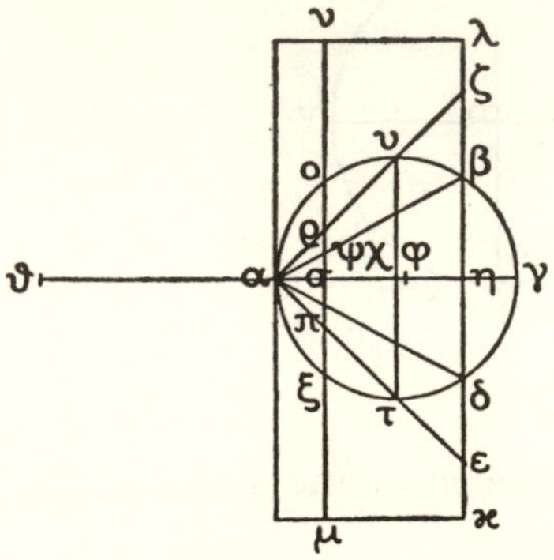

Fig. 132.

We also have

$$(\text{Cylinder, Cone } \alpha\varepsilon\zeta) = [\mathbf{T}(\eta\lambda), \tfrac{1}{3}\mathbf{T}(\eta\zeta)] .$$

In this

$$\eta\lambda = \alpha\gamma = \alpha\vartheta, \text{ and } \eta\zeta = \alpha\eta ,$$

whence

$$\tfrac{1}{3}\mathbf{T}(\eta\zeta) = \mathbf{O}(\tfrac{2}{3}\alpha\eta, \tfrac{1}{2}\alpha\eta) = \mathbf{O}(\tfrac{2}{3}\alpha\eta, \alpha\chi) = \mathbf{O}(\alpha\varphi, \alpha\chi) ,$$

if φ be a point on $\alpha\eta$ such that

$$\alpha\varphi = \tfrac{2}{3}\alpha\eta .$$

Thence

$$(\text{Cylinder, Cone } \alpha\varepsilon\zeta) = [\mathbf{T}(\alpha\vartheta), \mathbf{O}(\alpha\varphi, \alpha\chi)] \quad (2)$$

From (1) and (2) it follows *ex aequali* that

$$(\text{Cone } \alpha\varepsilon\zeta + \text{Segment, Cone } \alpha\varepsilon\zeta) = (\alpha\vartheta, \alpha\varphi) ,$$

whence

$$(\text{Segment, Cone } \alpha\varepsilon\zeta) = (\gamma\varphi, \alpha\varphi) = (\tfrac{3}{2}\gamma\varphi, \alpha\eta) . \quad (3)$$

We also have

$$[\text{Cone } \alpha\varepsilon\zeta, \text{Cone } \alpha\beta\delta] = [\mathbf{T}(\zeta\eta), \mathbf{T}(\beta\eta)] = [\mathbf{T}(\alpha\eta), \mathbf{T}(\beta\eta)] = (\alpha\eta, \gamma\eta) . \quad (4)$$

From (3) and (4) it follows *ex aequali* that

$$(\text{Segment, Cone } \alpha\beta\delta) = (\tfrac{3}{2}\gamma\varphi, \gamma\eta) .$$

In this, if R represent the radius of the sphere, we have

$$\tfrac{3}{2}\gamma\varphi = \tfrac{3}{2}(\alpha\gamma - \alpha\varphi) = 3R - \alpha\eta = R + \gamma\eta ,$$

so that

$$(\text{Segment, Cone } \alpha\beta\delta) = (R + \gamma\eta, \gamma\eta) .$$

The exact proof is to be found in **S.C. II, 2.**

In Prop. 8 it is merely stated that by this method it is also possible to find the volume of any segment of a spheroid (C.S. 29 and 31).

Proposition 9.

The centre of gravity of any segment of a sphere lies on the straight line which is the axis of the segment, which is so divided by it that the part towards the vertex of the segment is to the remaining part in the same ratio as the sum of the axis of the segment and four times the axis of the segment on the other side is to the sum of the axis of the segment and twice the axis contained in the segment on the other side.

In Fig. 133, the meaning of which will further be clear without any explanation, let χ be the required centre of gravity of the segment

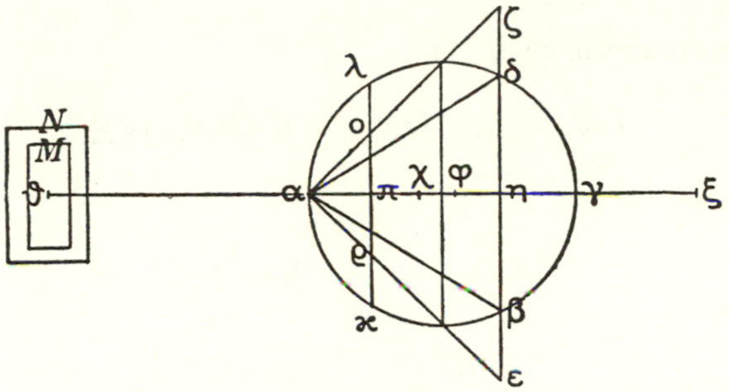

Fig. 133.

of the sphere $\alpha\beta\delta$, φ that of the cone $\alpha\varepsilon\zeta$ (so that $\eta\varphi = \frac{1}{4}\alpha\eta$). Just as in Prop. 6, it will be seen that the segment of the sphere and the cone $\alpha\varepsilon\zeta$ *suis locis* balance together the cone $\alpha\varepsilon\zeta$ in ϑ. If now the cone $\alpha\varepsilon\zeta$ in ϑ be replaced by the sum of two cylinders M and N (so that Cone $\alpha\varepsilon\zeta = M + N$), of which M balances in ϑ the cone $\alpha\varepsilon\zeta$ *suo loco*, N will balance in ϑ the segment *suo loco*.

We therefore have

$$(\text{Cylinder } N,\ \text{Segment}) = (\alpha\varphi,\ \alpha\vartheta)\,, \tag{1}$$

Further, if $\gamma\xi$ be equal to the radius of the sphere, we have

$$(\text{Segment},\ \text{Cone } \alpha\beta\delta) = (\eta\xi,\ \eta\gamma) \qquad \text{(Prop. 7)}$$

and

$$(\text{Cone } \alpha\beta\delta, \text{ Cone } \alpha\varepsilon\zeta) = [\mathbf{T}(\delta\eta), \mathbf{T}(\zeta\eta)]$$

$$= [\mathbf{O}(\alpha\eta, \gamma\eta), \mathbf{T}(\alpha\eta)] = (\gamma\eta, \alpha\eta) ,$$

whence *ex aequali*

$$(\text{Segment}, \text{ Cone } \alpha\varepsilon\zeta) = (\eta\xi, \alpha\eta) . \tag{2}$$

Further we have

$$(\text{Cone } \alpha\varepsilon\zeta, \text{ Cylinder } M) = (\alpha\vartheta, \alpha\varphi) = (\alpha\gamma, \alpha\varphi) ,$$

whence

$$(\text{Cone } \alpha\varepsilon\zeta, \text{ Cylinder } N) = (\alpha\gamma, \gamma\varphi) . \tag{3}$$

From (2) and (3) it follows that

$$(\text{Segment}, \text{ Cylinder } N) = [\mathbf{O}(\alpha\gamma, \eta\xi), \mathbf{O}(\alpha\eta, \gamma\varphi) ,$$

and by comparison with (1)

$$(\alpha\vartheta, \alpha\chi) = [\mathbf{O}(\alpha\gamma, \eta\xi), \mathbf{O}(\alpha\eta, \gamma\varphi)] ,$$

whence, since $\alpha\vartheta = \alpha\gamma$,

$$\mathbf{O}(\alpha\chi, \eta\xi) = \mathbf{O}(\alpha\eta, \gamma\varphi)$$

or

$$(\eta\xi, \alpha\eta) = (\gamma\varphi, \alpha\chi) .$$

In this

$$\eta\xi = \tfrac{1}{2}(\alpha\eta + \eta\gamma) + \eta\gamma = \tfrac{1}{2}\alpha\eta + \tfrac{3}{2}\eta\gamma$$

and

$$\gamma\varphi = \tfrac{1}{4}\alpha\eta + \eta\gamma .$$

Consequently we have

$$(\tfrac{1}{2}\alpha\eta + \tfrac{3}{2}\eta\gamma, \alpha\eta) = (\tfrac{1}{4}\alpha\eta + \eta\gamma, \alpha\chi) = (\tfrac{1}{4}\alpha\eta + \tfrac{1}{2}\eta\gamma, \eta\chi)$$

or

$$(\alpha\chi, \eta\chi) = (\tfrac{1}{4}\alpha\eta + \eta\gamma, \tfrac{1}{4}\alpha\eta + \tfrac{1}{2}\eta\gamma) = (\alpha\eta + 4\eta\gamma, \alpha\eta + 2\eta\gamma) .$$

In Prop. 10 it is merely stated that the theorem of Prop. 9 can be proved in the same manner for any segment of a spheroid; in Prop. 11 that the volume of any (right) segment of an amblyconoid can also be found (**C.S.** 25), and that the centre of gravity of such a segment divides the axis in such a way that the part towards the vertex is to the remaining part as the sum of three times the axis and eight times the axis produced (see III; 6.22) is to the sum of the axis and four times the axis produced.

8. In the last four propositions of the *Method* Archimedes deals with the volume of the so-called cylinder hoof, which is the first of the two solids referred to in the introductory letter.

Proposition 12.

If in a right prism with square bases a cylinder be inscribed which has its bases in the squares facing each other and the surface of which touches the four other faces, and a plane be drawn through the centre of the circle which is the base of the cylinder and a side of the opposite square, it is recognized by this method that the solid cut off by the plane thus drawn is one-sixth of the whole prism.

In the upper part of Fig. 134 let the rectangle $\alpha\beta$ be the intersection of the prism with the plane which bisects the side of the square upper base referred to in the proposition at right angles in β, γ the middle point of the base, consequently $\gamma\beta$ the intersection with the plane lateral face of the hoof, $\varepsilon\zeta$ the intersection with the plane which bisects the height $\gamma\delta$ at right angles, $\nu\upsilon$ the intersection with a variable plane at right angles to $\alpha\omega$. In the figure the intersection with the plane $\varepsilon\zeta$ is further shown. The variable plane $\nu\upsilon$ intersects the hoof in a rectangle whose sides are successively equal to $\tau\sigma$ and to $\nu\upsilon$, the cylinder in a rectangle whose sides are successively equal to $\tau\sigma$ and to $\omega\beta$.

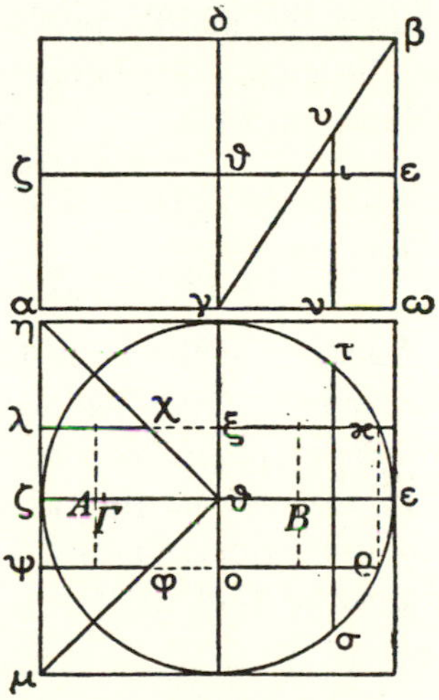

Fig. 134.

We now consider $\zeta\varepsilon$ as a balance with fulcrum ϑ.

The figure shows that

$$(\varepsilon\vartheta, \vartheta\iota) = (\omega\gamma, \gamma\nu) = (\omega\beta, \nu\upsilon) = [\mathbf{O}(\omega\beta, \tau\sigma), \ \mathbf{O}(\nu\upsilon, \tau\sigma)] .$$

Because $\vartheta\varepsilon = \vartheta\zeta$, this is equivalent to saying that the rectangular intersection of the plane $\nu\upsilon$ with the hoof, transferred to ζ, balances the rectangular intersection of the plane νv with the cylinder *suo loco*. We thus find as the result of Prop. 12 that the hoof in ζ balances half the cylinder $\gamma\omega\beta\delta$ *suo loco*.

The procedure is continued in Prop. 13. A prism with a triangular base is now considered (Fig. 134), of which $\mu\vartheta\eta$ is the section through the middle points of the sides and which is therefore one-fourth of the whole prism. A set of variable planes parallel to and equidistant on either side from the plane $\alpha\omega\beta$ intersects the triangular prism in rectangles, for each of which the sides are successively equal to $\lambda\chi$ $(=\psi\varphi)$ and to $\gamma\delta$, half the cylinder considered above in rectangles for each of which the sides are successively equal to $\varkappa\xi$ $(=o\varrho)$ and to $\gamma\delta$. The common centre of gravity of the first pair of rectangles is A, that of the second pair is B, A and B having been obtained by causing $\varepsilon\zeta$ to meet successively the line joining the middle points of $\lambda\chi$ and $\psi\varphi$ and the line joining the middle points of $\xi\varkappa$ and $o\varrho$.

Now we have

$$[\mathbf{O}(\lambda\chi, \delta\gamma),\ \mathbf{O}(\varkappa\xi, \delta\gamma)] = (\lambda\chi, \varkappa\xi) = (\pi\xi, \varkappa\xi) = (\varkappa\xi, \xi\Delta)$$

$$= (\varkappa\xi, \pi\xi + 2\vartheta\xi) = (\varkappa\xi, \lambda\chi + 2\chi\xi) = (2\vartheta B, 2\vartheta A) = (\vartheta B, \vartheta A)\,.$$

Apparently therefore the intersection of the prism and that of the cylinder, each *suo loco*, balance each other. Consequently the triangular prism *suo loco* also balances half the cylinder *suo loco*.

If we combine this with the result of Prop. 12, we see that the triangular prism $\vartheta\eta\mu$ *suo loco* balances the hoof in ε. Since the centre of gravity Γ of the prism lies on $\vartheta\zeta$ in such a way that

$$\vartheta\Gamma = \tfrac{2}{3}\vartheta\zeta\,, \qquad \text{(lemma 9)}$$

it follows from this that

$$\text{Hoof} = \tfrac{2}{3}.\text{Prism } \vartheta\eta\mu = \tfrac{1}{6}.\text{Prism } \alpha\omega\beta\,.$$

In Prop. 14 Archimedes deals once more with the volume of the cylinder hoof introduced in Prop. 12; he now avoids considerations of equilibrium, though he still makes use of the conception of a solid as being the sum of its parallel intersections. The proof therefore does not yet satisfy the requirement of exactness.

By way of explanation we give the reasoning in an analytical version. In Fig. 135 the square $\alpha\beta\gamma\delta$ represents the base of the prism; the plane lateral face of the hoof passes through the diameter $\varepsilon\eta$ of the base of the cylinder and through the point of the upper base of which ζ is the orthogonal projection on the lower base. Now it is intended to compare the hoof H with the triangular prism P, in which $\varepsilon\eta$, $\delta\gamma$, and the edge of the upper base of which $\delta\gamma$ is the orthogonal projection on the lower base are the vertical edges. This is done by comparing the areas of the intersections determined in both solids by a variable plane at right angles to $\varepsilon\eta$. If this plane intersects the lower base of the prism in $\mu\nu$ and the basic circle of the cylinder in ξ, it determines in the hoof and in the prism two similar triangles, in which $\mu\xi$ and $\mu\nu$ are homologous sides. From this it follows that

$$(\text{intersection of } P, \text{ intersection of } H) = [\mathbf{T}(\mu\nu),\ \mathbf{T}(\mu\xi)]\ .$$

Now construct on $\mu\nu$ a point λ, so that

$$[\mathbf{T}(\mu\nu),\ \mathbf{T}(\mu\xi)] = (\mu\nu,\ \mu\lambda)$$

or (III; 0.31)

$$(\mu\nu,\ \mu\xi) = (\mu\xi,\ \mu\lambda)$$

or

$$\mathbf{T}(\mu\xi) = \mathbf{O}(\mu\nu,\ \mu\lambda)\ .$$

If now σ be the projection of λ on $\vartheta\zeta$, this is equivalent to saying that

$$\mathbf{T}(\varkappa\zeta)-\mathbf{T}(\lambda\sigma) = \mathbf{O}(\varkappa\zeta,\ \varkappa\zeta-\zeta\sigma) = \mathbf{T}(\varkappa\zeta)-\mathbf{O}(\varkappa\zeta,\ \zeta\sigma)\ ,$$

whence

$$\mathbf{T}(\lambda\sigma) = \mathbf{O}(\varkappa\zeta,\ \zeta\sigma)\ .$$

The locus of λ therefore is an orthotome with vertex ζ, diameter $\zeta\vartheta$, and orthia $\varkappa\zeta$.

Now it is known that

$$(\text{intersection of } P, \text{ intersection of } H) = (\mu\nu,\ \mu\lambda)\ .$$

The volumes of the prism P and of the hoof H, being the sums of their parallel intersections, now are to each other as the sums of the parts $\mu\nu$ and $\mu\lambda$, *i.e.* as the areas of the rectangle $\varepsilon\delta\gamma\eta$ and the segment of the orthotome $\varepsilon\zeta\eta$. According to the main theorem of *Quadrature of the Parabola* this ratio is 3:2. From this it follows that the volume of the hoof H is two-thirds of that of the prism P, and consequently one-sixth of that of the regular quadrangular prism circumscribed about the cylinder.

In Prop. 15 the problem of the volume of the cylinder hoof is posited once more. Here, however, an exact geometrical discussion is given, in which the conception of a volume being the sum of areas of parallel intersections is abandoned and in which the method of the indirect limiting process is applied.

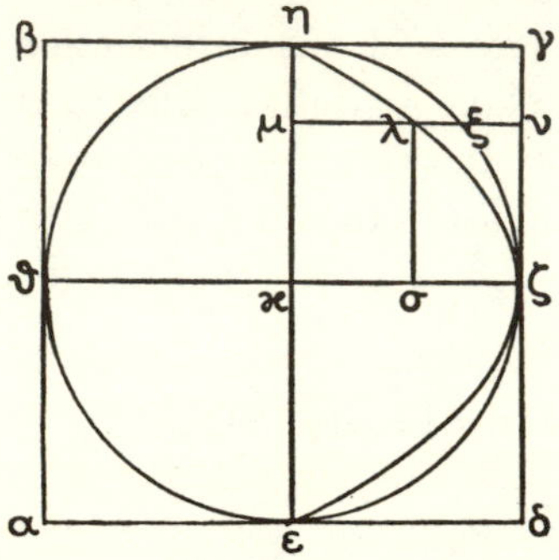

Fig. 135.

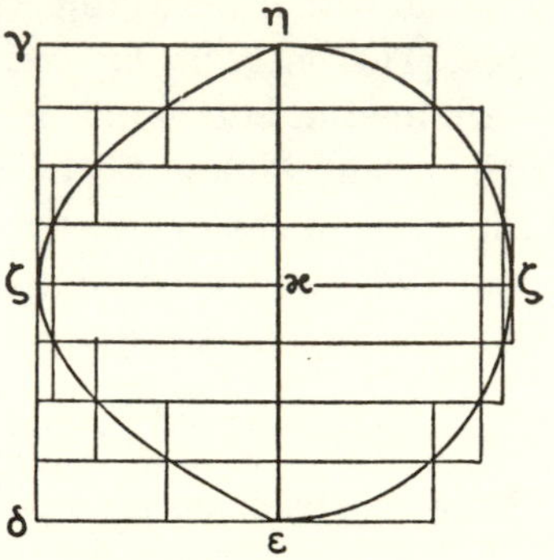

Fig. 136

Use is made of the difference form of the compression method (III; 8.21): in and about the hoof are constructed figures consisting of triangular prisms placed one behind the other, and it is first proved that the difference between the volumes of the circumscribed solid C_n and of the inscribed solid I_n can, by the choice of the number of frusta, be made less than any assigned volume.

In Fig. 136 the base of the hoof is shown once more. On the diameter $\varepsilon\eta$ dichotomy is applied, and through the points of division planes are drawn at right angles to $\varepsilon\eta$, each of which intersects the hoof in a triangle. The figure further shows for each of the inscribed and circumscribed triangular prisms the lateral face with which it is supported on the lower base of the cylinder. It is found that the difference $C_n - I_n$ is equal to the sum of the volumes of the two circumscribed prisms adjacent to $\varkappa\zeta$, which sum, by the choice of n, can be made less than any assigned magnitude. It is further evident that the planes drawn divide the prism P into equal triangular prisms p_n. Now suppose the volume of the hoof H not to be equal to two-thirds of the volume of the prism P. Then $H \gtrless \frac{2}{3}P$.

Case I. Suppose $H > \frac{2}{3}P$. Continue the dichotomy of $\varepsilon\eta$ until

$$C_n - I_n < H - \tfrac{2}{3}P ,$$

then, because $C_n > H$, we have

$$I_n > \tfrac{2}{3}P \quad \text{or} \quad P < \tfrac{3}{2}I_n \, .$$

Now it is proved in Prop. 14 for a cutting plane at right angles to $\varepsilon\eta$ (Fig. 135) that

$$(\mu\nu, \mu\lambda) = (\text{intersection of } P, \text{intersection of } H)$$
$$= (p_n, \text{inscribed prism of } H) \, . \qquad (1)$$

Now the straight lines which have been drawn through the points of division of $\varepsilon\eta$ at right angles to $\varepsilon\eta$ also determine on the orthotome O introduced in Prop. 14 points which are vertices, in the manner shown in Fig. 136 (where for convenience the curve has been drawn to the left of $\varepsilon\eta$), of circumscribed and inscribed rectangles to the orthotome. We call the figure of the circumscribed rectangles c_n, that of the inscribed rectangles i_n. Further the lines drawn divide the rectangle $\Phi(\varepsilon\delta\gamma\eta)$ into n equal rectangles φ_n. We now have

$$(\mu\nu, \mu\lambda) = (\varphi_n, \text{inscribed rectangle of } O) \, ,$$

whence by comparison with (1):

$$(p_n, \text{inscribed prism of } H) = (\varphi_n, \text{inscribed rectangle of } O) \, ,$$

whence (III; 7.21):

$$(P, I_n) = (\Phi, i_n) \, . \qquad (2)$$

Now it is known that

$$\Phi = \tfrac{3}{2}.\text{segment of the orthotome } \varepsilon\zeta\eta > \tfrac{3}{2}i_n \qquad (3)$$

However, we have

$P < \tfrac{3}{2}I_n$, so that it follows from (2) that

$\Phi < \tfrac{3}{2}i_n$,

which is contrary to (3).

In an analogous way it is found that also the supposition $H < \tfrac{2}{3}P$ results in a contradiction.

Here ends the preserved part of the *Method*, the manuscript of which already shows considerable gaps, which can, however, be easily filled up. The derivation of the theorem, mentioned in the introductory letter, on the solid determined by the intersection of two cylinders inscribed in a cube is missing. It appears not to be difficult to reconstruct it hypothetically, but we do not wish to go into this here[1]).

CHAPTER XI

QUADRATURE OF THE PARABOLA

1. The theorem on the volume of the cylinder hoof already enabled us to see how Archimedes arrived, from a surmise gained by mechanical means with the aid of the method of indivisibles, at a mathematical proof which satisfied all his requirements of exactness.

An even more beautiful example of such a logical confirmation of an intuitively gained insight is furnished by the theorem on the area of any segment of an orthotome, which formed the subject of the first proposition of the *Method*. In fact, it was to the mathematical proof of this theorem that Archimedes devoted a separate treatise, the *Quadrature of the Parabola*, in which he derives the already known result at great length in two different ways, namely, first with the aid of mechanical considerations and then purely geometrically. As we already observed above, this twofold character of the treatise may be deemed to furnish an argument in favour of the view that when Archimedes denies the demonstrative force of the mechanical method which he explains to Eratosthenes, he does not do so on account of its mechanical nature, but exclusively because it makes use of the method of indivisibles.

2. The treatise *Quadrature of the Parabola* opens with five propositions on properties of the orthotome, which we already incorporated in Chapter III. These are followed by eight propositions in which equalitities and inequalities about plane figures suspended on a balance are enunciated.

---

[1]) Such a reconstruction is to be found, *inter alia*, in T. L. Heath, *The Method of Archimedes* (Cambridge 1912), pp. 48 *et seq.*, and in E. Rufini, *Il "Metodo" di Archimede e le origini dell'analisi infinitesimale nell'Antichità* (*Per la Storia e la Filosofia delle Matematiche* No 4, Roma 1926), pp. 179–186.

In Props 6 and 7 (Fig. 137) there is suspended from one end A of a balance $A\Gamma$ supported in its middle point B a magnitude Z which balances a triangle $\Gamma H\Delta$, the side $H\Delta$ of which lies in the vertical of B (in Prop. 6, H is moreover placed in B). If Θ be the centre of gravity of the triangle, and if the vertical of Θ meet the straight line $B\Gamma$ in Π, then because $B\Pi = \frac{1}{3}B\Gamma$ we have

$$\triangle\, \Gamma H\Delta = 3Z\,.$$

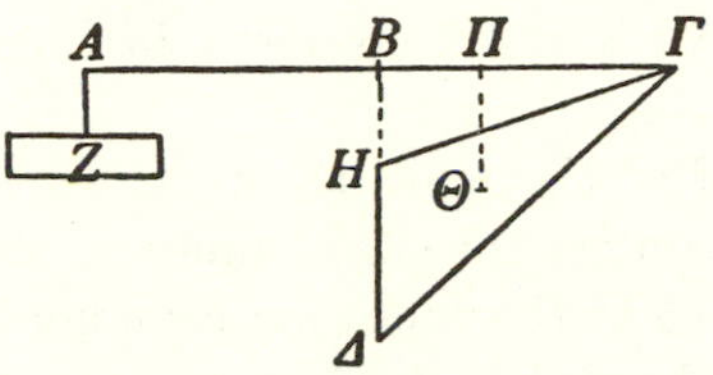

Fig. 137.

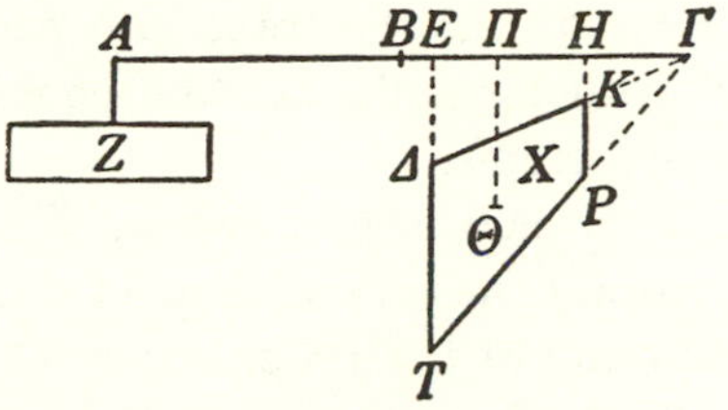

Fig. 138.

Archimedes here deviates from the method followed in *On the Equilibrium of Planes*, where the suspended magnitudes are attached to the balance in their centres of gravity; the centres of gravity are now at a level below that of the balance itself. It is not quite in order that he nevertheless applies the propositions from *On the Equilibrium of Planes*.

The propositions 8–12 are all particular cases of Prop. 13, so that we shall discuss the latter only (Fig. 138). In this case there is suspended from $B\Gamma$ a trapezium X ($TPK\Delta$), the vertical parallel sides of which, *viz.* ΔT and KP, meet $B\Gamma$ successively in E and H, while the other sides ΔK and TP converge towards Γ. The vertical through the centre of gravity Θ meets $B\Gamma$ in Π. Z and X balance. It results from this that

$$(AB, B\Pi) = (X, Z)\,. \qquad (1)$$

We now consider the magnitudes M and N, which would have to hang from A instead of Z in order to balance X, if X were so attached to $B\Gamma$ that the vertical of Θ passed successively through E and through H. M and N are determined by

$$(AB, BE) = (X, M)$$

$$(AB, BH) = (X, N)\,.$$

Since $BE < B\Pi < BH$, we find, by comparing these proportions with (1), the result

$$M < Z < N .$$

This is equivalent to saying that the magnitude at A which balances X grows smaller if X is displaced towards the fulcrum B, and larger if X is displaced in the direction away from the fulcrum B.

In Prop. 12, Δ fell in E and K in H, in Prop. 11 E fell in B, in Prop. 10 moreover Δ fell in E; in Props 8 and 9 P, K, H coincided in Γ, so that the trapezium passed into a triangle; the magnitude N (which would now be equal to X) is not considered in these propositions.

This part is followed by the proof of the main theorem, which occupies the propositions 14–17. For convenience we summarize the contents of the propositions 14 and 16 in the following argument:

3. A segment of an orthotome, a chord $B\Gamma$ of which is at right angles to the diameter (Fig. 139), is attached along the chord on the balance $A\Gamma$, which is supported in its middle point B.

The tangent at Γ meets the line, drawn through B parallel to the diameter, in Δ. $B\Delta$ is divided into equal segments BE, EH, etc., the number of which (n) is yet to be defined. The points of division are joined to Γ, and through the points in which these lines meet the curve are drawn straight lines parallel to the diameter, which meet $B\Gamma$ in the points $M, N \ldots$, and $\Gamma\Delta$ in the points $\alpha, \beta, \gamma \ldots$. The segments $BM, MN \ldots$ are now also equal (*vide* the

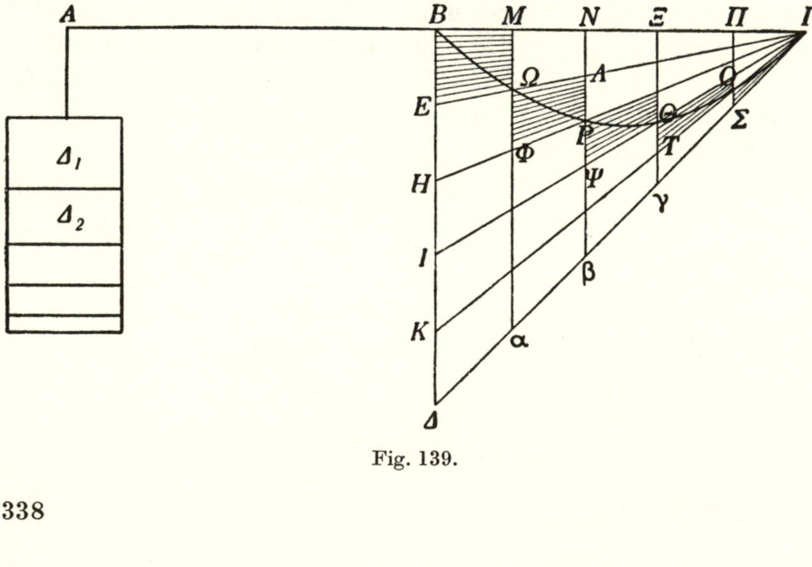

Fig. 139.

Note). In the figure several trapezia are thus formed, among which we distinguish three different types:

1. the trapezia $B\alpha$, $M\beta$, ..., which we call partial trapezia of $\triangle B\Gamma\Delta$; to be denoted by $d_1, d_2, \ldots d_n$.

2. the trapezia EM, ΦN, ..., which we call circumscribed trapezia to the segment; to be denoted by $c_1, c_2, \ldots c_n$; together they form the figure C_n.

3. the trapezia ΩN, $P\Xi$, ..., which we call inscribed trapezia to the segment; to be denoted by $i_2, i_3, \ldots i_n$; together they form the figure I_n. It is obvious that c_n and i_n are triangles with vertex Γ.

The derivation is given with the aid of the difference form of the compression method (III; 8.21); for this it is necessary that the difference $C_n - I_n$ can, by the choice of n, be made less than any assigned magnitude. Now $C_n - I_n$ is made up of the trapezia EM, $\Phi\Lambda$, ... With regard to these it is true that

$$\Phi\Lambda = \Omega N, \text{ etc.}$$

In fact, from $BE = EH$ it follows that $M\Omega = \Omega\Phi$ and $N\Lambda = \Lambda P$, etc.

The difference $C_n - I_n$ therefore is equal to $\triangle \Gamma BE$, so that it is indeed possible, by the choice of n, to make it less than any assigned magnitude.

It has now to be proved that the segment $B\Theta\Gamma$ is equal to $\frac{1}{3}\cdot\triangle B\Delta\Gamma$. This is done according to the principle of the compression method by proving the inequality

$$I_n < \tfrac{1}{3}\cdot\triangle B\Delta\Gamma < C_n\,. \tag{1}$$

Once this has been proved, what was required to be proved follows in the usual way, because the segment is then comprehended with $\frac{1}{3}\cdot\triangle B\Delta\Gamma$ between boundaries the difference between which can be made less than any assigned magnitude.

In order to prove the inequality (1), relations of equilibrium between the various trapezia distinguished above are derived (in Prop. 14). By the theorem **Q.P.** 5 (III; 2.7) we have

$$(\alpha\Omega, \Omega M) = (\Gamma M, BM)\,,$$

whence

$$(\alpha M, \Omega M) = (B\Gamma, BM) = (AB, BM)\,.$$

Similarly we have

$$(\beta N, PN) = (AB, BN)\,.$$

Now

$$(d_1, c_1) = (\Delta B + \alpha M, EB + \Omega M).$$

Since, however,

$$(\Delta B, EB) = (\alpha M, \Omega M) = (\Delta B + \alpha M, EB + \Omega M)$$

$$(AB, AH) = (X, N),$$

we also have

$$(d_1, c_1) = (\Delta B, EB) = (\alpha M, \Omega M).$$

Similarly it is seen that

$$(d_2, c_2) = (\alpha M, \Phi M) = (\beta N, PN), \text{ etc.}$$

However, we also have

$$(d_2, i_2) = (\alpha M, \Omega M) = (\beta N, \Lambda N), \text{ etc.}$$

From this it follows that

$$(d_1, c_1) = (d_2, i_2) = (\Delta B, EB).$$

Since it follows from

$$EB = EH, \text{ etc.}$$

that

$$BM = MN, \text{ etc.},$$

we have

$$(\Delta B, EB) = (\Gamma B, MB),$$

whence

$$(d_1, c_1) = (\Gamma B, MB) = (AB, BM).$$

Similarly

$$(d_2, c_2) = (d_3, i_3) = (AB, BN).$$

From this it appears that

c_1 at A balances d_1 at M

c_2 at A balances d_2 at N, etc.

and also

i_2 at A balances d_2 at M

i_3 at A balances d_3 at N, etc.

We now conceive to be suspended at A the magnitudes $\Delta_1, \Delta_2 \ldots$ which successively balance $d_1, d_2, \ldots$ *suis locis*. By the propositions 8, 10, and 12 we now have

$$c_k > \Delta_k \ (k = 1 \ldots n)$$
$$i_k < \Delta_k \ (k = 2 \ldots n) \, .$$

From this it follows that

$$i_2 + \ldots + i_n < \Delta_1 + \ldots + \Delta_n < c_1 + \ldots + c_n$$

or $$I_n < \Delta_1 + \ldots \Delta_n < C_n \, .$$

By Prop. 6, however, we have

$$3(\Delta_1 + \ldots + \Delta_n) = \triangle \, B\Delta\Gamma ,$$

from which follows the desired inequality

$$I_n < \tfrac{1}{3} . \triangle \, B\Delta\Gamma < C_n \, . \qquad (1)$$

The argument further proceeds automatically (in Prop. 16).

If the segment $B\Theta\Gamma$ is not equal to $\frac{1}{3} . \triangle \, B\Delta\Gamma$, it is either greater or smaller.

Case I. Suppose

$$\text{segment } B\Theta\Gamma > \tfrac{1}{3} . \triangle \, B\Delta\Gamma \, .$$

Now determine n such that

$$C_n - I_n < \triangle \, BE\Gamma < \text{segment } B\Theta\Gamma - \tfrac{1}{3} . \triangle \, B\Delta\Gamma ,$$

then *a fortiori*

$$\text{segment } B\Theta\Gamma - I_n < \text{segment } B\Theta\Gamma - \tfrac{1}{3} . \triangle \, B\Delta\Gamma ,$$

whence

$$I_n > \tfrac{1}{3} . \triangle \, B\Delta\Gamma ,$$

which is contrary to (1).

Case II. Suppose

$$\text{segment } B\Theta\Gamma < \tfrac{1}{3} . \triangle \, B\Delta\Gamma \, .$$

Now determine n such that

$$C_n - I_n < \triangle \, BE\Gamma < \tfrac{1}{3} . \triangle \, B\Delta\Gamma - \text{segment } B\Theta\Gamma ,$$

then *a fortiori*

$$C_n - \text{segment } B\Theta\Gamma < \tfrac{1}{3} . \triangle \, B\Delta\Gamma - \text{segment } B\Theta\Gamma ,$$

whence

$$C_n < \tfrac{1}{3} . \triangle \, B\Delta\Gamma ,$$

which is contrary to (1).

Note: It appears to be essential to the proof that from the equality of the segments BE, etc. of $B\Delta$ should follow the equality of the segments BM, MN, etc. of $B\Gamma$. Now Archimedes starts by assuming the latter equality in Prop. 14, in which he derives the inequality (1), and he then applies Prop. 14 in Prop. 16, in which not $B\Gamma$, but $B\Delta$ is divided into a number of equal segments, but he does not prove that by this the condition of Prop. 14 is also satisfied. This slight gap is naturally filled by observing that by **Q.P.** 5 (III; 2.7)

$$(\Gamma M, BM) = (\alpha\Omega, \Omega M) = (\Delta E, EB),$$

so that BM is found to be the nth part of $B\Gamma$, if BE is the nth part of $B\Delta$. Proceeding in this way, we find: $BM = MN$, etc.

In Prop. 15 the inequality (1) is proved for the case that the chord of the segment is not at right angles to the diameter. It is now attached to the balance in such a way that the diameter is vertical, one end of the chord lies at Γ, and the other in the vertical of B. The argument is scarcely altered by it. The only difference is that we have to use Props 7, 9, 11, and 13 instead of 6, 8, 10, and 12.

Finally in Prop. 17 the theorem is enunciated in the form in which Archimedes is accustomed to apply it:

Proposition 17.

This having been proved, it is manifest that the area of any segment which is comprehended by a straight line and an orthotome is greater by one-third than the triangle which has the same base as the segment and equal height.

Here (Fig. 140) "height" is meant to denote the distance from the chord $B\Gamma$ of the point Θ (vertex), where the tangent is parallel to the chord. Apparently we have

$$B\Delta = 2.ZE = 4.Z\Theta,$$

whence

$$\text{segment } B\Theta\Gamma = \tfrac{1}{3} \triangle B\Delta\Gamma = \tfrac{4}{3} \triangle B\Theta\Gamma.$$

Prop. 17 is followed by definitions of the terms base, height, and vertex of the segment, which, however, have already been used in Prop. 17. The height is now defined as the greatest distance of a point of the branch $B\Gamma$ of the curve from the base, the vertex as the point which has this greatest distance from the base. In Prop.

18 it is proved that the vertex is the point where the straight line through the middle point of the chord, parallel to the diameter, meets the curve.

4. With Prop. 18 the second part of the treatise has already started, in which Archimedes proves the theorem on the area of any segment of an orthotome once more by purely geometrical

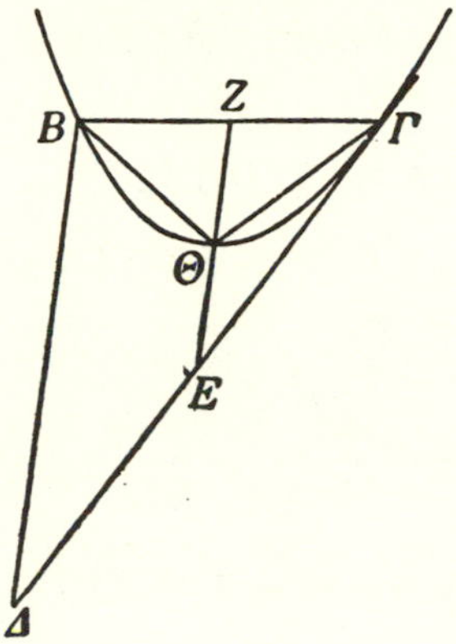

Fig. 140.

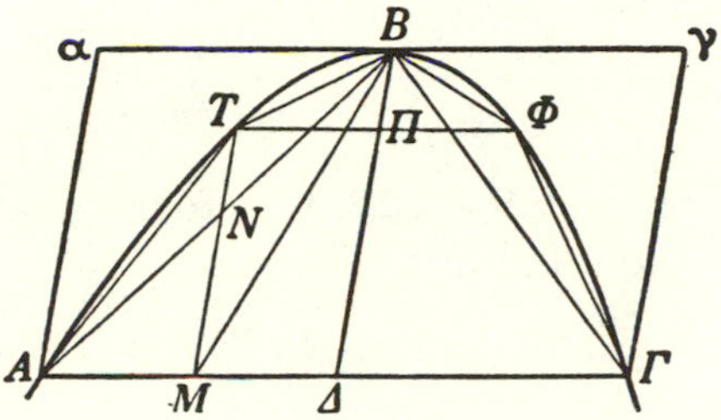

Fig. 141.

means. He here applies that form of indirect limiting process which we have referred to in the general discussion as approximation method (III; 8.30). He therefore inscribes in the segment a figure, the area of which is more than the half of that of the segment, treats the remaining part in the same way, and thus, by the lemma of Euclid, approximates the area to be found as accurately as is required.

In the following we summarize the contents of Props 19–24 (Fig. 141). Let Σ be the segment $AB\Gamma$ of the orthotome with base $A\Gamma$ and vertex B. Construct the parallelogram $A\Gamma\gamma\alpha$ with side $A\Gamma$, the opposite side of which passes through B, then it will appear in Prop. 20 that

$$\triangle\, AB\Gamma = \tfrac{1}{2}\, A\Gamma\gamma\alpha > \tfrac{1}{2}\Sigma\,.$$

Since we can deal with each of the segments AB, $B\Gamma$ in the same way, the lemma of Euclid applies (Prop. 20; Corollary). Now consider the triangle ATB inscribed in the segment AB(cf. III; 2.4). The vertex T lies on the straight line drawn through the middle point N of AB, parallel to the diameter, *i.e.* parallel to $B\Delta$; this straight line therefore will also pass through the middle point M of $A\Delta$. Join B and M. We will now compare the triangles $AB\Gamma$ and ABT.

To this end it is first proved (Prop. 19) that

$$B\Delta = \tfrac{4}{3}TM .$$

This follows from the symptom of the orthotome; for if the straight line through T parallel to $A\Gamma$ meet the straight line $B\Delta$ in Π, we have

$$(B\Delta, B\Pi) = [\mathbf{T}(A\Delta), \mathbf{T}(T\Pi)] = [\mathbf{T}(A\Delta), \mathbf{T}(M\Delta)] ,$$

whence

$$B\Delta = 4.B\Pi ,$$

therefore

$$TM = \Pi\Delta = \tfrac{3}{4}B\Delta .$$

Since $NM = \frac{1}{2}B\Delta$, it is found that $NM = 2.TN$.

From this it follows that

$$\triangle ABT = 2.\triangle BNT = \triangle BNM = \tfrac{1}{2}\triangle ABM = \tfrac{1}{8}\triangle AB\Gamma.$$

(Prop. 21)

If we now represent the area of $\triangle AB\Gamma$ by Z, the sum of the areas of the two triangles inscribed in the segments AB and $B\Gamma$ is: $Z_1 = \frac{1}{4}Z$; proceeding in this way, we find for the sum of the areas of the triangles which are inscribed in the remaining four segments: $Z_2 = \frac{1}{4}Z_1$, etc. Archimedes expresses this by introducing magnitudes H, Θ, I, etc., so that

$$H = \tfrac{1}{4}Z,\ \Theta = \tfrac{1}{4}H,\ I = \tfrac{1}{4}\Theta .$$

By this it has been proved (Prop. 22) that

$$\Sigma > Z + H + \Theta + I ,$$

or in general

$$\Sigma > Z + Z_1 + \ldots + Z_{n-1} .$$

In Prop. 23 a lemma on the sum of a geometrical progression with the proportion $\frac{1}{4}$ is then derived. By this lemma (discussed in III; 7.60) we have

$$Z + H + \Theta + I + \tfrac{1}{3}I = \tfrac{4}{3}Z \quad (1)$$

or in general

$$Z + Z_1 + \ldots + Z_{n-1} + \tfrac{1}{3}Z_{n-1} = \tfrac{4}{3}Z . \quad (2)$$

This is followed, in Prop. 24, by the proof of the main theorem. Let K be equal to $\frac{4}{3}Z$, then it has to be proved that $\Sigma = K$. Suppose this is not true, then either $\Sigma > K$ or $\Sigma < K$.

Case I. Suppose $\Sigma > K$. Now continue inscribing triangles in each of the segments obtained until the sum I_n of all the inscribed triangles satisfies the relation

$$\Sigma - I_n < \Sigma - K \qquad \text{(lemma of Euclid III; 0.5)},$$

whence

$$I_n > K .$$

Archimedes assumes this to be the case for $n = 4$, and then writes

$$Z + H + \Phi + I > K ,$$

which is contrary to (1).

More in general it is found that

$$Z + Z_1 + \ldots + Z_{n-1} > K , \qquad \text{which is contrary to (2).}$$

Case II. Suppose $\Sigma < K$. We cannot now, as with the compression method, argue on entirely analogous lines to Case I, because there is no sum C_n of circumscribed figures. Archimedes now continues inscribing triangles until the sum of the ultimately obtained areas satisfies the relation

$$I < K - \Sigma . \tag{3}$$

We further have

$$K - (Z + H + \Theta + I) = \tfrac{1}{3}I < I ,$$

whence

$$I > K - I_n > K - \Sigma , \qquad \text{which is contrary to (3).}$$

In general we get

$$Z_{n-1} < K - \Sigma . \tag{4}$$

Now since

$$Z + Z_1 + \ldots Z_{n-1} + \tfrac{1}{3}Z_{n-1} = K ,$$

we also have

$$K - (Z + Z_1 + \ldots + Z_{n-1}) = \tfrac{1}{3}Z_{n-1} < Z_{n-1} ,$$

whence

$$Z_{n-1} > K - I_n$$

and *a fortiori*

$$Z_{n-1} > K - \Sigma , \qquad \text{which is contrary to (4).}$$

CHAPTER XII

ON THE EQUILIBRIUM OF PLANES

Book II

The main object of this book is the determination of the centre of gravity of any segment of an orthotome, which object is attained in Prop. 8.

This determination is a combined application of the principles of the barycentric theory and the theory of the quadrature of the orthotome, dealt with in Book I; it is further based on various properties of this curve, which we have already discussed in Chapter III.

At first sight it seems rather strange to find the argument starting with a proposition which is nothing but a particular case of the lever principle proved in the general sense in Props 6 and 7 of Book I.

Proposition 1.

If two surfaces, comprehended by a straight line and an orthotome, which we can apply to a given straight line, do not have the same centre of gravity, the centre of gravity of the magnitude made up of these two will lie on the straight line joining their centres of gravity, while dividing the said straight line in such a way that the parts of it in inverted order are in the same ratio as the surfaces.

In this proposition the parenthetic clause "which we can apply to a given straight line" is not to be understood as a restrictive condition imposed on the segments used, but as a reminder of the possibility of the quadrature of such a segment, proved in *Quadrature of the Parabola.* It should therefore be read in the sense of: "which, as we know, can be transformed into a rectangle".

Now it appears that in this first proposition Archimedes furnishes a slightly simpler proof of the lever principle than in Book I.

In fact, in Fig. 142 let AB and $\Gamma\Delta$ be the two segments which are attached in their centres of gravity E, Z to a lever with fulcrum Θ. Supposition:

$$(AB, \Gamma\Delta) = (\Theta Z, \Theta E) .$$

It has to be proved that Θ is the common centre of gravity. As in Book I, Prop. 6, construct

$$ZH = ZK = \Theta E \quad \text{and} \quad E\Lambda = \Theta Z .$$

In view of this we have again:

$$\Theta Z = EH, \ \Lambda\Theta = K\Theta$$

and $$(\Lambda H, HK) = (\Theta Z, \Theta E) = (AB, \Gamma\Delta) . \qquad (1)$$

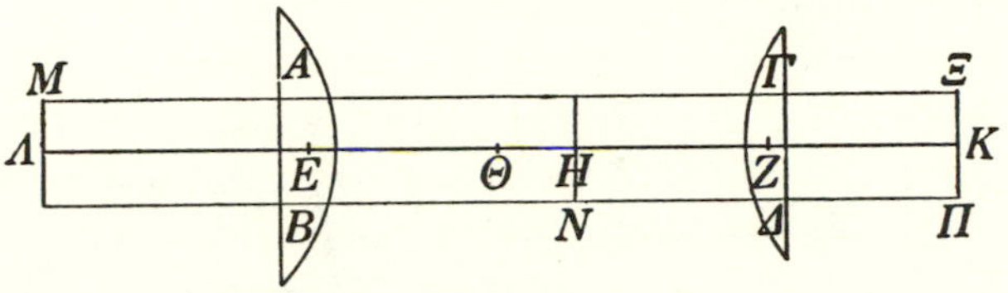

Fig. 142.

Now apply the segment AB (which according to **Q.P.** is four-thirds of its inscribed triangle) to the line segment ΛH and place the rectangle MN thus obtained in such a way that ΛH joins the middle points of two parallel sides. E is then the centre of gravity of MN. AB has thus been replaced by an equal magnitude in the same place (in the sense of postulate VI). Complete the rectangle MN with the rectangle $N\Xi$, in which HK joins the middle points of two parallel sides and which therefore has Z for centre of gravity.

Because of
$$(\Lambda H, HK) = (MN, N\Xi) ,$$

we then have in relation with (1)

$$N\Xi = \Gamma\Delta .$$

$\Gamma\Delta$, too, has therefore been replaced by an equal magnitude in the same place. Because $\Lambda\Theta = K\Theta$, the centre of gravity of the rectangle $M\Pi$ is Θ. There is thus equilibrium, and this equilibrium is maintained, according to postulate VI, if the rectangles MN and $N\Xi$ are replaced again by the segments AB and $\Gamma\Delta$.

The second proposition, in which the conception of a figure inscribed in the recognized manner in a segment of an orthotome is introduced, has been discussed in Chapter III; 2.5. We shall henceforth denote the figure in question by Ω.

Proposition 3.

If in each of two similar segments comprehended by a straight line and an orthotome a rectilinear figure be inscribed in the recognized manner, and the two inscribed rectilinear figures have an equal number of sides, the centres of gravity of the rectilinear figures divide the diameters of the segments in the same ratio.

With a view to the application which Archimedes is to make in Prop. 8 of Prop. 7 based on Prop. 3 it is important that for this theorem to be true the condition of similarity of the segments (III; 2.81) is not necessary.

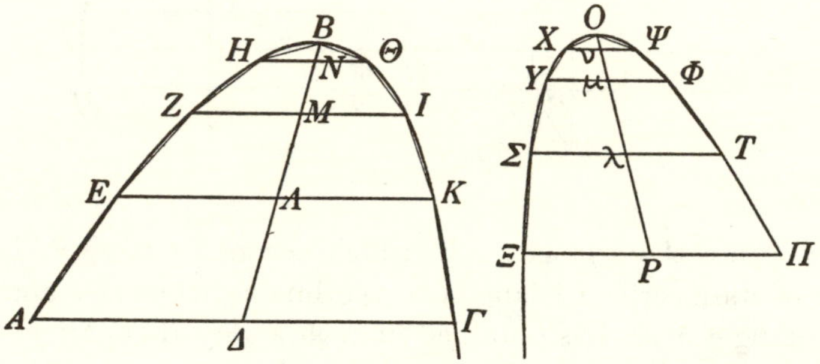

Fig. 143.

In the proof (Fig. 143) the figures $AE\ldots\Gamma$ and $\Xi\Sigma\ldots\Pi$ are considered, which are inscribed in the recognized manner. Now it is known (III; 2.5) that the successive parallel sides of the trapezia obtained ($H\Theta$, ZI, etc. and likewise $X\Psi$, $Y\Phi$, etc.) are to each other in the ratio of the successive natural numbers, while the heights of the successive trapezia are in the ratio of the successive odd numbers.

The areas of two corresponding trapezia (or of the two triangles $BH\Theta$ and $OX\Psi$ respectively) therefore have a constant ratio, while the centres of gravity divide the parts of the diameters inside the trapezia in the same proportion. From this it follows that also the centres of gravity of the whole figures $AE\ldots\Gamma$ and $\Xi\Sigma\ldots\Pi$ are similarly situated on the diameters.

Proposition 4.

The centre of gravity of any segment comprehended by a straight line and an orthotome lies on the diameter of the segment.

Suppose (Fig. 144) the centre of gravity E of the segment Σ ($AB\Gamma$) not to lie on the diameter $B\Delta$ and the straight line through E parallel to the diameter to meet the chord $A\Gamma$ in Z. Inscribe in the segment $AB\Gamma$ the triangle $AB\Gamma$, and now take a magnitude K such that

$$(\Gamma Z, \Delta Z) = (\triangle\ AB\Gamma, K)\ .$$

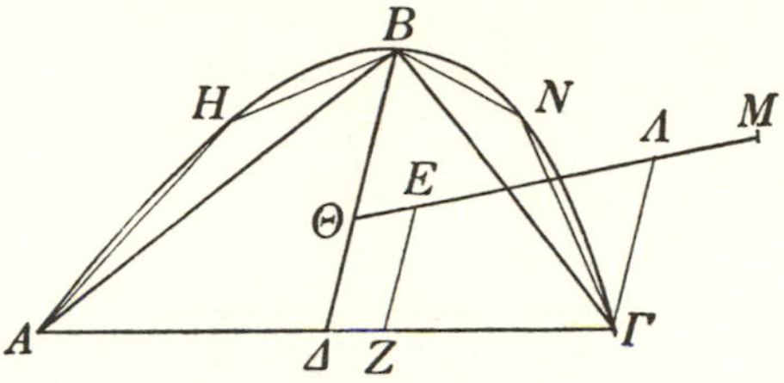

Fig. 144.

Now inscribe in the segment Σ in the recognized manner the figure Ω, such that the sum of the remaining segments $\Sigma - \Omega$ is less than K. Let Θ be the centre of gravity of Ω. Let ΘE be met in Λ by the straight line through Γ parallel to the diameter of the segment. Since $\Omega > \triangle\ AB\Gamma$ and $\Sigma - \Omega < K$, we have

$$(\Omega, \Sigma - \Omega) > (\triangle\ AB\Gamma, K) = (\Gamma Z, \Delta Z) = (\Lambda E, E\Theta)\ .$$

The centre of gravity M of $\Sigma - \Omega$ is now found from the centres of gravity E of Σ and Θ of Ω by means of the relation (Book I, Prop. 8):

$$(ME, \Theta E) = (\Omega, \Sigma - \Omega)\ .$$

Thence

$$(ME, \Theta E) > (\Lambda E, \Theta E), \text{ so that } ME > \Lambda E\ .$$

M is therefore on the opposite side of $\Gamma\Lambda$ to E, which is impossible because the whole segment lies on the same side of $\Gamma\Lambda$ with E.

Proposition 5.

If in a segment comprehended by a straight line and an orthotome a rectilinear figure be inscribed in the recognized manner, the centre of gravity of the whole segment is nearer to the vertex of the segment than the centre of gravity of the inscribed rectilinear figure is.

In Fig. 145 let $AB\Gamma$ be the given segment, in which $\triangle\ AB\Gamma$ has been inscribed in the recognized manner. Let E be the centre of gravity of $\triangle AB\Gamma$, Z the middle point of AB, H that of ΓB. Through Z and K have been drawn, parallel to $B\Delta$, the diameters ZK and $H\Lambda$ of the segments AB and ΓB; on these lie the centres of gravity Θ and I of those segments.

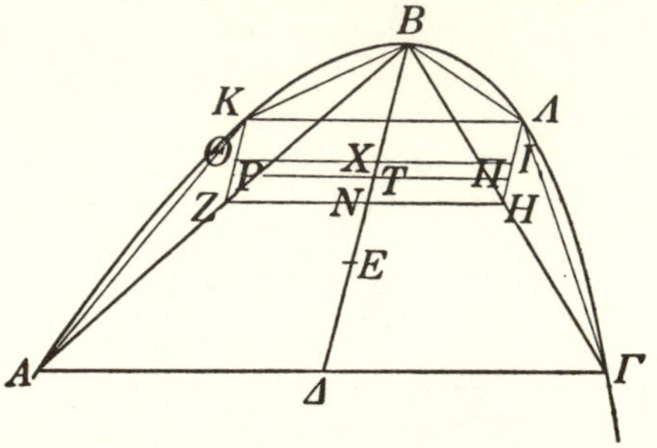

Fig. 145.

Archimedes now observes that ΘIHZ is a parallelogram; this is true, but it can be recognized only after Prop. 7. His conclusion that the common centre of gravity X of the segments AB and ΓB lies between N and B, and consequently that of the whole segment $AB\Gamma$ (which is made up of the segments AB and $B\Gamma$ with centre of gravity X and of the triangle $AB\Gamma$ with centre of gravity E) between B and E, however, remains true all the same, if only it is known (which is guaranteed by Prop. 4) that Θ lies on KZ and I on ΛH. The theorem has thus been proved for $\triangle AB\Gamma$ as figure Ω.

Now let P and Π successively be the centres of gravity of $\triangle AKB$ and $\triangle \Gamma\Lambda B$; we then know that Θ lies between K and P, I between Λ and Π. Let further T be the common centre of gravity of the triangles AKB and $\Gamma\Lambda B$, then T lies on $B\Delta$ between X and N.

The centre of gravity Ψ[1]) of segment $AB\Gamma$ now lies on XE in such a way that

$$(X\Psi, E\Psi) = (\triangle AB\Gamma, \text{ segment } AB + \text{segment } B\Gamma). \qquad (1)$$

The centre of gravity Φ of the figure Ω ($AKB\Lambda\Gamma$) lies on TE in such a way that

$$(T\Phi, E\Phi) = (\triangle AB\Gamma, \triangle AKB + \triangle \Gamma\Lambda B). \qquad (2)$$

[1]) This point Ψ and the point Φ to be presently introduced have not been given in the figure, for the sake of clarity.

A comparison of (1) and (2) shows that

$$(X\Psi, E\Psi) < (T\Phi, E\Phi)$$

or

$$(XE, E\Psi) < (TE, E\Phi) .$$

In this, $XE > TE$, therefore $E\Psi$ must be greater than $E\Phi$, so that Ψ is nearer to B than Φ is.

Continuing the argument (by now inscribing in each of the segments AB and ΓB a figure Ω with five sides, for which the theorem has just been proved), it is seen that the theorem is true for any figure Ω.

Nowadays we should furnish the proof by mathematical induction: imagine the theorem to be proved for a figure Ω with $(2n+1)$ sides; now inscribe such a figure Ω in each of the segments AB and ΓB; let Θ and I represent the centres of gravity of the segments again, P and Π those of the inscribed figures Ω, then the whole above argument applies. The theorem is therefore true for a figure Ω with $(2n+3)$ sides. It is true for $n=1$; it is therefore true for any value of n.

Proposition 6.

Given a segment comprehended by a straight line and an orthotome, it is possible to inscribe in the recognized manner in the segment a rectilinear figure such that the straight line between the centres of gravity of the segment and of the inscribed rectilinear figure is less than any assigned length.

In Fig. 146 let Σ be the segment $(AB\Gamma)$ with centre of gravity Θ and let Z be the assigned length. It is required to find a figure Ω such that, if E be its centre of gravity.

$$\Theta E < Z .$$

Imagine a surface X such that

$$(\triangle\ AB\Gamma, X) = (B\Theta, Z) .$$

Construct a figure Ω with a number of sides such that

$$\Sigma - \Omega < X .$$

If now $\Theta E \geqq Z$, we have

$$(\Omega, \Sigma - \Omega) > (\triangle\ AB\Gamma, X) = (\Theta B, Z) \geqq (\Theta B, \Theta E) .$$

Now determine H on $E\Theta$ produced such that

$$(\Omega, \Sigma - \Omega) = (\Theta H, \Theta E) .$$

From this it follows that

$$\Theta H > \Theta B .$$

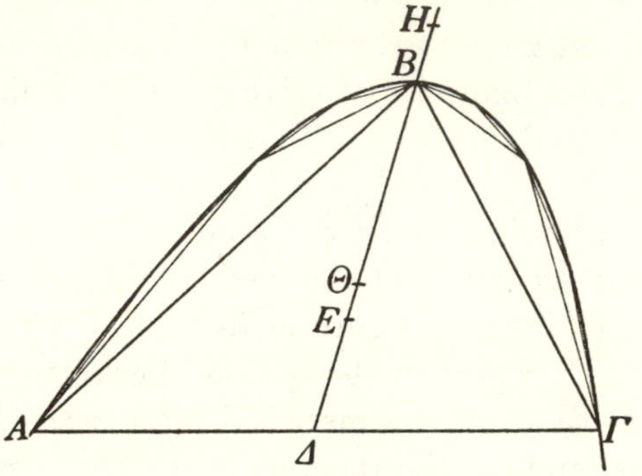

Fig. 146.

Since, however, E is the centre of gravity of Ω and Θ of Σ, H is the centre of gravity of the figure made up of the segments left over beyond Ω. But this is impossible, since all the segments lie on the same side of the straight line through B parallel to $A\Gamma$ with Ω, while H lies either at B or on ΘB produced.

Proposition 7.

The centres of gravity of two similar segments comprehended by a straight line and an orthotome divide their diameters in the same ratio.

In Fig. 147 let the similar segments $AB\Gamma$ and EZH with the successive centres of gravity K and Λ be given. It has to be proved that

$$(BK, K\Delta) = (Z\Lambda, \Lambda\Theta) .$$

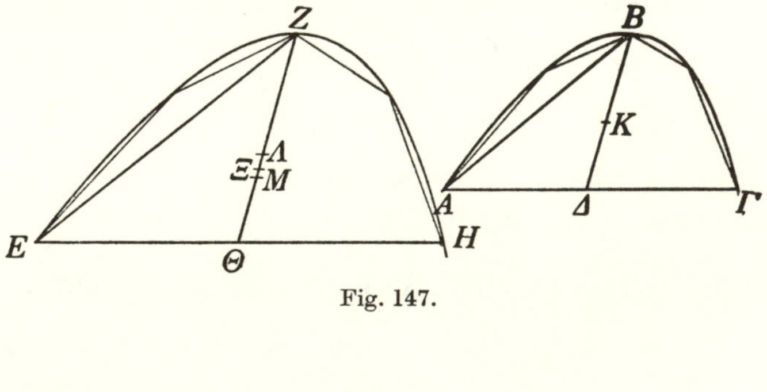

Fig. 147.

If this is not true, let M be so determined on $Z\Theta$ that

$$(BK, K\Delta) = (ZM, M\Theta) .$$

Then inscribe in EZH a figure Ω with a number of sides such that the distance between its centre of gravity Ξ and Λ is less than ΛM (Prop. 6). Then Ξ lies between Λ and M (because by Prop. 5 Ξ lies on that side of Λ where Θ lies and where M has also been taken; *vide* Note I). Now inscribe in $AB\Gamma$ a figure Ω with an equal number of sides; if the centre of gravity of this figure be P, we have by Prop. 3:

$$(BP, P\Delta) = (Z\Xi, \Xi\Theta) < (ZM, M\Theta) = (BK, K\Delta) ,$$

whence

$$BP < BK, \text{ which is contrary to Prop. 5.}$$

Note I. In the proof M has been taken between Λ and Θ; this is the case only if it be assumed that

$$(BK, K\Delta) > (Z\Lambda, \Lambda\Theta) .$$

If, however, it be assumed that

$$(BK, K\Delta) < (Z\Lambda, \Lambda\Theta) ,$$

we have

$$(Z\Lambda, \Lambda\Theta) > (BK, K\Delta) ,$$

and then the argument can be started with the segment $AB\Gamma$.

Note II. The supposition of the similarity of the segments has here been used no more than in Prop. 3. Archimedes accordingly applies the theorem just proved in Prop. 8 to any segment.

The above is followed by the main theorem:

Proposition 8.

The centre of gravity of any segment comprehended by a straight line and an orthotome divides the diameter of the segment in such a way that the part towards the vertex of the segment is half as large again as the part towards the base.

In Fig. 148 let $AB\Gamma$ be the given segment with chord $A\Gamma$ and vertex B. It has to be proved that the centre of gravity Θ so lies on the diameter $B\Delta$ that $B\Theta = \frac{3}{2}\Delta\Theta$. Of the segments AB and ΓB let K and Λ successively be the vertices, KZ and ΛH the diameters,

M and N the centres of gravity. Then $KZ \parallel B\Delta \parallel \Lambda H$, and $K\Lambda \parallel A\Gamma \parallel ZH$ (III; 2.5). $K\Lambda HZ$ therefore is a parallelogram. In this we have, by Prop. 7 applied to non-similar segments,

$$(KM, MZ) = (\Lambda N, NH) ,$$

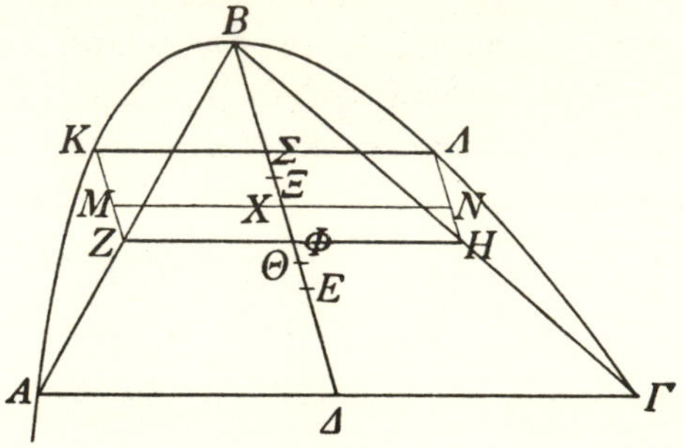

Fig. 148.

whence $MN \parallel ZH$. Now let $K\Lambda$, MN, ZH meet the diameter $B\Delta$ successively in Σ, X, Φ. X is then the centre of gravity of the segments AB and ΓB combined. Let further E be the centre of gravity of $\triangle\ AB\Gamma$.

Now we have

$$B\Delta = 4KZ ,$$

for $B\Delta = 4B\Sigma$ (III; 2.5) $= 2B\Phi$. Thence $B\Sigma = \Sigma\Phi = KZ = \frac{1}{4}B\Delta$.

Since by Prop. 7

$$(B\Theta, \Theta\Delta) = (KM, MZ), \text{ we also have}$$

$$\Theta\Delta = 4MZ = 4X\Phi ,$$

and therefore

$$B\Theta = 4\Sigma X .$$

(The two sets of three points B, Θ, Δ and Σ, X, Φ are therefore similar, with factor of similarity 4).

Since the segment $AB\Gamma$ consists of the triangle $AB\Gamma$ with centre of gravity E and the pair of segments AB, ΓB with centre of gravity X, for the centre of gravity Θ of $AB\Gamma$ the following relation applies:

$$(E\Theta, X\Theta) = (\text{segment } AB + \text{segment } \Gamma B,\ \triangle\ AB\Gamma) . \qquad (1)$$

Now by the main theorem (Prop. 17) of **Q.P.** we have

$$\text{segment } AB + \text{segment } \Gamma B = \tfrac{4}{3}(\triangle\, ABK + \triangle\, \Gamma B\Lambda) = \tfrac{1}{3}\,\triangle\, AB\Gamma\,,$$

whence

$$X\Theta = 3E\Theta\,.$$

Archimedes now determines a line segment which is equal to $E\Theta$. We have

$$B\Sigma + X\Theta = B\Theta - \Sigma X = 3\Sigma X\,,$$

so that

$$X\Theta = 3\Sigma X - B\Sigma\,.$$

Now determine on ΣX a point Ξ such that $B\Sigma = 3\Sigma\Xi$, then

$$X\Theta = 3\Sigma X - 3\Sigma\Xi = 3X\Xi\,,$$

whence

$$X\Xi = E\Theta\,,$$

and consequently

$$\Xi E = \Xi X + X\Theta + \Theta E = 5E\Theta\,.$$

We also have

$$B\Xi = B\Sigma + \Sigma\Xi = \tfrac{4}{3}B\Sigma = \tfrac{1}{3}B\Delta = \Delta E\,,$$

whence

$$\Delta E = \Xi E = 5\Theta E\,.$$

From this it follows that

$$\Delta\Theta = 6\Theta E,\ BE = 10\Theta E,\ B\Theta = 9\Theta E\,,$$

so that indeed

$$B\Theta = \tfrac{3}{2}\Theta\Delta\,.$$

In the algebraic form the derivation would of course be much simpler.

If $B\Delta = a$ and $B\Theta = \lambda a$,

$$KM = \tfrac{1}{4}\lambda a \text{ and } BX = \tfrac{1}{4}a(1+\lambda)\,,$$

whence

$$X\Theta = \lambda a - \tfrac{1}{4}a(1+\lambda) \text{ and } E\Theta = a(\tfrac{2}{3} - \lambda)\,.$$

The relation (1) now gives for λ the equation

$$\frac{\frac{2}{3} - \lambda}{\lambda - \frac{1}{4}(1+\lambda)} = \tfrac{1}{3}\,,$$

from which it follows that

$$\lambda = \tfrac{3}{5}\,.$$

The two remaining propositions 9 and 10, as they stand, are about the most indigestible thing in all Greek mathematics. In Prop. 9 a reduction of a given expression, occurring in Prop. 10, is formulated as a theorem and proved synthetically before we have any idea as to the motives for dealing precisely with this complicated question. These motives do not become known until the likewise perfectly synthetical argument of Prop. 10 is found to lead to the expression concerned. Add to this that in the course of the proof a detour is made which might have been avoided.

In the following pages we shall summarize the two propositions in one analytical argument (Fig. 149).

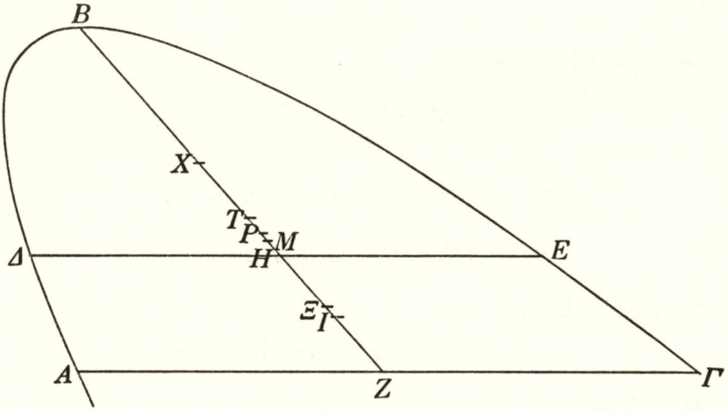

Fig. 149.

The object is to determine the position of the centre of gravity of the figure cut off from any segment of an orthotome on the side of the base by a chord parallel to the base, *i.e.* of the frustum $A\Delta E\Gamma$, which we shall represent by Σ. Imagine X and P to be the centres of gravity of the segments ΔBE and $AB\Gamma$, then by Prop. 8 the following relations apply:

$$XH = \tfrac{2}{5}BH \text{ and } PZ = \tfrac{2}{5}BZ\ ,$$

in which H and Z are the middle points respectively of ΔE and $A\Gamma$.

If I be the centre of gravity of Σ, I therefore has to satisfy the relation

$$(XP, PI) = (\Sigma, \text{segment } \Delta BE)\ . \tag{1}$$

Now we have

$$\text{segment } AB\Gamma = \tfrac{4}{3} \triangle AB\Gamma$$

$$\text{segment } \varDelta BE = \tfrac{4}{3} \triangle \varDelta BE ,$$

from which it follows that

$$(\text{segment } AB\Gamma, \text{ segment } \varDelta BE) = [\mathbf{O}(A\Gamma, BZ), \ \mathbf{O}(\varDelta E, BH)]$$

$$= \mathbf{T\Lambda}(AZ, \varDelta H) ,$$

because

$$(BZ, BH) = [\mathbf{T}(AZ), \ \mathbf{T}(\varDelta H)] .$$

Now the method of application of areas demands that we represent the triplicate ratio of AZ and $\varDelta H$ by the ratio of two line segments. For this purpose imagine the points Ξ and T to be constructed such that

$$BZ, B\Xi, BH, BT$$

form a series of line segments in continued proportion (a geometrical progression).

We now have

$$(BZ, BH) = \mathbf{\Delta\Lambda}(BZ, B\Xi) ,$$

whence

$$(AZ, \varDelta H) = (BZ, B\Xi) ,$$

$$(BZ, BT) = \mathbf{T\Lambda}(BZ, B\Xi) = \mathbf{T\Lambda}(AZ, \varDelta H) .$$

Consequently

$$(BZ, BT) = (\text{segment } AB\Gamma, \text{ segment } \varDelta BE) ,$$

and therefore

$$(TZ, BT) = (\Sigma, \text{ segment } \varDelta BE) .$$

For the determination of I we therefore have, in connection with (1), the relation

$$(XP, PI) = (TZ, BT) . \qquad (2)$$

In order to know the position of I relative to the straight boundary lines of Σ, the ratio (IZ, HZ) has to be determined.

Now $IZ = PZ - PI$, in which $PZ = \tfrac{3}{5} BZ$.

Now consider a line segment BM such that $PI = \tfrac{3}{5} BM$, then $IZ = \tfrac{3}{5} MZ$.

From (2) it follows that

$$(PI, BT) = (XP, TZ),$$

in which

$$XP = BP - BX = \tfrac{3}{5}HZ,$$

whence

$$(PI, BT) = (\tfrac{3}{5}HZ, TZ). \qquad (3)$$

Since

$$BT, BH, B\Xi, BZ$$

form a geometrical progresssion, the differences between successive terms

$$TH, H\Xi, \Xi Z$$

will also do so, with the same ratio.

The subsequent derivation is mainly based on a property of a geometrical progression and the progression of its differences, which may be formulated generally as follows:

If the geometrical progression be

$$t_1, t_2, \ldots t_n,$$

the progression of the differences

$$v_1, v_2, \ldots v_{n-1},$$

we have

$$\frac{\lambda_1 v_1 + \ldots \lambda_{n-1} v_{n-1}}{\mu_1 v_1 + \ldots \mu_{n-1} v_{n-1}} = \frac{\lambda_1 t_1 + \ldots \lambda_{n-1} t_{n-1}}{\mu_1 t_1 + \ldots \mu_{n-1} t_{n-1}} = \frac{\lambda_1 t_2 + \ldots \lambda_{n-1} t_n}{\mu_1 t_2 + \ldots \mu_{n-1} t_n} \; (A),$$

in which $\lambda_1 \ldots \lambda_{n-1}$, $\mu_1 \ldots \mu_{n-1}$ are any constants.

In view of this it is then found that

$$(\lambda_1 = 0,\ \lambda_2 = \lambda_3 = 1;\ \mu_1 = \mu_2 = \mu_3 = 1)$$

$$(HZ, TZ) = (B\Xi + BZ, BH + B\Xi + BZ),$$

so that (3) may be written

$$(PI, BT) = [\tfrac{3}{5}(B\Xi + BZ),\ BH + B\Xi + BZ],$$

and because $PI = \frac{2}{5}BM$,

$$(BM, BT) = [3(B\Xi + BZ),\ 2(BH + B\Xi + BZ)]. \qquad (4)$$

If in this, for greater clarity, we put

$$BZ = 1,\ B\Xi = r,\ BH = r^2,$$

the above is equivalent to

$$\frac{BM}{BT} = \frac{3(1+r)}{2(1+r+r^2)},$$

whereas Archimedes uses a relation which, in the same algebraic symbolism, has in the second member the proportion

$$\frac{3+6r+3r^2}{2(1+r^3)+4(r+r^2)}.$$

This fraction, however, may be written

$$\frac{3(1+r)^2}{2(1+r)(1+r+r^2)} = \frac{3(1+r)}{2(1+r+r^2)}.$$

Because Archimedes neglects to carry out this reduction, his calculation becomes too complicated. If we imagine it to have been made, we find from (4), by going through the synthesis given in Prop. 9 in the reverse direction:

$$(TM, BT) = [B\Xi - BH + BZ - BH,\ 2(BH + B\Xi + BZ)]$$
$$= [H\Xi + HZ,\ 2(BH + B\Xi + BZ)]\ .$$

Further we have by (A)

$$(BT, TH) = (BH, H\Xi) = (BH + B\Xi, HZ) = (2BH + B\Xi, H\Xi + HZ),$$

whence *ex aequali*

$$(TM, TH) = [2BH + B\Xi,\ 2(BH + B\Xi + BZ)]\ .$$

From this it follows that

$$(HM, TH) = [B\Xi + 2BZ,\ 2(BH + B\Xi + BZ)]\ ,$$

or by (A)

$$(HM, TH) = [BH + 2B\Xi,\ 2(BT + BH + B\Xi)]$$
$$= (BH + 3B\Xi + 2BZ,\ 2BT + 4BH + 4B\Xi + 2BZ)\ .$$

We also have by (A)

$$(TH, HZ) = (BH, B\Xi + BZ) = (BT, BH + B\Xi)$$
$$= (2BH + BT,\ BH + 3B\Xi + 2BZ)\ ,$$

whence *ex aequali*

$$(HM, HZ) = (2BH + BT, 2BT + 4BH + 4B\Xi + 2BZ),$$

whence

$$(MZ, HZ) = (3BT + 6BH + 4B\Xi + 2BZ, 2BT + 4BH + 4B\Xi + 2BZ),$$

whence, because $IZ = \frac{2}{5}MZ$,

$$(IZ, HZ) = (3BT + 6BH + 4B\Xi + 2BZ, 5BT + 10BH + 10B\Xi + 5BZ),$$

which is enunciated (in words) in Prop. 10 and proved with the aid of the derivation—given in Prop. 9—of the expression for (BM, BT), which we have reduced to (4). The reading of the original is rendered even more difficult because the letters used in Prop. 9 are quite different from those in Prop. 10, in which the result obtained is applied.

We finally reproduce the result in algebraic symbolism.

If $BZ = a$, $BH = b$, then

$$\frac{IZ}{a-b} = \frac{3b\sqrt{b} + 6b\sqrt{a} + 4a\sqrt{b} + 2a\sqrt{a}}{5b\sqrt{b} + 10b\sqrt{a} + 10a\sqrt{b} + 5a\sqrt{a}}.$$

CHAPTER XIII

THE SAND-RECKONER

1. The work *The Sand-Reckoner*, though meant by the author as a contribution to Greek arithmetic, owes its historical interest not only to what it contains as such; it is no less valuable as a document of Archimedes' astronomical activity. It was of course to be expected that he engaged in astronomy, though he has not left any work exclusively devoted to it: astronomy and mathematics in his day were scarcely distinguished as two different branches of science; his father Pheidias had already cultivated this science, and his friend Conon had even gained a great reputation in it. He himself also enjoyed a great astronomical reputation in antiquity. *Unicus spectator caeli siderumque* he is called by Livy[1]), and several other authors quote his observations and theories. Thus Hipparchus

[1]) Livy, *Ab urbe condita* XXIV, 34.

(whose remarks on this point have been preserved by Ptolemy[1])) mentions his determination of the length of the year; with reference to the same problem he is quoted by Ammianus Marcellinus[2]), while Macrobius reports on his theory concerning the mutual distances of the sun, the moon, and the planets, and their positions in relation to the sphere of the fixed stars[3]). His practical skill in the matter of astronomy is proved not only by his construction of a planetarium, but also by a remarkable determination of the apparent diameter of the sun, with which we shall become acquainted in the treatise now to be discussed.

2. By way of introduction it may first be explained how a fundamentally purely arithmetical problem could give rise to a work in which both theoretical discussions and practical observations in the field of astronomy are contained.

The problem concerned is the dual question, of such fundamental importance to Greek mathematics, of the reading and writing of large numbers, a question therefore which lies partly in the sphere of linguistics and partly in that of mathematical symbolism. The linguistic difficulty consisted in that for the successive powers of the base of the numerical system only three names were in use *ἑκατόν* = one hundred; *χίλιοι* = one thousand; *μύριοι* = ten thousand). For numbers exceeding 10^4 in the first place the number of myriads (tens of thousands) always had to be given, and this involved considerable difficulties already for numbers upwards of 10^8. For the mathematical notation the problem was no less serious. In the usual alphabetical numerical system numbers up to 10^4 could be denoted by the small letters of the alphabet (for the thousands with accents added). For 10^4 the symbol M or $\overset{Y}{\mathrm{M}}$ was in use, but above this it was necessary again to denote the number of myriads

---

1) The statement by Hipparchus was to be found in his work *Περὶ τῆς μεταπτώσεως τῶν τροπικῶν καὶ ἰσημερινῶν σημείων* quoted by Ptolemy, *Syntaxis Mathematica* III, 1. Ed. J. L. Heiberg (Leipzig 1898) I, 194.

2) Ammianus Marcellinus (Latin historian of the 4th century) refers to him as excelling among the *periti mundani motus et siderum. Ammiani Marcellini rerum gestarum libri qui supersunt.* XXVI; 1, 8. Ed. E. U. Clark, II (Berlin 1915), 391.

3) Macrobius was a Latin writer of African origin at the beginning of the 5th century. *Ambrosii Theodosii Macrobii Commentarius in Somnium Scipionis* II; 3, 13. Ed. F. Eyssenhardt (Leipzig 1868), 584.

in the number to be expressed with the aid of the other symbols, which gave rise to very complicated expressions.

Archimedes now wants to propose so efficacious a solution of this problem that even the largest numbers that may ever occur in the natural sciences shall be susceptible of being expressed shortly. In order to demonstrate the force of his system he chooses an example of astronomical dimensions: to determine the number of grains of sand which the sphere of the fixed stars could hold. Since this number depends directly on the dimensions in the universe, he has to start with considerations about astronomical distances.

3. We are reproducing the introductory part of the work in a literal translation:

There are some, king Gelon[1]) *who think that the number of grains of sand is infinite. I mean not only of the sand which is present in Syracuse and the rest of Sicily, but also of that which is found all over the world, whether inhabited or uninhabited. Others, indeed, do not assume that it is infinite, but they think that no such large expressible number exists that it exceeds its multitude. It is clear that if those who hold this view imagined a volume of sand as large as would be the volume of the earth if all the seas and hollows in it were filled up to a height equal to that of the highest mountains, they would be even less inclined to believe that any number could be expressed which exceeds the number of grains of this sand. But I will try to show by means of geometrical proofs which you will be able to follow that the numbers named by us, as published in the work destined for Zeuxippus, include some which exceed not only the number of grains of the sand which, as stated, has a volume equal to that of the earth filled up in the way described, but also of the sand which has a volume equal to that of the cosmos. You know that 'cosmos' is the name given by most astronomers to the sphere whose centre is the centre of the earth and whose radius is equal to the distance between the centre of the sun and the centre of the earth ... Aristarchus of Samos has, however, enunciated certain hypotheses in which it results from the premisses that the universe is much greater than that just mentioned. As a matter of fact, he supposes*[2]) *that the fixed stars and the sun do not move,*

[1]) Gelon was the son and co-governor of king Hieron II.

[2]) This is the most ancient and at the same time the most authoritative evidence of the existence of a heliocentric system in antiquity. The author, Aristarchus of Samos, lived in the early part of the 3rd century B.C., *i.e.*

but that the earth revolves in the circumference of a circle about the sun, which lies in the middle of the orbit, and that the sphere of the fixed stars, situated about the same centre as the sun, is so great that the circle in which the earth is supposed to revolve has the same ratio to the distance of the fixed stars as the centre of a sphere to its surface. Now it is obvious that this is impossible; for since the centre of a sphere has no magnitude, it cannot be conceived to bear any ratio to the surface of the sphere. It is, however, probable that Aristarchus meant the following: since we conceive the earth to be, as it were, the centre of the universe, he assumes that the ratio which the earth bears to what we call the cosmos is the same as the ratio which the sphere containing the circle in which the earth is conceived to revolve bears to the sphere of the fixed stars[1]*). For his proofs of the phenomena fit in with this supposition, and in particular he appears to suppose the magnitude of the sphere in which he represents the earth as moving to be equal to what we call the cosmos. We now say that, if a sphere were made up of sand of such a magnitude as Aristarchus supposes the sphere of the fixed stars to be, the numbers named in the Principles*[2]*) would still include some which exceed the number of grains of the sand which has a volume equal to that of the said sphere, the following assumptions being made:*

---

after Euclid and before Archimedes. Of the work in which he explained his system nothing has come down to us. The only thing that has been preserved is a treatise on the sizes of sun and moon, *Περὶ μεγεθῶν καὶ ἀποστημάτων ἡλίου καὶ σελήνης*, edited with an English translation by T. L. Heath, *Aristarchus of Samos, the ancient Copernicus* (Oxford 1913).

[1]) It is more likely that Aristarchus wanted to express that the orbit of the earth is "like a point" with regard to the sphere of the fixed stars. In fact, he also speaks in this way about the earth with regard to the sphere of the moon.

Hypothesis 2: *τὴν γῆν σημείου τε καὶ κέντρου λόγον ἔχειν πρὸς τῆς σελήνης σφαῖραν*. (Ed. Heath 352).

In the same way Ptolemy expresses himself when comparing the earth with the sphere of the sun (*Syntaxis Mathematica* I, 6; ed. J. L. Heiberg (Leipzig 1898) I, 20.

[2]) This translation is uncertain. In the Heiberg edition it says *τῶν ἐν ἀρχᾷ ἀριθμῶν τῶν κατονομαξίαν ἐχόντων*, which he translates by *numerorum denominatorum quos supra significavimus* (*Opera* II, 221; line 5). A variant reading, however, has *τῶν ἐν Ἀρχαῖς κ.τ.λ.*, *i.e.* "*in the Principles*". Here *Principles* would therefore be the title of the work sent to Zeuxippus.

I. *that the perimeter of the earth is three hundred myriads of stadia and not greater*[1]) ...

II. *next that the diameter of the earth is greater than the diameter of the moon and the diameter of the sun is greater than the diameter of the earth* ...

III. *next that the diameter of the sun is thirty times the diameter of the moon and not greater*[2]) ...

IV. *next that the diameter of the sun is greater than the side of the chiliagon inscribed in a greatest circle of the cosmos* (*i.e. a circle whose radius is equal to the distance from the earth to the sun*) ...

4. Of the last assumption a mathematical proof is first given, which is based on a measurement of the apparent diameter δ of the sun. Archimedes explains that it is difficult to give an exact figure for this, *because neither the eye, nor the hands, nor the instruments that have to be used are reliable in finding the exact value*[3]). It will, however, be sufficient for him to give an upper and a lower limit, which have been obtained by means of the following appliance.

Along a horizontal rod mounted on a vertical base a small vertical cylinder can be displaced. The rod is pointed in the direction of the sun when the latter is just above the horizon[4]), the eye is placed at one end, and the cylinder is then displaced in the direction of the sun until the latter becomes just visible to the right and to

---

[1]) Archimedes comments on this that someone (probably Dikaiarchus of Messina, a geographer of the 2nd half of the 4th century) tried to prove that the perimeter is 30 myriads (3.10^5) of stadia, but that he assumes as a maximum a value ten times greater, *viz.* 300 myriads of stadia (3.10^6). In this 1 stadium = 190 m; the real value of the perimeter is about 2.10^5 stadia).

Other values assumed by the Greeks are: Aristotle: 4.10^5 stadia; Eratosthenes, Hipparchus, Strabo: $2.5 . 10^5$ stadia; Poseidonius: $2.4 . 10^5$ stadia; Ptolemy: $1.8 . 10^5$ stadia. Archimedes exaggerates his supposition deliberately, in order to find a value which is sure not to be too small for the dimensions of the sphere of the fixed stars.

[2]) At this, other suppositions about this ratio are quoted. According to Eudoxus it was 9, according to Pheidias (this is the passage referred to in Chapter I, § 1) 12, according to Aristarchus more than 18 and less than 20. *But I, exaggerating this as well, suppose the diameter of the sun to be thirty times the diameter of the moon, in order that my proposition may be proved beyond dispute.*

[3]) *Opera* II, 222; lines 12–14.

[4]) The ancients preferably made their observations of the sun in the horizon, because they had no means for reducing the sunlight.

the left of the cylinder. If the pupil of the eye were a point, the angle included by the tangents drawn from this point to the base of the cylinder would apparently be a lower limit for δ. The extension of the pupil of the eye disturbs this argument. In fact, if in Fig. 150, drawn in the horizontal plane of the rod directed at the

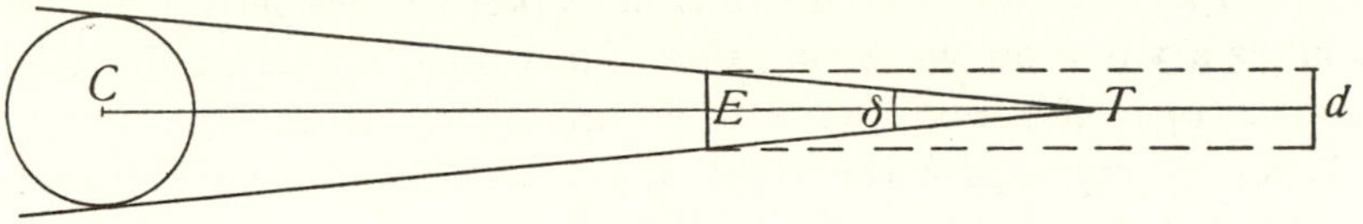

Fig. 150.

centre of the sun, the circle C be the base of the cylinder, the tangents drawn to this be rays from the rims of the sun, and T their meeting point, an eye in the form of a point situated between C and T can never receive sunlight, but an eye the diameter of whose pupil is d *can* receive it, provided its centre be between T and the point E where the distance d just fits into the angle included by the rays of the sun; in fact, in that case the diameter of the cast shadow in this place is smaller than the pupil of the eye.

Now a lower limit for δ is found by placing in the spot where the eye just sees a little of the sun round the cylinder a horizontal circular disc, whose diameter is not smaller than the diameter d of the pupil of the eye, and then measuring the angle φ which the common tangents to this disc and the circle C include[1]).

In order to find an upper limit for δ, Archimedes finds the position in which the eye no longer sees anything. This is between C and E. The tangents drawn from such a point to the circle C include an angle ψ, which is greater than δ, and which differs more

[1]) Properly speaking, therefore, one ought to measure the angle included by the tangents to the circle C from the extremities of the diameter of the disc at right angles to CT. This results, however, in so small a difference that it was not necessary to take it into account. In order to determine the diameter of the pupil of the eye the following experiment is carried out. Two small cylindrical rods of equal thickness and length, one of which is white, the other is coloured, are combined to one rod; the coloured end of this is held against the cornea of the eye. If the diameter of the rod is less than that of the pupil, the eye will still see something of the white rod. The experiment is then repeated with rods of increasing diameters, until for the first time the eye no longer sees anything of the white rod; the diameter is now an upper limit for the diameter of the pupil of the eye.

from δ according as the point in question can be taken further away from C. It would of course be possible to lower the upper limit here by finding the angle included by the common tangents to C and the above-mentioned disc. Archimedes does not do this, probably because the inaccuracy which is committed by measuring the angle included by the tangents drawn from a point to circle C can never produce a value on the wrong side of δ, which might, on the other hand, happen if φ were found in an analogous manner[1]).

When R represents the right angle, the result found may be reproduced as follows:

$$\varphi > \frac{R}{200}, \quad \psi < \frac{R}{164},$$

whence *a fortiori*

$$\frac{R}{200} < \delta < \frac{R}{164}.$$

In a different notation this amounts to

$$27' < \delta < 32'56''$$ [2]).

5. It can now be proved that the diameter of the sun is greater than the side of the regular chiliagon inscribed in a greatest circle of the cosmos[3]).

In Fig. 151 the plane of the paper is the plane through the eye of the observer, Δ, and the centres Θ, K, of earth and sun. The cosmos is cut in a circle through K with centre Θ.

$\Delta\Xi$, through Δ, touching the sun in T, is the intersection of the plane of the paper with the horizon of Δ. Further, from Δ the tangent $\Delta\Lambda$ is drawn to the sun; the point of contact is N. From Θ are also drawn two tangents to the sun, ΘO with point of contact X and point of intersection with the cosmos B, ΘM with point of contact P and point of intersection with the cosmos A.

---

[1]) The difference between the methods of finding φ and ψ is indicated by the fact that Archimedes refers to ψ as γωνία ἐν στίγῳ (angle in a point).

[2]) As is known, δ lies in reality between 31′28″ and 32′32″.

[3]) If Archimedes had dared to assume that the tangent cone to the sun from the eye touches the sun in a greatest circle, and if he had also dared to neglect the parallax, this proposition would be a direct consequence of the consideration that the arc subtended by a chord of a regular chiliagon is $\frac{R}{250}$, *i.e.* certainly smaller than δ.

The line joining the centres Θ and K meets the earth in E and Y, the sun in Σ and H, the line AB in Φ.

Because the sun is above the horizon and $\angle\,\Theta\Delta O$ is right, $\angle\,\Theta\Delta K$ is obtuse, so that $\Theta K > \Delta K$. Consequently the apparent diameter of the sun for Θ, *i.e.* $\angle\,M\Theta O$, is less than that for Δ, *i.e.* $\angle\,\Lambda\Delta\Xi$.

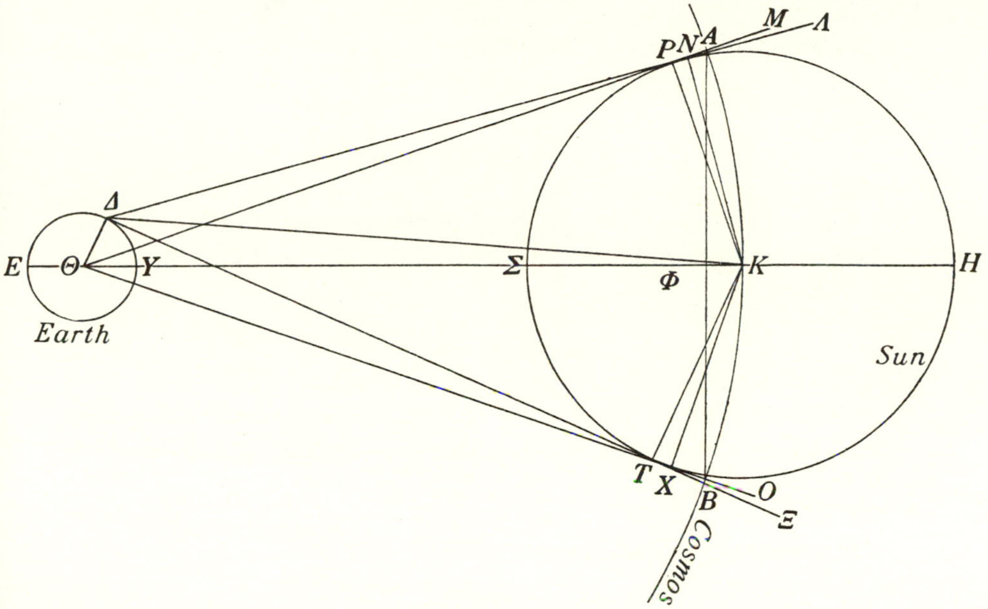

Fig. 151.

Now it is known that

$$\frac{R}{200} < \angle\,\Lambda\Delta\Xi < \frac{R}{164},$$

whence certainly

$$\angle\,M\Theta O < \frac{R}{164},$$

so that AB is less than the chord subtending $\frac{1}{656}$th of the circumference of the circle AKB. Since the perimeter of the regular 656-sided polygon bears to the radius ϱ of the cosmos a ratio which is less than 44:7 (the ratio of the circumference of the circle to the radius already being less than this), we have accordingly

$$(AB, \varrho) < \tfrac{1}{656} \cdot \tfrac{44}{7} = \tfrac{11}{1148} < \tfrac{1}{100},$$

whence

$$AB < \tfrac{1}{100}\varrho .$$

Now AB is equal to the diameter d of the sun ($\triangle\, \Theta KX \cong \triangle\, \Theta B\Phi$, whence $B\Phi = KX = \frac{1}{2}d$), whence

$$d < \tfrac{1}{100}\varrho .$$

Now the diameter of the earth $EY < d$, whence

$$\Theta Y + K\Sigma < \tfrac{1}{100}\Theta K \quad \text{or} \quad Y\Sigma > \tfrac{99}{100}\Theta K ,$$

whence

$$(\Theta K,\ Y\Sigma) < (100, 99) .$$

Now $\Theta K > \Theta P$, but $Y\Sigma < \Delta T$ (because $Y\Sigma$ is the shortest distance from a point of the earth to a point of the sun).

Thence *a fortiori*

$$(\Theta P, \Delta T) < (100, 99) .$$

If we now compare the triangles ΘKP and ΔKT, which are both right-angled and in which $KP = KT$, $\Theta P > \Delta T$ (because $\Theta K > \Delta K$), then by a lemma widely used in Greek mathematics[1]) we have

$$(\angle\, K\Delta T,\ \angle\, K\Theta P) < (\Theta P, \Delta T) < (100, 99) .$$

Consequently we also have

$$(\angle\, \Lambda\Delta\Xi,\ \angle\, M\Theta O) < (100, 99) .$$

Since $\angle\, \Lambda\Delta\Xi > \dfrac{R}{200}$, we find

$$\angle\, M\Theta O > \frac{99}{100} \cdot \frac{R}{200},$$

---

[1]) In modern formulation this lemma states for $\frac{\pi}{2} > \alpha > \beta$:

$$\frac{\alpha}{\beta} < \frac{\operatorname{tg}\alpha}{\operatorname{tg}\beta}.$$

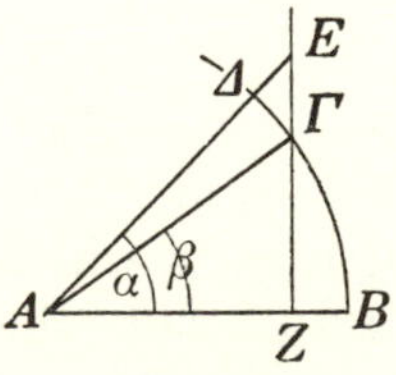

Fig. 152.

The proof may be furnished as follows: In Fig. 152, suppose $\angle\, \Delta AB = \alpha$, $\angle\, \Gamma AB = \beta$. Then we have $(\angle\, \Delta A\Gamma,\ \angle\, \Gamma AB)$ = (sector $\Delta A\Gamma$, sector ΓAB) $<$ ($\triangle\, EA\Gamma$, $\triangle\, \Gamma AZ$), whence $(\angle\, \Delta AB,\ \angle\, \Gamma AB) < (\triangle\, EAZ,\ \triangle\, \Gamma AZ) = (EZ, \Gamma Z)$.

and since

$$\frac{99}{20,000} < \frac{1}{203}:$$

$$\angle M\Theta O > \frac{R}{203}\ ^{1)}.$$

AB therefore is certainly greater than the side of the regular 812-sided polygon, thence *a fortiori* it is greater than the side of the regular chiliagon inscribed in a greatest circle of the cosmos.

6. If we represent the diameters of moon, earth, sun, and cosmos by D with an index, we now know that

$$D_{\text{sun}} = 30D_{\text{moon}} < 30D_{\text{earth}}.$$

Since further the perimeter of the chiliagon inscribed in the section of the cosmos is greater than $3D_{\text{cosmos}}$ (in fact, the perimeter of the regular hexagon is already $3D$), and at the same time, by § 5, less than $1,000D_{\text{sun}}$, we also have

$$3D_{\text{cosmos}} < 1,000D_{\text{sun}} < 30,000D_{\text{earth}},$$

and since the perimeter of the earth is $3 \cdot 10^6$ stadia and its diameter therefore less than 10^6 stadia,

$$D_{\text{cosmos}} < 10^{10} \text{ stadia}^{2)}.$$

7. Suppositions as to the size of the grains of sand now have to be made. These naturally have to be exaggerated on the small side in order to strengthen the evidence for the possibility of expressing large numbers. It is assumed that a volume no greater than a poppy-seed contains not more than a myriad of grains, while the diameter of such a seed is not less than $\frac{1}{40}$th of a finger-breadth. The latter limit is based on a measurement: 25 seeds, placed side by side, filled more than a finger-breadth. The diameter is therefore greater than $\frac{1}{25}$th of a finger-breadth. The limit of $\frac{1}{40}$th is thus exaggerated on the small side.

---

[1]) The result of the whole laborious reduction, which is equivalent to a correction for daily parallax, thus consists in the replacement of the lower limit $\frac{R}{200}$ for δ, measured from Δ, by $\frac{R}{203}$ for δ, measured from Θ.

[2]) We give this result in modern notation for clearness' sake. In the Greek text it says ἁ τοῦ κόσμου διάμετρος ἐλάττων ἐστὶν ἢ σταδίων μυριάκις μυριάδες ρ̄ (less than ten thousand times one hundred myriads of stadia).

8. Before the calculation can be started, the new system for expressing large numbers must be explained.

The names of the numbers from one to ten thousand are known, and by stating the number of myriads we may go up to ten thousand myriads (*μύριαι μυριάδες*, 10^8) without any difficulty. The numbers thus obtained are called first *numbers* (*πρώτοι ἀριθμοί*[1])). We now take 10^8 as the new unit, and by means of this form the *second numbers*, *viz.* 10^8, 2.10^8 ... up to $10^8.10^8$, *i.e.* 10^{16}. The number 10^{16} now becomes the unit of the *third numbers*, which end with 10^{24}.

For clearness' sake we shall call

$1, 2 \ldots 10^8$	the numbers of the first order
$10^8, 2.10^8 \ldots 10^{16}$	the numbers of the second order
$10^{8(i-1)}, 2.10^{8(i-1)} \ldots 10^{8i}$	the numbers of the ith order.

The numbers $1, 10^8 \ldots 10^{8(i-1)}$ we call the units of the 1st, 2nd, ith order.

This procedure may be continued up to $i = 10^8$. We thus obtain a number A, which Archimedes describes as ten thousand myriads of the myriad-myriad-th numbers, and which we may refer to as 10^8 times the unit of the order 10^8.

The value is

$$A = 10^{8.10^8}.$$

Although for all practical purposes the numbers hitherto named suffice, Archimedes extends the system even further.

The numbers from 1 to A form the *first period*. A now becomes the unit of the *second period*, which, just like the first, has its successive orders. The last number of the 2nd period is therefore A^2; this becomes the unit of the *third period*, which runs up to A^3. The procedure is continued until 10^8 periods have been obtained. The last concrete number is called in Greek

μυριακισμυριοστᾶς περιόδου μυριακισμυριοστῶν ἀριθμῶν μύριαι μυριάδες

i.e. literally

a myriad-myriad units of the myriad-myriad-th order of the myriad-myriad-th period.

---

[1]) The same name which otherwise denotes prime numbers.

This means therefore: 10^8 times the unit of the (10^8)th order of the (10^8)th period. The number obtained is

$$A^{10^8} = (10^{8.10^8})^{10^8} = 10^{8.10^{16}} .$$

9. In order to simplify slightly more the expression of the numbers occurring in the calculation, which are all multiples of powers of ten, Archimedes now considers the series of the successive powers of 10 with non-negative whole exponents, which he divides into groups of eight, the octads:

$$\underbrace{1,\ 10,\ 10^2 \ldots 10^7}_{\text{1st octad}},\ \underbrace{10^8 \ldots 10^{15}}_{\text{2nd octad}},\ \underbrace{10^{16} \ldots 10^{23}}_{\text{3rd octad}},\ \text{etc.}$$

It is easily seen that the first octad contains exclusively numbers of the first order, the second octad exclusively numbers of the second order, etc. The ith octad $(10^{8i-8}, 10^{8i-7} \ldots, 10^{8i-1})$ thus contains the unit of the ith order, its decuple, its centuple, etc. up to its 10^7-fold. The unit of the ith order is also called unit of the ith octad. The numbers 1, 10^8, 10^{16}, etc. are termed the units of the successive octads.

In addition Archimedes now gives a rule of multiplication for numbers of a geometrical progression the first term of which is 1:

If among any numbers which are in continued progression starting from unity there are some which multiply each other[1]), *the product will occur in the same series in a place which is as far removed from the greater of the numbers that have multiplied each other as the smaller from unity, and it will be distant from unity by one place less than the sum of the numbers which denote how far the numbers that have multiplied each other are removed from unity.*

This rule therefore makes it possible to find in the geometrical progression

$$A, B, \Gamma, \Delta, E, Z, H, \Theta, I, K, \Lambda, \text{ in which } A = 1 ,$$

for the product $\Delta . \Theta$ at once Λ, because Λ is as far removed from Θ as Δ from A. It further states that the number of Λ is found by reducing the sum of the numbers of Δ and Θ by 1.

In modern symbolism the first part of the above rule signifies nothing but the property

[1]) That is actually the term which we have corrupted to "multiplying by each other".

$$a^m . a^n = a^{m+n} ,$$

while the second part states that in the geometrical progression

$$t_1 = 1, t_2, t_3 \ldots$$

the product of t_i and t_j is the term t_{i+j-1}.

10. After all this preparatory work the calculation can now begin; of course it is given entirely in words, and thus occupies many pages. As far as it is given here, we shall also reproduce it in words.

A sphere with a diameter of one finger-breadth contains no more than sixty-four thousand seeds (40^3), consequently not more grains of sand than six units of the second octad with four thousand myriads added, in any case therefore less than ten units of the second octad. A sphere with a diameter of one hundred finger-breadths therefore contains less than one hundred myriad times ten units of the second octad, *i.e.* less than the product of the seventh and the tenth number of the above series of the successive powers of ten; this product is the sixteenth number of the series of one thousand myriad units of the second octad. A sphere with a diameter of ten thousand finger-breadths, *i.e.* a stadium, contains one thousand myriad times this number again, thus less than the twenty-second number of the series (10^{21}). Proceeding in this way, we find upper limits for the numbers of grains of sand in spheres whose diameters are each time one hundred times greater than that of the preceding one; the numbers in the series of powers of ten leap six units each time. For a sphere with a diameter of ten times one hundred thousand myriads of stadia we thus arrive at the fifty-second number, which is one thousand myriad units of the seventh octad.

In modern notation, for the number of grains of sand in the last-mentioned sphere, which has a diameter of 10^{10} stadia, the limit

$$(10^{10})^3 . 10^{21} = 10^{51}$$

is found at once.

The number last obtained is therefore an upper limit for the number of grains in the cosmos. Since by hypothesis[1]) the diameter

[1]) In fact, according to the interpretation of Aristarchus' estimation as suggested by Archimedes the ratio of the (diameters of the) sphere of the fixed stars and the cosmos is the same as that of the (diameters of the) cosmos and the earth, which latter ratio was found in § 6 to be less than 10,000.

of the sphere of the fixed stars is less than one myriad cosmos diameters, it is found that the number of grains of sand in the sphere of the fixed stars is less than one thousand myriad units of the eighth octad (*i.e.* 10^{63}).

Archimedes thus succeeds in naming this number without even requiring the whole of the first period. He then concludes the discussion with the following words:

I conceive, king Gelon, that these things will appear incredible to the numerous persons who have not studied mathematics; but to those who are conversant therewith and have given thought to the distances and the sizes of the earth, the sun, and the moon, and of the whole cosmos, the proof will carry conviction. It is for this reason that I thought it would not displease you either to consider these things.

CHAPTER XIV

FLOATING BODIES

1. As we saw in Chapter II, the Greek text of this work has only been known since 1899; however, it still exhibits considerable lacunae. Before that time scholars always had to make shift with a Latin translation by William of Moerbeke; the latter is still used to supply the undecipherable or missing parts of the Greek text.

2. Book I of the work opens with a postulate the precise meaning of which may be subject to doubt and which on that account we quote in the original.

ὑποκείσθω τὸ ὑγρὸν φύσιν ἔχον τοιαύταν, ὥστε τῶν μερέων αὐτοῦ τῶν ἐξ ἴσου κειμένων καὶ συνεχέων ἐόντων ἐξωθεῖσθαι τὸ ἧσσον θλιβόμενον ὑπὸ τοῦ μᾶλλον θλιβομένου, καὶ ἕκαστον δὲ τῶν μερέων αὐτοῦ θλίβεσθαι τῷ ὑπεράνω αὐτοῦ ὑγρῷ κατὰ κάθετον ἐόντι, εἴ κα μὴ τὸ ὑγρὸν ᾖ καθειργμένον ἔν τινι καὶ ὑπὸ ἄλλου τινὸς θλιβόμενον.

Let it be granted that the fluid is of such a nature that of the parts of it which are at the same level and adjacent to one another that which is pressed the less is pushed away by that which is pressed the more, and that each of its parts is pressed by the fluid which is vertically above it, if the fluid is not shut up in anything and is not compressed by anything else.

We will first adopt this postulate without any commentary or criticism; its meaning will have become clearer as soon as we have seen it applied in the proofs of the propositions, and it is not until then that we shall be able to deal successfully with a number of questions arising in connection with it.

The first of the propositions now following is purely geometrical in character; it enunciates that a surface which is cut by all the planes through a given point in circles having that point for its centre is a sphere. This result is applied in

Proposition 2.

The surface of any fluid which is so located that it remains motionless will have the form of a sphere which has the same centre as the earth.

Figure 153 represents an intersection with a plane through the centre of the earth K, which cuts the surface of the fluid in $ABGD$.

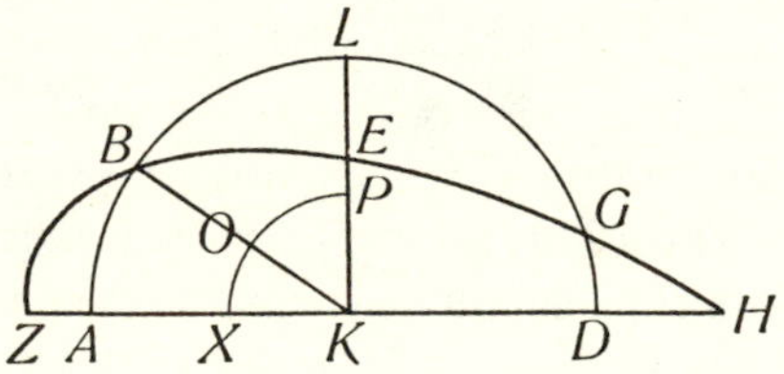

Fig. 153.

If $ABGD$ is not a circle, a radius vector KB is chosen such that the radii vectores of K to the curve $ABGD$ include some which are greater as well as some that are smaller, while a circle $ZBEH$ is described about K as centre and with radius KB. Further make $\angle ZKB$ equal to $\angle BKL$, and about K as centre describe a circle with radius $KX < KE$. Now consider the equal surfaces XO and OP, which are at the same level and adjacent to each other. XO is compressed by the fluid in $ABOX$, OP by the fluid in $OBLP$. It is evident that the latter quantity is greater than the former, so that by the postulate the fluid cannot be at rest. If therefore it is at rest, $ABGD$ must be a circle with centre K. Since the cutting plane through K is arbitrary, it results from this by Prop. 1 that the surface of a fluid at rest is a spherical surface, the centre of which coincides with the centre of the earth.

Proposition 3.

Of solids those which are of equal weight with the fluid will sink down in the fluid until the surfaces no longer project above that of the fluid, and they will not be driven down any further.

Figure 154 represents an intersection of a fluid (AMD) and a solid ($EZTH$) by a plane through the centre of the earth K. Suppose the portion $EZGB$ of the solid to project above the fluid. KLM is (the section of) a pyramid whose base LM lies in the surface of the fluid, KMN is a pyramid equal and similar thereto, XOP is a part of a spherical surface with centre K. $RYSC$ is a volume of fluid equal and similar to $BGTH$. Now XO is compressed by the fluid between XO, LK, and MK and beyond $BGTH$, as also by the solid $EZTH$; OP is compressed by the fluid $MNPO$, of which the portion beyond $RYCS$ is of equal weight with the portion of $XOML$ beyond $BGTH$, while $RYCS$ has the same weight as $BGTH$. XO is therefore compressed more than OP, *viz.* in addition by $EZGB$; hence the fluid cannot be at rest, which is contrary to the supposition. This contradiction can be removed by supposing EZ to fall in LM. There is, however, no reason why the solid should sink down any further, for even as it is there is already equality on equal adjacent surfaces at the same level.

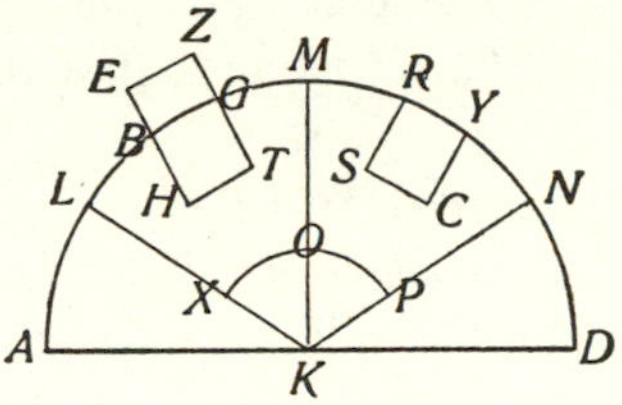

Fig. 154.

Proposition 4.

Of solids one which is lighter than the fluid, when thrown into the fluid, will not sink down altogether, but a portion of it will project above the surface of the fluid.

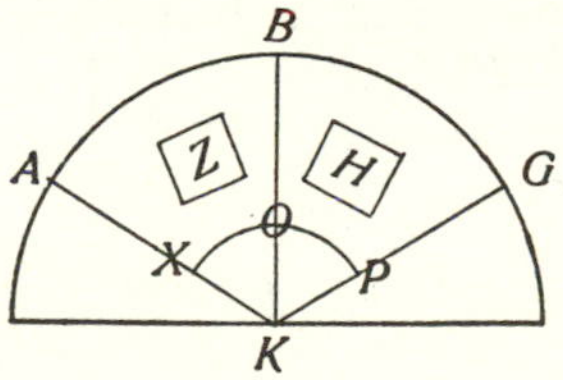

Fig. 155.

By means of the same considerations as in Prop. 3 it is proved (Fig. 155) that the pressures on XO and OP would be unequal, if Z were a solid equal and similar to a portion of fluid H, but lighter than the latter.

Proposition 5.

Of solids one which is lighter than the fluid, when thrown into the fluid, will sink down until a volume of the fluid equal to the volume of the immersed portion has the same weight as the whole solid.

In Fig. 154 let *EZTH* now be the given solid which is specifically lighter than the fluid, *RYCS* a volume of the fluid equal and similar to the immersed portion *BGTH*. If the fluid is at rest, the weight of *EZTH* must be equal to that of *RYCS*.

Proposition 6.

Solids which are lighter than the fluid, when forcibly immersed in the fluid, are thrust upwards by a force[1]) *equal to the weight by which a volume of the fluid equal to the solid exceeds that solid.*

In Fig. 156 let A be the given solid with weight B, while $B+G$ is the weight of a quantity of fluid the volume of which is equal to that of A. D is a solid with weight G. The composite solid $A+D$ weighs less than an equal volume of fluid, because its weight, *viz.* $B+G$, is equal to that of a smaller volume of fluid, *viz.* that of A. By Prop. 4 it will sink down until A is just immersed, because it has the same weight as a fluid the volume of which is equal to that of A. The force by which A is forced upwards is therefore equal to the weight of D, *i.e.* G, *i.e.* the difference between the weights of a volume of fluid equal to that of A and of A itself.

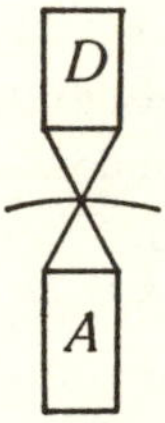

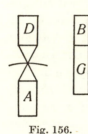

Fig. 156.

Proposition 7.

Solids heavier than the fluid, when thrown into the fluid, will be driven down as far as they can sink, and they will be lighter in the fluid by the weight of a portion of the fluid having the same volume as the solid.

In Fig. 157 let A be the given solid with weight $B+G$, while an equal volume of fluid weighs B. Take a solid D with weight B so that an equal volume of fluid has a weight $B+G$. Now combine A and D to one solid with weight $2B+G$. An

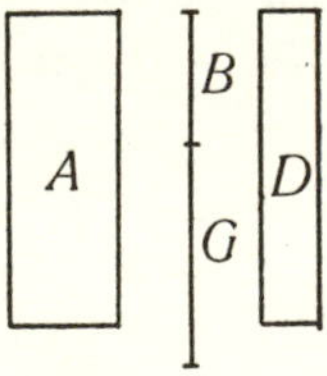

Fig. 157.

[1]) It is to be noted that this is not what in present-day physics is called the upward thrust exerted on a solid immersed in a fluid (which is equal to the weight of the displaced fluid), but the amount by which this force exceeds the weight of the solid.

equal volume of fluid also weighs $2B+G$. A and D will therefore be suspended together; D alone would by Prop. 6 be forced upwards by a force equal to G. G thus also stands for the force by which A alone would be driven down, *i.e.* the weight of A in the fluid. A is therefore lighter in the fluid by an amount B than out of it.

It is now time to raise the question as to the real meaning of the hydrostatic argument of Archimedes. This question cannot be answered at once; for that, in spite of its appearance of perfect strictness, the whole argument is built on too vague a basis, and is also too remote from the physical reality, to which after all it must relate. There is at once one difficulty: all the above propositions deal with natural phenomena constantly manifesting themselves before our eyes on the earth; but what has become of the earth itself? And the answer has to be: it has been thought away, because it does not play any part in those natural phenomena. The only thing that is required is its centre, or rather the centre of the universe, which coincides with that of the earth. It is about this centre that, in the absence of the earth, the second heaviest element, water, accumulates, and in this water the hydrostatic phenomena to be studied take place.

All this is a perfectly Aristotelian conception: gravity is a constant tendency of heavy solids towards a certain point in the universe where the earth indeed has its centre, but which remains the goal of the tendency of everything that is heavy even if the earth is conceived to be removed from its central position.

Now imagine that this has been done, and instead of the earth let all the water have accumulated in a sphere about the centre of the universe. Conceive the surface of this sphere to be divided into a large number of equal parts, which are the bases of conical sectors of the sphere having the centre for their vertex. In all these sectors the water tends downwards, and at the centre a position of equilibrium has been created under the influence of all these conflicting tendencies. What will happen if in one of these sectors the downward tendency becomes more intense because a solid is added? The answer can be deduced with sufficient clarity from the arguments conducted above in the propositions: if in one sector the pressure increases, in another sector the fluid must yield.

But why in one sector, why not in all? And if it happens in one only, in which? Archimedes avoids a direct answer to these embarrassing questions in that, returning to some extent to the reality denied so completely in the above, he considers not the pressures at the centre, but those on a spherical surface with the same centre, of which a horizontal surface in a fluid filling a vessel may be considered to form part. On this spherical surface he takes two adjacent equal surfaces *a* and *b*, and he now seems to imagine that if the pressure on *a* is greater than the pressure on *b*, the fluid in the sector having *a* for its section will force the fluid in the sector in which *b* lies upwards *via* the centre of the universe. We do not hear what happens to the fluid in the other sectors, whether this is assumed to be motionless or whether it is also displaced; the postulate is explicitly confined to adjacent surfaces at the same level.

The interpretation here given of the hydrostatic postulate of Archimedes is also supported by the following consideration: In all the other investigations in which statics plays a part (notably in the *Method*, in *On the Equilibrium of Planes*, and in *Quadrature of the Parabola*) it is naturally assumed that the verticals along which the heavy bodies tend downwards are parallel: consequently, in the proper sense of the word there is no longer a centre of the universe, nor is it referred to in any of the works in question. But in Book I of the work *Floating Bodies* this is all at once different: the verticals are convergent and meet in the centre of the earth; this supposition, however, is made exclusively in Book I, in which the fundamental theorems are derived, while in Book II, where they are applied, the verticals are parallel again. What may have induced the author to make this difference (which also involves a complication) if not the desire to make it clear how phenomena taking place in a column of fluid can effect a column adjacent to the first? It should be borne in mind that gravity tends vertically downwards; if this alone is known and if all verticals are considered to be parallel, it cannot well be understood how (Fig. 158) a difference in pressure at the same level in *a* and *b* can cause motion, not even if both columns are conceived to be continued beneath the bottom of the vessel.

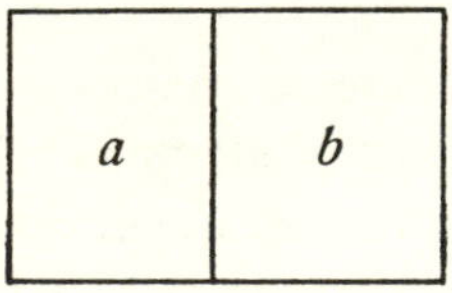

Fig. 158.

If, on the other hand, the verticals are considered to be convergent, the fact that motion is caused by the difference in pressure is indeed not immediately clear either, but it is suddenly understood if, both columns being continued downwards, they are seen to be connected in the centre of the universe. It remains an enigma how the same phenomenon is brought about if they are not connected in that point, but are bounded by the bottom of a vessel, but then such a boundary has been explicitly excluded in the condition formulated in the postulate. As long as the postulate is applied (*i.e.* in Props 2–5), the fluid is conceived to extend to the centre of the universe. In later propositions it is only the results obtained as to the behaviour of solids of different specific gravity in a fluid that are applied. Actually Archimedes ought to have postulated that the hydrostatic phenomena are the same in a vessel with a bottom and in a sphere of water about the centre of the universe.

The above discussion may have served to elucidate the postulate on which the work *Floating Bodies* is based somewhat further. It is true that the first part was always understandable, even though in the first edition of Heiberg the text of it was evidently mutilated. The second part, however, was altogether unintelligible in that edition: it was stated that each portion of the fluid was compressed by the fluid above it, *if* the fluid was shut up in a vessel and was compressed by something else. This *if*, however, has been replaced in the second edition, on the strength of the text of the Jerusalem palimpsest, by *if* . . . *not*, a striking instance of the extent to which the study of the history of mathematics and physics depends on the work of the philologists, who have to furnish us with the sources for the investigation.

This *if* . . . *not* apparently is not to be taken in the sense that the absence of the circumstance in question is a necessary condition for the correctness of the enunciation; in order to make this clear, we have avoided *unless* in the translation: it is merely postulated that the condition is sufficient.

We finally observe that, as appears from the whole of the above considerations, the concept of the hydrostatic pressure, which has a constant magnitude in a given point of a fluid and acts perpendicularly on any plane through that point, in any position whatever, is perfectly alien to Archimedes; *a fortiori* the hydrostatic paradox, the knowledge of which used sometimes to be attributed to him, was unknown to him[1]).

[1]) Attention was already drawn to this by P. Duhem, *Archimède connaissait-il le paradoxe hydrostatique?* Bibl. Math. (3) I (1900), 15–19.

3. We have now come to that part of Archimedes' work which deserves the highest admiration of the present-day mathematician, both for the high standard of the results obtained, which would seem to be quite beyond the pale of classical mathematics, and for the ingenuity of the argument. We are referring to the investigations on the different positions in which a right segment of an orthoconoid (III; 6.21) can float in a fluid, which investigations form the subject of ten propositions in the second book of the work *Floating Bodies*.

In the sequel we shall replace the method of exposition of the original, which is frequently very complicated and inconvenient, by a shorter argumentation of a more analytical character, while we shall also make the connection between the various propositions clearer by bringing uniformity in the notation. A disadvantage of this consists in that not all the figures have been taken over with the same letters, so that a comparison with the original becomes somewhat difficult.

In the following figures, which invariably represent an axial cross-section of the solid, we call:

of the solid itself: the vertex O, the diameter of the base in the plane of the paper AB, the middle point of this Γ, the centre of gravity Z (so that $OZ = 2Z\Gamma$. **Meth.** 5).

of the portion immersed in the fluid, in so far as this is itself a segment of an orthoconoid: the vertex (*i.e.* the point with a horizontal tangent plane) T, the intersection of the base with the plane of the paper ΔE, the middle point of this N, the centre of gravity M (so that $TM = 2MN$).

half the orthia (III; 2.0) of the orthotome AOB in which the plane of the paper cuts the solid: Π.

the ratio between the weights of the solid and of an equal volume of fluid (*i.e.* the ratio between the specific gravities): s; s is invariably supposed to be less than 1.

K: a point on the line segment OZ such that $ZK = \Pi$.

H: the point of TN the projection of which on the axis is K.

I: the projection of T on the axis.

Λ: the point in which the line touching the orthotome AOB in T meets the produced part of the axis (so that $OI = O\Lambda$; III; 2.2).

We further put $O\Gamma = h$, $TN = k$.

4. The theory to be reproduced is mainly based on two fundamental theorems, one of which is enunciated in Book I as a postulate, while the other forms the subject of the first proposition of Book II.

The postulate, which succeeds Prop. 7 of Book I, reads:

ὑποκείσθω, τῶν ἐν τῷ ὑγρῷ ἄνω φερομένων ἕκαστον ἀναφέρεσθαι κατὰ τὰν κάθετον τὰν διὰ τοῦ κέντρου τοῦ βάρεος αὐτοῦ ἀγμέναν.

Let it be granted that any body which is thrust upwards in the fluid is thrust upwards along the vertical drawn through its centre of gravity.

The reference is to the centre of gravity of the portion immersed in the fluid.

The proposition concerns a solid which floats in a fluid:

Proposition 1.

If a solid lighter than the fluid be thrown into the fluid, its weight will be to that of the fluid in the same ratio as the immersed portion to the whole.

Proof (Fig. 159): Of the solid with volume $A+F$ let A be the volume of the immersed portion. Consider a quantity of fluid the volume $N+I$ of which is equal to that of the solid in such a way that $N=F$, $I=A$. Let B be the weight of the solid, O that of the fluid N, R that of I. We then have:

$$[\text{weight }(F+A),\ \text{weight }(N+I)] = (B, O+R).$$

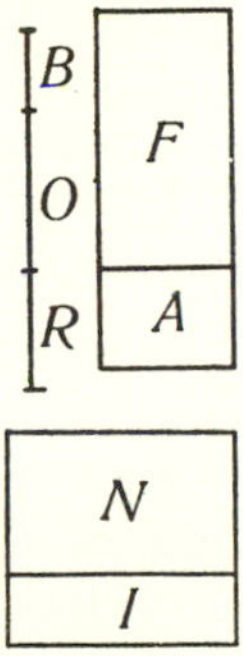

Fig. 159.

By Prop. 5 of Book I:

$$R = B\,.$$

Thence:

$$[\text{weight }(F+A),\ \text{weight }(N+I)] = (R, O+R) = (I, N+I)$$

$$= (A, F+A)\,.$$

5. In order to find the positions in which the solid can float in stable equilibrium, it is first stated as a necessary condition that the centres of gravity Z and M should lie in the same vertical; further the solid should be so far immersed that the weight of the displaced fluid is equal to that of the solid.

If the immersed portion is a (generally) oblique segment of the orthoconoid, the volumes of this portion and of the whole solid are in the ratio of the squares of the diameters k and h (C.S. 24), and consequently, by Prop. 1 of Book II, we must have the relation

$$s = \frac{k^2}{h^2}. \tag{1}$$

In order to test the stability, Archimedes gives the axis an inclination such that M no longer lies in the vertical of Z, and he conceives the solid to be so far immersed that the diameter TN of the portion below the surface of the fluid is still equal to the value k determined by the relation (1). The solid is now acted upon by a couple consisting of the upward thrust along the vertical through M upwards and the weight of the solid along the vertical through Z downwards. In Archimedes' view it is formed by the difference between the weights of the displaced fluid and the immersed portion of the solid, acting along the vertical through M upwards, and the weight of the non-immersed portion of the solid along the vertical through Z upwards. The question is now whether the centre of gravity M does or does not lie on the opposite side of the vertical through Z to the part ZO of the axis of revolution[1]).

The behaviour of the floating segment depends on its geometrical properties as well as on the value of the ratio s. Moreover, a distinction has to be made between equilibria with a vertical and those with an inclined axis.

6. In the first place it can now be ascertained in what case the solid can float in stable equilibrium with the axis vertical for any value of s (but naturally always less than 1). Z and M then already lie in the same vertical. If now the axis is made to incline to the left at the top and the above-mentioned depth of immersion is chosen (the diameter of the immersed segment therefore being k), it is

[1]) It may also be asked whether the point in which the vertical through M meets the axis lies above or below Z. This point was afterwards to be called metacentre. For a formulation of some of Archimedes' propositions on floating paraboloid segments in modern terminology see: Ch. Bonny, *Les Oeuvres d'Archimède et les progrès de la construction navale*. Revue et Bulletin Technique de l'Association des Ingénieurs sortis des Ecoles Spéciales de Louvain et de l'Union des Ingénieurs navals de Belgique. 1945 No 2, pp. 41–68.

necessary and sufficient for stability that M should lie to the left of the vertical through Z. This is undoubtedly the case when the vertex T of the immersed oblique segment lies already to the left of this vertical. This position is shown in Fig. 160. By a well-known property of the subnormal of an orthotome (III; 2.24) we now have

$$I\Omega = \Pi .$$

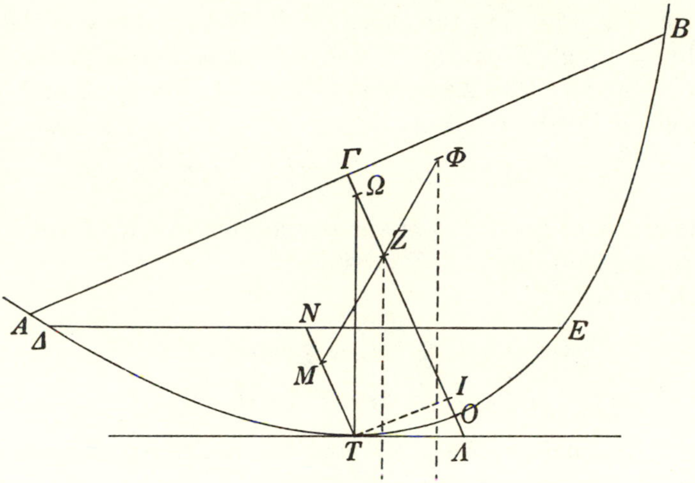

Fig. 160.

In order that T may lie to the left of the vertical through Z it is necessary and sufficient that

$$O\Omega > OZ .$$

For this it is sufficient that

$$I\Omega \geqq OZ$$

or

$$\Pi \geqq \tfrac{2}{3}h$$

or

$$h \leqq \tfrac{3}{2}\Pi .$$

In this way it has become understandable how Archimedes can have arrived at the first proposition about the floating orthoconoid segment, which he enunciates as follows in Book II:

Proposition 2.

If a right segment of an orthoconoid has an axis which is not greater than one and a half times the line-as-far-as-the-axis[1]*), and if it is so placed in the fluid that its base does not touch the surface of the fluid, it will, when given an inclined position, not remain in this inclined position, but will return to the vertical position, whatever may be the ratio of its gravity to that of the fluid*[2]*).*

Proof (Fig. 160): Let the chord ΔE lie in the surface of the fluid. The line through T parallel to the axis passes through the middle point N of ΔE (III; 2.33). The centres of gravity Z and M are determined by the relations

$$OZ = 2Z\Gamma, \qquad TM = 2MN.$$

The centre of gravity Φ of the portion projecting above the surface of the fluid lies in MZ produced.

Now we have

$$\tfrac{3}{2}OZ = O\Gamma \leqq \tfrac{3}{2}\Pi, \qquad \text{whence } OZ \leqq \Pi.$$

Because of $\quad I\Omega = \Pi$, we have $OZ \leqq I\Omega < O\Omega$,

so that the angle between ZT and the horizontal tangent $T\Lambda$ is acute. The vertical through Z therefore meets the horizontal tangent at T in a point between T and Λ. Now also conceive the verticals through M and Φ to be drawn.

The portion lying outside the fluid will now be driven downwards along the vertical through Φ. In fact, it has been assumed that any heavy solid is driven downwards along the vertical through the centre. The portion in the fluid, however, since it is lighter than the fluid, will be thrust upwards, along the vertical through M. Since the portions of the solid are not compressed against each other along the same vertical, the solid will not keep its vertical position, but the portion on the side of A will be thrust upwards, that on the side of B downwards, and this will continue until the solid has reached the vertical position.

By the condition enunciated in the proposition, *viz.* that the base of the segment should not touch the surface of the fluid, it is ensured that the immersed portion is also an (oblique) segment of

---

[1]) This refers to half the orthia; *vide* III; 1.4.

[2]) What is naturally meant here is that the solid should be specifically lighter than the fluid.

the orthoconoid. The condition can always be satisfied by taking the inclination of $O\Gamma$ in relation to the vertical sufficiently small.

In Prop. 3 the theorem is proved once more, this time for the case that the solid floats with the top up and the base is completely immersed in the liquid.

7. In Prop. 2 the condition

$$h \leqq \tfrac{3}{2}\Pi$$

had been imposed on the axis of the segment. The axis therefore must not be too long if the solid is to float in stable equilibrium in a vertical position for any value of s (<1). This condition is now waived, but stable equilibrium with vertical axis is still required. Now the value of s will begin to play a part.

We therefore imagine $h > \tfrac{3}{2}\Pi$ and stipulate that M in Fig. 161 shall lie to the left of the vertical through Z. Now we have

$$I\Omega = ZK = \Pi .$$

If therefore $KH \perp O\Gamma$, H will lie on the vertical of Z (the triangles $TI\Omega$ and HKZ being equal and similar). We now have

$$s = \frac{k^2}{h^2} = \frac{TM^2}{OZ^2} > \frac{TH^2}{OZ^2} = \frac{\Omega Z^2}{OZ^2} = \frac{(\tfrac{3}{2}\Omega Z)^2}{O\Gamma^2}. \qquad (2)$$

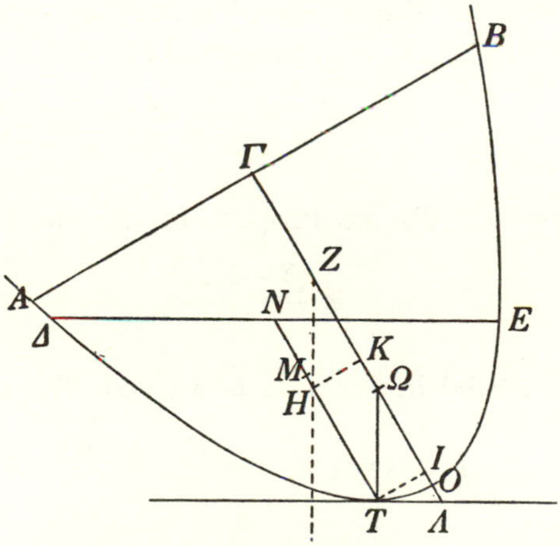

Fig. 161.

Now we have

$$\tfrac{3}{2}\Omega Z = \tfrac{3}{2}(OZ - O\Omega) < \tfrac{3}{2}(OZ - I\Omega) = h - \tfrac{3}{2}\Pi .$$

The relation (2) is therefore satisfied in any case if

$$s \geqq \frac{(h-\frac{3}{2}\Pi)^2}{h^2}.$$

For the case $h > \frac{3}{2}\Pi$ this is therefore a sufficient condition for stability with vertical axis. The result obtained is enunciated by Archimedes in

Proposition 4.

If a right segment of an orthoconoid is lighter than the fluid and has an axis which is greater than one and a half times the line-as-far-as-the-axis, and if its weight has to that of an equal volume of the fluid a ratio which is not less than the ratio which the square of the amount by which the axis exceeds one and a half times the line-as-far-as-the-axis bears to the square of the axis, and if it is so placed in the fluid that its base does not touch the surface of the fluid, it will, when given an inclined position, not remain in this inclined position, but will return to the vertical position.

We shall omit the synthetic proof of this, the contents of which do not differ from the above analytical derivation, and also skip Prop. 5, which deals with the analogous case that the solid floats with the top up, while the base is completely immersed. The result in this case is that for stable equilibrium with vertical axis, if the condition $h > \frac{3}{2}\Pi$ holds, it is sufficient that

$$s \leqq \frac{h^2-(h-\frac{3}{2}\Pi)^2}{h^2}.$$

8. We will now first discuss Prop. 8. In this the condition that

$$s \geqq \frac{(h-\frac{3}{2}\Pi)^2}{h^2},$$

which has been stated in Prop. 4, is waived. The suppositions are now therefore

$$h > \tfrac{3}{2}\Pi \text{ and } s < \frac{(h-\frac{3}{2}\Pi)^2}{h^2}.$$

It appears that the solid, which is now specifically lighter than before, can no longer float in stable equilibrium with vertical axis. In fact, the figure of Prop. 4 (Fig. 161) now has to be changed in such a way that

$$\frac{k^2}{h^2} = s < \frac{(h-\frac{3}{2}\Pi)^2}{h^2}, \text{ whence } k < h-\tfrac{3}{2}\Pi$$

or

$$\tfrac{2}{3}k < \tfrac{2}{3}h - \Pi$$

i.e.

$$TM < OZ - \Pi$$

or, because $I\Omega = KZ$,

$$TM < OZ - KZ = OK\,.$$

If TM becomes even less than IK, then $TM < TH$, so that M lies to the right of the vertical through Z. This, however, takes place already at a slight inclination of the axis relative to the vertical. In fact, when the solid floats vertically, we also have $\frac{2}{3}k < OK$. If therefore $\frac{2}{3}k = OK - \varepsilon$, however small ε may be, an inclination of the axis will always be possible so that $OI < \varepsilon$.

But then we have

$$TM < OK - OI = IK\,.$$

The axis will not therefore return from its inclined position to the vertical position. The question is now whether in this case equilibrium with inclined axis is possible.

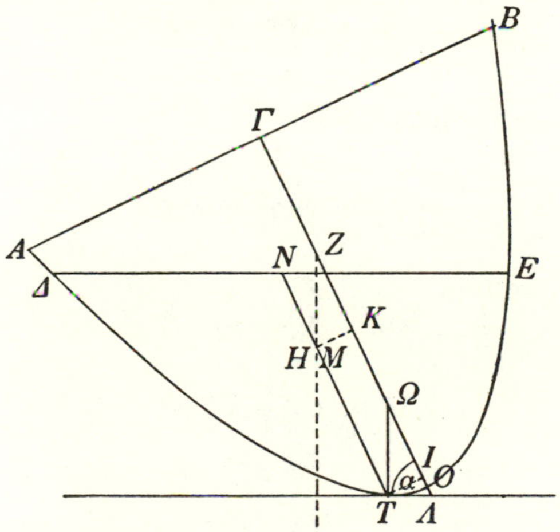

Fig. 162.

In order to ascertain this, we first stipulate, without regarding the stability, that M shall lie on the vertical of Z. In Fig. 162 we therefore conceive M to coincide with H.

We now have as above

$$s = \frac{k^2}{h^2} = \frac{TN^2}{O\Gamma^2} = \frac{TH^2}{OZ^2} = \frac{IK^2}{OZ^2} < \frac{OK^2}{OZ^2} = \frac{(\frac{2}{3}h - \Pi)^2}{(\frac{2}{3}h)^2} = \frac{(h - \frac{3}{2}\Pi)^2}{h^2}.$$

This condition, already mentioned above, is therefore necessary for equilibrium with inclined axis. Further, if we call the angle of the axis with the horizon (which is equal to $\angle\ \Omega TI$) α, we have

$$\cot^2\alpha = \frac{IT^2}{\Pi^2} = \frac{2\Pi . OI}{\Pi^2} = \frac{OI}{\frac{1}{2}\Pi},$$

whence

$$\tfrac{1}{2}\Pi . \cot^2\alpha = OI = O\Omega - I\Omega = OZ - TH - I\Omega = \tfrac{2}{3}(h-k) - \Pi . \quad (3)$$

By this the value of α, at which equilibrium is possible, is determined. Now it has to be ascertained whether this equilibrium is stable.

For this purpose conceive α to be replaced by an angle $\alpha_1 > \alpha$. We now have

$$\cot^2\alpha_1 < \cot^2\alpha ,$$

whence

$$\frac{\Lambda I^2}{TI^2} < \frac{(\frac{2}{3}h - k) - \Pi}{\frac{1}{2}\Pi}.$$

The first member is equal to

$$\frac{4 . OI^2}{2\Pi . OI}, \text{ and consequently to } \frac{OI}{\frac{1}{2}\Pi}$$

and we therefore find

$$OI < \tfrac{2}{3}(h-k) - \Pi ,$$

whence

$$IK = OZ - KZ - OI$$

$$= \tfrac{2}{3}h - \Pi - OI > \tfrac{2}{3}h - \Pi - \tfrac{2}{3}(h-k) + \Pi = \tfrac{2}{3}k = TM, \text{ so that}$$

$$TM < TH .$$

M therefore lies to the right of the vertical through *Z*, and the axis consequently returns to the position of equilibrium. The same is found to be the case, if it is supposed that $\alpha_1 < \alpha$, so that *TM* becomes greater than *TH*.

What we have found here by means of an analysis expressed in modern symbols appears to be in complete agreement with what Archimedes enunciates in Prop. 8. The only difference is that here the length of the axis is subjected to a condition to which we shall have to revert.

Proposition 8.

If a right segment of an orthoconoid has an axis which is greater than one and a half times the line-as-far-as-the-axis, but not so great that it should be to the line-as-far-as-the-axis as fifteen to four, and if further its weight has to that of the fluid a ratio less than that which the square of the amount by which the axis exceeds one and a half times the line-as-far-as-the-axis bears to the square of the axis, it will, when so placed in the fluid that the base does not touch the surface of the fluid, not return to the vertical position and not remain in the inclined position either, except when its axis includes with the surface of the fluid an angle equal to one to be defined more in detail.

Archimedes begins by constructing the angle in question. He takes a line segment $\beta\delta = h$, divides this at $\varkappa$ in such a way that $\beta\varkappa = 2\varkappa\delta$, makes $\varkappa\varrho$ equal to Π and $\beta\tau$ equal to $\frac{3}{2}\beta\varrho$, so that therefore $\tau\delta = \frac{3}{2}\varkappa\varrho$. A second line segment $\varphi+\chi$ (φ being equal to 2χ) is now determined so that

$$[\mathbf{T}(\varphi+\chi), \mathbf{T}(\beta\delta)] = s < [\mathbf{T}(h-\tfrac{3}{2}\Pi), \mathbf{T}(h)],$$

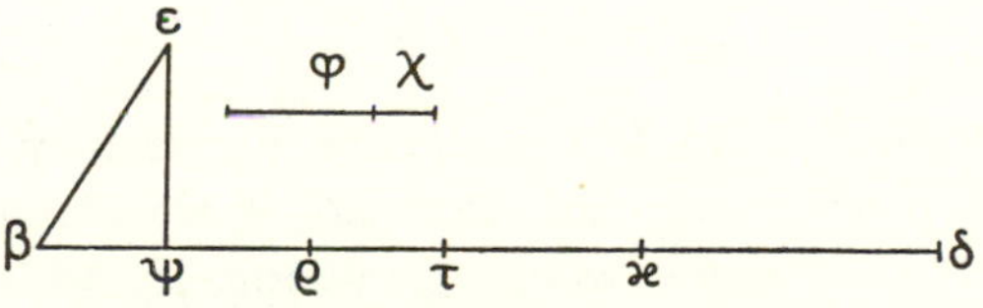

Fig. 163.

from which it follows that

$$\varphi+\chi < h-\tfrac{3}{2}\Pi = \beta\delta-\delta\tau = \beta\tau,$$

whence

$$\varphi < \tfrac{2}{3}\beta\tau = \beta\varrho.$$

Now make $\varrho\psi$ equal to φ and construct $\varepsilon\psi$ at right angles to $\beta\delta$, so that

$$\mathbf{T}(\psi\varepsilon) = \tfrac{1}{2}\mathbf{O}(\varkappa\varrho, \beta\psi),$$

then $\angle\, \varepsilon\beta\psi$ is the angle which the axis of the segment has to make with the horizon if the solid is to float in stable equilibrium.

Evidently this angle is none other but the one called α above. In fact, we here have

$$\frac{(\varphi+\chi)^2}{h^2} = s = \frac{k^2}{h^2},$$

whence

$$\varphi+\chi = k\,.$$

Consequently

$$\beta\psi = \beta\varrho - \tfrac{2}{3}k = \tfrac{2}{3}h - \Pi - \tfrac{2}{3}k = \tfrac{2}{3}(h-k) - \Pi\,,$$

whence

$$\varepsilon\psi^2 = \tfrac{1}{2}\Pi[\tfrac{2}{3}(h-k) - \Pi]\,,$$

and therefore

$$\cot^2 \angle\, \varepsilon\beta\psi = \frac{\tfrac{2}{3}(h-k) - \Pi}{\tfrac{1}{2}\Pi},$$

which is in agreement with the relation (3) for α.

It is seen that, short of using the word cotangent, Archimedes determines the required angle entirely with the aid of trigonometry. The rest of the proof is now merely an inversion of the analysis given above.

9. It now remains to ascertain what is the significance of the condition

$$h < \tfrac{15}{4}\Pi$$

in all this, for in the proof this relation apparently has not been used. Yet the proof would not be valid, if this relation were not satisfied, for the argument would go wrong, if the immersed portion should no longer be an oblique segment of the orthoconoid, *i.e.* if the base were partly immersed in the fluid (indeed, in that case one could no longer write $s = \frac{k^2}{h^2}$, etc.). A condition therefore has to be made which is sufficient to prevent this. This is done in Prop. 6, in which it is proved that for $h < \frac{15}{4}\Pi$ (and invariably $> \frac{3}{2}\Pi$) the solid will return from the position in which the circular base touches the surface of the fluid to the vertical position, so that the edge will project entirely above the fluid.

We again give the analysis of the proof (Fig. 164).

Let the base of the orthoconoid touch the surface of the fluid at A, then the solid will return to the vertical position if in this position M lies to the left of ZH. This is the case when $TH < TM$, or in other words: $\frac{TN}{TH} > \frac{3}{2}$.

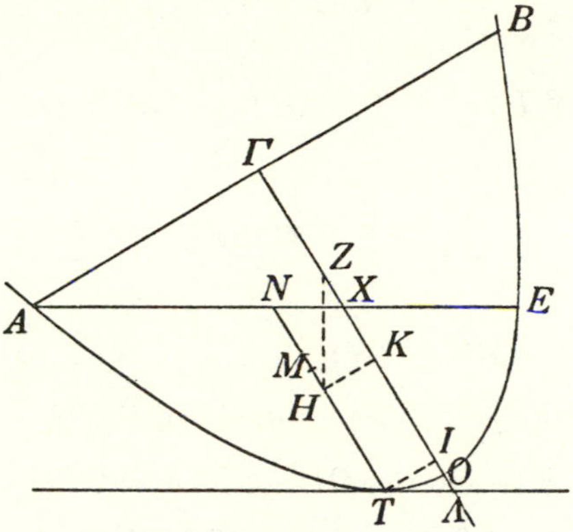

Fig. 164.

We transfer this proportion to the axis. By a lemma to be discussed below we have

$$\frac{TN}{TH} \geqq \frac{\Gamma K}{OK}.$$

The condition in question is therefore undoubtedly satisfied if

$$\frac{\Gamma K}{OK} > \frac{3}{2} \quad \text{or} \quad \frac{OK}{O\Gamma} < \frac{2}{5}.$$

We then have: $OK < \frac{2}{5}h$, and since $OZ = \frac{2}{3}h$, $\Pi = ZK > \frac{4}{15}h$.

Starting from this condition, the synthetic proof gives the argument inversely.

The proof of the lemma in question, which probably was contained together with other lemmas in an appendix to the work, has been lost. It may have been given as follows.

It has to be proved that

$$(TN, TH) \geqq (\Gamma K, OK)$$

or

$$\mathbf{O}(TN, OK) \geqq \mathbf{O}(\Gamma K, TH).$$

If the axis meets the horizontal lines through T and N in Λ and X respectively, it can be proved instead that

$$\mathbf{O}(\Lambda X, OK) \geqq \mathbf{O}(\Gamma K, IK) .$$

In order to determine the position of X on $O\Gamma$, we note that

$$(\Gamma X, I\Lambda) = (\Lambda\Gamma, TI) ,$$

whence

$$[\mathbf{T}(\Gamma X), \mathbf{T}(I\Lambda)] = (O\Gamma, OI) = [\mathbf{O}(O\Gamma, OI), \mathbf{T}(OI)] ,$$

and, because $\Lambda I = 2OI$,

$$\mathbf{T}(\Gamma X) = 4\mathbf{O}(O\Gamma, OI) .$$

Now we have

$$\begin{aligned} &\mathbf{O}(\Lambda X, OK) - \mathbf{O}(\Gamma K, IK) \\ &\quad = \mathbf{O}(OX + OI, OK) - \mathbf{O}(O\Gamma - OK, OK - OI) \\ &\quad = \mathbf{T}(OK) - \mathbf{O}(OK, O\Gamma - OX) + \mathbf{O}(O\Gamma, OI) \\ &\quad = \mathbf{T}(OK) - \mathbf{O}(OK, \Gamma X) + \tfrac{1}{4}\mathbf{T}(\Gamma X) = \mathbf{T}(OK - \tfrac{1}{2}\Gamma X) \\ &\quad \text{or } \mathbf{T}(\tfrac{1}{2}\Gamma X - OK) . \end{aligned}$$

In any case therefore

$$\mathbf{O}(\Lambda X, OK) \geqq \mathbf{O}(\Gamma K, IK) ,$$

with which that which was required to be proved has indeed been proved.

Prop. 7 contains the analogous theorem for a segment floating with the top up.

9. Finally in the very extensive Proposition 10 the case that

$$h > \tfrac{15}{4}\Pi$$

is dealt with, which leads to a distinction of six possibilities. Archimedes starts from the position shown in Fig. 165: the circular base of the segment touches the surface of the fluid and the solid is so far immersed that the diameter k of the immersed segment satisfies the relation

$$s = \frac{k^2}{h^2}.$$

In order to permit a clear view of the different possibilities, the diameter AB has been drawn in the horizontal position, while the section of the surface of the fluid AE revolves about A; E thus describes the orthotome AOB (I). If N is again the middle point of AE, the line through N parallel to $O\Gamma$ meets the curve in the point T, where the tangent is horizontal (parallel to the surface of the

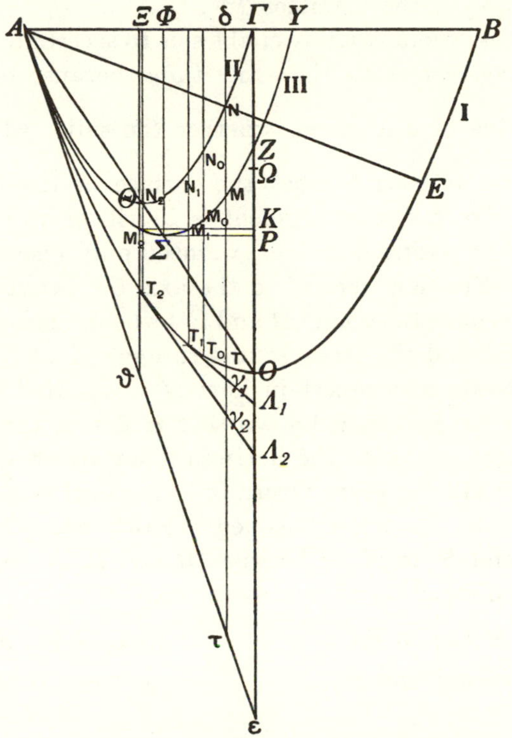

Fig. 165.

fluid AE). If AE revolves about A, the diameter $TN = k$ is displaced parallel to the axis. Since $AN = \frac{1}{2}AE$, it is clear that N describes an orthotome $\Gamma\Theta A$ (II), which passes through A and the middle point Γ of AB and the vertex Θ of which is the middle point of AO. If then the centre of gravity M is determined each time on the variable diameter TN of the immersed segment by means of the relation $TM = 2MN$, a simple calculation shows that M describes an orthotome (III), which passes through A and through Z, and the vertex Σ of which lies on AO in such a way that its distance $\Sigma\Phi$ from AB is $\frac{2}{3}h$. M describes the arc $Z\Sigma A$ of this curve.

As always, let ZK be equal to Π, and let P be the projection of Σ on $O\Gamma$. Then $PZ=(\frac{3}{5}-\frac{1}{3})\,h=\frac{4}{15}h$, and consequently, in connection with the hypothesis $h>\frac{15}{4}\Pi$, we have $ZP>ZK$. The line through K at right angles to $O\Gamma$ therefore meets the curve III in two points M_1 and M_2, which lie successively on the positions T_1N_1 and T_2N_2 of the diameter TN.

The figure now immediately enables us to ascertain, when a value of s or—which comes to the same thing because of the relation $s=\frac{k^2}{h^2}$—a value of k is given, whether the solid, when so placed that the circular base touches the surface of the fluid and the diameter of the immersed segment is k, will or will not continue floating in that position. In fact, we need only insert the line segment k in a direction parallel to the axis $O\Gamma$ between the curves I and II, determine the points M and H[1]) where it meets successively the curves III and the straight line through K at right angles to $O\Gamma$, and subsequently ascertain how M is situated in relation to T, H, and N. In fact, it is known that H lies on the vertical of Z (HZ is at right angles to the horizontal surface of the fluid AE). The axis $O\Gamma$ will therefore return to its vertical position, keep its position or take up a less inclined position according as M lies between N and H, in H, or between H and T. To begin with, this makes for a distinction of five cases, viz.

	If M lies on III	then M lies on TH
1.	between Z and M_1	between H and N
2.	in M_1	in H
3.	between M_1 and M_2	between H and T
4.	in M_2	in H
5.	between M_2 and A	between H and N.

10. We will now first explain how all this is dealt with by Archimedes. He starts by introducing the orthotomes II and III, postulating that the segments which AB cuts off from them, *i.e.* $A\Theta\Gamma$ and $A\Sigma Y$, shall be similar to the segment AOB (*i.e.* that the chords $A\Gamma$, AY, AB shall be proportional to the diameters $\Theta\Xi$, $\Sigma\Phi$, $O\Gamma$), while II should have its vertex Θ in the middle point of AO and III in a point Σ of AO so situated that the projection P of Σ on $O\Gamma$ has a distance $\frac{4}{15}h$ from Z.

[1]) H is not shown in Fig. 165.

It now has to be proved therefore that

1. N is the middle point of AE.
2. The curve III passes through Z.
3. $TM = 2MN$.

These proofs either are not supplied or are sketched only cursorily. They may have been given as follows:

ad 1). The curves I and II have the tangent in common at A; indeed, the line which touches I at A passes through a point ε, so situated on ΓO produced that $\varepsilon O = O\Gamma$, and it therefore meets $\Xi\Theta$ produced in ϑ in such a way that $\Xi\Theta = \Theta\vartheta$; thence it also touches II at A. Application of **Q.P.** 5 (III; 2.7) successively to I and to II now gives

$$(T\tau, T\delta) = (A\delta, B\delta), \text{ whence } (\tau\delta, T\tau) = (AB, A\delta)$$

$$(N\tau, N\delta) = (A\delta, \Gamma\delta), \text{ whence } (\tau\delta, N\tau) = (A\Gamma, A\delta)\,.$$

Since $AB = 2A\Gamma$, it follows from this that $N\tau = 2T\tau$.

By a second application of **Q.P.** 5 (to the segment AOE of I) we find

$$(T\tau, TN) = (AN, EN), \text{ whence } (\tau N, T\tau) = (AE, AN)\,,$$

from which it follows, because $\tau N = 2T\tau$, that

$$AE = 2AN\,.$$

Archimedes merely observes that AN and AE are homologous line segments in the similar segments $A\Theta\Gamma$ and AOB.

ad 2) The theorem **Q.P.** 4 (III; 2.6), when read in the inverse order, formulates a condition which is sufficient to conclude that a point lies on an orthotome. In order that Z may lie on III it is therefore required that

$$(O\Gamma, OZ) = (A\Phi, \Phi\Gamma)\,.$$

Now we have

$$\Sigma\Phi = P\Gamma = h - OP = h - (\tfrac{2}{3}h - \tfrac{4}{15}h) = \tfrac{3}{5}h\,,$$

whence, because of the similarity of the segments,

$$AY = \tfrac{3}{5}AB, \text{ whence } A\Phi = \tfrac{3}{10}AB\,.$$

Further we have

$$\Phi\Gamma = (\tfrac{1}{2}-\tfrac{3}{10})AB = \tfrac{2}{10}AB, \text{ whence}$$

$$(A\Phi, \Phi\Gamma) = (3,2) = (O\Gamma, OZ)\,.$$

ad 3) As above, we see that $A\varepsilon$ also touches III at A. Application of **Q.P.** 5 to I, III, II now gives successively

$$(\tau T, T\delta) = (A\delta, B\delta), \text{ whence } (\tau\delta, T\tau) = (AB, A\delta)$$

$$(\tau M, M\delta) = (A\delta, Y\delta), \text{ whence } (\tau\delta, M\tau) = (AY, A\delta)$$

$$(\tau N, N\delta) = (A\delta, \Gamma\delta), \text{ whence } (\tau\delta, N\tau) = (A\Gamma, A\delta)\,.$$

We therefore have

$$\mathbf{O}(AB, T\tau) = \mathbf{O}(AY, M\tau) = \mathbf{O}(A\Gamma, N\tau)\,,$$

whence

$$\mathbf{O}(AB, M\tau - MT) = \mathbf{O}(AY, M\tau)\,,$$

whence

$$\mathbf{O}(M\tau, BY) = \mathbf{O}(MT, AB) \quad \text{or} \quad (MT, M\tau) = (BY, AB)\,.$$

In the same way $(M\tau, MN) = (A\Gamma, \Gamma Y)\,.$

The ratio (MT, MN) is therefore compounded of

$$(BY, AB) \text{ and } (A\Gamma, \Gamma Y)\,.$$

Now it was found above that

$$AY = \tfrac{3}{5}AB, \text{ whence } BY = \tfrac{2}{5}AB \text{ and } \Gamma Y = \tfrac{1}{10}AB\,.$$

We therefore have

$$(BY, AB) = (2, 5) \text{ and } (A\Gamma, \Gamma Y) = (5, 1)\,,$$

so that

$$(MT, MN) = (2, 1)\,.$$

Archimedes does not speak of the way in which a line segment TN of assigned length k is inserted between the curves I and II each time in the direction of $O\Gamma$. It is probable that this has to be conceived of as a νεῦσις construction (III; 9).

If we now trace in Archimedes the cases distinguished by him, which are obtained by imposing certain conditions on s, we find that he has not five, but six cases. This is due to the fact that the position of equilibrium for the case that

$$s \geqq \frac{(h - \frac{3}{2}\Pi)^2}{h^2}$$

is already known from Prop. 4. In fact, in the latter proposition it was exclusively assumed with regard to h that $h > \frac{3}{2}\Pi$, and if the condition $h > \frac{15}{4}\Pi$ of Prop. 10 applies, this assumption is satisfied.

We now introduce $h - \frac{3}{2}\Pi$ into the figure as the distance from O to a point Ω on $O\Gamma$ (Archimedes does this by taking $K\Omega$ equal to $\frac{1}{2}OK$; this is equivalent to: $O\Omega = \frac{3}{2}OK = \frac{3}{2}(\frac{2}{3}h - \Pi) = h - \frac{3}{2}\Pi$).

The position of Ω in relation to Z depends again on h. We have $O\Omega \gtreqless OZ$ for $h \gtreqless \frac{9}{2}\Pi$; in the figure $O\Omega < OZ$.

For

$$s = \frac{k^2}{h^2} \geqq \frac{O\Omega^2}{h^2},$$

i.e. for $k \geqq O\Omega$, the solid will therefore return to the vertical position and float in stable equilibrium with vertical axis. If we now imagine TN in the position T_0N_0, in which $T_0N_0 = O\Omega$, we can obtain the distinction of the five cases dealt with anew in Prop. 10 by causing M to move between M_0 and A. If we call the angles which the tangents make with ΓO at T_1 and T_2 successively γ_1 and γ_2, we can summarize the theorems enunciated by Archimedes as follows:

If

		in the position of stable equilibrium the perimeter of the base of the segment will meet the surface of the fluid in	and in that position the axis will incline relative to the horizon at an angle α which satisfies the relation
I.	$\frac{O\Omega^2}{O\Gamma^2} > s > \frac{T_1N_1^{\,2}}{O\Gamma^2}$	0 points	$\alpha > \gamma_1$
II.	$s = \frac{T_1N_1^{\,2}}{O\Gamma^2}$	1 point	$\alpha = \gamma_1$
III.	$\frac{T_1N_1^{\,2}}{O\Gamma^2} > s > \frac{T_2N_2^{\,2}}{O\Gamma^2}$	2 points	
IV.	$s = \frac{T_2N_2^{\,2}}{O\Gamma^2}$	1 point	$\alpha = \gamma_2$
V.	$\frac{T_2N_2^{\,2}}{O\Gamma^2} > s$	0 points	$\alpha < \gamma_2$

Archimedes proves these propositions in great detail by the synthetic method in figures in which the surface of the fluid is drawn in horizontal position.

Since, however, his syntheses are invariably the complete inversions of the analyses given above, we shall not record them here.

In conclusion we observe that in the propositions 8 and 9 of Book I Archimedes has already enunciated, with regard to a segment of a sphere, theorems of the same nature as those in Book II on the floating segment of an orthoconoid. It is proved that in the condition of stable equilibrium the axis is vertical, both when the flat boundary plane, turned upwards, projects completely above the surface of the fluid and when, turned downwards, it lies completely beneath the surface of the fluid. The method is identical with that applied in the case of the orthotome. The truth of the result follows immediately from the fact that the centre of gravity of the immersed portion (which is also a spherical segment) lies in the vertical of the centre.

In both propositions the surface of the fluid has been conceived of as spherical about the centre of the universe.

CHAPTER XV

MISCELLANEOUS

In this last chapter we will discuss some minor treatises by Archimedes, which have only been preserved either in fragments or by references in other writers.

I. *The Cattle Problem*[1]).

This problem[2]), which was already famous in antiquity, is contained in an epigram which, as appears from the heading, was sent

[1]) *Opera* II, 528–534. We already saw in Chapter II that the text was first published in 1773 by G. E. Lessing.

[2]) In the Scholia to Plato's *Charmides* 165e it is mentioned as a subject from logistic (quoted by Heiberg, *Opera* II, 528). Likewise by Heron, *Definitiones, Heronis Opera* IV, 98. When Cicero speaks on two occasions (*Epistulae ad Atticum* XII, 4 and XIII, 28) of a *πρόβλημα Ἀρχιμήδειον* in order to indicate something extremely difficult, he is probably also thinking of the cattle problem.

by Archimedes to Eratosthenes with instructions to submit it to the Alexandrian mathematicians. It is considered improbable by philologists that it was written by Archimedes himself in the form in which it has come down to us[1]). This, however, does not exlude the possibility that the problem may quite well originate from him. Hypotheses[2]) have even been drawn up as to the purpose for which he may have set the problem: it is suggested to have been a *tour de force* of Archimedes, intended to put to shame Apollonius, who had calculated the proportion of the circumference and the diameter of a circle more accurately than had been done in the *Measurement of the Circle*, and who had been inspired by the *Sand-Reckoner* to write also a treatise on the notation of large numbers. It is naturally impossible to test such hypotheses as to the motives that may have prompted Greek mathematicians to write their works.

We here translate the formulation of the problem from the Ionian distichs in which it is stated into algebraic symbolism.

The bulls and cows of Helios are grazing in the island of Sicily in four herds of different colours: white, black, dappled, and yellow. If we call the numbers of bulls in these herds respectively W, Z, P, B and the numbers of cows similarly w, z, p, b, the following relations between these numbers are given:

$$\left.\begin{array}{ll} W = (\frac{1}{2}+\frac{1}{3})\,Z+B & w = (\frac{1}{3}+\frac{1}{4})\,(Z+z) \\ Z = (\frac{1}{4}+\frac{1}{5})\,P+B & z = (\frac{1}{4}+\frac{1}{5})\,(P+p) \\ P = (\frac{1}{6}+\frac{1}{7})\,W+B & p = (\frac{1}{5}+\frac{1}{6})\,(B+b) \\ & b = (\frac{1}{6}+\frac{1}{7})(W+w) \end{array}\right\} \quad \text{(I)}$$

In addition it is required[3]) that

---

[1]) The philological side of the question is discussed in detail by B. Krumbiegel, *Das Problema bovinum des Archimedes*. Zeitschr. f. Math. u. Phys. XXV (1880). Hist.-litt. Abt. 121–136.

[2]) Hultsch in Pauly-Wissowa, *Real-Encyclopädie der classischen Altertumswissenschaft, s.v.* Archimedes, col. 534*b*–535*a*.

[3]) This requirement is made by giving the figure formed by the bulls and cows upon orderly disposition. With regard to the white and black bulls it is stated that they stood *ἰσόμετροι εἰς βάθος εἰς εὖρός τε*. If this is taken in the sense that there were equal numbers lengthwise and broadwise, the requirement $W+Z=n^2$ is arrived at. If, however, it is taken to mean that they filled a square, then, since a bull is longer than it is broad, there cannot have been equal numbers lengthwise and broadwise. In that case the requirement is: $W+Z=mn$ (m and n to be positive integers). The problem is greatly simplified in this way, but this view seems hardly ten-

(II) $W+Z$ be a square number, *i.e.* of the form n^2
(III) $P+B$ be a triangular number, *i.e.* of the form

$$\frac{m(m+1)}{2}.$$

} n and m to be positive integers

If only the seven homogeneous equations (I) with eight unknowns are considered, the values of the eight unknowns can be written as multiples of an auxiliary variable t, *viz.*[1]):

$$\begin{aligned}
W &= 2.3.7.53.4657\,t = 10{,}366{,}482\,t\\
Z &= 2.3^2.89.4657\,t = 7{,}460{,}514\,t\\
P &= 2^2.5.79.4657\,t = 7{,}358{,}060\,t\\
B &= 3^4.11.4657\,t = 4{,}149{,}387\,t\\
w &= 2^3.3.5.7.23.373\,t = 7{,}206{,}360\,t\\
z &= 2.3^2.17.15.991\,t = 4{,}893{,}246\,t\\
p &= 2^2.3.5.7.11.761\,t = 3{,}515{,}820\,t\\
b &= 3^2.13.46.489\,t = 5{,}439{,}213\,t
\end{aligned}$$

By II it is required that

$$W+Z = 2^2.3.11.29.4657\,t \text{ be the square of an integer.}$$

For this it is necessary and sufficient that

$$t = 3.11.29.4{,}657x^2 \qquad (x \text{ to be a positive integer})$$

By III it is further required that

$P+B=7.353.4{,}657t=3.7.11.29.353.4{,}657^2x^2$ be of the form $\frac{1}{2}y(y+1)$ (y to be a positive integer).

If we put $2y+1=u$, we have

$$u^2-1 = 4y(y+1) = 2^3.3.7.11.29.353(4{,}657x)^2$$

---

able. In fact, in order that the bulls may form a square, there must be between m and n another relation which is connected with the proportion of length and breadth of a bull. Assuming that length and breadth are commensurable, we get the condition $W+Z=\lambda n^2$ (n to be a positive integer, λ a positive rational number), owing to which the simplification is almost completely lost again. For the solution of the simplified problem consult Heath, *Archimedes*, 320.

[1]) The solution of the problem here given has been taken from A. Amthor, *Das Problema bovinum des Archimedes*. Zeitschr. f. Math. u. Phys. XXV (1880). Hist. litt. Abt. 153–171.

or, if we represent 2.4,657x by v:

$$u^2 - \lambda v^2 = 1 \qquad (\lambda = 2.3.7.11.29.353)\,.$$

The problem therefore leads to a Pellian equation. It is required to find the smallest solution for v which is divisible by 2.4,657.

This equation has been solved with the aid of the continued fraction development of $\sqrt{\lambda} = \sqrt{4729494}$, which has a period of 91 numbers.

It was found that W was a number of 206,545 digits, beginning with 1,598, and that the whole number of the bulls and cows of Helios was expressed by a number of 206,545 digits, beginning with 7,766.

It is unlikely that Archimedes solved this problem completely.

Heiberg's edition contains a scholium[1]) to the cattle problem, in which a set of solutions for the eight unknowns is given. The eight values given are found to satisfy the relations (I); they are the values obtained by putting $t = 80$ in the above expressions for W, Z, P, B, as also the corresponding values for w, z, p, and b. The sum $Z + W$, however, is not a square number, nor is the sum $P + B$ a triangular number.

II. *Lemmas*.

As has already been mentioned, this work has only been preserved in an Arabic version, which is accessible in a Latin translation. The possibility of its being an original treatise by Archimedes in the form in which it has been handed down is excluded, for his name is quoted twice (Props 4 and 14), while moreover there is a reference to a treatise on quadrilaterals as *noster tractatus de figuris quadrilateris*, which is nowhere mentioned among his works.

In his preface[2]) the Arabian editor Thābit b. Qurra says on the authority of one doctor Almochthasso that the work is to be attributed to Archimedes, and that it contains only a few, but very elegant propositions, which ought to be studied between the reading of Euclid and that of the Almagest. For his edition he made use of a work by Abusahal Alkuhi[3]) entitled *Ordinatio libri Archimedis de*

[1]) *Opera* II, 532 *et seq.* The values there given are $W = 829{,}318{,}560$; $w = 576{,}508{,}800$; $Z = 596{,}841{,}120$; $z = 391{,}459{,}680$; $P = 588{,}644{,}800$; $p = 281{,}265{,}600$; $B = 331{,}950{,}960$; $b = 435{,}137{,}040$.

[2]) Quoted *Opera* II, 511 Note.

[3]) *i.e.* Abu Sahl.

assumptis. Apparently therefore the Arabian writers also changed the wording somewhat. Probably, however, it was a compilation of important planimetric lemmas and results already in the original Greek version[1]), so that it is no longer possible to ascertain in how far the contents originate from Archimedes.

The work comprises fifteen propositions of widely varied quality, which we shall not mention in full. We will confine ourselves to the discussion of propositions 4–6 and 14, in which the figures *ἄρβηλος* and *σάλινον* are dealt with, with the express mention that Archimedes is the author.

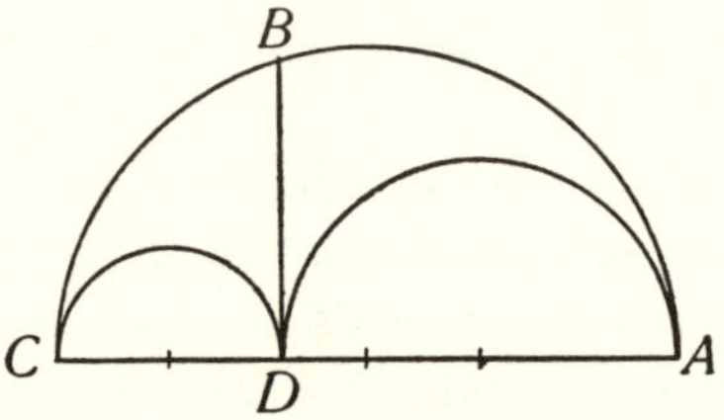

Fig. 166.

In Prop. 4 (Fig. 166) the figure *ἄρβηλος* is introduced as follows:

On a line segment AC a point D is chosen, upon which on one side of AC are described the semicircles having successively AC, AD, CD as diameters. Now *ἄρβηλος*[2]) is the name of the figure included between the circumferences of the three semicircles. With regard to this, in Prop. 4 the property is proved that its area is equal to that of the circle whose diameter is the semi-chord DB which is perpendicular to AC in D. This follows at once from

$$\mathbf{T}(AC) = \mathbf{T}(AD)+\mathbf{T}(CD)+2\mathbf{O}(AD, CD)$$
$$= \mathbf{T}(AD)+\mathbf{T}(CD)+2\mathbf{T}(BD)\,,$$

whence $\frac{1}{2}[\mathbf{T}(AC)-\mathbf{T}(AD)-\mathbf{T}(CD)] = \mathbf{T}(BD)\,,$

which relation also exists between the areas of the circles on AC, AD, CD, and BD as diameters.

In Prop. 5 it is proved that the circles in each of the parts into which the internal common tangent of the two smaller circles divides the *ἄρβηλος* are equal. Each of the inscribed circles touches the outer circle, the common tangent of the two inner circles, and one of these two circles themselves. Their construction (one of the Apollonian tangency problems) is not discussed.

[1]) Heath, *Archimedes, Introduction* XXXII, Note, points to the fact that the compiler has probably drawn from the same source as Pappus, judging from the large number of substantially identical propositions in the *Liber Assumptorum* and the *Collectio*.

[2]) The word denotes a shoemaker's knife.

In Fig. 167 let the ἄρβηλος be formed by the circles **K**(AB), **K**(AC), and **K**(BC). The circles, inscribed in the parts into which the semi-chord which is perpendicular to AB in C divides the ἄρβηλος, touch the outer circle, one of the inner circles, and the chord respectively in F, G, E and in N, M, L. If then EH is the diameter of the first-mentioned circle, by a lemma proved in

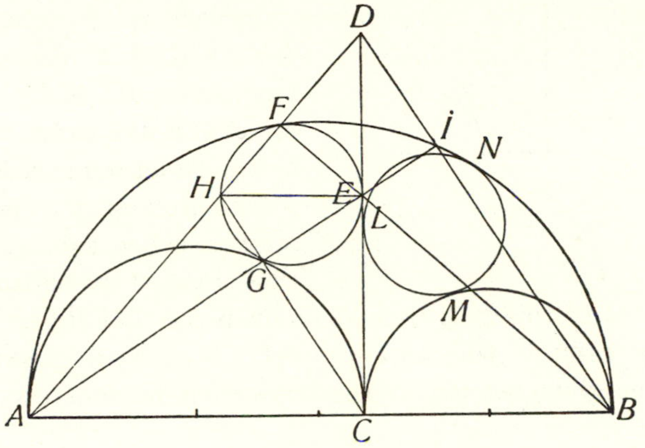

Fig. 167.

Prop. 1 the sets of three points F, H, A and F, E, B are collinear. In $\triangle\, ABD$, E is the orthocentre and consequently BD and AE meet in I on the circumference of **K**(AB); it is further obvious that AE and HC pass through G(G is the internal centre of similitude of **K**(AC) and **K**(HE)).

Apparently $CH \parallel BD$, whence

$$(AB, BC) = (AD, DH) = (AC, HE)\,,$$

whence

$$\mathbf{O}(AB, HE) = \mathbf{O}(AC, BC)\,,$$

by which the diameter HE is determined. Since the expression is symmetrical in AC and BC, the diameter of the circle NLM appears to be equal to HE.

In Prop. 6 the diameter of a circle inscribed in an ἄρβηλος (*i.e.* touching the outer circle internally and the two inner circles externally) is found for the case when the ratio of the diameters of the two inner circles is $3:2$.

Additional theorems on the *ἄρβηλος* are found in Pappus and in several later authors[1]).

In Props 7–13 several lemmas, partly of a rather elementary kind, are derived, which we will pass over in silence. Prop. 14 then deals with the figure *σάλινον*[2]) (*Salinum*), which is formed as follows (Fig. 168):

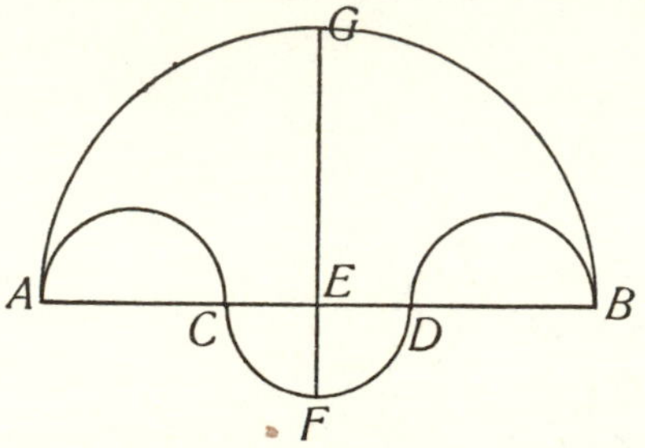

Fig. 168.

On the diameter AB of a semicircle AGB we make $AC = BD$, and then describe semicircles on AC and BD as diameters on the same side of AB as the given semicircle, and also a semicircle on CD as diameter on the other side. The figure thus formed is called Salinon. With regard to this it is proved in Prop. 14 that its area is equal to that of a circle whose diameter FG is equal to the sum of the radii of the given semicircle and the semicircle constructed on the other side.

Proof:

$$\mathbf{T}(AC)+\mathbf{T}(AD) = 2[\mathbf{T}(AE)+\mathbf{T}(ED)] \qquad \text{(Euclid II, 10)} .$$

Because $AB=2AE$, $CD=2ED$ we also have

$$\mathbf{T}(AB)+\mathbf{T}(CD) = 4[\mathbf{T}(AE)+\mathbf{T}(ED)] ,$$

whence, since $AD=FG$,

$$\mathbf{T}(AB)+\mathbf{T}(CD) = 2[\mathbf{T}(AC)+\mathbf{T}(FG)] .$$

---

[1]) Pappus, *Collectio* IV, 14 *et seq.*; 208 *et seq.*

A. Lidonnici, *Gli Arbeli*. Period. d. Matem. (4) 12 (1932), 253–269.

D. J. E. Schrek, *De sikkel van Archimedes.* Nieuw Tijdschrift voor Wiskunde 30 (1942–43), 1–13.

[2]) Opinions differ as to the meaning of the word. A detailed discussion is to be found in Heath, *Archimedes*, Intod. XXXIII, Note. Probably the name of *σάλινον* was applied by a later author to the figure introduced by Archimedes. Heath takes it to be a Graecized form of the Latin word *salinum*, salt-cellar; the name is then supposed to be due to the resemblance of the lower part of the figure to this part of the domestic equipment, which was hightly essential in antiquity. Cantor connects it with *σάλος*, oscillation; the translation might then be *wave-line*. Heiberg thinks of an Arabic corruption of *σέλινον*, *apium*, while Gow translates it by *sieve*.

We therefore also have

$$\tfrac{1}{2}[\mathbf{K}(AB)+\mathbf{K}(CD)-\mathbf{K}(AC)-\mathbf{K}(BD)] = \mathbf{K}(FG).$$

In Prop. 15 an isolated planimetric theorem is given.

As may be seen, the Arbelos and the Salinon only occupy a minor part of the work. The propositions relating to these figures, however, are most in the nature of a result attained, while all the others are true lemmas, which only derive their importance from the applications that can be made of them.

III. *Semi-regular Polyhedra.*

According to Pappus[1]) Archimedes discovered thirteen[2]) of the solids which later became known as semi-regular or Archimedian polyhedra. They are all of the kind now called semi-regular polyhedra of the first type, a class which is defined by abandoning among the various conditions for the regularity of a polyhedron the requirement of congruence of the faces. The faces therefore are still regular polygons, and they still form congruent polyhedral angles at the vertices, but they do not all have the same number of sides and consequently the polyhedral angles are no longer regular.

Pappus, not quite fully, defines the solids in question as *figures contained by equilateral and equiangular, but not similar, polygons*[3]).

In the sequel we give a survey of the polygons mentioned by Archimedes. The significance of the letters is as follows:

Z: the number of faces.
B: the nature of the boundary, in the sense that $(m_i, n_k \ldots)$ in-

---

[1]) Pappus, *Collectio* V, 34 *et seq.*; 352 *et seq.*

[2]) It is wrongly stated by Heron, *Definitiones* 104, *Heronis Opera* IV, 64, that to the five known (regular) polyhedra Archimedes added eight more, so that in all he knew thirteen polyhedra with circumscribed spheres. He also records that two of the solids discussed by Archimedes were already known to Plato, *viz.* a polyhedron with 14 faces, which is bounded by eight equilateral triangles and six squares (No II of the above list) and another polyhedron with 14 faces, which is bounded by eight squares and six equilateral triangles (which, however, does not appear among the solids discussed by Archimedes).

[3]) The shortest formulation of the definition is: polyhedra with regular, non-congruent faces and irregular, congruent polyhedral angles. The semi-regular polyhedra of the second type have irregular, congruent faces and regular, non-congruent polyhedral angles.

dicates that the solid is bounded by m regular polygons of i sides, n regular polygons of k sides, etc.

P: the number of sides of each polyhedral angle.

S: the constitution of each polyhedral angle, in the sense that $(i, k\ldots)$ indicates that at each vertex a polygon of i sides, a polygon of k sides, etc. meet.

H: the number of vertices, calculated from B with the aid of P

$$\left(H = \frac{mi+nk+\ldots}{P}\right)$$

R: the number of edges, calculated from B

$$\left(R = \frac{mi+nk+\ldots}{2}\right)$$

V: the manner in which the figure is produced; the significance of the notation is dealt with below; bracketed figures indicate that the construction method given is not attested by a classical text.

Archimedian polyhedra

No	Z	B	P	S	H	R	V
I	8	$4_3, 4_6$	3	3,6,6	12	18	2
II	14	$8_3, 6_4$	4	3,4,3,4	12	24	1
III	14	$6_4, 8_6$	3	4,6,6	24	36	2
IV	14	$8_3, 6_8$	3	3,8,8	24	36	3
V	26	$8_3, 18_4$	4	3,4,4,4	24	48	(4)
VI	26	$12_4, 8_6, 6_8$	3	4,6,8	48	72	(4)
VII	32	$20_3, 12_5$	4	3,5,3,5	30	60	(1)
VIII	32	$12_5, 20_6$	3	5,6,6	60	90	(2)
IX	32	$20_3, 12_{10}$	3	3,10,10	60	90	(3)
X	38	$32_3, 6_4$	5	3,3,3,3,4	24	60	
XI	62	$20_3, 30_4, 12_5$	4	3,4,4,5	60	120	(4)
XII	62	$30_4, 20_6, 12_{10}$	3	3,6,10	120	180	(4)
XIII	92	$80_3, 12_5$	5	3,3,3,3,5	60	150	

With this the list of the semi-regular polyhedra of the first type is almost complete. The only polyhedra not appearing in the list are the two series which are usually, but somewhat deceptively, called Archimedian prisms and anti-prisms[1]).

The method by which the polyhedra described are produced is explained in a scholium to Pappus[2]), which, however, has unfortunately been preserved only in part; it breaks off in the discussion of the solid No V. From what is left it appears that for the construction of the solids I–IV Archimedes started from the five regular polyhedra, and that he applied to these the three procedures which Stevin in his study on the semi-regular polyhedra[3]) has styled successively truncation 1) *per laterum media*, 2) *per laterum tertias*, and 3) *per laterum divisiones in tres partes*.

In this construction the edges of the regular polyhedra are divided successively into two equal parts, into three equal parts, and into three parts, the middle one of which is to the two outermost as the diagonal of a face to the side. In all three cases the polyhedron is truncated at every vertex by a plane which of each edge bounded by said point contains the point of division closest to that point.

Thus according to the Scholium the solid I is formed from the tetrahedron by construction 2), II from the hexahedron[4]) by 1), III from the octahedron by 2), IV from the hexahedron by 3). The fragment breaks off before it has been said how V can be obtained, which is much to be regretted, since here for the first time a different construction is required from that used with the four pre-

---

[1]) Archimedian prisms are all regular prisms whose height is equal to the edge of the base $(Z = n + 2 . B = (2_n, n_4) . P = 3 . S = (4, 4, n) . H = 2n . R = 3n)$. Archimedian anti-prisms are all prismoids whose bases are congruent regular polygons with n sides, while all the faces are equilateral triangles. The upper and the lower base are turned through $\frac{\pi}{n}$ relative to each other from the position they have in a regular prism, while their distance is such that each lateral edge is equal to an edge of the base $(Z = 2n + 2 . B = (2n_3, 2_n) . P = 4 . S = (3, 3, 3, n) . H = 2n . R = 4n)$.

[2]) Pappus, *Collectio* 1170. Reprinted *Opera* II, 538.

[3]) Simon Stevin, *Problematum Geometricorum Libri* V. Antwerp (1583). A detailed study on this work, which has been used in the above discussion, is: N. L. W. A. Gravelaar, *Stevin's Problemata geometrica.* Nieuw Archief voor Wiskunde (2) V (1902), 106 *et seq.* The reader may also consult E. J. Dijksterhuis, *Simon Stevin.*'s-Gravenhage 1943. p. 99.

[4]) II is also formed from the octahedron by construction 1.

ceding solids. In fact, by means of the truncations described it is only possible to obtain VII from the dodecahedron or the icosahedron (construction 1), VIII from the icosahedron (construction 2), and IX from the dodecahedron (construction 3). Of course it is highly probable that Archimedes did indeed construct the three solids in question in this way, but it cannot be said with certainty how he constructed the six remaining solids. We can merely suspect that for the solids V, VI, XI, and XII he probably made use of the method of double truncation (4), which consists in that first, parallel to each edge, a plane is constructed which divides the edges meeting at its ends in a given proportion, and that the solid thus obtained is subsequently truncated at some of the vertices in a suitable manner[1]).

IV. *The Stomachion.*

1. Of this work two fragments have been preserved, one of which, in Greek, appears in the palimpsest discovered at Jerusalem, which contains the *Method*[2]), while the other is a fragment of an Arabic translation[3]). Together they are still insufficient to give an idea of

---

[1]) Thus V, for example, is formed from a hexahedron as follows. Divide each of the edges into three parts, of which the middle one is to each of the two outermost as a diagonal of a face to an edge. Parallel to each edge construct a plane which passes through the points of division of the edges meeting at its ends which are closest to it. The solid obtained has eight vertices on the diagonals of the hexahedron. Now truncate it at each of these points by a plane through the three vertices of the squares formed in the faces of the hexahedron which are closest to it.

In order to construct VI from a hexahedron, each edge is divided into five parts, of which the middle one is to each of the four others as a diagonal of a face to an edge. In the same way as above, at each edge a triangular prism is cut off. In each of the faces of the hexahedron a square is thus left. Each edge of this is divided into three parts in the proportion mentioned above. Upon this, at each of the vertices lying on the diagonals of the hexahedron a hexagonal pyramid is cut off. In an analogous manner V and VI can also be obtained from the octahedron, XI and XII from the dodecahedron and the icosahedron.

We will not discuss the construction of the solids X and XIII in this place. For this one may consult: M. Brückner, *Vielecke und Vielfläche. Theorie und Geschichte.* (Leipzig 1900), p. 138.

[2]) *Opera* II, 416.

[3]) *Vide* Chapter II. In Suter the stomachion is called συντεμάχιον (= combination of cut-off bits).

the end which Archimedes may have had in view when writing his treatise. We shall therefore have to confine ourselves to giving a summary of their contents.

First a few words are to be devoted to the Stomachion[1]) itself. This is a kind of game, played with bits of ivory in the form of simple planimetrical figures, the object being to fit these bits together in such a way that various shapes of human beings, animals or different objects were imitated. A number of passages in literature are known, where this game is referred to and from which its meaning has been inferred. We are mentioning the following:

α) Ausonius[2]) compares a form of poetry in which all sorts of different metres are used promiscuously to a game which the Greeks called *ostomachia*[3]) and which was played with fourteen bits of ivory in the form of isosceles or equilateral triangles, right-angled or oblique-angled triangles. As examples of what figures could be composed from these bits he mentions: an elephant, a boar, a flying goose, an armed gladiator, a squatting huntsman, a barking dog, a tower, a tankard; he adds that the composition of these figures by adepts might be called marvellous, whereas it became ridiculous when done by inexperienced people.

β) Ennodius[4]) entitles one of his Carmina *De ostomachio eburneo*[5]). The drift of the poem is somewhat mysterious, but the first two lines, *viz.*:

Sollicitata levi marcescunt corda virorum
Tormento: fas est ludere virginibus.

---

[1]) The name Stomachion is translated by Heiberg (*Eine neue Archimedeshandschrift*, Hermes 42 (1907), 240) by *Neckspiel, das einen ärgert und erregt* (cf. Latin stomachari).

[2]) *Decimi Magni Ausonii Burdigalensis Opuscula*, rec. R. Peiper (Leipzig 1886), p. 208. Ausonius was a Roman poet and statesman in the fourth century.

[3]) Thus in the edition quoted. Heiberg (*loc. cit.*), however, notes that the best mss. read *stomachion*, and that this word has wrongly been connected with *ὀστέον* and *μαχία*. In that case *ostomachia* would have to mean something like "combat of bones", which would be a queer name for a game to be played with bits of ivory.

[4]) *Magni Felicis Ennodii Opera*, rec. F. Vogel. Monumenta Germ. Hist. Auct. antiq. tomus VII (Berlin 1885), p. 249. Ennodius (474–521) was bishop of Ticino.

[5]) *Vide* Note 33. Here again the best ms. has stomachio.

(to be rendered by: through a light torment the hearts of men succumb; playing befits girls) may testify to the annoyance to which the stomachion puzzle could give rise, while the next two, *viz.*:

Frangunt Marmaricis elefans quod misit ab arvis
Per micas sparsum mox solidatur opus.

(to be translated by: they break into pieces what the elephant from the regions of Marmarica has brought; the work, scattered in pieces, is soon made into a whole) apparently refer in a rather elaborate way to the composition of figures from bits of ivory.

γ) The Stomachion is connected with the name of Archimedes, under the style of *loculus Archimedius* (Archimedian box), by Marius Victorinus[1]) and by Attilius Fortunatianus[2]), from whose statements we infer that it consisted of fourteen bits of ivory of different forms, that these bits together formed a square, that it was possible to compose from them all sorts of figures (a ship, a sword, a tree, a helmet, a dagger, a column), and that this game was considered very instructive for children, because it strengthened the memory.

2. The name *loculus Archimedius* by no means proves that the game, which has been sufficiently described by the above quotations (the idea of it still survives in toys of our own day), was an invention of Archimedes. The attribute *Archimedius* may have the meaning of difficult, as in *πρόβλημα Ἀρχιμήδειον*[3]) or it may indicate that he studied the game from a mathematical point of view. This latter possibility is corroborated by the introductory sentences of the treatise he devotes to it, in which he calls it necessary to discuss some of the properties of the so-called Stomachion. From the mutilated text it may be inferred that he wanted to ascertain, *inter alia*, which pairs of angles of the figures were together equal to two right angles, either exactly or approximately, and also whether it sometimes happened that two of the figures, taken together, were equal and similar to another figure or to a combination of two of them.

In the Greek fragment, however, we do not find much about this investigation. A square *ΓBZE* (Fig. 169) is considered, in which

---

[1]) Quoted *Opera* II, 417, Note.

[2]) Quoted *Opera* II, 417, Note.

[3]) *Vide* Note 2 of p. 398.

the line joining Γ and the middle point K of ZE meets BZ produced in Δ. ΓK meets BE in H, X is the middle point of ΓH. It is proved that $\angle\ \Gamma XB$ is obtuse and consequently $\angle\ BXH$ is acute, but here the text already breaks off. What follows is a fragment of a second proposition, which is too small to enable us to infer anything as to the object of the argument.

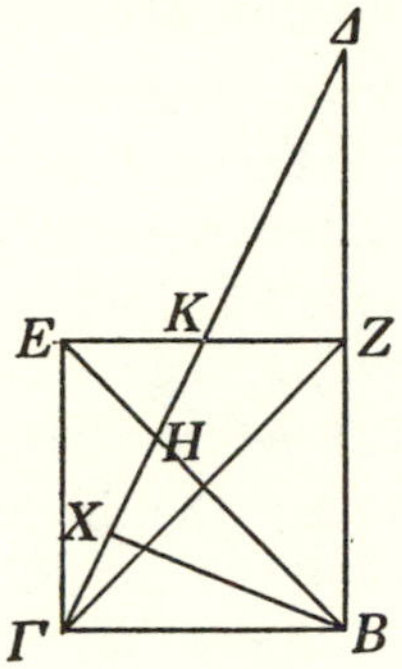

Fig. 169.

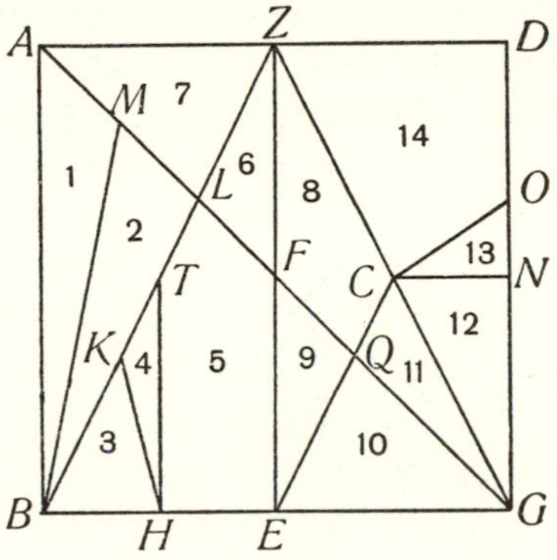

Fig. 170.

3. The fragment in Arabic is more interesting than that in Greek in so far that here the fourteen bits of which the Stomachion consisted are actually produced by division of a square.

The proposition which is enunciated about the figure, however, has no ascertainable connection with the object of the investigation as defined above. In fact, it is proved only that all the fourteen fragments obtained are commensurable with the square. We are not going to give the values of all these proportions, and will confine ourselves to performing the division.

In Fig. 170 let $ABGD$ be a square, and E the middle point of BG, Z the middle point of AD. AG meets EZ in F, BZ in L. M is the middle point of AL, H that of BE. Now draw through H a line perpendicular to BE, which meets BZ in T, and a line HK, which, when produced, passes through A and meets BZ in K. Join B and M. The rectangle AE is now divided into seven parts. In the rectangle ED the middle point C of GZ is joined with E and with the middle point N of GD, and OC is drawn such that, when produced, it passes through B. This rectangle, too, is now divided into seven parts. The whole square is now divided into fourteen parts, which are proved to be commensurable.

It can no longer be ascertained whether this result was the object aimed at or whether it played a part (and if so, what part) in the investigation as originally announced.

V. *Area of the Triangle.*

According to a statement by the Arabian mathematician al-Bîrûnî[1]) Archimedes is the author of the expression—usually attributed to Heron—of the area of a triangle in terms of the sides, which is nowadays rendered by the formula $\sqrt{s(s-a)(s-b)(s-c)}$.

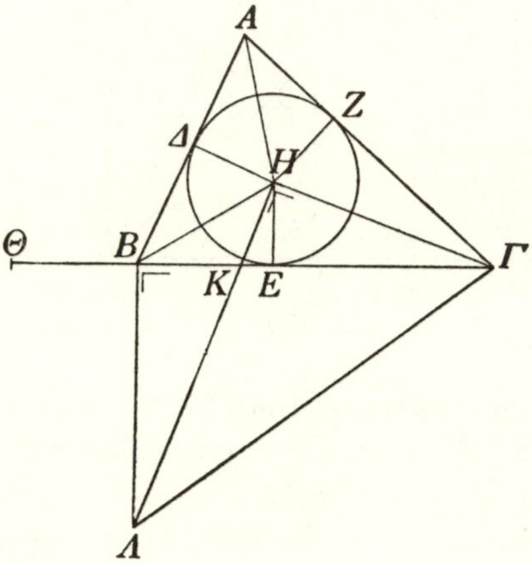

Fig. 171.

In Heron[2]) this proposition is enunciated and proved as follows: *If the sides of a triangle be given, to find its area.*

In Fig. 171 let the triangle $AB\Gamma$ be given; the inscribed circle with centre H touches the sides successively in Δ, E, Z. If Π represents the circumference of $AB\Gamma$, we have the familiar relation

$$\mathbf{O}(\Pi, HE) = 2.\triangle\ AB\Gamma\,.$$

Now produce ΓB with $B\Theta = A\Delta$, then we have $\Gamma\Theta = \frac{1}{2}\Pi$, whence

$$\mathbf{O}(\Gamma\Theta, HE) = \triangle\ AB\Gamma \qquad (1)$$

What now follows is a step which was not usual in classical Greek mathematics, and which raises some doubt whether the proof in the form in which it is quoted by Heron can really originate from Archimedes. In fact, the rectangle $\mathbf{O}(\Gamma\Theta, HE)$ (in Greek τὸ ὑπὸ τῶν $\Gamma\Theta$, HE) in turn is taken as a side of an entity of a higher order, which cannot, however, occur in the three-dimensional space

[1]) *Das Buch der Auffindung der Sehnen im Kreise* von Abū'l Raihān Muhammad-al-Bīrūnī. Übersetzt von H. Suter. Bibl. Math. (3) XI (1910–1911), p. 39. In the same work (page 37) the calculation of the perpendiculars of a triangle and of the parts into which they divide the sides is also attributed to Archimedes.

[2]) Heron, *Metrica*, I, 8. *Heronis Opera* III, 18–24.

of Greek geometry. This shows that the expression τὸ ὑπὸ τῶν $\Gamma\Theta$, HE has lost its direct geometrical meaning, and, just like $\Gamma\Theta$ and HE themselves, is looked upon as a dimensionless magnitude (or number), which can be squared in its turn. In accordance with this we may write the conclusion drawn from (1)

$$\mathbf{T}(\Gamma\Theta).\mathbf{T}(HE) = \triangle\, AB\Gamma.\triangle\, AB\Gamma \tag{2}$$

Now draw through H a line perpendicular to ΓH and through B a line perpendicular to ΓB, which two perpendiculars may meet in Λ. Now $\Gamma HB\Lambda$ is a cyclic quadrilateral, so that we have

$$\angle\, \Gamma HB + \angle\, \Gamma\Lambda B = 2R\,.$$

However, we also have

$$\angle\, \Gamma HB + \angle\, AH\Delta = 2R\,,$$

from which it follows that

$$\angle\, \Gamma\Lambda B = \angle\, AH\Delta\,.$$

Through this we know that

$$\triangle\, \Gamma\Lambda B \sim \triangle\, AH\Delta\,,$$

whence

$$(B\Gamma, B\Lambda) = (\Delta A, \Delta H) = (B\Theta, EH)$$

or

$$(B\Gamma, B\Theta) = (B\Lambda, EH)$$

or, if K is the point of intersection of $B\Gamma$ and $H\Lambda$,

$$(B\Gamma, B\Theta) = (BK, EK)\,,$$

whence

$$(\Gamma\Theta, B\Theta) = (BE, EK)$$

or

$$[\mathbf{T}(\Gamma\Theta),\ \mathbf{O}(\Gamma\Theta, B\Theta)] = [\mathbf{O}(BE, \Gamma E),\ \mathbf{O}(EK, \Gamma E)]$$

$$= [\mathbf{O}(BE, \Gamma E),\ \mathbf{T}(EH)]\,.$$

From this it follows, with abandonment of the geometrical representation,

$$\mathbf{T}(\Gamma\Theta).\mathbf{T}(EH) = \mathbf{O}(\Gamma\Theta, B\Theta).\mathbf{O}(BE, \Gamma E)\,,$$

whence, because of (2),

$$\triangle\, AB\Gamma.\triangle\, AB\Gamma = \mathbf{O}(\Gamma\Theta, B\Theta).\mathbf{O}(BE, \Gamma E)\,.$$

If we now put $\Gamma\Theta = s$ and the sides of the triangle a, b, c, this is equivalent in modern notation to

$$(\triangle AB\Gamma)^2 = s(s-a)(s-b)(s-c)\ .$$

VI. *Construction of a Regular Heptagon.*

According to Arabian tradition Archimedes has written a work *On the Heptagon in a Circle.* For some decades past we have been acquainted with a text by Thābit b. Qurra[1]), which he announces as the *Book of Archimedes.* This contains, after a number of theorems on triangles and without any connection therewith, a construction of a regular heptagon. It is based on the following theorem:

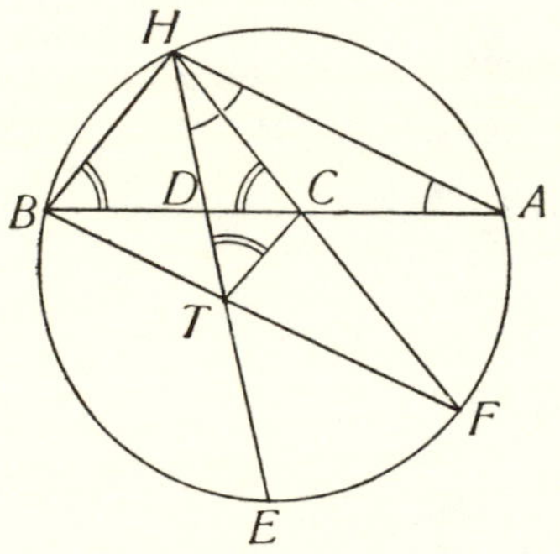

Fig. 172.

In Fig. 172 let there be given a straight line segment AB, and on it two points C and D, so that

$$\mathbf{O}(AD, CD) = \mathbf{T}(BD) \qquad (1)$$

$$\mathbf{O}(CB, BD) = \mathbf{T}(AC) \qquad (2)$$

Now construct $\triangle DHC$ in such a way that $HD = BD$ and $HC = AC$.

Now construct a circle through A, H, B, then in this circle BH is a side of the inscribed regular heptagon.

Proof: Let HC and HD produced meet the circle successively in F and in E, and let BF meet HE in T.

From (1) it follows:

$$(AD, BD) = (BD, CD)\ ,$$

whence $$(AD, DH) = (HD, DC)\ ,$$

whence $$\triangle ADH \backsim \triangle HDC\ ,$$

whence

$$\angle HAD = \angle CHD, \text{ so that arc} BH = \text{arc} FE = \text{arc} AF$$

$$(\text{because } CH = CA)\ .$$

---

[1]) *Vide* p. 49, Note 3. The *Book of Archimedes* is rendered in full by J. Tropfke, *Die Siebeneckabhandlung des Archimedes.* Osiris I (1936), 636–651.

Now therefore $\angle BTH = \angle BCH$, so that $BHCT$ is a cyclic quadrilateral.

Because $DH = DB$, we also have $DC = DT$, whence $HT = BC$.

From (2) it follows:

$$(CB, AC) = (AC, BD),$$

for which we may now write

$$(HT, HC) = (HC, HD),$$

so that $$\triangle HTC \sim \triangle HCD,$$

whence $\angle HTC = \angle HCD$, so that $\angle HBA = \angle HCD = 2 . \angle BAH$.

Consequently

$\text{arc}\, AH = 2 . \text{arc}\, HB$, from which follows that which it was required to prove.

In fact, $\text{arc}\, HB = \frac{1}{2}\text{arc}\ HA = \text{arc}\, AF = \text{arc}\, FE = \frac{1}{2}\text{arc}\, BE = \frac{1}{7} . 360°$.

The regular heptagon can therefore be constructed, if C and D can be so found on AB that the relations (1) and (2) are satisfied. This is possible with the aid of a νεῦσις, which is, however, less simple than the insertions of line segments of known length which we have so far encountered.

In Fig. 173 let $ABCD$ be a square (AB is not the line segment AB from Fig. 172). Through D draw a line which meets CB in T, CA in E, and BA produced in F, so that

$$\triangle DTC = \triangle AEF$$ [1].

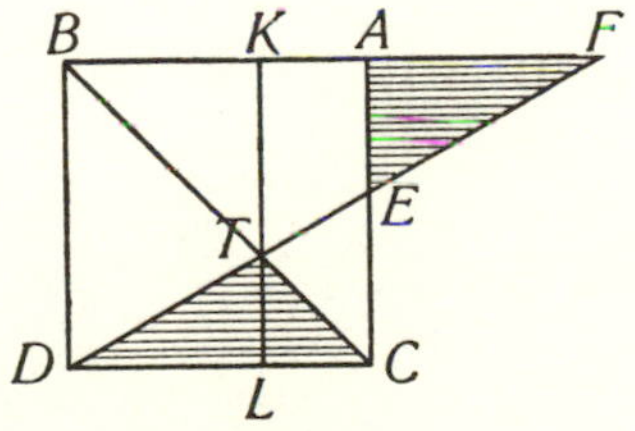

Fig. 173.

If the line through T parallel to BD meets the sides DC and BA successively in L and K, we have

$$\mathbf{O}(DC, TL) = \mathbf{O}(AF, AE)$$

or, because $TL = LC = AK$,

$$\mathbf{O}(AB, AK) = \mathbf{O}(AF, AE),$$

[1]) It does not appear how this νεῦσις is performed. It may of course have been done by means of conics.

from which it follows in connection with $AB > AE$ that

$$AF > AK\,.$$

Further we have $(AB, AF) = (AE, TL) = (AF, LD)$,

whence $$\mathbf{T}(AF) = \mathbf{O}(AB, BK)\,. \tag{1}$$

We also have $$(TL, TK) = (LD, KF)$$

or $$(AK, KB) = (KB, KF)\,,$$

whence

$$\mathbf{T}(KB) = \mathbf{O}(AK, KF)\,.$$

It therefore appears from (1) and (2) that BF has been divided as required. This division may be transferred by means of oblique projection to the line segment AB from Fig. 172, where with B, K, A, F, there correspond respectively B, D, C, A.

BIBLIOGRAPHY

With regard to a number of works which are repeatedly quoted, the title and the abbreviation employed in the text are given below.

Euclides Opera Omnia edd. J. L. Heiberg et H. Menge. I–IV. *Elementa* (Leipzig 1883–1885). VII. *Optica, Opticorum Recensio Theonis, Catoptrica, cum scholiis antiquis* (Leipzig 1895). VIII. *Phaenomena et Scripta Musica* (Leipzig 1916). To be quoted as: *Euclidis Opera*; propositions from the *Elements* are referred to by a mention of the Book and the Proposition, as follows:

Euclid X, 1 = *Elements*, Book X, Prop. 1.

E. J. Dijksterhuis, *De Elementen van Euclides*. I (Groningen 1929). II (Groningen 1930). To be quoted as: *Elements of Euclid*.

Archimedis Opera Omnia cum Commentariis Eutocii iterum edidit J. L. Heiberg. I (Leipzig 1910). II (Leipzig 1913). III (Leipzig 1915). To be quoted as: *Opera*.

T. L. Heath, *The Works of Archimedes, edited in modern notation with introductory chapters* (Cambridge 1897). To be quoted as: Heath, *Archimedes*.

Paul Ver Eecke, *Les Oeuvres Complètes d'Archimède, traduites du grec en français avec une introduction et des notes* (Paris, Bruxelles, 1921). To be quoted as: Ver Eecke, *Archimède*.

Apollonii Pergaei quae Graece exstant cum commentariis antiquis, ed. J. L. Heiberg. I (Leipzig 1891). II (Leipzig 1893). To be quoted as: Apollonius, *Conica*.

Heronis Alexandrini Opera quae supersunt omnia. II, Fasc. 1. *Mechanica et Catoptrica* edd. L. Nix et W. Schmidt (Leipzig 1900). III. *Rationes Dimentiendi et Commentatio Dioptrica* ed. H. Schöne (Leipzig 1903). IV. *Definitiones cum variis collectionibus. Geometrica* ed. J. L. Heiberg (Leipzig 1912). V. *Stereometrica. De Mensuris* ed. J. L. Heiberg (Leipzig 1914). To be quoted as: *Heronis Opera*. II, 1 as *Mechanica*. III as *Metrica*.

Pappi Alexandrini Collectionis quae supersunt ed. F. Hultsch. I (Berlin 1875) *Librorum* II, III, IV *Reliquiae*. II (Berlin 1877) *Librorum* VI *et* VII *Reliquiae*. III (Berlin 1878) *Libri* VIII *Reliquiae. Supplementa*. To be quoted as: Pappus, *Collectio*, as follows:

Collectio VIII, 5; 1030=Liber VIII, caput 5, page 1030 of the quoted edition with continuous pagination.

T. L. Heath, *A History of Greek Mathematics*. I. II (Oxford 1921). To be quoted as: Heath, *Greek Mathematics*.

Propositions from Archimedes are quoted:

α) in general: by reference to the work, the book (if any), and the number of the proposition; for the works a system of abbreviations is used which is mentioned at the end of Chapter II, *e.g.*: **S.C.** II, 4 = *On the Sphere and Cylinder*, Book II, Proposition 4.

β) in the course of the discussion of the work in which they occur: by mere reference to the book (if any) and the number of the proposition.

Chapter III of this work is quoted as: III, followed by a decimal fraction indicating the subdivision.

ARCHIMEDES AFTER DIJKSTERHUIS: A GUIDE TO RECENT STUDIES

BY WILBUR R. KNORR

E. J. Dijksterhuis (1892–1965) gained wide respect for his historical studies in the exact sciences. These included major contributions on the work of the sixteenth-century Flemish physicist-engineer Simon Stevin (see Dijksterhuis [1955; 1970]), and most notably, his expansive survey of the history of physical science from antiquity to Newton, *The Mechanization of the World Picture* (1950; 1961). (For further particulars on Dijksterhuis' life and work, I refer the reader to the admirable portraits by Struik [1986] and Hooykaas [1966].)

In these works Dijksterhuis criticizes the ancients for building mechanical science on the foundation of formal geometry, thus elevating considerations of logic and rigor, but neglecting matters of practical computation and physical application. It is surprising, then, that Dijksterhuis himself devoted particular energy to the study of ancient geometry, producing a Dutch translation with commentary of Euclid's *Elements* (2 vols., 1929–30) and a series of essays on the work of Archimedes (in Dutch, 1938–44). The latter, in English translation, collected and revised, appeared in 1956 as *Archimedes*, the volume here being reissued by Princeton University Press.

In so pursuing ancient mathematics, Dijksterhuis was elucidating the tradition that furnished the prevalent model for the exact sciences throughout antiquity and the Middle Ages. If mechanical science in the sixteenth and seventeenth centuries took greater account of computational and experimental methods, it did not at all reject the ancient technical precedent. The leading scientists—one may think of Stevin, Galileo, Kepler, Huygens, and Newton—were firmly grounded in the study of the ancients. Galileo's *Two New Sciences* and Newton's *Principia*, for instance, drew from the Euclidean-Archimedean model the essential elements of their expository form. While thus retaining the ancient conception of *knowledge*, as secured through precise deductive reasoning, the moderns emphasized the search for more powerful methods for the *discovery* of mathematical and physical truths.

Viewed in this light, Archimedes is a "modern." In the preface to his *Method* he speaks with evident pride of his "mechanical method," applying the physical principle of the balance toward the discovery of geometric propositions. Such an analysis can be useful, Archimedes maintains, even though it might lack the rigor of formal geometric proof. His writings thus offer a rich reward for the student interested in getting behind formal exposition to grasp the originating mind at work. Unlike Euclid and Apollonius, whose extant writings fall primarily within the category of textbooks, Archimedes' work is dominated by research letters, communicating in precise terms his own relatively recent findings. We are thus close to the stage of discovery—close enough, one often feels, to retrieve a sense of the actual process of discovery.

As critic and commentator, Dijksterhuis shares Archimedes' estimation of discovery. In his *Archimedes* he undertakes with skill and insight the task of reconstructing the heuristic thought behind Archimedes' formal proofs and so carries the reader to a deeper understanding of Archimedes' achievement. Thirty years after its publication, Dijksterhuis' *Archimedes* remains the best survey available in English. Only T. L. Heath's *Works of Archimedes* (1897; with supp., 1912) attempts a comparably detailed overview; but this work is compromised by being based on Heiberg's older edition of the text (1880–81), entirely superseded by the second edition (1910–15; see Section II below). By adopting modern notation Heath also risks misrepresenting the thought-line of Archimedes' proofs. Literal translations based on Heiberg's revised text have been produced in modern languages other than English (see References, Section A). As Dijksterhuis notes, however, such renditions of the "dry-as-dust" formal style can be taken in by modern readers only with considerable difficulty. By introducing a new notational format, Dijksterhuis hoped to provide a full transcription of Archimedes' proofs while avoiding the cumbersome expository style typical of ancient formal writing. At the same time, he incorporated findings drawn from the secondary literature on Archimedes and so pursued the examination of general conceptual issues more fully than had the older editions.

Thus, his study has become, and remains, an important starting point for contemporary discussion of Archimedes. But scholarship has inevitably outdistanced it on many fronts. It thus seems appropriate to accompany the reissue of Dijksterhuis' volume with a brief overview of Archimedes study over the last three decades. This project is considerably facilitated by the appearance of I. Schneider's *Archimedes* in 1979, providing a thorough bibliographical survey of Archimedes scholarship in this century, and so both updating Dijksterhuis and filling in gaps in his coverage of earlier work. Thus, the present survey can be selective for the period covered by Schneider and concentrate on efforts during the last ten years. No doubt some worthy contributions will be omitted through inadvertence; for this I apologize.

The following sections adopt the order of topics in Dijksterhuis and follow the numbering of his chapters. For convenience, I will cite Dijksterhuis' book by the abbreviation "EJD," and refer to the works in the Archimedean corpus by the acronyms there set out (see EJD, 46–47).

I. *Archimedes' Life and Work*

Full-length Archimedes studies which have appeared since Dijksterhuis' include Kagan [1955]; Schneider [1979a]; Zhitomirskii [1981]. Among articles and chapters devoted to Archimedes, one may note van der Waerden [1954, chap. 7]; Dedron and Itard [1959, vol. 1, chap. 4.2]; Rybnikov [1960]; Itard [1963]; Aaboe [1964, chap. 3]; Becker [1966]; Onicescu [1967]; Stamatis [1968c, 108–49]; Bashmakova [1970]; Clagett [1970]; Krafft [1971]; Bernhardt [1975]; Edwards [1979, chap. 2]; Reixach i Vilà [1982]; Zhitomirskii [1982]; Knorr [1986a, chap. 5]. Extensive bibliographies, including editions, translations, and secondary literature, are given by Schneider [1979a] (emphasizing twentieth-century titles), by Procissi [1981, 17–33], and by Stamatis [1970–74, 1(A):309–52; 2:565–72]. A useful overview of current directions in research on ancient mathematics, including notes on Archimedes study, is given by Berggren [1984].

To supplement the biographical account presented by Dijksterhuis (EJD, chap. I), one should consider the more general cultural and political context of the Hellenistic world, as surveyed by Tarn [1952]; see also contributions in the *Cambridge Ancient History* [1928; 1930; 1984]. For the specific situation of Sicily, see Finley [1968], who includes observations on Archimedes, and the monograph of Berve [1959]. Ptolemaic Alexandria provides a useful focus for examining the social and intellectual currents, and is surveyed with especial thoroughness by Fraser [1972]. One should note the reservations expressed by Toomer [1976a, 2n], however, who emphasizes the diversity of intellectual centers in this period. For an account of the general setting of Hellenistic science, see Lloyd [1973]; on the literary side, see Aujac [1979–80].

Our oldest and best witness to the life of Archimedes is Polybius, whose account is augmented by Livy and Plutarch (see EJD, chap. I; Schneider [1979a, chap. 1]). The ancient testimonia are assembled by Stamatis [1970–74, vol. 1(A)], who gives the original Greek or Latin, with modern Greek translation. On specific questions relating to Archimedes' biography, see Tronquart [1966] and Haury [1980]. The later literary use of Archimedes as a type is discussed by Courcelle [1959]; Quacquarelli [1960–61]; Gigon [1973]; Dragoni [1975]. The striking mosaic which Dijksterhuis reproduces as his frontispiece illustrates the death of Archimedes (see EJD, 32). More recent analysts assign the work to a Renaissance artist, perhaps imitating a lost ancient mosaic; see van der Waerden [1954, 210]; Schneider [1979a, 19].

In both the ancient and modern popular imagination, the fame of Archimedes rests on his remarkable mechanical contrivances. The legend of his boasting that he could move the earth itself, were he given a place to stand, epitomizes the limitless potential that came to be assigned to Archimedes' feats (see EJD, 14–18). Drachmann [1958] suggests that Archimedes performed the related feat of singlehandedly launching King Hieron's three-master by means of a combination of triple pulley, windlass, and endless screw. For a textual criticism of the passages, see Ver Eecke [1955]. Overall, the mechanical element of Archimedes' activity is deemed central in the accounts by Schneider [1979a, esp. chap. 3; 1979b; 1982]. A similar emphasis is placed on mechanics by Zhitomirskii [1981] and Kozlov [1984].

In the following, I will take up the several mechanisms and instruments attributed to Archimedes and note treatments of them; these references should be supplemented by discussions and bibliography in Schneider [1979a].

Although the mechanism underlying the boast may be a form of compound pulley, some ancient witnesses envision a *charistiōn*, or steelyard (the so-called Roman balance). That Archimedes invented a device of the latter sort for hydrostatic measurements (e.g., in connection with the problem of the "wreath"; see EJD, 18–21) is attested in Arabic sources (see Stamatis [1979]). For a review of textual and archeological evidence of such balances, see Knorr [1982c, app. A] and Jenemann [1984]. On the physical solution of the wreath problem, see Hoddeson [1972], who brings in the method proposed by Galileo. Some ramifications of the legend in the Renaissance are discussed by Gavagna [1979].

The Archimedean invention of the *cochlias*, or "snail" (a form of endless screw) for water lifting is doubted by Dijksterhuis (EJD, 21–23), but defended by Drachmann [1963b, 152–54]; see also Oleson [1984]. The tradition of Archimedean water clocks is expansive among writers in Arabic; see Hill [1974; 1978; 1984]. On the Vitruvian hodometer, see Sleeswijk [1979], who suggests its Archimedean provenance. For surveys of ancient mechanical devices, not restricted to those of Archimedes, see Drachmann [1963b]; Landels [1978]; Hill [1984]. The survey in Hill [1984] also includes coverage of medieval developments.

Archimedes' construction of mammoth catapults for the defense of Syracuse is firmly rooted in the ancient biographical tradition (see EJD, 26–28). On these and related military engines, see Drachmann [1963b]; Soedel and Foley [1979]. The ancient treatises, with translation and commentary, are assembled by Marsden [1970; 1971]. Hero of Alexandria's account of catapult design includes a cubic relation between throw weight and linear scale; the *neusis* he describes for its solution is assigned an Archimedean provenance by Knorr [1986a, 188–91]. The alternative solution discussed by Hart [1982] is better known as the "pseudo-Platonic" method of cube duplication; its Eudoxean invention is proposed by Knorr [1986a, 57–61].

The tradition of Archimedes' burning mirrors is, by contrast, far less secure. The reviews of the ancient sources by Dijksterhuis (EJD, 28–29) and Schneider [1969] are strongly skeptical. But there are still ardent modern advocates, notably Stamatis and Sakkas [1973]; recent notices are collected in Stamatis [1982]. On technological aspects of the question, see Simms [1977], who also supplies ample textual and bibliographical data; on historical and geometric aspects, see Knorr [1983]. The history of the legend is examined by Middleton [1961] and Baltrušaitis [1978]. For briefer accounts, see Africa [1975] and Thuillier [1979].

Archimedes' practical work in optics is represented by his design and analysis of a sighting device in the *Sand Reckoner* (cf. EJD, 364–66). A translation of the principal passage is given by Shapiro [1975], who styles it "the earliest extant Greek account of an astronomical measurement." The instrument and Archimedes' use of it are examined in detail by Delsedime [1970a] and Lejeune [1947]; for a summary, see Schneider [1979a, 92–95], and for additional references see Shapiro [1975]. There survive testimonia to Archimedean efforts in theoretical optics, touching specifically on the geometry of reflected and refracted rays and the study of mirrors (catoptrics); for the ancient witness, see Schneider [1979a, 96–97]; Rome [1932]. (Mugler [1957] notes an Archimedean optical aspect to one ancient conception of the straight line.) But the correctness of the ancient attributions is questioned by Knorr [1985b], who argues their confusion with the pseudo-Euclidean *Catoptrics*.

The ancients held in especially high esteem Archimedes' planetarium, a spherical device which reproduced mechanically the motions of stars, sun, moon, and planets (see EJD, 23–25). Archimedes' own account of the device is lost, but references to it survive in several ancient authors, particularly Cicero. These passages are surveyed by Gigon [1973] with special emphasis on their philosophical and theological context. Technical and astronomical aspects of the design are surmised by Price [1974, 55–59] in the light of the Antikythera mechanism (a geared calculator for solar and lunar cycles). For an extensive discussion of Archimedes' planetarium, with proposals for a reconstruction, see Zhitomirskii [1978]. Although well familiar with Aristarchus' heliocentric cosmology, Archimedes adopted a geocentric design for this device. Among the forms of planetary model he might have used are the homocentric spheres, earlier introduced by Eudoxus and Callippus, and the eccentric and epicyclic circles, later studied by Apollonius. The ancient sources do not make clear which of these models, if any, Archimedes adopted. In another context, Zhitomirskii [1977; 1983] favors assigning to Archimedes a partial eccentric model. The question appears to deserve further examination. On the early Greek planetary schemes, one may consult Neugebauer [1975, ID; IIIC]. Other aspects of Archimedes' astronomical work, including his figures for the dimension of the cosmos, arise in connection with his *Sand Reckoner* (see below).

II. *The Archimedean Corpus: Text and Transmission*

The Archimedean works extant in Greek have come down to us in three textual lines. Six codices from the fifteenth and sixteenth centuries were used by Heiberg to reconstruct their prototype (designated A). A Nuremberg codex, associated with Regiomontanus, is in the same line; although not used by Heiberg, some of its readings have been noted by Toomer [1976a, app. A]. The prototype was a Byzantine manuscript already available to European scholars in the thirteenth century; it passed into the hands of the Renaissance humanist G. Valla, thence to be copied before being lost. Heiberg's first edition of Archimedes [1880–81] is based almost exclusively on this textual line; for *C.F.*, lacking in A, Heiberg resorted primarily to the Latin version by Tartaglia (1543), which he thought was based on a lost Greek manuscript (see Heiberg 1880–81, II:359n]), but which Clagett [1978, 553–56] has since shown depended on Willem of Moerbeke's Latin translation (1269). After the appearance of his edition, Heiberg came upon the Moerbeke translation and established that, while it for the most part reflected the A tradition, in several of the mechanical writings it referred to a second manuscript. The Latin (designated B) was so literal in style that Heiberg could use it as further evidence of the text. Later, through an extraordinary find, Heiberg identified the Archimedean contents of a certain tenth-century palimpsest manuscript and inspected it at Constantinople in 1906. This text (designated C) provided a new line of the text; not only did it compensate for scribal deficiencies of A and offer practically the whole of *C.F.*, but also it preserved most of the text of the *Method*, unique in its witness to Archimedes' heuristic techniques.

Thus, for his second edition, Heiberg [1910–15] collated A, B, and C to obtain a substantially improved Greek text. The basic discussion of the manuscripts is given by Heiberg in his "Prolegomena critica" [ibid., vol. III], and summarized in detail by Dijksterhuis (EJD, chap. II), as also by Heath [1897; with supp., 1912]; cf. also Schneider [1979a, chap. 5]. On the history of the Valla manuscript, see Rose [1977]. For an account of Moerbeke's activity, with the full Latin text and translation of B, see Clagett [1976]. The description of the discovery and nature of C is given by Heiberg [1907] with a first edition of the text of the *Method*; for a German translation and commentary on the *Method*, see Heiberg and Zeuthen [1906–7]. Codex C was removed from Istanbul in the 1920s and has not been available for scholarly examination since then.

The medieval Latin tradition, including translation and commentaries from the twelfth to the sixteenth centuries, is covered in the five volumes of Clagett's *Archimedes in the Middle Ages* (1964–84); cf. also the mechanical tracts in Moody and Clagett [1952] and Clagett [1959a]. Clagett has noted how the older Latin versions relating to *D.C.* and *S.C.*, made from Arabic or derivative Greek versions by twelfth-century translators, persisted through the Renaissance, long after the appearance of the direct, Greek-based versions by Moerbeke and Jacob of Cremona; cf. Clagett [1978, 1225–46]. For briefer surveys of the Archimedean element in the Middle Ages, see

Clagett [1959b; 1969]. The Arabic-based versions of *D.C.* have been used for reconstructing phases of the text; see Sato [1979] and Knorr [1986b]. A reconstruction of aspects of the lost Greek prototype underlying Latin and Arabic renditions of materials related to Archimedes' *S.C.* I is given in Knorr [1987a].

On the medieval Arabic tradition of Archimedes, see Sezgin [1974, 121–36]; for notes on the general interest of the Arabic mathematical tradition for evidence of the Greek, see Toomer [1984] and Berggren [1984]. Several Archimedean efforts survive only through medieval Arabic translations. These include the *Lemmas* and the *Construction of the Regular Heptagon* (see EJD, chap. XV. 2 and 6). For the text of the latter, with discussion of Arabic elaborations, see Hogendijk [1984]. Another such work is the tract *On Mutually Tangent Circles* (not discussed in EJD), edited by Dold-Samplonius [1973; 1975]. Texts of these Arabic versions, with Greek translation, appear in Stamatis [1970–74, vol. 3].

Valuable fragments testifying to lost Archimedean works, particularly in mechanics, survive in Arabic (see Drachmann [1963a]). The mechanical tract on the balance by Thābit ibn Qurra (ed. Jaouiche [1976]), for instance, according to Knorr [1982c], continues the line of an Archimedean work. The Arabic also preserves complete translation of *S.C.* and *D.C.* (for references, see Sezgin [1974, 128–31]). These were made in the ninth century, that is, before the oldest of the extant Greek manuscripts was copied, and thus stand to provide valuable collateral testimony to the text. To date, however, the matter has not been systematically explored. A study of the Archimedean tradition in Arabic, including editions of treatises by Thābit ibn Qurra and ibn al-Haytham, is in preparation by R. Rashed. The revised Arabic version of *S.C.* and *D.C.* produced by al-Ṭūsī in the thirteenth century has been published [1939–40, vol. 2, no. 5]; his commentary is discussed by Kubesov [1969]. On a contribution to the later Archimedean tradition in Arabic among the Ottomans, see Berggren [1987].

The Archimedean element in medieval Hebrew includes versions of *D.C.* and *S.C.* made from the Arabic. For bibliographic details, see Steinschneider [1893, 502–3]. The works and methods of the Hebrew translators are examined by Sarfatti [1968, chap. 7], who includes a medieval Hebrew text of *D.C*, prop. 1 [ibid., 213–14]. The potential of the Hebrew materials for text studies of the ancient works is suggested in Knorr [1986c]; but it has not yet been significantly examined. For an account of one medieval Hebrew tract in the Archimedean tradition, see Rabinovitch [1974].

Archimedes' importance for the exact sciences of the early modern period was enormous. Extensive data on the Archimedean element in the work of leading sixteenth-century figures are collected in Clagett [1978]. A systematic survey of Archimedes in the seventeenth century has yet to be undertaken. Extensions of Archimedean limiting methods are discussed by Whiteside [1961, chap. 9]. Among worthy studies of individual figures in the Archimedean line, one may note Dijksterhuis [1970] on Stevin; Napolitani [1982] on Valerio; Giusti [1980] and Andersen [1985] on Cavalieri; Roero [1983] on Jakob Bernoulli.

The editions and translations of Archimedes since the Renaissance are surveyed by Dijksterhuis (EJD, chap. II); cf. also Heath [1897, chap. 2] and Procissi [1981, 17–33]. Schneider [1979a, 177–79] provides bibliographical citations of texts and translations since Heiberg. For terminological studies of the Archimedean corpus in relation to ancient mathematics generally, Mugler [1958–59] is a convenient tool.

Commentaries on five of Archimedes' books (*S.C.* I–II, *D.C.*, *PL.AE.* I–II) are extant from Eutocius of Askalon (sixth century). These were attached to the main Archimedes codices and are included in the editions by Heiberg [1910–15], Ver Eecke [1960], and Mugler [1970–72]. Save for the commentary on *D.C.*, these were rendered into Latin by Moerbeke along with his Archimedes translations; see Clagett [1976]. The commentaries on *S.C.* and *D.C.* were included in the Arabic of Archimedes and were exploited by al-Ṭūsī in his recension of these works; apparently, only the long fragment on cube duplication from book II survives in Arabic (see Sezgin [1974, 128–30, 188]). On Eutocius, see Tannery [1884]; Mogenet [1956]; Bulmer-Thomas [1971]; Knorr [1982a; 1986a, chap. 8].

Valuable insights into the development of the Archimedean corpus are provided by writers of late antiquity. Hero of Alexandria (first century A.D.) incorporated Archimedean materials into his extensive writings on practical geometry and mechanics; see Drachmann [1948; 1972]. The *Collection* of Pappus of Alexandria (mid-fourth century) and the Ptolemy commentaries by Pappus and Theon of Alexandria (late fourth century) also include Archimedes among their sources (cf. Sections VI and VIII below). On Pappus, see Bulmer-Thomas [1974], Jones [1986], and Knorr [1986a, chap. 8]; on Theon, see Toomer [1976b].

For the chronological ordering of Archimedes' writings, Dijksterhuis, like most commentators, follows the scheme of Heath (see EJD, 46–47). On at least two important points, however, the list is dubious: the late placement of *D.C.* results from a simple scribal error by Heath, while the early placement of *METH.* was already admitted by its advocate Zeuthen to be conjectural; see Knorr [1978e]. Revised orders have been argued by Kierboe [1913–14]; Arendt [1913–14]; Itard [1963]; Neuenschwander [1974, 121n]; Knorr [1978e]; Schneider [1979a, chap. 2.2]. The issue is crucial for any account which attempts to sketch out the development of Archimedes' ideas, as Schneider [1979a] and Knorr [1978e] do. But it is immaterial for Dijksterhuis, whose survey follows the order of Heiberg's edition, an essentially arbitrary sequence inherited from the order of Codex A.

III. *General Technical Elements*

Dijksterhuis consolidates the discussion of technical preliminaries within one section (EJD, chap. III) under the heading of "elements." This is useful on pedagogical and expository grounds; but it may be misleading if taken literally in the historical sense, for it could suggest that Archimedes depended specifically on the *Elements* of Euclid (cf. EJD, 49). Archimedes' activity came only a generation after Euclid; and according to Archimedes himself, his own father Pheidias was an astronomer, who, like Eudoxus and Aristarchus, had proposed a value for the solar diameter (EJD, 10, 364n). One would reasonably suppose that Pheidias, like the other astronomers named, included observational and theoretical elements in his work and was well versed in the geometry required for the study of heavenly motions. He would thus have had ready access to the chief works in the pre-Euclidean technical literature and, without doubt, instructed his son Archimedes by their means. In principle, then, Archimedes could preserve valuable evidence of the pre-Euclidean tradition. In several of the areas that follow (in particular, in proportion theory and convergence techniques) this possibility deserves serious consideration.

Dijksterhuis criticized the tendency, evident in Heath [1897], to render ancient materials too freely into modern format and notation; in his accounts of Euclid and Archimedes, Dijksterhuis devised his own alternative format, which in his view both avoided anachronism and offered a more supple text for the modern reader than a literal rendering could (see EJD, 7–8, 49–54). This notation, already implemented in Dijksterhuis' Euclid (1929–30), has been adopted also by Mueller [1981]. It provides a means for reducing the modernizing tendency inherent in the notion of "geometric algebra," espoused by Zeuthen [1886] and almost universally adopted since. Even Dijksterhuis, for all his reservations about notation, admits the term (see EJD, 51). But the conception of an ancient "geometric algebra" has been sharply criticized by Unguru [1975; 1979], who in turn drew the fire of van der Waerden [1975], Freudenthal [1977], and Weil [1978]. While regretting the hostility of the debate, one can admit the elements of justice on both sides. Indeed, this debate is but a special instance of the problem of the status of translations. Purists may insist that texts can be properly dealt with only in the original, while others may stress the pragmatic need to accommodate the contemporary audience in accordance with forms familiar to them. One can see in Dijksterhuis' ploy a vessel as worthy as any for navigating between the perilous extremes.

To supplement Dijksterhuis' review of proportion theory (EJD, 51–54) one may note Euclid's key definition: "magnitudes are said to be 'in the same ratio' . . . whenever equimultiples of the first and third alike exceed, or alike equal, or alike are exceeded by, equimultiples of the second and fourth, according to any multiplication whatever" (*Elem.* V, def. 5). That is, $A{:}B = C{:}D$ if and only if, for arbitrary integers $m, n, mA \rangle = \langle nB$ entails that $mC \rangle = \langle nD$ (relations in corresponding order). Euclid commentaries show how this definition parallels the real-number concept of Dedekind and Weierstrass (see, e.g., Mueller [1981, chap. 3]). Archimedes gives clear evidence of following this technique only in *SPIR.*, prop. 1. (Frajese [1972b] maintains, however, that even here Archimedes adopts a technique different from that of Euclid.)

Antecedent to this Euclidean technique was an alternative method based on *anthyphairesis*, that is, the "Euclidean division algorithm" (equivalent to the continued fraction expansion). Here, $A{:}B = C{:}D$ if the sequence of quotients in the expansion for A, B is identical to that for C, D. For integers and commensurable magnitudes, the sequence will be finite; otherwise, it will be infinite. For a survey of this form of proportion theory, see Knorr [1975, chap. 8, app. B] and Fowler [1979]. There is no indication that Archimedes applied this alternative method, although in *D.C.*, prop. 3, he appears to use *anthyphairesis* for the reduction of specific numerical ratios (see Knorr [1976]; Fowler [1982, 198]).

Yet in one remarkable instance—the proof of the balance principle in *PL.AE.* I, 6 and 7—Archimedes follows another form of proportion theory. From other ancient examples, one can generalize the scheme thus: to establish that stated magnitudes satisfy the proportionality $A{:}B = C{:}D$, one first proves the claim for the case where A, B are commensurable with each other. For the incommensurable case, an indirect argument is followed: if $A{:}B \neq C{:}D$, then let $A{:}X = C{:}D$ for $X \neq B$ (that is, either $X \langle B$ or $X \rangle B$). One then introduces a magnitude Y commensurable with A and intermediate between X and B; applying the commensurable case, already proved, one finally obtains a contradiction. Archimedes' proof of the balance principle follows this procedure in a converse form, as is appropriate to the particular formulation of his theorem: in *PL. AE.* I, 6, the case for commensurable weights A, B is established; in I, 7, the incommensurable case. For further details on this technique, with exposition of Archimedes' proof and citations of other specimens from Pappus and Theodosius, see Knorr [1978b]; it is there argued that this technique, an alternative equivalent to that expounded in *Elements* V, owed its origin to Eudoxus. One may observe that the mere possibility of Archimedes' adopting such an alternative technique must call into question the harsh assessments typically made of the proof in *PL. AE.* I, 7 (cf. EJD, 305–6).

Archimedes assumes or proves many theorems on conic sections in *C.S.*, *PL. AE.*, *Q.P.*, *C.F.*, and *METH.*, and also in the appendix to *S.C.* II relating to the solid construction assumed in its prop. 4 (cf. EJD, 195–205). Writing a generation before Apollonius, Archimedes preserves valuable evidence for the older theory of conics, of which Dijksterhuis provides an ample survey (EJD, 55–118). Archimedes adheres to the archaic manner of forming conics by planes cutting orthogonal cones at right angles to a generator; Dijksterhuis' derivations of the standard algebraic expressions for parabola, hyperbola, and ellipse (EJD, 55–63) may be compared with those given by Zeuthen [1886, chap. 2]. Further observations on the older theory are made by Toomer [1976a, 9–15], who incorporates newly discovered evidence from Diocles and notes that a fully comprehensive overview of the Archimedean conics is still needed. For suggestions relating to the earliest phases of the study of conics, see Zeuthen [1886, chap. 21] and Knorr [1982b; 1986a, chap 3].

Dijksterhuis provides a handy taxonomy of Archimedes' convergence methods (EJD, 130–33). For the most part, Archimedes follows a "compression" technique, in which a given curvilinear figure is bounded above and below by two *similar* rectilinear figures; as the number of sides of the bounding figures increases (usually by successive doubling), two sequences are obtained *simultaneously* converging toward each other, hence *a fortiori* toward the intermediate curvilinear figure. This method takes two forms: in one form, the *ratio* of the similar figures can be made arbitrarily close to *1:1* (cf. *S.C.* I, 3–5); in the other, the *difference* between them is made arbitrarily small (cf. *C.S.* 19–20). But Archimedes also knows an alternative "approximative" method (in Dijksterhuis' phrase), in which the difference between a curvilinear figure and a rectilinear figure is made arbitrarily small by successively doubling the number of sides of the latter. This is the technique employed in Euclid's book XII, representing, one may suppose, the method of Eudoxus. In Archimedes' *Q.P.* 24 only inscribed rectilinear figures are considered, as is the case for Euclid. But Dijksterhuis errs in maintaining that a "compression by difference" method is used in *D.C.* 1. For in fact, Archimedes here makes two independent applications of the "approximative" method, first for inscribed figures, then for circumscribed figures; cf. Knorr [1978e; 1986a, 153–56; 1986b]. The issue alters one's view of the relation of *D.C.* to *S.C.* I, by setting *D.C.* firmly in the technical orbit of the older elementary methods, and so could have important implications for one's view of the Archimedean chronology.

For a review of the related issue of Archimedes' differential methods, see Bashmakova [1964–65].

In his account of Archimedes' *neusis* (or sliding ruler) constructions from *SPIR.* (EJD, 133–40), Dijksterhuis emphasizes Archimedes' acceptance of the *neusis* in itself as a valid construction procedure. He thus differs from the widely held view, that one should have expected some alternative procedure—e.g., the "Euclidean" technique of compass and straightedge. Dijksterhuis' view is sustained in Knorr [1978d]. For an overview of *neusis* constructions, see Knorr [1986a].

IV–V. Sphere and Cylinder *I–II*

Dijksterhuis presents his discussion of the much-debated "Archimedean axiom" in the context of *S.C.* I, where it enters as the fifth of the introductory postulates (EJD, 145–49). His account concurs with Hjelmslev's, save on one major point. For Dijksterhuis, Archimedes' purpose in formulating the axiom is to exclude the case of infinitesimal magnitudes, such as arise in the heuristic analyses of *METH.* (see below). The associated axiom of Euclid (or Eudoxus), namely, *Elements* V, def. 4, specifies when given magnitudes have a ratio, but does not mention the case of their *difference*, as does Archimedes' formulation. The latter case, in Dijksterhuis' view, could give rise to infinitesimals unless expressly ruled out.

Although the range of interpretations of the axiom is diverse, most, I think, would subscribe to Dijksterhuis' basic position: that Archimedes perceived subtle difficulties not covered by the Euclidean formulation and so introduced his axiom; moreover, that the two axioms are different, and that Archimedes intended his axiom to be supplementary to Euclid's.

This position captures important *technical* nuances; but, like many analyses of this and related issues, it overlooks the central *textual* aspects. For when one considers the texts of the two axioms, it is clear that Archimedes is not working from Euclid. The Euclidean axiom, framed as a definition, reads thus: "magnitudes are said to 'have a ratio to each other' which can, when multiplied, exceed each other" (*Elem.* V, def. 4). That is, the given magnitudes A, B must be such that there exist integers m, n, m', n' such that $mA > nB$ but $m'A < n'B$. This manner of transcribing Euclid's statement (which is also adopted by Mueller [1981, 144–45]), although not the usual one (contrast EJD, 51n), conforms to the text; moreover, it is precisely what the applications of the Euclidean proportion theory require, as one can see through comparison with the definition of proportion given in V, def. 5 (cited above). The earliest extant appearance of Archimedes' axiom, on the other hand, reads thus: "of unequal areas the excess by which the greater exceeds the lesser can, added itself to itself, exceed any preassigned finite area" (*Q.P.*, pref.). Dijksterhuis singles out for discussion the somewhat different form adopted in post. 5 of the later work *S.C.* I. Later still, however, in *SPIR.*, pref., Archimedes follows the same form as in *Q.P.* The template for the wording in *Q.P.* and *SPIR.* (i.e., "excess by which . . . exceeds . . .") is not the Euclidean definition of *Elements* V, but rather the step of the convergence arguments in *Elements* XII where the bisection principle is invoked. In *Q.P.*, pref., Archimedes, speaking of his own lemma, notes that former geometers assumed "this very lemma" in their treatment of the circle and sphere theorems (cf. *Elem.* XII, 2, 18) and also for the pyramid and cone theorems (XII, 5, 10). In Euclid the circle, cone, and pyramid theorems use the bisection principle to reduce a hypothesized finite difference, while the sphere theorem adopts the same bisection principle in a markedly different manner (see XII, 16). Further, Euclid *proves* convergence by bisection (in X, 1) on the basis of an assumption comparable to his definition (V, def. 4.) These facts indicate that Archimedes must be working with a text differing in certain important details from the extant Euclid. In affirming that his own sources, as he says, "assumed" the convergence step, Archimedes must surely be taken to mean that no formal proof (such as X, 1) was there proposed—indeed, that the step was not even articulated as an explicit postulate.

The issue can hardly be considered closed. But any satisfactory interpretation must go beyond the purely technical aspects to accommodate fully the textual constraints. The classic discussion of the axiom is Stolz [1883]; in addition to Dijksterhuis' account, one may also consult Onicescu [1961]; Stamatis [1968b]; Neuenschwander [1974, 110–13]; Knorr [1978b; 1978e]; Schneider [1979a, 46–54]; Mueller [1981, 138–45].

A general overview of Archimedes' proofs of the sphere theorems in *S.C.* I is given by van der Waerden [1973, chap. 2]. Suggestions on the manner of their discovery by Archimedes are offered by Frajese [1972a; 1975]. Lorent [1955] develops a generalization of the principal results, to the effect that for a sphere inscribed in a solid of revolution of the Archimedean type (cf. EJD, 170), the surfaces and volumes of the two solids have the same ratio. A deficiency in the proof of I, 3, is examined in Berggren [1977]. The origin of a derivative form of I, 34–that the sphere is $^{11}\!/_{21}$ the cube of its diameter–is traced in Knorr [1986d]. On Archimedes' equivalent of fractional exponents (cf. *S.C.* II, 8), see Stamatis [1967a].

Eutocius' commentaries on *S.C.* contain important materials on cube duplication (ad *S.C.* II, 1) and on the use of conics by Archimedes, Dionysodorus, and Diocles for solving a certain solid problem (ad *S.C.* II, 4). On the cube duplication texts, see Knorr [1986a]. The solid problem is discussed in EJD (195–205), while Diocles' text has been retrieved from the Arabic by Toomer [1976a]; see also Knorr [1986a, chap. 5]. As noted above, the commentaries on *S.C.* were included in the Arabic translation of Archimedes. In his recension, al-Ṭūsī substitutes for Eutocius' methods of cube duplication an alternative construction related to that of Apollonius (see Knorr [1986a, 307–8]); but he rehearses all three solutions of the solid problem given by Eutocius. These solving methods appear to influence some of the Arabic solutions for Archimedes' heptagon problem (see Section XV below; Hogendijk [1984]; Knorr [1986a, 181–85]). They also bear on the methods adopted for solving cubics by Omar al-Khayyam (see Youschkevitch [1976, chap. 14]). Links between Archimedes' solid problem and Tartaglia's solution of the cubic are discussed by Schultz [1984].

On the medieval Latin tradition of *S.C.*—in particular, the tracts *Verba filiorum* and *De curvis superficiebus*—see Clagett [1964, chaps. 4, 6]; Sato [1985]; and Knorr [1987a].

VI. Dimension of the Circle

The short tract on the circle survives in a version far removed from its Archimedean prototype (see EJD, 222). With reference to the first proposition, Sato [1979] has argued that the medieval version (specifically, the Latin of Gerard of Cremona) derives from a better text than the extant Greek; through comparisons with versions in the Greek commentators Knorr [1986b] argues a fifth-century date for the extant *D.C.*

On the ancient variant forms for *D.C.*, prop. 2—e.g., that "14 circles equal 11 squares of the diameter"—see Knorr [1986d].

In *D.C.*, prop. 3, Archimedes calculates that the ratio of the circumference of the circle to its diameter is less than $3\frac{1}{7}$ but greater than $3\frac{10}{71}$. The numerical method is explicated in Dedron and Itard [1959, vol. 2, chap. 7], Stamatis [1970–74, 1(B):458–68], and Knorr [1976]. The error against which Dijksterhuis warns (EJD, 228n), of describing Archimedes' procedure in terms of a sequence of harmonic and geometric means, still detracts from some accounts, such as that by Phillips [1981]. On Archimedes' probable use of "Hero's rule" for the approximation of square roots, see EJD (229–34); for a different proposal, see Paev [1965]. A survey of the scribal variants in the manuscript transmission of the numerical figures is given by Fowler [1987, chap. 7.3a]. On the basis of a passage from Hero of Alexandria (cf. EJD, 239–40), Knorr [1976; 1986a, 157] argues that Archimedes went on to establish closer bounds, namely, that the ratio is less than $377/120$, but greater than $333/106$. This view is reported favorably in Schneider [1979a, 147–49], but is attacked by Bruins [1979].

The rate of convergence of Archimedes' algorithm is probed by Phillips [1981], who considers the use of linear combinations of upper and lower bounds to improve convergence. Such procedures are discussed also by Bruins [1976] and Knorr [1976], who note the extensive efforts in this connection by Huygens. One of these linear combinations is derivable from an area rule presented by Hero (*Metrica* I, 32): that any circular segment is greater than $4/3$ the triangle having the same altitude and base as the segment. The Archimedean provenance of this rule is argued in Knorr [1978e; 1986a, 168–69].

Doubtless the most extensively debated of Archimedean questions is how he derived the bounds for $\sqrt{3}$ introduced in *D.C.*, prop. 3 (namely, $265/153 < \sqrt{3} < 1351/780$). Dijksterhuis provides a detailed review (EJD, 229–38); the discussions by Heath and Hofmann (cited in EJD, 230n) include extensive bibliographies. More recent attempts are surveyed by Schneider [1979a, 145–47]. Recursive algorithms of the "side and diameter" type are proposed by Stamatis [1955; 1970–74, 1(B):468–77; 1980, 118–21), Gazis and Herman [1959], and Knorr [1976]. Weil [1984, chap. 1.8] notes possible connections with the study of second-order diophantine equations. Developing a suggestion by D. T. Whiteside, Fowler [1979; 1982] has observed that the convergent fractions obtained from the continued fraction for $\sqrt{27}$, when divided by 3, yield an alternating sequence of bounds for $\sqrt{3}$: $5/3$, $26/15$, $265/153$, $1351/780$. This method thus directly yields Hero's value ($26/15$) as well as Archimedes', and avoids the arbitrariness characteristic of most other accounts. Thus, even accepting that the extant evidence does not permit certainty, one may well wonder what a more satisfactory solution of the question could possibly look like.

VII. Conoids and Spheroids

Matters of terminology in *C.S.* are raised by Kierboe [1913–14] and summarized by Schneider [1979a, 30–32]. Dijksterhuis' account of the content and technique of this Archimedean work has not been superseded. On the Arabic tradition of this material, see Rashed [1981].

VIII. Spiral Lines

Dijksterhuis' overview of *SPIR.* may be supplemented by Schneider [1979a, chap. 4.2]. Archimedes' tangent method (see *SPIR.* 18–20) is examined, along with other examples of differential methods, in Bashmakova [1964–65]. The neuses, auxiliary to the tangent theorems, are discussed in Knorr [1978d]. On the arithmetic lemma on sums of square numbers (*SPIR.* 10–11; cf. EJD, 118–30), see Stamatis [1975]. Pappus reports an alternative method for the area of the spiral, involving its correlation with a procedure for the volume of the cone; the Archimedean provenance of this method is argued in Knorr [1978c].

IX–XII. *The Mechanical Writings*

Fundamental for Archimedes' mechanical writings is the concept of center of gravity. Despite this, center of gravity is nowhere explicitly defined within the extant works. The question is examined in detail by Dijksterhuis (EJD, 295–304), particularly in connection with the postulates which open *PL.AE.* I. In contrast to the formalist view of Stein and others, that Archimedes here was developing an implicit definition of center of gravity, Dijksterhuis favors a naive quasi-physicalist position, that Archimedes allowed himself to assume certain basic mechanical properties of idealized bodies. The collateral evidence from Hero and Pappus which Dijksterhuis incorporates into his discussion has been further elaborated in Drachmann [1963a]; for a survey, see Schneider [1979a, chap. 3.1]. A more recent analysis appears in Souffrin [1980].

On the formal structure of *PL.AE.* I, see Schmidt [1975] and Suppes [1981]. Beisenherz [1981] gives an account of its principal conceptions in the light of the contemporary philosophical critique of physics called "Protophysik." The formal validity of the proof of the balance principle in I, 6–7, called into question by Mach (see EJD, 289–95), is defended by Goe [1972] and Beisenherz [1981]; cf. also Suppes [1981, 212n3], Knorr [1978b], and Child [1921]. Berggren [1977] calls attention to anomalies of text and technique which in his view render the authenticity of *PL.AE.* I suspect, and Souffrin [1980] follows Berggren's view as the basis of his own elaborations. Some of the principal objections, however, have been answered in Knorr [1978b, 185–86; 1978e, 245–47], among these the technical difficulties lodged against the proof of I, 7 (see above).

The *Method* occupies a special place in Archimedes studies, since it provides unique insight into his heuristic methods. Recent studies include Rufini [1961]; Babini [1966]; García de la Sienra [1983]. Upon its rediscovery and publication by Heiberg (see Section II above), this writing stirred up a flurry of excited notice in scholarly and popular science journals. It is thus extraordinary that fully eight decades later there is still no adequate translation of this work into English; for the version by Heath [1912 supp. to 1897] is based on Heiberg's 1907 text, subsequently superseded by the edition of 1912. The ample discussion by Dijksterhuis (EJD, chap. X) may be supplemented by Schneider [1979a, chap. 4.1], who cites some alternative applications of Archimedes' method from van der Waerden [1973; 2nd ed. 1968], Frajese [1972a], and Gould [1955]. Gould [1955, 474] notes that merely by substituting squares for the circular cross sections in the figure for the sphere measurement in *METH.* 2, one obtains the volume of the cylindrical "hoof" of *METH.* 12–15. This very conception is basic for Chinese measurements of the sphere and the "hoof" (or "box lid"); cf. Wagner [1978] and Lam and Shen [1985]. But Archimedes develops an alternative procedure for this solid (cf. EJD, 331–35). Other accounts which explore Archimedes' procedures in the *Method* include Child [1921] and Edwards [1979, 68–74].

The "mechanical method" leads to the determination of areas, volumes, and centers of gravity of specified figures. It conceives of the given figure as set in balance with another figure of known measure, in such a way that each indivisible constituent element of the first is in balance with a corresponding indivisible of the second. Archimedes takes care to claim only heuristic, not formal, validity for this procedure. Dijksterhuis maintains that the appeal to infinitesimals is its invalidating assumption (EJD, 318–21); Schneider [1979a, 114–16] concurs, noting that Archimedes could mean by "mechanical" the "material" (i.e., atomistic) aspect of bodies, adopted within some contemporaneous physical theories. But doubts are expressed by Becker [1957] and Knorr [1982d], who locate the difficulty in the assumption of the physical properties of the balance.

Interesting parallels link Archimedes' use of indivisibles in the *Method* with certain later Chinese studies of the sphere—indeed, both effect the measurement of the same cylindrical solid (the so-called hoof; cf. EJD, 331–36); see Wagner [1978] (summarized in van der Waerden [1983]) and Lam and Shen [1985]. A survey of the ancient use of indivisibles is presented in Knorr [1987b].

In *Q.P.* Archimedes provides two proofs of the area of the parabolic segment: one in the manner of the heuristic procedure in *METH.* 1, but eliminating indivisibles; the other following an unrelated geometric method along standard Euclidean lines. Dijksterhuis outlines both fully (EJD, chap. XI). Souffrin [1980, 15–19] also compares these proofs; he maintains that Archimedes would designate both proofs as "geometric," restricting the term "mechanical" to the alternative heuristic procedure followed in *METH.* 1; but in this he is contradicted by Archimedes' clear usage in *Q.P.* pref. (ed. Heiberg, II:264–66). For an alternative account of the implications of the double proving procedure, see Knorr [1978e].

Mugler [1973] proposes a textual emendation and interpretation of Archimedes' puzzling phrase "section of the whole cone" in *Q.P.*, pref. Stamatis [1963] discusses a generalization of Archimedes' geometric progression in *Q.P.* 23.

A number of lost mechanical writings can be inferred both from remarks by Archimedes himself and from allusions in Pappus and Hero. (a) The *Mechanics* of Hero, extant only in Arabic, includes references to *On Balances*, *On Columns*, and other lost works; the passages are assembled and discussed by Drachmann [1963a], whose tentative recommendations on the character of the lost works are expanded by Krafft [1970]; for a summary, see Schneider [1979a, chap. 3.2]. But important aspects of these views are questioned in Knorr [1982c]; it is there argued that certain medieval Arabic and Latin writings on the balance—in particular, Thābit ibn Qurra's tract on the so-called *qarasṭūn* (Latin: *karaston*, the steelyard)—depend on a Greek prototype (*On the Charistiōn*) descended from Archimedes' lost tract *On Balances*. This view has the effect of minimizing the dichotomy of static ("Archimedean") and dynamic ("Aristotelian") approaches commonly assumed by historians of early mechanics; for both approaches are assignable to the lost Archimedean prototype. (b) An incorrect analysis of the weight distribution problem, ostensibly drawn by Hero from Archimedes' *On Columns*, is examined by Schneider [1979c]. (c) The content and method of the lost *Equilibria*, on the center of gravity of solids, are reconstructed in Knorr [1978a]; the reconstructions worked out by Commandino and Galileo are presented in Knorr [1978a] and Drake [1974, app.], respectively. For different proposals on this and others of the lost mechanical writings, see Sato [1981].

XIII. Sand Reckoner

In contrast with the other works in the corpus, the *Sand Reckoner* is a specimen of popular writing, addressed by Archimedes not to a scientific colleague, but to his friend and patron Gelon, coregent of Syracuse. It reads, one might imagine, like the transcript of a lecture, amplified with technical notes in its published version. In it Archimedes incorporates ingenious bits of arithmetic, geometry, and astronomy, both theoretical and applied, for the purpose of challenging a popular misconception about the infinite: even the number of grains of sand which could fill the universe is finite, he maintains, and it is possible easily to express numbers of this magnitude according to simple arithmetical rules.

Despite its nonspecialist audience, the *Sand Reckoner* provides unique insight into Archimedes' scientific work, for Archimedes here applies his geometric expertise in ways nowhere else seen in the corpus. Only here does he address issues of numerical notation; his notion of the arithmetic infinite and his novel scheme for the representation of large numbers are discussed in Delsedime [1970b]. Only here does he reveal his activity in observational optics; on his design and use of an optical sighting instrument for measuring the sun's apparent angular diameter, see Section I above; for a reconstruction of the derivation of his reported figures for the apparent diameter of the sun, see Shapiro [1975]. Within the extant corpus, only here do we learn of his astronomical studies; for discussions of his report of Aristarchus' heliocentric system, the primary surviving ancient testimony of that system, see Rosen [1978]; Derenzini [1974]; Stamatis [1971].

The values Archimedes here proposes for the size of the cosmos are merely rough estimates; a more elaborate Archimedean scheme of dimensions, including figures for the planets, is reported by later witnesses, perhaps on the authority of the lost writing on the planetarium; see Zhitomirskii [1977; 1983]; Osborne [1983]; Klimenko [1980]; Neugebauer [1975, 647–51]. Archimedes here also provides valuable insight into the evolution of trigonometrical computations, through his application of the tangent and chord inequalities (cf. EJD, 368n); on the ancient and medieval traditions of the texts of these inequalities, see Knorr [1985a; 1986c]; for further remarks on Archimedes' relation to ancient trigonometry, see Section XV below.

XIV. Floating Bodies

As with his account of the *Method*, Dijksterhuis' account of the two books of *Floating Bodies* stands as the most satisfactory account in English, in view of Heath's reliance on the superseded first edition of Heiberg's text. In addition, in his examination of the contents of book II, Dijksterhuis adopts an "analytic" approach: the principal conditions which Archimedes stipulates for the synthetic demonstrations of his propositions on floating paraboloids are derived by Dijksterhuis from geometric conditions on the centers of gravity which suffice for stability. This avoids the opaqueness of expositions in the formal synthetic manner, like that of Heath.

The two books of *Floating Bodies*—particularly the first on general principles of hydrostatics—were influential in the Renaissance through the Latin versions of Tartaglia and Commandino, and spawned numerous adaptations and commentaries, such as those of Stevin, Guidobaldo, and Galileo. For texts and discussion, see Dijksterhuis [1955; 1970]; Drake and Drabkin [1969]; Clagett [1978]; Drake [1981].

XV. *Miscellaneous*

On the *Cattle Problem*: Mita [1951]; Fowler [1982, 203–4]; for modern computational efforts, see Fowler [1981] and Lawrence (Livermore) Laboratories [1981].

On the semiregular polyhedra: Papadatos [1978]; for a detailed popularized account of technical aspects, including suggestions for models, see Cundy and Rollett [1961, chap. 3.7].

To the ancient testimonia of Archimedes' board game, presented in the *Stomachion*, one may add the passage from Lucretius identified by Rose [1956].

On Archimedes' rule for the measurement of triangles: a reconstruction of the Archimedean proof is proposed by Taisbak [1980].

On the construction of the regular heptagon: the Arabic text of Archimedes' tract has been edited, in conjunction with a set of original Arabic contributions to the solution of this problem, by Hogendijk [1984]; the text of ibn al-Haytham's construction has been edited by Rashed [1979]; see also the discussions in Knorr [1986a, chap. 5] and Stamatis [1970–74, 3: 88–101]. A back-translation of the text into the ancient Doric dialect has been composed by Stamatis [1968a; 1970–74, 3:82–86; 1973].

The text of the tract *On Mutually Tangent Circles*, extant only in Arabic translation (not discussed in EJD), has been edited by Dold-Samplonius [1973; 1975]. The result in its prop. 15 is identical with what al-Bīrūnī calls the "lemma of Archimedes" and uses toward his own trigonometric computations. Against the view of Tropfke, however, Toomer [1973, 21–23] argues that Archimedes himself did not apply this lemma to such ends.

Postscript

Who was Archimedes? The wonderworker of popular legend serves as focus for the portrait by Schneider [1979a], who maps out Archimedes' career first as engineer, then only belatedly as pure mathematician after exposure to a Pythagorean-oriented natural philosophy. For Luria [1945], Archimedes becomes a type of dialectical materialist, inspired by the atomist doctrines of Democritus. Delsedime [1970b] discerns an Aristotelian, rather than a Pythagorean, element in Archimedes' conception of the infinite (cf. Section XIII above). Virieux-Reymond [1979] argues a Platonist influence behind the mechanical method; but the claim is doubted by Gardies [1980]. Platonist? Aristotelian? Stoic? Epicurean? Schneider's survey of the question [1979a, 54–57] is frustrated by Archimedes' virtually complete silence on philosophical issues; indeed, in Schneider's view intellectual defensiveness becomes Archimedes' virtual hallmark [ibid., chap. 2.4].

If Archimedes' philosophical affiliation proves elusive, perhaps one can pin down his scientific method. A common assumption is that some form of physical experimentation underlies not only the mechanical studies, e.g., of balance and hydrostatics (see Drachmann [1967]), but the geometric applications of the "mechanical method" as well (see Gould [1955]; Schneider [1979a, chap. 4.1.]). Contemporary conceptions of scientific method provide the framework for schematizations of Archimedes' method, as in Frajese

[1972a; 1974, 23–25; 1975] and Giorello [1975] (both cited in this context by Schneider [1979a, 120–21, 152]); van der Waerden [1973, chap. 1]; Pogrebysski [1968]. But the geometric writings withhold comment on heuristic method, the mechanical writings are silent on the form of concomitant physical trials, and the documentation for the practical works is merely anecdotal, not technical. All efforts to codify Archimedes' scientific procedures are thus conjectural.

What, then, were Archimedes' leading scientific roles and the keys to his scientific genius? Dijksterhuis' volume ends abruptly, without even raising such general questions, let alone volunteering answers. He has surveyed the corpus, a monument of formal mathematics, which (with the partial exception of the *Method*) offers virtually no insight at all into motives, heuristic methods, or philosophical inclinations. Does Dijksterhuis thus mean to imply that through the corpus Archimedes presents himself foremost as a geometer and mathematical scientist? But in the light of Dijksterhuis' own silence, one perhaps does best to lay conjecture to rest.

REFERENCES

A. *Editions and Translations*

Heiberg, J. L. 1880–81. *Archimedis opera.* 1st ed. 3 vols. Leipzig: Teubner (vol. III contains "Prolegomena critica" and Eutocius' Commentaries).

———. 1910–15. *Archimedis opera.* 2d ed. 3 vols. Leipzig: Teubner. Repr. 1972. Stuttgart.

Czwalina, A. 1922–25. *Archimedes Werke.* 5 vols. Leipzig: Teubner (selected works in German trans.; see Schneider [1979a, 177–78] for individual entries). Repr. 1963. Darmstadt: Wissenschaftliche Buchgesellschaft (includes *Kreismessung*, trans. F. Rudio, and *Methodenlehre*, trans. J. L. Heiberg).

Frajese, A. 1974. *Archimede*: *Opere* [Italian]. Turin: Unione tipografico-editrice torinese.

Heath, T. L. 1897. *The Works of Archimedes.* Cambridge. Repr. (n.d.) with *A supplement, The Method of Archimedes, of 1912.* New York.

Mugler, C. 1970–72. *Les oeuvres d'Archimède.* 4 vols. Paris: Budé (Heiberg's text with French trans.; includes Eutocius' Commentaries).

Stamatis, E. S. 1970–74. *Archimēdous Hapanta.* 3 vols. in 4. Athens: Technikon epimelētērion tēs Hellados (Heiberg's text with modern Greek trans.; includes ancient testimonia [vol. 1, pt. 1] and versions of the Arabic-based tradition [vol. 3]).

Ver Eecke, P. 1960. *Archimède: Les oeuvres complètes.* 2d ed. 2 vols. Liège: Vaillant-Carmanne (includes Eutocius' Commentaries). 1st ed. 1921. 1 vol. Paris, Brussels: Desclée, de Brouwer (without Eutocius).

Veselovskii, I. N., and Rosenfeld, B. A. 1962. *Works of Archimedes* [Russian]. Moscow: Gos. izd. phyziko-matemat. lit.

B. *Studies on or Related to Archimedes*

Aaboe, A. 1964. *Episodes from the early history of mathematics*. New York: Random House. Repr. 1975. Mathematical Association of America.

Africa, T. W. 1975. Archimedes through the looking glass. *Classical World* 68:305–8.

AHES = *Archive for History of Exact Sciences*.

AIHS = *Archives internationales d'histoire des sciences*.

Andersen, K. 1985. Cavalieri's method of indivisibles. *AHES* 31:291–367.

Arendt, H. 1913–14. Zu Archimedes. *Bibliotheca mathematica*, 3d ser. 14:289–311.

Aujac, G. 1979–80. La lettre à teneur scientifique à l'époque alexandrine. *Bulletin de la Société toulousaine d'études classiques* 179–80:79–102.

Babini, J. 1966. *Arquímedes: el Método* [Spanish]. Buenos Aires: Editorial universitaria.

Baltrušaitis, J. 1978. *Le miroir: Essai sur une légende scientifique*. Paris: Elmayan.

Bashmakova, I. G. 1963. Some problems of the history of ancient mathematics [Russian]. *Istoriko-matematicheskie issledovaniya* 15:37–50.

———. 1964–65. Les méthodes différentielles d'Archimède. *AHES* 2:87–107.

———. 1970. Hellenistic lands and Roman Empire [Russian]. In *Istoriya matematiki*, ed. A. P. Iushkevich, vol. 1, pt. 1, chap. 5, 106–53. Moscow: Izdatel'stvo Nauka.

Becker, O. 1957. Review of E. J. Dijksterhuis, *Archimedes*. *Gnomon* 29:329–32.

———. 1966. *Das mathematische Denken in der Antike*. 2d ed. Göttingen: Vandenhoeck & Ruprecht.

Beisenherz, H. G. 1981. Archimedes und die Protophysik. *Philosophia naturalis* 18:438–78.

Berggren, J. L. 1976–77. Spurious theorems in Archimedes' *Equilibria of planes*, Book I. *AHES* 16:87–103.

———. 1977. A lacuna in Book I of Archimedes' *Sphere and cylinder*. *HM* 4:1–5.

———. 1984. History of Greek mathematics: A survey of recent research. *HM* 11:394–410.

———. 1987. Archimedes among the Ottomans (in press).

Bernhardt, H. 1975. Archimedes. In *Biographien bedeutender Mathematiker*, ed. H. Wussing and W. Arnold, 33–42. Berlin: Volk und Wissen. Repr. 1978. Cologne: Aulis Verlag Deubner. 2d ed. Berlin, 1983; Cologne, 1985.

Berve, H. 1959. *König Gelon II*. Bayerische Akademie der Wissenschaften, *Abhandlungen* (phil.-hist. kl.), n.s. 47.

Bruins, E. M. 1976. The division of the circle and ancient arts and sciences. *Janus* 63:61–84.

———. 1979. On interpretation in the history of mathematics. *Janus* 66:83–129.

BSSM = *Bollettino di storia delle scienze matematiche*.

Bulmer-Thomas, I. 1971. Eutocius of Askalon. *DSB* 4:488–91.

———. 1974. Pappus of Alexandria. *DSB* 10:293–304.

Cambridge ancient history. 1928. VI: *The Hellenistic kingdoms*, ed. S. A. Cook et al. Cambridge: Cambridge University Press.

———. 1930. VIII: *Rome and the Mediterranean 218–133* B.C., ed. S. A. Cook et al. Cambridge: Cambridge University Press.
———. 1984. VII, pt. I: *The Hellenistic world*, ed. F. W. Walbank et al. 2d ed. Cambridge: Cambridge University Press.
Child, J. M. 1921. Archimedes' principle of the balance and some criticisms upon it. In *Studies in the history and method of science*, ed. C. Singer, 2:490–520. Oxford: Clarendon Press. Repr. 1975. New York: Arno.
Clagett, M. 1959a. *The science of mechanics in the Middle Ages*. Madison, Wis.: University of Wisconsin Press.
———. 1959b. The impact of Archimedes on medieval science. *Isis* 50:419–29.
———. 1964. *Archimedes in the Middle Ages*. I: *The arabo-latin tradition*. Madison, Wis.: University of Wisconsin Press.
———. 1969. Leonardo da Vinci and the medieval Archimedes. *Physis* 11:100–51.
———. 1970. *Archimedes*. *DSB* 1:213–31.
———. 1976. *Archimedes in the Middle Ages*. II: *The translations from the Greek by William of Moerbeke*. 2 vols. Philadelphia: American Philosophical Society.
———. 1978. *Archimedes in the Middle Ages*. III: *The fate of the medieval Archimedes 1300–1565*. 3 vols. Philadelphia: American Philosophical Society.
———. 1980. *Archimedes in the Middle Ages*. IV: *A supplement on the medieval Latin traditions of conic sections*. 2 vols. Philadelphia: American Philosophical Society.
———. 1984. *Archimedes in the Middle Ages*. V: *Quasi-Archimedean geometry in the thirteenth century*. 2 vols. Philadelphia: American Philosophical Society.
———, and Moody, E. A.: see Moody, E. A., and Clagett, M.
Courcelle, P. 1959. Le souvenir d'Archimède en occident chrétien. In *Convivium dominicum*, 287–96. Catania: Università di Catania.
Cundy, H. M., and Rollett, A. P. 1961. *Mathematical models*. 2d ed. Oxford University Press. 1st ed. 1951.
Czwalina, A. 1922–25. *Archimedes Werke*: see Section A above.
Dedron, P., and Itard J. 1959. *Mathématiques et mathématiciens*. Paris: Magnard. Trans. 1974 by J. V. Field as *Mathematics and mathematicians*. 2 vols. London: Transworld.
Delsedime, P. 1970a. Uno strumento astronomico descritto nel corpus archimedeo: la dioptra di Archimede. *Physis* 12:173–96.
———. 1970b. L'infini numérique dans l'Arénaire d'Archimède. *AHES* 6:345–59.
Derenzini, G. 1974. L'Eliocentrismo di Aristarco da Archimede a Copernico. *Physis* 16:289–308.
Dijksterhuis, E. J. 1955. *Principal works of Simon Stevin*. Vol. 1. Amsterdam: Swets and Zeitlinger.
———. 1961. *The mechanization of the world picture*. Oxford: Clarendon Press (trans. of 1st Dutch ed., 1950). Repr. 1986. Princeton: Princeton University Press.
———. 1970. *Simon Stevin: Science in the Netherlands around 1600* (abr. trans. of 1943 Dutch ed.). The Hague: Nijhoff.
Dold-Samplonius, Y. 1973. Archimedes: Einander berührende Kreise. *Sudhoffs Archiv* 57:15–40.
———. 1975. *Archimedis opera omnia*. Vol. 4: *Über einander berührende Kreise*. Stuttgart: Teubner.

Drabkin, I. E.: see Drake, S., and Drabkin, I. E.

Drachmann, A. G. 1948. *Ktesibios, Philon and Heron: a study in ancient pneumatics.* Copenhagen: Munksgaard.

———. 1958. How Archimedes expected to move the Earth. *Centaurus* 5:278–82.

———. 1963a. Fragments from Archimedes in Heron's *Mechanics. Centaurus* 8:91–146.

———. 1963b. *The mechanical technology of Greek and Roman antiquity.* Copenhagen: Munksgaard.

———. 1967. Archimedes and the science of physics. *Centaurus* 12:1–11.

———. 1972. Hero of Alexandria. *DSB* 6:310–4.

Dragoni, G. 1975. Introduzione allo studio della vita e delle opere di Eratostene. *Physis* 17:41–70.

Drake, S. 1974. *Galileo Galilei: Two new sciences.* Madison, Wis.: University of Wisconsin Press (trans., introduction, notes; includes appendix: On centers of gravity).

———. 1981. *Cause, experiment and science: A Galilean dialogue incorporating a new English translation of Galileo's "Bodies that stay atop water, or move in it."* Chicago: University of Chicago Press.

——— and Drabkin, I. E. 1969. *Mechanics in sixteenth century Italy.* Madison, Wis.: University of Wisconsin Press.

DSB = *Dictionary of scientific biography*, ed. C. C. Gillispie. 16 vols. New York: Scribner's, 1970–80.

Edwards, C. H., Jr. 1979. *The historical development of the calculus.* New York, Heidelberg, Berlin: Springer.

Finley, M. I. 1968. *Ancient Sicily.* New York: Viking Press.

Foley, V.: see Soedel, W., and Foley, V.

Fowler, D. H. 1979. Ratio in early Greek mathematics. *Bulletin of the American Mathematical Society*, n.s. 1:807–46.

———. 1981. Archimedes' Cattle problem and the pocket calculating machine. Preprint. Coventry: University of Warwick Mathematics Institute.

———. 1982. Book II of Euclid's *Elements* and a pre-Eudoxan theory of ratio. Part 2: Sides and diameters. *AHES* 26:193–209.

———. 1987. *The mathematics of Plato's academy: A new reconstruction.* Oxford: Oxford University Press (forthcoming).

Frajese, A. 1972a. Como trovò Archimede il volume della sfera? *Archimede* 24:281–89.

———. 1972b. Da Eudosso a Euclide. *Scientia* 66:563–68 (English trans., ibid., 569–73).

———. 1974. *Archimede: Opere:* see Section A above.

———. 1975. Archimedea. *Cultura e Scuola* 14:190–96.

Fraser, P. M. 1972. *Ptolemaic Alexandria.* 3 vols. Oxford: Clarendon Press.

Freudenthal, H. 1977. What is algebra and what has it been in history? *AHES* 16:189–200.

García de la Sienra, A. 1983. El Método de Arquímedes. *Dianoia* 29:53–80.

Gardies, J. L. 1980. La méthode mécanique et le platonisme d'Archimède. *Revue philosophique* 170:39–43.

Gardner, M., and Fisher, L. E. (illus.). 1965. *Archimedes, mathematician and inventor.* New York: Macmillan.

Gavagna, R. 1979. Cusano e Alberti a proposito del "De architectura" di Vitruvio. 1: Cusano, Alberti e un esperimento di Archimede. 2: l'Igrometro. *Rivista critica di storia della filosofia* 34:162–76.

Gazis, D. C. and Herman, R. 1959. Square roots, geometry and Archimedes [Greek with English summary]. *Platon* 11:357–70 (English trans., *Scripta mathematica* 25 [1960]: 228-41).

Gigon, O. 1973. Posidoniana—Ciceroniana—Lactantiana [German]. In *Romanitas et christianitas: Studia J. H. Waszink . . . oblata*, ed. W. den Boer et al., 145–80. Amsterdam: North Holland.

Giorello, G. 1975. Archimede e la metodologia dei programmi di ricerca. *Scientia* 69:111–23 (English trans., ibid., 125–35).

Giusti, E. 1980. *Bonaventura Cavalieri and the theory of indivisibles*. Rome: Edizioni Cremonese (separate publication of introduction to repr. ed. of B. Cavalieri, *Exercitationes geometricae sex*).

Goe, G. 1972. Archimedes' theory of the lever and Mach's critique. *Studies in History and Philosophy of Science* 2:329–45.

Gould, S. H. 1955. The Method of Archimedes. *American Mathematical Monthly* 62:473–76.

Hart, V. G. 1982. A Greek cube-root device. *Mathematical Gazette* 66 (no. 438): 294–96.

Haury, A. 1980. Cicerone giudice della genialità di Archimede. *Ciceroniana*, n.s. 4:115–20.

Heath, T. L. 1897. *The works of Archimedes*: see Section A above.

Heiberg, J. L. 1880–81; 1910–15. *Archimedis opera*: see Section A above.

———. 1907. Eine neue Archimedeshandschrift. *Hermes* 42:234–303.

———, and Zeuthen, H. G. 1906–07. Eine neue Schrift des Archimedes. *Bibliotheca mathematica*, 3d ser. 7:321–63.

Hill, D. R. 1974. *The book of knowledge of ingenious mechanical devices of al-Jazarī*. Dordrecht: Reidel.

———. 1978. al-Jazarī. *DSB* 15:253–55.

———. 1984. *A History of engineering in classical and medieval times*. La Salle, Ill.: Open Court.

HM = Historia mathematica.

Hoddeson, L. H. 1972. How did Archimedes solve King Hiero's crown problem?—An unanswered question. *Physics Teacher* 10:14–19.

Hogendijk, J. 1984. Greek and Arabic constructions of the regular heptagon. *AHES* 30:197–330.

Hooykaas, R. 1966. In memoriam E. J. Dijksterhuis. *AIHS* 19:138–40 (cf. *Isis* 58 [1967]: 223–25).

Itard, J. 1963. Archimedes. In *Ancient and medieval science*, ed. R. Taton, 279–88. New York: Basic Books (trans. of 1957 French ed.).

———, and Dedron, P.: see Dedron, P., and Itard, J.

Jaouiche, K. 1976. *Le livre du Qarasṭūn de Thābit ibn Qurra*. Leiden: Brill.

Jenemann, H. R. 1984. Eine römische Waage mit nur einer Schale und festem Gegengewicht. *Archäologisches Korrespondenzblatt* 14:81–96.

Jones, A. 1986. *Pappus of Alexandria: Book 7 of the Collection*. 2 vols. New York, Berlin, Heidelberg, Tokyo: Springer.

Kagan, V. F. 1955. *Archimedes: Sein Leben und sein Werk*. Leipzig: Fachbuchverlag. Trans. of 2d Russian ed., 1951. Moscow: Gos. izd. tekhn.-teor. lit.

Keller, A. 1971. Archimedean hydrostatic theorems and salvage operations in 16th century Venice. *Technology and Culture* 12:602–17.

Kierboe, T. 1913–14. Bemerkungen über die Terminologie des Archimedes. *Bibliotheca mathematica*, 3d ser. 14:33–40.

Klimenko, A. V. 1980. On the origin of the results, mentioned by Aristotle and by Archimedes, on the determination of the measurement of the earth [Russian]. *Istoriko-astronomicheskie issledovaniya* 15:189–97.

Knorr, W. R. 1975. *Evolution of the Euclidean Elements*. Dordrecht: Reidel.

———. 1976. Archimedes and the measurement of the circle: A new interpretation. *AHES* 15:115–40.

———. 1978a. Archimedes' lost treatise on centers of gravity of solids. *Mathematical Intelligencer* 1:102–8.

———. 1978b. Archimedes and the pre-Euclidean proportion theory. *AIHS* 28:183–244.

———. 1978c. Archimedes and the spirals: The heuristic background. *HM* 5:43–75.

———. 1978d. Archimedes' *neusis*-constructions in *Spiral lines*. *Centaurus* 22:77–98.

———. 1978e. Archimedes and the *Elements*: Proposal for a revised chronological ordering of the Archimedean corpus. *AHES* 19:211–90.

———. 1982a. The hyperbola-construction in the *Conics*, Book II. *Centaurus* 25:253–91.

———. 1982b. Observations on the early history of the conics. *Centaurus* 26:1–24.

———. 1982c. *Ancient sources of the medieval tradition of mechanics: Greek, Arabic and Latin studies of the balance. Annali dell' Istituto e Museo di Storia della Scienza,* supp. monograph no. 6.

———. 1982d. Infinity and continuity: The interaction of mathematics and philosophy in antiquity. In *Infinity and continuity in ancient and medieval thought*, ed. N. Kretzmann, 112–45. Ithaca, N.Y.: Cornell University Press.

———. 1983. The geometry of burning-mirrors in antiquity. *Isis* 74:53–73.

———. 1985a. Ancient versions of two trigonometric lemmas. *Classical Quarterly* 35:362–91.

———. 1985b. Archimedes and the pseudo-Euclidean *Catoptrics*. *AIHS* 35:28–105.

———. 1986a. *The ancient tradition of geometric problems*. Boston, Basel, Stuttgart: Birkhäuser.

———. 1986b. Archimedes' *Dimension of the circle*: A view of the genesis of the extant text. *AHES* 35:281–324.

———. 1986c. The medieval tradition of a Greek mathematical lemma. *Zeitschrift für Geschichte der arabisch-islamischen Wissenschaften* (in press).

———. 1986d. On two Archimedean rules for the circle and the sphere. *BSSM* 6:145–58.

———. 1987a. The medieval tradition of Archimedes' *Sphere and cylinder*. In *Mathematics and its applications to science and natural philosophy in the Middle Ages: Essays in honor of Marshall Clagett*, ed. E. Grant and J. E. Murdoch. Cambridge: Cambridge University Press (forthcoming).

———. 1987b. Before and after Cavalieri: The method of indivisibles in ancient geometry (in progress).
Kozlov, B. I. 1984. Archimedes and the genesis of technical knowledge [Russian with English summary]. *Voprosy istorii estestvoznaniya i tekhniki*, 1984 (no. 3): 18–32.
Krafft, F. 1970. *Dynamische und statische Betrachtungsweise in der antiken Mechanik*. Wiesbaden: Steiner.
———. 1971. Archimedes. In *Die Grossen der Weltgeschichte*, ed. K. Fassmann, 726–44. Zurich.
Kubesov, A. 1969. On Naṣīr al-Dīn al-Ṭūsī's commentary on Archimedes' *Sphere and cylinder* [Russian]. *Voprosy istorii estestvoznaniya i tekhniki* 27:23–28.
Lam Lay-Yong and Shen Kangsheng. 1985. The Chinese concept of Cavalieri's principle and its applications. *HM* 12:219–28.
Landels, J. G. 1978. *Engineering in the ancient world*. London: Chatto & Windus.
Lawrence Laboratories (Livermore, Calif.). 1981. Remarks on Archimedes' Cattle Problem (by research group under H. L. Nelson). Reported in *Scientific American* 244 (no. 6): 84.
Lejeune, A. 1947. La dioptre d'Archimède. *Annales de la Société scientifique de Bruxelles* 61:27–47.
Lloyd, G. E. R. 1973. *Greek science after Aristotle*. London: Chatto & Windus; New York: Norton.
Lorent, H. 1955. Sur les traces des pas d'Archimède. *Mathesis* 64, supp. 1.
Luria [Lur'ye], S. I. 1945. *Archimedes* [Russian], Moscow. Leningrad: Akademiya Nauk. German trans. 1948. Vienna.
Marsden, E. W. 1970. *Greek and Roman artillery: Historical development*. Oxford: Clarendon Press.
———. 1971. *Greek and Roman artillery: Technical treatises*. Oxford: Clarendon Press.
Middleton, W. E. K. 1961. Archimedes, Kircher, Buffon and the burning-mirrors. *Isis* 52:533–43.
Mita, H. 1951. Problema bovinum of Archimedes [Japanese]. *Jap. journ. hist. sci.* 18:16–28.
Mogenet, J. 1956. *l'Introduction à l'Almageste*. Académie royale de Belgique, *Mémoires* 51, fasc. 2.
Moody, E. A. and Clagett, M. 1952. *The medieval science of weights (scientia de ponderibus)*. Madison, Wis.: University of Wisconsin Press.
Mueller, I. 1981. *Philosophy of mathematics and deductive structure in Euclid's Elements*. Cambridge: MIT Press.
Mugler, C. 1957. Sur l'histoire de quelques définitions de la géométrie grecque et les rapports entre la géométrie et l'optique. I: la ligne droite. *Antiquité classique* 36:331–45.
———. 1958–59. *Dictionnaire historique de la terminologie géométrique des Grecs*. 2 vols. Paris: Klincksieck.
———. 1970–72. *Les oeuvres d'Archimède:* see Section A above.
———. 1973. Sur un passage d'Archimède. *Revue des études grecques* 86:45–47.
Napolitani, P. D. 1982. Metodo e statica in Valerio, con edizione di due sue opere giovanili. *BSSM* 2:3–173.

Neuenschwander, E. 1974. Die stereometrischen Bücher der *Elemente* Euklids. *AHES* 14:91–125.

Neugebauer, O. 1975. *A history of ancient mathematical astronomy*. 3 vols. Berlin, Heidelberg, New York: Springer.

Oleson, J. P. 1984. *Greek and Roman mechanical water-lifting devices: The history of a technology*. *Phoenix,* supp. 16.

Onicescu, O. 1961. La science des grandeurs dans l'oeuvre d'Archimède. *Acta Logica* (*Analele*, Univ. Bucharest) 4:113–16. Repr. In Syracuse [1962–65, vol. 4].

———, et al. 1967. *Figuri ilustre ale antichitatii: Archimede, Pitagora* . . . . Bucharest.

Osborne, C. 1983. Archimedes on the dimensions of the cosmos. *Isis* 74:234–42.

PAA = *Praktika tēs Akadēmias Athēnōn*.

Paev, M. E. 1965. Approximate calculation of square roots in ancient Greece [Russian]. *Istoriko-matematicheskie issledovaniya* 16:219–33.

Papadatos, I. S. 1978. *Archimedes' 13 semiregular polyhedra* [Greek]. Athens.

Phillips, G. M. 1981. Archimedes, the numerical analyst. *American Mathematical Monthly* 88:165–69.

Pogrebysski, J. B. 1968. Structures mathématiques et théories physiques depuis Archimède jusqu' à Lagrange. *Revue scientifique* 89:247–56.

Price, D. de S. 1974. *Gears from the Greeks: The Antikythera mechanism—a calendar from ca. 80 B.C.* American Philosophical Society, *Transactions*, n.s. 64, pt. 7. Repr. 1975. New York: Science History Publications.

Procissi, A. 1981. Bibliografia della matematica greca antica. *BSSM* 1:1–151.

Quacquarelli, A. 1960–61. La fortuna di Archimede nei retori e negli autori cristiani antichi. Messina semin. matem. dell'Univ., *Rendiconti* 5:10–50. Repr. in Syracuse [1962–65, vol. 4].

Rabinovitch, N. L. 1974. An Archimedean tract of Immanual Tov-Elem (14th cent.). *HM* 1:13–27.

Rashed, R. 1979. La construction de l'heptagone régulier par ibn al-Haytham. *Journal for History of Arabic Science* 3:309–87.

———. 1981. Ibn al-Haytham et la mesure du paraboloide. *Journal for History of Arabic Science* 5:191–262.

Reixach i Vilà, P. 1982. Arquímedes de Siracusa. La ciencia entra en la praxis. *Atlantida* (Caracas: Univ. Simon Bolivar) 9 (25) :45–56.

Roero, C. S. 1983. Jakob Bernoulli attento studioso delle opere di Archimede. *BSSM* 3:77–125.

Rome, A. 1932. Notes sur les passages des *Catoptriques* d'Archimède conservés par Théon d'Alexandrie. *Annales de la Société scientifique de Bruxelles* 52 (2), ser. A, 30–41.

Rose, H. J. 1956. Lucretius II. 778–783. *Classical review*, n.s. 6:6–7.

Rose, P. L. 1977. For the history of Codex A of Archimedes: Notes on the Estense, Carpi and Ridolfi Libraries. *Manuscripta* 21:180–183.

Rosen, E. 1978. Aristarchus of Samos and Copernicus. *Bulletin of the American Papyrological Society* 15:85–93.

Rufini, E. 1961. *Il "Metodo" di Archimede e le origini del calcolo infinitesimale nell'antichità*. 2d ed. Milan: Feltrinelli. 1st ed. 1926. Rome: Stock.

Rybnikov, K. A. 1960. Infinitesimal calculus: Archimedes. In *History of mathematics* [Russian] I, chap. 5, 25–78. Moscow.

Sakkas, J. 1973. Report of burning-mirror experiments [Greek]. *Technika chronika*, Sept. 1973, 771–79. Repr. in Stamatis [1970–74, 3:309–12].

Sarfatti, G. B. 1968. *Mathematical terminology in Hebrew scientific literature of the Middle Ages* [Hebrew with English summary]. Jerusalem: Hebrew University.

Sato, T. 1979. Archimedes' *On the measurement of a circle* proposition 1: An attempt at a reconstruction. *Japanese Studies in the History of Science* 18:83–99.

———. 1981. Archimedes' lost works on the center of gravity of solids, plane figures, and magnitudes. *Historia scientiarum* 20:1–41.

———. 1985. Quadrature of the surface area of the sphere in the early Middle Ages—Johannes de Tinemue and Banū Mūsā. *Historia scientiarum* 28:61–90.

Schmidt, O. 1975. A system of axioms for the Archimedean theory of equilibrium and center of gravity. *Centaurus* 19:1–35.

Schneider, I. 1969. Die Entstehung der Legende um die kriegstechnische Anwendung von Brennspiegeln bei Archimedes. *Technikgeschichte* 36:1–11.

———. 1979a. *Archimedes: Ingenieur, Naturwissenschaftler und Mathematiker.* Darmstadt: Wissenschaftliche Buchgesellschaft.

———. 1979b. Archimedes: Wissenschaft und Technik im Spannungsfeld der Politik. *Kultur und Technik*, 1979 (no. 3): 4–11.

———. 1979c. Archimedes unfehlbar? Ein Fehler in der nach Heron rekonstruierten Schrift *Über Stützen*. In *Arithmos-Arrhythmos . . . Festschrift für J. O. Fleckenstein*, ed. K. Figala and E. H. Berninger, 235–43. Munich: Minerva.

———. 1982. Technik in der Sicht der exakten Naturwissenschaften am Beispiel von Archimedes, Christiaan Huygens und Carl Friedrich Gauss. *Kultur und Technik*, 1982 (no. 1): 21–41.

Schultz, P. 1984. Tartaglia, Archimedes and cubic equations. *Australian Math. Soc. Gazette* 11:81–4.

Sezgin, F. 1974. *Geschichte des arabischen Schrifttums.* V: *Mathematik bis ca. 430 H.* Leiden: Brill.

Shapiro, A. E. 1975. Archimedes's measurement of the sun's apparent diameter. *Journal for the History of Astronomy* 6:75–83.

Shen Kangsheng: see Lam Lay-Yong and Shen Kangsheng.

Simms, D. L. 1977. Archimedes and the burning mirrors of Syracuse. *Technology and Culture* 18:1–24.

Sleeswijk, A. W. 1979. Vitruvius' waywiser. *AIHS* 29:11–22.

Soedel, W. and Foley, V. 1979. Ancient catapults. *Scientific American* 240 (no. 3): 150–60.

Souffrin, P. 1980. Trois études sur l'oeuvre d'Archimède. *Cahiers d'histoire et de philosophie des sciences* 14:1–33.

Stamatis, E. S. 1955. Geometric proof of Archimedes' arithmetic approximation of the square root of 3 [Greek]. *PAA* 30:255–62 (cf. *Platon* 7 [1955]: 305–11).

———. 1963. Generalization of the theorem of Archimedes [Greek with German summary]. *Platon* 15:165–68.

———. 1967a. Powers with fractional exponents in Archimedes [Greek with English summary]. *Platon* 19:111–17.

———. 1967b. Archimedeia I [Greek with English summary]. *Platon* 19:150–153.
———. 1968a. Archimedes on the construction of the side of the regular heptagon inscribed in the circle [Greek]. *Bull. Math. Soc. Greece (Deltion tēs Hellēn. Mathēm. Hetaireias)*, n.s. 9 (2): 9–24.
———. 1968b. On the axiom of continuity [Greek]. *Platon* 20:144–47.
———. 1968c. *Greek science* [Greek]. Athens.
———. 1970–74. *Archimēdous Hapanta*: see Section A above.
———. 1971. Das heliozentrische System der alten Griechen [Greek with German summary]. *PAA* 46:64–83 (cf. The heliocentric system of Greeks, offpr. from Contrib. Res. Cent. Astr. and Appl. Math., Acad. Athens, 1973, ser. 1, no. 32).
———. 1973. The regular heptagon of Archimedes [Greek]. *Platon* 25:274–77.
———. 1975. The summation of square numbers [Greek with English summary]. *Platon* 27:103–8.
———. 1979. *Archimedes' balance* [Arabic text with Greek and English trans.]. Athens (cf. *Platon* 31 [1979]: 265–67).
———. 1980. *History of Greek mathematics: Arithmetic—The beginnings of Greek geometry* [Greek]. 2d ed. Athens (includes bibliog. of author's works).
———. 1982. *The burning mirrors of Archimedes* [Greek]. Athens (reprints of reports and journalistic notices, in Greek, English, and Russian; includes general bibliog. of author's works).
Steinschneider, M. 1893. *Die hebraeischen Übersetzungen des Mittelalters und die Juden als Dolmetscher*. Berlin. Repr. 1956. Graz: Akad. Druck- u. Verlagsanstalt.
Stolz, O. 1883. Zur Geometrie der Alten, insbesondere über ein Axiom des Archimedes. *Mathematische Annalen* 22:504–19.
Struik, D. J. 1986. Foreword to repr. ed. of E. J. Dijksterhuis, *The mechanization of the world picture*. Princeton: Princeton University Press.
Suppes, P. 1981. Limitations of the axiomatic method in ancient Greek mathematical sciences. In *Proceedings of the 1978 Pisa conference*, ed. J. Hintikka et al., 1:197–213. Dordrecht: Reidel.
Syracuse (Città di Siracusa). 1962–65. *Celebrazioni archimedee del secolo XX, 11–16 aprile 1961*. 4 vols. (see Procissi [1981,26] for contents).
Taisbak, C. M. 1980. An Archimedean proof of Heron's formula for the area of a triangle. *Centaurus* 24:110–16.
Tannery, P. 1884. Eutocius et ses contemporains. *Bull. sci. math. astr.*, ser. 2. 8:315–29. Repr. 1912 in *Mémoires scientifiques* II:118–36. Paris: Gauthier-Villars; Toulouse: Privat.
Tarn, W. W. 1952. *Hellenistic civilization*. 3d ed. London: St. Martin's Press. Repr. 1961. New York: New American Library.
Thuillier, P. 1979. Une énigme. Archimède et les miroirs ardents. *La recherche* 10:444–53.
Toomer, G. J. 1973. The chord table of Hipparchus and the early history of Greek trigonometry. *Centaurus* 18:6–28.
———. 1976a. *Diocles: On burning mirrors—The Arabic translation of the lost Greek original*. Berlin, Heidelberg, New York: Springer.
———. 1976b. Theon of Alexandria. *DSB* 13:321–25.
———. 1984. Lost Greek mathematical works in Arabic translation. *Mathematical Intelligencer* 6 (no. 2): 32–38.

Tronquart, G. 1966. Quelques réflexions sur Archimède et sa mort. *Bulletin de l'Association G. Budé,* 4th ser., 1966 (3): 299–308.

al-Ṭūsī, Naṣīr al-Dīn. 1939–40. *Rasā'il.* 2 vols. Hyderabad: Osmania University Press.

Unguru, S. 1975. On the need to rewrite the history of Greek mathematics. *AHES* 15:67–114.

———. 1979. History of ancient mathematics: Some reflections on the state of the art. *Isis* 70:555–65.

Ver Eecke, P. 1955. Note sur une interprétation erronée d'une sentence d'Archimède. *Antiquité classique* 24:132–33.

———. 1960. *Archimède: Les oeuvres complètes:* see Section A above.

Veselovskii, I. N., and Rosenfeld, B. A. 1962. *Works of Archimedes* [Russian]: see Section A above.

Virieux-Reymond, A. 1979. Le platonisme d'Archimède. *Revue philosophique* 169:189–92.

van der Waerden, B. L. 1954. *Science awakening.* Groningen: Noordhoff (rev. trans. of 1951 Dutch ed.). 2d ed. 1961. 3d ed. 1971. New York: Oxford University Press.

———. 1973. *Einfall und Überlegung: Beiträge zur Psychologie des mathematischen Denkens.* 3d ed. Basel: Birkhäuser. 1st ed. 1954. 2d ed. 1968. Chaps. 1–3 originally in *Elemente der Mathematik* 8 (1953): 121–29; 9 (1954): 1–9, 49–56.

———. 1975. Defence of a "shocking" point of view. *AHES* 15:199–210.

———. 1983. *Geometry and algebra in ancient civilizations.* Berlin, Heidelberg, New York: Springer.

Wagner, D. B. 1978. Liu Hui and Tsu Keng-chih on the volume of a sphere. *Chinese Science* 3:59–79.

Weil, A. 1978. Who betrayed Euclid? *AHES* 19:91–93.

———. 1984. *Number theory.* Basel, Stuttgart, Boston: Birkhäuser.

Whiteside, D. T. 1961. Patterns of mathematical thought in the later seventeenth century. *AHES* 1:179–388.

Youschkevitch [Iushkevich], A. P. 1976. *Les mathématiques arabes (viiie–xve siècles).* Paris: Vrin (based on pt. III of 1964 rev. German trans.). 1st Russian ed. 1961.

Zarankiewicz, K. 1957. *Z dziejow mechaniki: Archimedes, Galileusz, Newton.* Warsaw.

Zeuthen, H. G. 1886. *Die Lehre von den Kegelschnitten im Altertum.* Copenhagen: Fischer-Benzon. Repr. 1965. Hildesheim: Olms.

———. 1906–7: see Heiberg, J. L., and Zeuthen, H. G.

Zhitomirskii, S. V. 1977. Astronomical works of Archimedes [Russian]. *Istoriko-astronomicheskie issledovaniya* 13:319–37.

———. 1978. The celestial globe of Archimedes [Russian]. *Istor.-astron. issled.* 14:271–302.

———. 1981. *Archimedes* [Russian]. Moscow: Prosveshcheniye.

———. 1982. *The sage of Syracuse, Archimedes* [Russian]. Moscow: Molodaya gvardiya.

———. 1983. Ancient ideas on the dimensions of the world [Russian]. *Istor.-astron. issled.* 16:291–326.

INDEX OF NAMES

ERRATA

Page 8, line 28: *For* Purely *read* Pure
Page 10, line 5: *For* Silus *read* Silius
Page 10, note 2: *For* Silus *read* Silius
Page 13, note 1: *For* Tar' īkh alhukamā *read* Ta'rīkh al-ḥukamā'
Page 13, note 1: *For* al-Qiftî *read* al-Qifṭî
Page 17, line 15: *For* Thabīt *read* Thābit
Page 22, line 3: *For* Byzantium *read* Alexandria
Page 27, note 3: *For* Silus *read* Silius
Page 29, note 3: For 'Επιτομή read 'Επιτομὴ
Page 34, note 7: *For* A.D. *read* B.C.
Page 40, line 4: *For* Book II *read* Book I
Page 43, line 6: *For* Arabian *read* Arabic
Page 43, line 14: *For* Maii *read* Maius
Page 44, line 6: *For* Maii *read* Maius
Page 46, line 5: For σφαίρασ read σφαίρας
Page 46, line 22: For -ισμός read -ισμὸς
Page 47, line 2: For προς read πρὸς
Page 48, line 17: *For* Arabian *read* Arabic
Page 48, note 6: *For* Greek historian and alchemist (fl. about A.D. 400) *read* Platonist philosopher (second half of sixth century)
Page 48, note 8: *For* al-Qiftī *read* al-Qifṭī
Page 49, line 4: *For* Arabian *read* Arabic
Page 49, line 9: *For* Arabian *read* Arabic
Page 49, note 3: For *Raihân Muh. ibn Ahmad* read *Raiḥân Muḥ. ibn Aḥmad*
Page 50, line 2: *For* collection *read* collections
Page 51, line 14: *For* non *read* now
Page 65, line 6: For ἄν read ἂν
Page 70, line 12: For *orthotome* read *orthotome, and not meeting the diameter inside the section,*
Page 70, note 1: *For* double *read* equal
Page 79, line 33: *For* ΛΑΖ *read* ΔΑΖ
Page 87, line 24: *For* ΓΔ *read* ΓΛ
Page 99, line 20: *For*). *read*)].
Page 110, line 20: *For* 11a. *read* 11α.

Page 123, line 21: *For* [OΘ *read* O[Θ
Page 123, line 22: *For* [OΘ *read* O[Θ
Page 127, line 8: *For* ΓΔ +) *read* ΓΔ) +
Page 134, diagram: *For* Π *read* P
Page 165, line 4: *For* ΔZ·AH *read* ΔZ and AH
Page 165, line 8: *For* AE *read* ΔE
Page 182, note: *For* TΔ *read* TΛ
Page 182, line 8: *For* (**S, X**) *read* (**X, S**)
Page 197, line 30: *For* HK *read* HK_0
Page 201, note 1: *For* 100 *read* 200
Page 209, line 14: *For* [ZK, ZB] *read* (ZK, ZB)
Page 209, line 17: *For*) . . . *read*).
Page 249, line 20: *For* (EΛ, *read* (EΛ),
Page 264, line 17: *For* etc.) the *read* etc.); the
Page 271, line 11: *For* radius φ. *read* radius ϱ.
Page 283, line 3: *For* $(i - 1)$. *read* $(i - 1, 1)$.
Page 299, line 9: *For* third *read* fourth
Page 329, line 20: For αφ, read αχ,
Page 330, line 10: *For*), *read*)],
Page 331, diagram: points π and Δ, respectively the upper and lower endpoints of the vertical diameter of the circle, have not been marked
Page 361, note 1: For τροπίκῶν read τροπικῶν
Page 369, line 2: *For* $<$ *read* $>$
Page 401, line 28: *For* Arabian *read* Arabic
Page 401, note 3: *For* Abu *read* Abū
Page 402, line 1: *For* Arabian *read* Arabic
Page 406, line 28: *For* 3,6,10 *read* 4,6,10
Page 412, line 5: *For* Arabian *read* Arabic
Page 412, note 1: *For* Raihān *read* Raiḥān
Page 412, note 1: *For* Muhammad *read* Muḥammad
Page 414, line 5: *For* Arabian *read* Arabic
Page 417, line 3: For *Euclides* read *Euclidis*
Page 417, line 30: *For* 1875 *read* 1876
Page 453, line 22: *For* Milete *read* Miletus
Page 454, line 6: *For* Maii *read* Maius
Page 454, line 26: *For* Silus . . . , 10 *read* Silius . . . , 10, 27
Page 454, line 37: *For* Byzantius *read* Alexandrinus

Library of Congress Cataloging-in-Publication Data

Dijksterhuis, E. J. (Eduard Jan), 1892–1965
Archimedes.

Bibliography: p.
Includes index.
1. Archimedes. 2. Mathematics, Greek. I. Title.

QA31.D4813 1987 510′92′4 86-43144
ISBN 0-691-08421-1
ISBN 0-691-02400-6 (pbk.)